NUMERICAL METHODS IN BIOMEDICAL ENGINEERING

This is a volume in the
ACADEMIC PRESS SERIES IN BIOMEDICAL ENGINEERING

Joseph Bronzino, Series Editor
Trinity College—Hartford, Connecticut

Titles in this series include:

Numerical Methods in Biomedical Engineering, by Stanley M. Dunn, Alkis Constantinides, and Prabhas V. Moghe, January 2006

Introduction to Biomedical Engineering, 2 Ed. edited by John Enderle, Susan Blanchard, and Joseph Bronzino, March 2005

Circuits, Signals and Systems for Bioengineering by John Semmlow, March 2005

Bioelectrical Signal Processing in Cardiac and Neurological Applications by Leif Sörnmo and Pablo Laguna, June 2005

Diagnostic Ultrasound Imaging: Inside Out by Thomas Szabo, September 2004

Clinical Engineering Handbook edited by Joseph Dyro, August 2004

Modeling Methodology for Physiology and Medicine by Claudio Cobelli and Ewart Carson, October 2000

Handbook of Medical Imaging edited by Isaac Bankman, September 2000

Introduction to Applied Statistical Signal Analysis, 2 Ed. by Richard Shiavi, April 1999

NUMERICAL METHODS IN BIOMEDICAL ENGINEERING

STANLEY M. DUNN

Department of Biomedical Engineering
Rutgers, The State University of New Jersey

ALKIS CONSTANTINIDES

Department of Chemical and Biochemical Engineering
Rutgers, The State University of New Jersey

PRABHAS V. MOGHE

Department of Biomedical Engineering &
Department of Chemical and Biochemical Engineering
Rutgers, The State University of New Jersey

AMSTERDAM • BOSTON • HEIDELBERG • LONDON
NEW YORK • OXFORD • PARIS • SAN DIEGO
SAN FRANCISCO • SINGAPORE • SYDNEY • TOKYO
ACADEMIC PRESS IS AN IMPRINT OF ELSEVIER

Elsevier Academic Press
30 Corporate Drive, Suite 400, Burlington, MA 01803, USA
525 B Street, Suite 1900, San Diego, California 92101-4495, USA
84 Theobald's Road, London WC1X 8RR, UK

 This book is printed on acid-free paper.

Library of Congress Cataloging-in-Publication Data

Numerical methods in biomedical engineering / Stanley M. Dunn, Alkis
Constantinides, Prabhas V. Moghe.
p. ; cm.
Includes bibliographical references and index.
ISBN-13: 978-0-12-186031-8 ISBN-10: 0-12-186031-0 (casebound : alk. paper)
1. Biomedical engineering--Mathematics. 2. Biomedical engineering--Mathematical models.
I. Dunn, Stanley Martin. II. Constantinides,
A. III. Moghe, Prabhas V.
[DNLM: 1. Biomedical Engineering--methods. 2. Mathematical Computing.
3. Models, Theoretical. QT 36 N977 2005]
R857.M34N86 2005
610'.28--dc22

2005020031

British Library Cataloguing in Publication Data

A catalogue record for this book is available from the British Library

ISBN-13: 978-0-12-186031-8
ISBN-10: 0-12-186031-0

For all information on all Elsevier Academic Press publications,
visit our Web site at www.books.elsevier.com

Printed in the United States of America
08 09 10 9 8 7 6 5 4

Contents

Chapter 3 Concepts of Numerical Analysis 65

Part II: Steady-State Behavior

Chapter 4 Linear Models of Biological Systems 85

Appendices

Preface

The purpose of this textbook is to serve as an introductory overview of computational tools to solve numerical problems in the rapidly emerging discipline of biomedical engineering. Despite the popularity of bioengineering as a major in engineering, only a handful of textbooks have been written primarily for the instruction of undergraduates in bioengineering, none of which are in the area of numerical methods in biomedical engineering. Addressing this void was one of the driving forces for the current effort.

This book is intended as the primary text for an undergraduate course in biomedical engineering. The authors have adopted the book for the Fall semester junior course on Numerical Methods in the Department of Biomedical Engineering at Rutgers University—the book could be easily adopted for either semester of the junior year as well as for the senior year in BME. If the bioengineering concepts are somewhat de-emphasized and the calculus offerings are adjusted, it could also be adopted in the sophomore class. The book assumes that students have prerequisite skills in Calculus (I-IV), freshman Chemistry and Physics, General Biology, and an Introduction to Biomedical Engineering. The Numerical Methods course using this book may be offered in parallel with the treatment of junior topics such as Biomedical Transport Phenomena, Biomedical Thermodynamics/Kinetics, Biomechanics, and Bioinstrumentation. This book is well suited to train bioengineers interested in all major subfields within biomedical engineering, because it addresses a wide range of biosystems topics. The book can additionally be used as a text for quantitative biology curriculum aimed at life scientists (cell biologists, biomaterials scientists, and biochemists).

Fig. 1 illustrates our philosophy of the role that computing and numerical methods plays in the education of modern-day biomedical engineers. Placed early in the junior year, the course serves as a focal point for integrating fundamentals and problem-solving skills in the context of biomedical applications. The course satisfies two major goals: assimilating computing tools within the student's tool-kit, and applying these tools to a wide range of numerical models encountered in modern biomedical engineering.

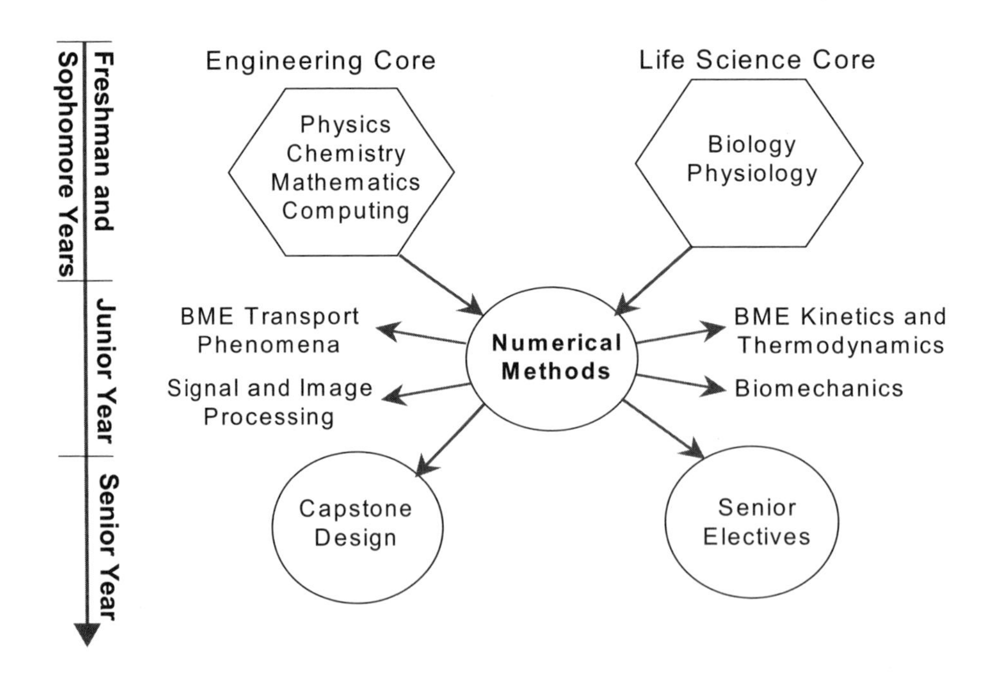

Figure 1 Numerical methods as an enabling pathway from engineering principles to biomedical applications.

A major challenge in biomedical engineering is the immense scope of biomedically relevant problems, which range from the molecular scale to the macroscopic scale and have classically been treated by different "breeds" of engineers: chemical engineers handling the molecular and tissue engineering problems; mechanical engineers tackling the biomechanics aspects; and electrical engineers treating imaging and measurement problems. We have sought to unify these traditionally disparate approaches into an integrated overview of the major problems encountered by bioengineers. Several classes of problems that may require the application of numerical methods for their solution are tabulated in Table 1.

Table 1 Bioengineering problems from a diverse range of system scales can be effectively treated through the use of the numerical and modeling techniques that are discussed in this book.

	Physiological Systems	Cell and Tissue Systems	Molecular Systems
Nonlinear Systems: Root Finding	Blood pressure Friction factor in perfusion	Cell migration Cell ablation	Receptor-ligand binding Enzyme-binding kinetics
Numerical Linear Algebra	Biomedical image reconstruction	Cell migration Musculoskeletal biomechanics	Acid-base equilibria Metabolic mass balances
Integration and Differentiation	Arrhythmia detection Heart rate variability	Cell migration Oxygen uptake in growing cells	Pharmacokinetic analysis Melting curve analysis
Ordinary Differential Equations	Membrane/blood oxygenation	Capillary transport Cell biomechanics Membrane and action potentials	Metabolic pathways Enzymatic reactions
Partial Differential Equations	Impedance imaging Capillary transport	Cell migration on biomaterials Laser ablation	Drug transport Metabolic pathways
Modeling Tools	Peritoneal dialysis Evoked potentials	Enzyme reactor parameters Heparin conversion Cell migration	Ligand binding

Organization and Outline of the Book

The organization of the book is illustrated in Table 2. The book is divided into four parts, followed by the Appendices, which should serve as a comprehensive resource.

Part I: Fundamentals: Introduces the student to the nature and behavior of physiological systems, and how to apply mathematical modeling techniques to these systems in order to develop models that may be used to simulate their behavior. This section also discusses the different types of models and relates them to numerical methods that may be used for their solution. The use of computing languages and computer programs to solve such problems is explained.

Part II: Steady-State Behavior: Treats the analysis of systems that result in algebraic models (both linear and nonlinear), develops the numerical methods necessary for solution of such systems, and applies the numerical methods to several examples drawn from physiological, cell, and molecular systems. The classical numerical methods, such as Gauss elimination, Gauss-Jordan reduction, and Gauss-Seidel substitution methods for linear systems, and linear interpolation, Newton-Raphson, and Newton's methods for simultaneous nonlinear systems, are presented to the reader, discussed, and applied.

Part III: Dynamic Behavior: Concentrates on the analysis of dynamic and multidimensional systems whose models contain ordinary and partial differential equations. Finite difference methods are used to develop the integration and differentiation algorithms. The classical techniques, such as quadrature and Newton-Cotes formulas for integration and Euler and Runge-Kutta methods for ordinary differential equations, are presented and applied. The stability of several finite difference methods applied to the solution of partial differential equations are analyzed and discussed.

Part IV: Modeling Tools and Applications: The solution of the mathematical models of complex biosystems requires a combination of several of the methods discussed in the previous sections of the book.

Appendices: Appendix A offers a tutorial introduction to MATLAB. This appendix is strongly recommended for students who need a review of the language and capabilities of MATLAB. **Appendix B** focuses on the Simulink environment within MATLAB. **Appendix C** reviews the foundations of linear algebra, which are integral to the numerical methods discussed in Chapters 4 and 6. **Appendix D** offers analytical approaches for solutions of ordinary and partial differential equations—a summary that will be helpful to the students and instructors to allow validation of the numerical techniques discussed in Chapters 7 and 8. **Appendix E** offers an overview of topics related to numerical stability.

Table 2 Overview of Numerical Methods in Biomedical Engineering.

Part I: Fundamentals

Chapter 1. Modeling Biosystems
- Biomedical Engineering
- Fundamental Aspects of Biomedical Engineering
- Constructing Engineering Models
- Examples of Solving Biomedical Engineering Models by Computer
- Overview of the Text

Chapter 2. Introduction to Computing
- Introduction
- The Role of Computers in Biomedical Engineering
- Programming Language Tools and Techniques
- Fundamentals of Data Structures for MATLAB
- An Introduction to Object-Oriented Systems
- Analyzing Algorithms and Programs

Chapter 3. Concepts of Numerical Analysis
- Scientific Computing
- Numerical Algorithms and Errors
- Taylor Series, Keeping Errors Small
- Floating-Point Representation in MATLAB

Part II: Steady-State Behavior

Chapter 4. Linear Biological Systems
- Examples of Linear Biological Systems
- Simultaneous Linear Algebraic Equations
  - Gauss elimination, Gauss-Jordan reduction, iterative methods
- Iterative Approach for Solution of Linear Systems
  - Jacobi and Gauss-Seidel methods
- Applications
  - Force balance in biomechanics
  - Biomedical imaging and image processing
  - Metabolic engineering and cellular biotechnolgy

Chapter 5. Nonlinear Biological Systems
- General Form of Nonlinear Equations
- Examples of Nonlinear Equations in Biomedical Engineering
- The Method of Successive Substitution
- The Method of False Position (Linear Interpolation)
- The Newton-Raphson Method
- Newton's Method for Simultaneous Nonlinear Equations
- Applications:
  - Friction factor in a catheter
  - Michaelis-Menten kinetics
  - Ventricular pressure measurements
  - Receptor-ligand dynamics

Part III: Dynamic Behavior

Chapter 6. Finite Difference Methods: Interpolation, Differentiation, and Integration
- Symbolic operators
- Backward, Forward, and Central Differences
- Interpolating of Equally Spaced Points
  - Gregory-Newton, Stirling's interpolation
- Interpolating Polynomials
- Interpolation of Equally Spaced Points
  - Gregory-Newton interpolation
- Interpolation of Unequally Spaced Points
  - Lagrange and spline
- Integration Formulas
  - Newton-Cotes Formulas of Integration

Chapter 7. Dynamic Systems: Ordinary Differential Equations
- Classification of Ordinary Differential Equations
- Transformation to Canonical Form
- Nonlinear Ordinary Differential Equations
  - Euler, Runge-Kutta methods
- Linear Ordinary Differential Equations
  - Eigenvalue methods
- Steady State Solutions and Stability Analysis
- Numerical Stability
- Applications:
  - Pharmacokinetics: the drug absorption problem
  - Enzyme catalysis reactions
  - Glycolysis pathways of living cells
  - Dynamics of membrane and nerve cell potentials
  - Dynamics of stem cell differentiation
  - Tissue engineering: cell migration

Chapter 8. Dynamic Systems: Partial Differential Equations
- Multidimensional distributed systems
- Classification of Partial Differential Equations
- Initial and Boundary Conditions
- Solution of Partial Differential Equations: elliptic, parabolic, hyperbolic equations
- Polar coordinates
- Stability analysis
- Applications:
  - Dynamics of the basilar membrane
  - Migration of human leukocytes
  - Drug diffusion
  - Laser ablation

Part IV: Modeling Tools and Applications

Chapter 9. Measurements, Models and Statistics
- The Role of Numerical Methods
- Measurements, Errors and Uncertainty
- Descriptive Statistics, Inferential Statistics
- Least Squares Modeling, Curve Fitting
- Fourier Transforms
- Applications:
  - Computing statistics of MRI and CT image intensities
  - Hypothesis testing in DNA microarray analysis
  - Analysis of mass spectra data
  - Separating EEG frequency components

Chapter 10. Modeling Biosystems: Applications
- Numerical Modeling of Bioengineering Systems
- PhysioNet, PhysioBank, and Physio Toolkit
- Signal Processing
- Applications:
  - ECG simulation; EEG simulation; Model of glucose regulation
  - Diabetes and insulin regulation; Renal clearance; Motion of rigid body
- PHYSBE Simulations
  - Normal PHYSBE operation; Simulink model of coarctation of the aorta; Simulink model of aortic valve stenosis; Ventricular septal defect; Left ventricular hypertrophy; Pressure-volume loops

Part V: Appendices

A. Introduction to MATLAB
B. Introduction to Simulink
C. Review of Linear Algebra and Related MATLAB Commands
D. Analytical Solutions of Differential Equations
E. Numerical Stability and Other Topics

Several worked examples are given in each chapter to demonstrate the numerical techniques. Most of these examples require computer programs to be solved. These programs were written in the MATLAB language and are compatible with MATLAB 7.0 or higher. In most of the examples, we tried to present a general MATLAB function that implements the method and may be applied to the solution of other problems that fall in the same category of application as the worked example.

All the MATLAB programs that appear in the text are on the book's Web site (http://books.elsevier.com/companions/0121860310) and may be downloaded as needed. The reader is strongly encouraged to visit this Web site for updates, errata and author contact information.

Acknowledgments: The authors are grateful to the Whitaker Foundation, and Dr. John Linehan in particular, for reviewing, endorsing, and supporting the curriculum proposal as part of the Teaching Materials Program of the Foundation. The timelines for this activity were instrumental in enabling the authors to complete this ambitious project within two eventful years. The authors thank their colleagues at Rutgers for help with identifying current numerical problems in bioengineering, including: John Li, Charlie Roth, David Shreiber, and Martin Yarmush. Excellent new books in biomedical engineering enabled the authors to draw upon a wider range of problems—credits have been cited in the book where appropriate. The Rutgers biomedical engineering junior classes of 2003 and 2004, and in particular, Devin Fensterheim, Ronn Friedlander, Kelly Horn, Jessica Nikitchuk, Nayyereh Rajaei, Jonathon Scarpa, Ashley Winter and Jillian Zaveloff, contributed significantly to Chapter 10 and helped to test the manuscript version of the book—many thanks to them for their valuable feedback.

Stanley M. Dunn
Alkis Constantinides
Prabhas V. Moghe

Chapter 1
Modeling Biosystems

1.1 Biomedical Engineering

Biomedical Engineering (BME) is the branch of engineering that is concerned with solving problems in biology and medicine. This text is an introduction to *Numerical Methods* for biomedical engineers. Numerical methods are mathematical techniques for performing accurate, efficient and stable computation, by computer, to solve mathematical models of biomedical systems. Numerical methods are the tools engineers use to realize computer implementation of analytic models of system behavior.

Biomedical engineers use principles, methods, and approaches drawn from the more traditional branches of electrical, mechanical, chemical, materials, and computer engineering to solve this wide range of problems. These methods include: principles of Electrical Engineering, such as circuits and systems; imaging and image processing; instrumentation and measurements; and sensors. The principles from Mechanical Engineering include fluid and solid mechanics; heat transfer; robotics and automation; and thermodynamics. Principles from Chemical Engineering include transport phenomena; polymers and materials; biotechnology; drug design; and pharmaceutical manufacturing.

Biomedical engineers apply these and other principles to problems in the life sciences and healthcare fields. That means the biomedical engineer must also be familiar with biological concepts of anatomy and physiology at the system, cellular, and molecular levels. Working in healthcare requires familiarity with the cardiovascular system, the nervous system, respiration, circulation, kidneys, and body fluids.

Other terms, such as bioengineering, clinical engineering, and tissue engineering, are used to identify subgroups of biomedical engineers. Bioengineering refers to biomedical engineers who focus on problems in biology and the relationship between biological and physiological systems. Clinical engineering is a term used to refer to biomedical engineers who solve problems related to the clinical aspects of healthcare delivery systems and patient care. Tissue engineering is the subspecialty in which engineering is used to design and create tissues and devices to replace structures with lost or impaired function. Tissue engineers use a combination of cells, engineering materials, and suitable biochemical factors to improve or replace biological functions in an effort to effect the advancement of medicine.

Although the most visible contributions of biomedical engineering to clinical practice involve instrumentation for diagnosis, therapy, and rehabilitation, there are examples in this text drawn from both the biological and the medical arenas to show the wide variety of problems that biomedical engineers can and do work on.

The field of biomedical engineering is rapidly expanding. Biomedical engineers will play a major role in research in the life sciences and development of devices for efficient delivery of healthcare. The scope of biomedical engineering ranges from bionanotechnology to assistive devices, from molecular and cellular engineering to surgical robotics, and from neuromuscular systems to artificial lungs. The principles presented in this text will help prepare biomedical engineers to work in this diverse field.

There are a number of good histories of biomedical engineering; each has its own beginning date of the field and the significant milestones. The beginning of biomedical engineering can be traced to either the 17th, 18th, or 19th century, the choice depending on the definition used for biomedical engineering. The reader is referred to Nebeker (2002) for a comprehensive history of how biomedical engineers build diagnostic (data that characterizes the system) or therapeutic (replace or enhance lost function) devices as solutions to problems in healthcare or the life sciences.

1.2 Fundamental Aspects of Biomedical Engineering

Any biomedical engineering device includes one or more *measurement, modeling,* or *manipulation* tasks. By measurement, we mean sensing properties of the physical, chemical, or biological system under consideration. Realizing that no property can be measured exactly is an important concept for biomedical engineers to understand. For this reason, an appreciation of statistics is required and will be covered later in this text. Principles of measurement also include an understanding of variability and sources of noise; sensing instruments and accuracy; and resolution and reproducibility as characterizations of measurements.

Manipulation in this engineering sense means interacting with a system in some way. For the most part, biomedical engineers will interact with the human body or biological system by constructing diagnostic or therapeutic interventions. The process of developing an intervention system starts with requirements and constraints, which, in turn, lead to specifications for the particular intervention and then to design, fabrication, and testing.

The task of engineering modeling is the process by which a biomedical engineer expresses the principles of physics, chemistry, and biology in a mathematical statement that describes the phenomena or system under consideration. The mathematical model is a precise statement of how the system interacts with its environment. *A model is a tool that allows the engineer to predict how the system will react to changes in one of the system parameters.*

Engineering models are mathematical statements using one or more of four areas of continuous mathematics: algebra, calculus, differential equations and statistics. Other than the pure modeling tasks, any of the other significant biomedical engineering projects highlighted above use models, signal processing, or control systems that are implemented in a computer system. In these cases, one must describe the behavior algebraically, as an integral, differential equation or as an expression of the variability in analogs of these continuous mathematical models; the challenge is to preserve the accuracy and resolution to the highest degree possible and perform stable computation. This is the purpose of numerical methods.

1.3 Constructing Engineering Models

Practicing engineers are asked to solve problems based on physical relationships between something of interest and the surrounding environment. Because there is quite a wide range of phenomena that will occur in biosystems, it is important to have a common language that can be used to describe and model bioelectric, biomechanical, and biochemical phenomena.

1.3.1 A framework for problem solving

The problem-solving framework is based on first identifying the conservation law that governs the observed behavior. There are four steps to developing the solution to a problem in biomedical engineering:

1. **Identify the system to be analyzed:** A *system* is any region in space or quantity of matter set aside for analysis; it's the part of the universe in which we are interested. The *environment* is everything not inside the system. The system *boundary* is an infinitesimally thin surface, real or imagined, that separates the system from its environment. The boundary has no mass, and merely serves as a delineator of the extent of the system.

2. **Determine the extensive property to be accounted for:** An *extensive property* does not have a value at a point, and its value depends on the extent or size of the system; e.g., it is proportional to the mass of the system. The amount of an extensive property for a system can be determined by summing the amount of extensive property for each subsystem that comprises the system. The value of an extensive property for a system only depends upon time. Some examples that we are familiar with are mass and volume.

   There is scientific evidence that suggests that the property can neither be created nor destroyed. An extensive property that satisfies this requirement is called a *conserved property*.

   There are many experiments reported in the literature that support the idea that charge, linear momentum and angular momentum are conserved. Mass and energy, on the other hand are conserved under some restrictions:

   a. If moving, the speed of the system is significantly less than the speed of light.
   b. The time interval is long when compared to the time intervals of quantum mechanics.
   c. There are no nuclear reactions.

   It is very unlikely that conditions a and b will be violated by any biomedical system that will be studied. In the time scale of biology, the shortest event is on the order of 10^{-8}, which is still longer than nuclear events. However, emission tomography and nuclear medicine are systems where nuclear reactions do occur and one will have to be careful of assuming that mass or energy is conserved in these systems.

   Conservation of mass, charge, energy, and momentum can be very useful in developing mathematical models for analysis of engineering artifacts. In addition to conserved properties, there are other extensive properties for which we know limits on the generation/consumption terms. The classic example of this is the *Second Law of Thermodynamics* and its associated property, *entropy*. When written as a conservation equation, it is easy to see that entropy can only be produced within a system. Furthermore, in the limit of an internally reversible process, the entropy production rate reduces to zero.

3. **Determine the time period to be analyzed:** When a system undergoes a change in state, we say that the system has undergone a *process*. It is frequently the goal of engineering analysis to predict the behavior of a system, i.e., the path of states that result, when it undergoes a specified process. Processes can be classified in three ways based on the time interval involved: steady-state, finite-time, and transient processes.

4. **Formulate a mathematical expression of the conservation law:** Experience has taught us that the amount of an extensive property within a system may change with time. This change can only occur by two mechanisms:
   - Transport of the extensive property across the system boundary
   - Generation (production) or consumption (destruction) of the extensive property inside the system

   Thus, the change of an extensive property within a system can be related to the amount of the extensive property transported across the boundary and the amount of the extensive property generated (and/or consumed) within the system. This "accounting principle" for an extensive property is known to engineers as a *balance equation*. Although this principle can be applied to a system for any extensive property, it will be especially useful for properties that are conserved. There are two forms of a balance equation: the accumulation form and the rate form.

1.3.2 Formulating the mathematical expression of conservation

The accumulation form

In the accumulation form, the time period used in the analysis is finite and is therefore used to formulate equations that model steady-state or finite-time processes. When accounting for the input and output, the total amount that enters the system (during the time period) is computed and the total amount that exits in the same time is subtracted. The accounting statement is:

$$\left\langle \begin{matrix} \text{Net amount} \\ \text{accumulated} \\ \text{inside the system} \end{matrix} \right\rangle = \left\langle \begin{matrix} \text{Net amount} \\ \text{transported} \\ \text{into the system} \end{matrix} \right\rangle + \left\langle \begin{matrix} \text{Net amount} \\ \text{generated} \\ \text{inside the system} \end{matrix} \right\rangle$$

or, for a conserved property P:

$$P_{inside}^{final} - P_{inside}^{initial} = (P_i - P_o) + (P_G - P_C)$$

where the left-hand side is the net amount accumulated inside; the first difference term on the right-hand side is the net amount transported into the system (input – output); and the second difference term on the right is the net amount of the property generated (generated – consumed).

The advantage of using an accumulation form of the conservation or accounting laws is that the mathematical expression is in the form of either algebraic or integral equations. The disadvantage of the accumulation form of the law is that it is not always possible to determine the amount of the property of interest entering or exiting from the system.

The rate form

The rate form of a balance equation is similar to the accumulation form, except that the time period is infinitesimally small, so in the limit the net amounts become rates of change. The mathematical relationship between the accumulation form and the rate form can easily be developed by dividing the accumulation form through by the time interval Δt and taking the limit as $\Delta t \to 0$. That is:

$$\left\langle \begin{array}{c} \text{Rate of change} \\ \text{inside the system} \\ \text{at time t} \end{array} \right\rangle = \left\langle \begin{array}{c} \text{Transport rate} \\ \text{into the system} \\ \text{at time t} \end{array} \right\rangle + \left\langle \begin{array}{c} \text{Generation rate} \\ \text{into the system} \\ \text{at time t} \end{array} \right\rangle$$

or, for a conserved property P:

$$\frac{dP}{dt} = (\dot{P}_i - \dot{P}_o) + (\dot{P}_G - \dot{P}_C)$$

where the first term on the right-hand side is the difference in transport rates across the system boundary, and the second term is the difference in the rates at which the property is generated inside the system. The sum of these two terms is the rate of change of the property, inside the system, with respect to time. The advantage of the rate form of the law is that the laws of physics generally make it easy to find the rates at which things are happening. The disadvantage of the rate form is that it generates differential equations.

Although the accounting principle can be applied for any extensive property, it is most useful when the transport and generation/consumption terms have physical significance. The most useful applications of this principle occur when something is known *a priori* about the generation/consumption term.

For conserved extensive properties the equations that apply the accounting principle are significantly simpler. In the accumulation form, the equations become:

$$P_{inside}^{final} - P_{inside}^{initial} = (P_i - P_o).$$

1.3.3 Using balance equations

The mathematical formulation of problems that biomedical engineers solve is based on one or more of the conservation laws from chemistry and physics. The solution to a problem becomes obvious when one has the right formulation. The approach used herein is that the solution to a problem can be formulated using one or more of the forms of the following conservation laws.

The model formulation and problem-solving framework presented in Section 1.3.1 shows that there are two forms for each of the conservation laws: the accumulation (or sum) form and the rate form. The former is used in solutions of steady-state or finite-time problems, whereas the latter is used in solving problems with transient behavior.

Example 1.1 How conservation laws lead to the Nernst equation.

Show how Fick's law, Ohm's law and the Einstein relationship can be derived from the conservation laws in Section 1.3.1. Show how these three conservation models lead to the Nernst equation. This problem is derived from Section 3.4 of Enderle et al. (2000).

The rate form of the principle of conservation of mass is:

$$\frac{dm_{sys}}{dt} = \sum_{inlets} \dot{m}_i - \sum_{outlets} \dot{m}_e$$

meaning that the overall time rate of change of mass in the system is equal to the difference between the sum of the rates of change of mass into the system and the sum of the rates of change of mass out of the system.

For the derivation of the Nernst equation, the system being considered is a cell membrane surrounded by extracellular fluid, with ion flow across the membrane. Assume that one of K^+, Na^+, or Cl^- ions flow across the membrane, with the positive direction taken to be from the outside, across the membrane, to inside the cell. The rate form above leads to the following three relationships:

1. *Fick's Law* is a model that relates the flow of ions due to diffusion and the ion concentration gradient across the cell membrane. It is a law that expresses the influence of chemical force on the conservation of mass in the system. The left-hand side of the conservation law is the flow rate of mass, that is, the mass flux in the system, which is the flow due to diffusion. The right-hand side is the sum of the rates of change of mass through the system inlets and outlets, that is, the flow through the ion channels. The rate form of the principle of conservation of mass leads to:

$$J_{diffusion} = -D\frac{dI}{dx}$$

that is, the flow of ions due to diffusion is equal to the ion concentration gradient across the membrane scaled by the diffusivity constant D. That is, there is ion flow in response to an ion concentration gradient. The negative sign indicates that the ion flow is in the opposite direction of the gradient.

2. *Ohm's Law* is a model for the influence of electrical force on ion flow across the cell membrane. It is still based on conservation of mass, but the driving force on the right-hand side is mass flow due to an electric field induced by other charged particles. An electric field $\vec{E}$ is applied, creating a rate of change of potential, dv/dx. The right-hand side is the ion flow due to this potential, and leads to:

$$J_{drift} = -\mu Z[I]\frac{dv}{dx}$$

in which J_{drift} is the ion flux due to the electric field, μ is the mobility, Z is the ion valence, $[I]$ is the ion concentration, and v is the voltage.

3. *The Einstein relationship* is a form of conservation of momentum that expresses a relationship between diffusion and ion mobility. The electric field induces a force on the ions that causes a flow that is balanced by osmotic pressure. The conservation law,

$$\frac{dP_{sys}}{dt} = \sum F_{ext} + \sum_{inlets} \dot{P}_i - \sum_{outlets} \dot{P}_o$$

leads to the form:

$$D = \frac{KT\mu}{q}$$

in which K is Boltzmann's constant (1.38×10^{-23} J/K), T is absolute temperature in Kelvin, and q is the magnitude of the electric charge (1.61×10^{-19} C).

Both Fick's Law and Ohm's Law express conservation of mass and can be combined by the engineering principle of superposition, yielding an expression for total flow:

$$J = J_{diffusion} + J_{drift} = -D\frac{d\left[K^+\right]}{dx} - \mu Z\left[K^+\right]\frac{dv}{dx}$$

which, in the steady state, is 0 (that is, no net transport). The conservation of momentum principle, Einstein's relationship, relates the two rates on the right-hand side:

$$0 = J_{diffusion} + J_{drift} = -\frac{KT}{q}\mu\frac{d\left[K^+\right]}{dx} - \mu Z\left[K^+\right]\frac{dv}{dx}$$

for K^+, Z=1 and after integrating potential across the cell boundary:

$$\int_{v_o}^{v_i} dv = -\frac{KT}{q}\int_{K_o^+}^{K_i^+}\frac{d\left[K^+\right]}{\left[K^+\right]}$$

which yields:

$$v_i - v_o = -\frac{KT}{q}\ln\frac{\left[K^+\right]_i}{\left[K^+\right]_o}$$

known as the Nernst equation, expressing the potential difference across a cell membrane as a function of chemical and electrical forces driving ion transport.

1.4 Examples of Solving Biomedical Engineering Models by Computer

The materials presented in the previous section are the basic tools that a biomedical engineer will use to solve a problem or design a new diagnostic or therapeutic device. However, an analytic solution is not sufficient; the device will be designed and/or controlled using a computer. This means that one must know how to transform the model from continuous mathematics to discrete mathematics (numerical analysis) and then write a computer program to implement the new equation (numerical methods). This section shows some examples of how biomedical engineers use numerical analysis and numerical methods.

1.4.1 Modeling rtPCR efficiency

PCR is an acronym that stands for polymerase chain reaction, a technique that will allow a short stretch of DNA (usually fewer than 3,000 base pairs) to be amplified about a million fold so that one can determine its size and nucleotide sequence.

The particular stretch of DNA to be amplified, called the *target sequence*, is identified by a specific pair of DNA *primers*, which arc usually about 20 *nucleotides* in length. For convenience, these four nucleotides are called dNTPs.

There are three major steps in a PCR reaction, which are repeated for 30 or 40 cycles (Hsu et al., 1996). This process is done on an automated cycler, which can heat and cool the tubes with the reaction mixture in a very short time. The three steps of a PCR reaction (illustrated in Fig. 1.1) are:

1. **Denaturation** at 94°C: During the denaturation, the double strand melts open to single-stranded DNA and all enzymatic reactions stop.
2. **Annealing** at 54°C: Bonds are constantly formed and broken between the single-stranded primer and the single-stranded template. When primers fit exactly, the bonds are more stable and last a bit longer. On that piece of double-stranded DNA (template and primer), the polymerase can attach and starts copying the template. Once there are a few bases built in, the ionic bond is so strong between the template and the primer that it does not break.
3. **Extension** at 72°C: The primers, where there are a few bases built in, already have a strong attraction to the template. Primers that are on positions with no exact match get loose again and don't give an extension of the fragment. The bases complementary to the template are coupled to the primer on the 3' side.

Because both strands are copied during PCR, there is an **exponential** increase of the number of copies of the gene.

Real-time PCR (rtPCR) is used to determine gene expression over and above size and sequence information. In rtPCR, it is assumed that the efficiency is constant; however, analysis of data shows that the efficiency is *not* constant, but is instead a function of cycle number (Gevertz et al., 2005). Based on this observation, it seems reasonable that rtPCR quantification techniques can be improved upon by understanding the behavior of rtPCR efficiency. A mathematical model of rtPCR will provide this understanding and will lead to more accurate methods to quantify gene expression levels from rtPCR data.

The mathematical model of annealing and extension is as follows: after the double-stranded DNA is denatured, there are two single strands of DNA, called T1 and T2, for template 1 and template 2, respectively. Once the temperature is cooled to 54°C (signifying the starting of the annealing stage), we expect that primer 1 (P1) will anneal with T1 to form a hybrid (H1) and we also expect that P2 and T2 will anneal to form a hybrid (H2). These two reactions can be represented by the chemical equations:

$$P1 + T1 \leftrightarrow H1 \tag{1.1}$$

$$P2 + T2 \leftrightarrow H2 \tag{1.2}$$

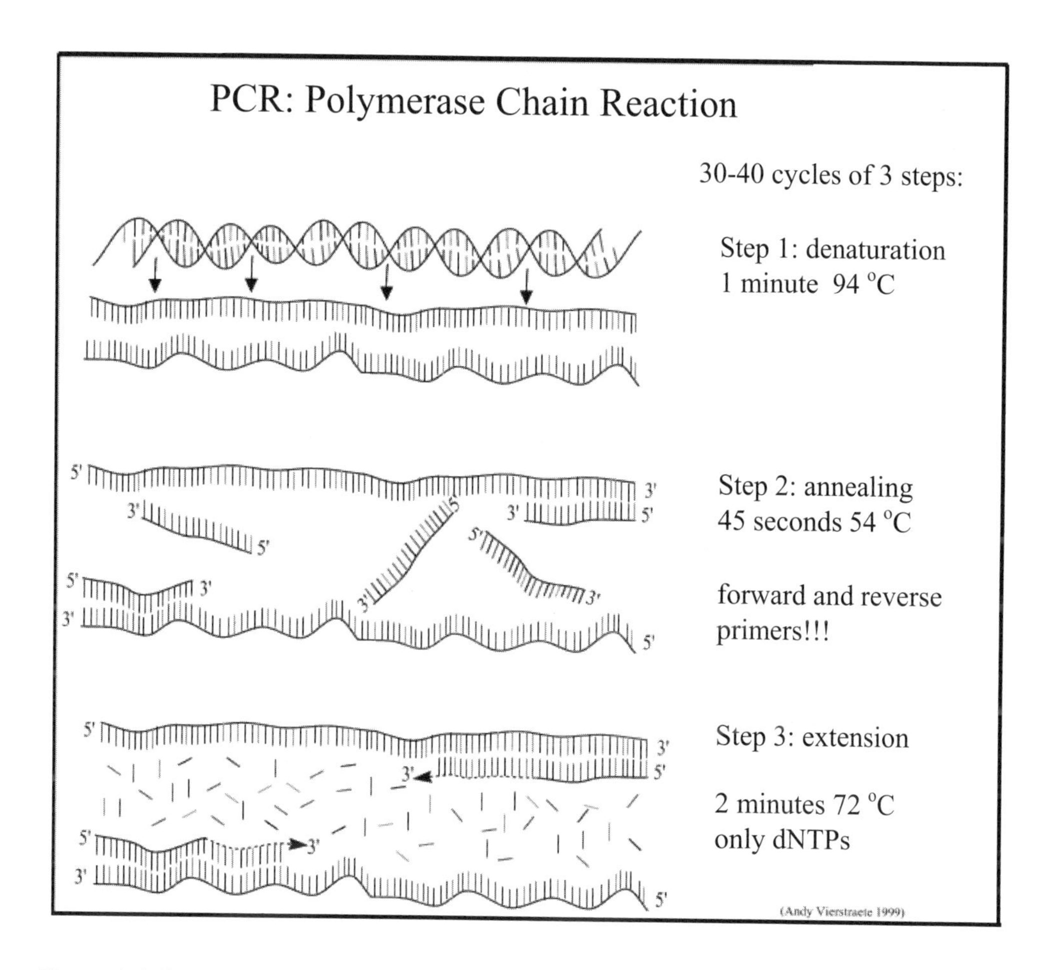

Figure 1.1 The exponential amplification of the gene in PCR. From Vierstraete (1999).

Unfortunately, other reactions can occur during the annealing stage of PCR. The template strands, T1 and T2, are complementary and can re-anneal upon contact, forming a template hybrid (HT) as represented by the following reaction:

$$T1 + T2 \leftrightarrow HT \tag{1.3}$$

Finally, depending on primer design, primer-dimers (D) can form as well:

$$P1 + P2 \leftrightarrow D \tag{1.4}$$

Hence, four reactions can occur simultaneously during the primer annealing stage. These reactions can be modeled using either thermodynamic (steady state) or kinetic (transient) equations.

The steady-state thermodynamic model tracks the total concentration of all products performing active roles during the annealing stage of PCR, a mass balance in performed on reactions (1.1) to (1.4). This procedure results in the following four equations:

$$\begin{aligned}
[\mathrm{P1}]_\mathrm{T} &= [\mathrm{P1}]+[\mathrm{H1}]+[\mathrm{D}] \\
[\mathrm{P2}]_\mathrm{T} &= [\mathrm{P2}]+[\mathrm{H2}]+[\mathrm{D}] \\
[\mathrm{T1}]_\mathrm{T} &= [\mathrm{T1}]+[\mathrm{H1}]+[\mathrm{HT}] \\
[\mathrm{T2}]_\mathrm{T} &= [\mathrm{T2}]+[\mathrm{H2}]+[\mathrm{HT}]
\end{aligned} \tag{1.5}$$

where $[\mathrm{X}]_\mathrm{T}$ is a parameter denoting the total concentration of product X.

To complete the thermodynamic model, we need to introduce four more parameters into the system—the equilibrium constants of each reaction. We define K_{H1}, K_{H2}, K_{HT}, and K_D to be the equilibrium constants of (1.1) to (1.4), respectively. Since the equilibrium constant is, by definition, the fixed ratio of reactant to product concentrations, these allow us to introduce four more equations into the model:

$$\begin{aligned}
K_{H1} &= \frac{[\mathrm{P1}]\cdot[\mathrm{T1}]}{[\mathrm{H1}]} \\
K_{H2} &= \frac{[\mathrm{P2}]\cdot[\mathrm{T2}]}{[\mathrm{H2}]} \\
K_{HT} &= \frac{[\mathrm{T1}]\cdot[\mathrm{T2}]}{[\mathrm{HT}]} \\
K_{D} &= \frac{[\mathrm{P1}]\cdot[\mathrm{P2}]}{[\mathrm{D}]}
\end{aligned} \tag{1.6}$$

The efficiency of the annealing stage, $\varepsilon_{ann}(n)$, can be calculated by comparing the amount of hybrids after the nth annealing stage to the total amount of template present throughout the nth annealing stage:

$$\varepsilon_{ann}(n) = 0.5\left(\frac{[\mathrm{H1}]}{[\mathrm{T1}]_\mathrm{T}} + \frac{[\mathrm{H2}]}{[\mathrm{T2}]_\mathrm{T}}\right) \tag{1.7}$$

in which [H1] and [H2] can be found by solving the nonlinear system of eight equations with eight unknowns in terms of the eight free parameters.

This deceivingly simple system of eight equations can be solved analytically either by hand or using a computer program for solving the system of equations symbolically. The numerical solution to this problem can be obtained using the techniques described in Chapter 4.

1.4.2 Modeling transcranial magnetic stimulation

Transcranial magnetic stimulation (TMS), the stimulation of cortical tissue by magnetic induction, is potentially a new diagnostic and therapeutic tool in clinical neurophysiology. The magnetic fields are delivered to the tissue, by placing coils of the surface of the skull; TMS shows the promise of being useful for brain mapping and appears to show potential for treating brain disorders (Hallett 2000). The advantage of techniques like TMS and EEG (electroencephalogram) (see Chapter 10) is high temporal resolution; a disadvantage of TMS is that the 3D spatial resolution and depth penetration is not very good. Poor spatial resolution is due to the face that the magnetic fields cannot easily be focused at a particular point in the brain.

Norton (2003) proposed a different method of stimulating cortical tissue, which potentially may allow deeper penetration and better focusing in cortical tissue. His idea is to create an electrical current by propagating an ultrasonic wave in the presence of a strong DC magnetic field. His results (Norton 2003) show that the amplitude of the magnetic field generated in this fashion is less than that of a traditional TMS approach, but there are some distinct advantages in both spatial and temporal characteristics of the induced field.

As one can imagine, numerical methods play a large role in calculating the magnitude of an electric field throughout the neural tissue in the brain. Briefly, let the cortical tissue be represented in a cylindrical 3D coordinate system (r, ϕ, z) and assume that the ultrasonic beam is collimated and propagating in the z direction. Furthermore, assume that the profile is axially symmetric and will be represented by $p(r)$. Norton (2003) modeled the particle velocity by:

$$v(\mathbf{r}) = v_0 p(r) e^{ik_0 z}\hat{z} \tag{1.8}$$

where v_0 is the peak velocity of the particles, $\hat{z}$ is the unit vector in the z direction and the wave number $k_0 = \omega/c_0$, i.e., the frequency of the ultrasonic wave divided by the speed of the ultrasonic wave.

The components of the magnetic field $\mathbf{E}_s$ induced in the brain are given by:

$$\begin{aligned}
E_r(r,\phi,z) &= \mathbf{B}_0 v_0 \left[\frac{d^2A(r)}{dr^2}\right] e^{ik_0 z}\sin\phi \\
E_\phi(r,\phi,z) &= \mathbf{B}_0 v_0 \left[\frac{1}{r}\frac{dA(r)}{dr}\right] e^{ik_0 z}\cos\phi \\
E_z(r,\phi,z) &= \mathbf{B}_0 v_0 \left[ik_0\frac{dA(r)}{dr}\right] e^{ik_0 z}\sin\phi
\end{aligned} \tag{1.9}$$

where

$$A(r) = K_0(k_0 r)\int_0^r I_0(k_0 r')p(r')r'dr' + I_0(k_0 r)\int_r^\infty K_0(k_0 r')p(r')r'dr' \quad (1.10)$$

and both $I_0(\cdot)$ and $K_0(\cdot)$ are functions that model the ultrasonic wave propagation.

This mathematical model is used to predict the distribution of the induced electric field in the brain when generated by an ultrasonic wave. In order for this set of equations to be used, they have to be solved numerically. But, in order to solve the three equations, one has to solve Eq. (1.10) for $A(r)$. Except for a very ideal case the integrals in Eq. (1.10) and the differential equations in Eq. (1.9) have to be evaluated using the numerical integration techniques described in Chapter 6 and the differential equation solvers of Chapter 7.

1.4.3 Modeling cardiac electrophysiology

Cardiac arrythmias are a leading cause of death in the United States and abroad. Computer simulations are rapidly becoming a powerful tool for modeling the factors that cause these life-threatening conditions (see Chapter 10). High accuracy simulations require fine spatial sampling and time-step sizes at or below a microsecond. To further complicate matters, there are many factors such as heat transport, fluid flow and electrical activity to model. The heart does not have a simple geometry and is not composed of one type of tissue. All of these factors mean that a complete simulation will require fast processors and lots of memory.

Chapter 10 shows examples of modeling fluid flow and heat transport in the heart. Pormann et al. (2000) developed a simulation system for the flow of electrical current. This simulation is based on a set of partial differential equations called the *Bidomain Equations*, which are a widely used model of cardiac electro-physiology:

$$\begin{aligned}
\nabla \cdot \sigma_i \nabla \Phi_i &= \beta C_m \frac{\partial V_m}{\partial t} + I_{ion}(V_m, q) \\
\nabla \cdot \sigma_e \nabla \Phi_e &= -\beta C_m \frac{\partial V_m}{\partial t} - I_{ion}(V_m, q) \\
\frac{dq}{dt} &= M(V_m q)
\end{aligned} \quad (1.11)$$

where Φ_i and Φ_e are the intra- and extra-cellular potentials respectively, V_m is the transmembrane potential ($V_m = \Phi_i - \Phi_e$), and q is a set of state variables which define the physiological state of the cellular structures. I_{ion} and M are functions that approximate the cellular membrane dynamics, C_m is a transmembrane capacitance and σ_i and σ_e are conductivities.

This system of equations can be used to model 1-D nerve fibers, 2-D sheets of tissue, or 3-D geometries of the heart. The conductivities σ_i may be inhomogeneous (to model dead or diseased tissue). Different I_{ion} and M functions simulate nerve, atrial, or ventricular dynamics. The model parameters can be varied spatially to simulate diseased tissue or to study the effects of a channel-blocking drug on electrical conductivity.

The Bidomain Equations are solved for the potential V_m at each point in the 1-D, 2-D, or 3-D space, depending on the problem to be solved. In Pormann et al.(2000), the user has a choice of ten numerical integration methods to solve this set of partial differential equations. Many of these methods (explicit, semi- and fully implicit time integrators, adaptive time steppers and Runge-Kutta methods) are described in Chapter 7.

1.4.4 Using numerical methods to model the response of the cardiovascular system to gravity

Since the beginning of the space program, a problem has been to understand the response of the cardiovascular system to returning to a normal gravitational environment. Astronauts returning to earth may experience *postspaceflight orthostatic intolerance* (OI) - the inability to stand after returning to normal gravity. OI is an active area of modeling research, including: explaining observations seen during space flight, simulating the cardiovascular response from experiments in earth gravity and modeling interventions to the cardiovascular problems caused by return to gravity. A review of a number of models for OI is given by Melchior et al. (1992).

Modeling the response to gravity can best be illustrated with the work of Heldt et al. (2002), in which is a single cardiovascular model is used to simulate the steady-state and transient response to ground-based tests. The authors compared their modeling results compared with population-averaged hemodynamic data and found that their predicted results compared well with subject data. Their model provides a framework with which to interpret experimental observations and to study competing physiological hypotheses of the cause of OI.

The hemodynamics are modeled in terms of an electrical network; Fig. 1.2 shows the model for a single compartment. Assuming that the devices behave linearly, the model is a set of first-order differential equations. Although the equations are in terms of electrical units, the assumption of linearity allows one to use the model for hemodynamics. The flow rates, q, across the resistors, R, and capacitor, C, expressed in terms of the pressures, P, are given by:

$$\begin{aligned} q_1 &= (P_{n-1} - P_n)/R_n \\ q_2 &= (P_n - P_{n+1})/R_{n+1} \\ q_3 &= \frac{d}{dt}\left[C_n(P_n - P_{bias})\right] \end{aligned} \tag{1.12}$$

The subscripts are defined in Fig. 1.2.

Applying conservation of charge to the node at P_n yields $q_1 = q_2 + q_3$. The rate form of the conservation law yields:

$$\frac{dP_n}{dt} = \frac{P_{n+1} - P_n}{C_n P_{n+1}} + \frac{P_{n-1} - P_n}{C_n R_n} + \frac{P_{bias} - P_n}{C_n} \cdot \left(\frac{dC_n}{dt} \right) + \frac{dP_{bias}}{dt} \tag{1.13}$$

The entire compartmental model is shown in Fig. 1.3. The peripheral circulation is divided into upper body, renal, splanchnic, and lower extremity sections; the intrathoracic superior and inferior vena cavae and extrathoracic vena cava are separately identified. The model thus consists of 10 compartments, each of which is represented by a linear resistance, R and a capacitance, C. There are many similarities with the PHYSBE system in Chapter 10, but the reader will find PHYSBE to be a more detailed model of the cardiovascular system.

The model of Heldt et al. is described by 12 differential equations similar to Eq. (1.13). The authors used an adaptive step-size fourth-order Runge-Kutta integration routine, described in Chapter 7, to solve the system of differential equations numerically. The results reported by Heldt et al. use integration steps that range from 6.1×10^{-4} to 0.01 s with a mean step size of 5.6×10^{-3} s. The initial pressures, required to start the solution, were computed by a linear algebraic solution of a hemodynamic system where all pressures are assumed constant. These methods are described in Chapter 4.

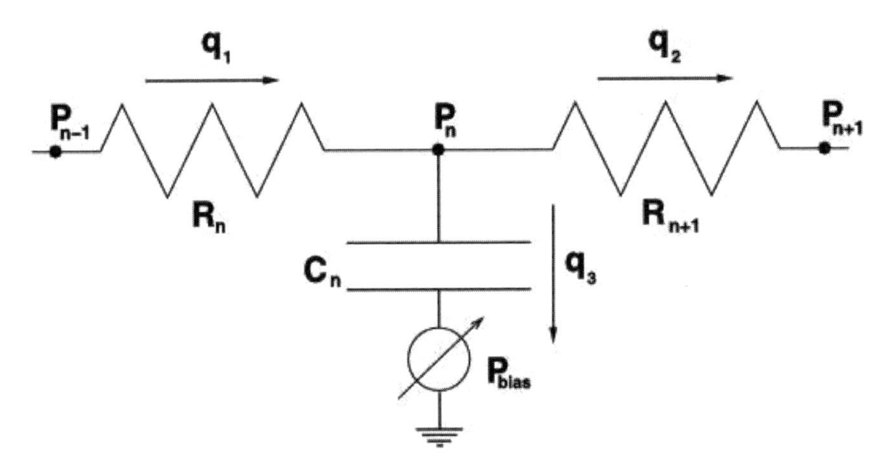

Figure 1.2 Single-compartment circuit representation. P is pressure; R is resistance; C is vascular compliance; q1, q2, and q3 are blood flow rates; n-1, n, n+1 are compartment indexes; Pbias, external pressure. From (Heldt 2002).

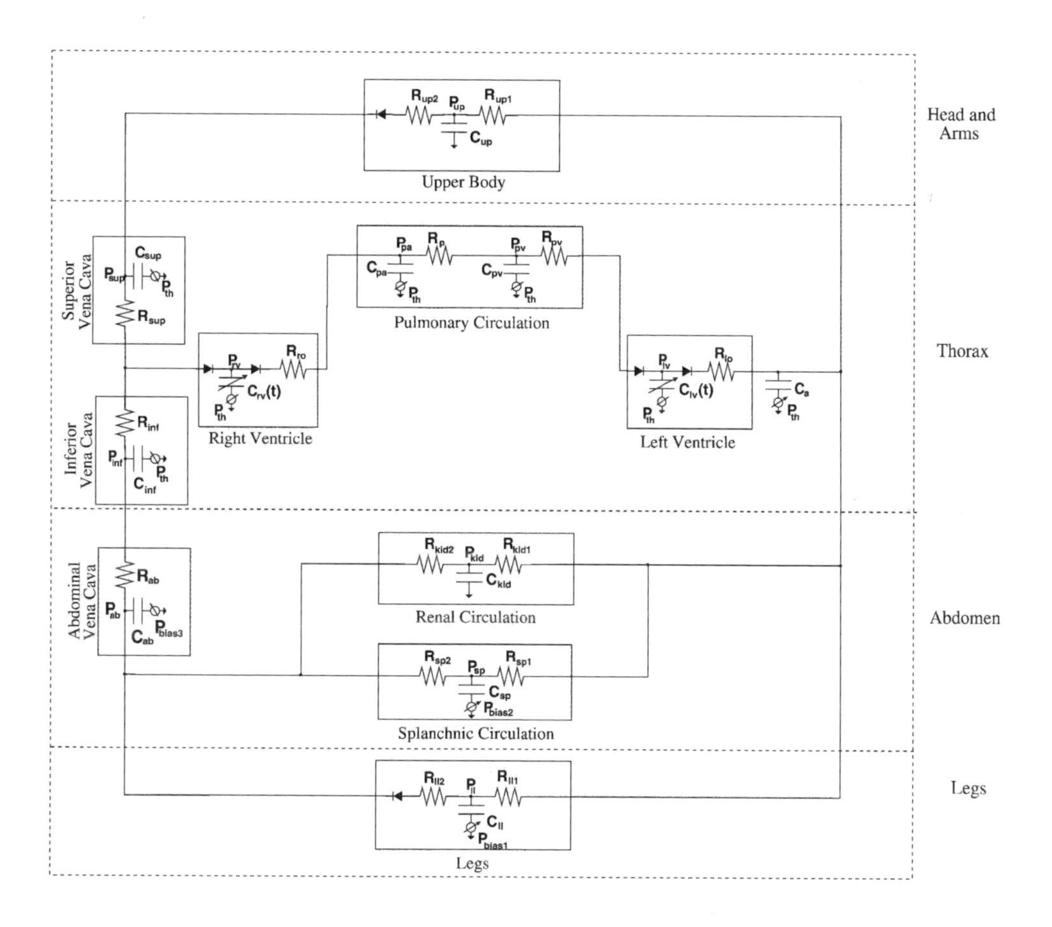

Figure 1.3 Circuit diagram of the hemodynamic system. lv, left ventricle; a, arterial; up, upper body; kid, kidney; sp, splanchnic; ll, lower limbs; ab, abdominal vena cava; inf, inferior vena cava; sup, superior vena cava; rv, right ventricle; p, pulmonary; pa, pulmonary artery; pv, pulmonary vein; ro, right ventricular outflow; lo, left ventricular outflow; th, thoracic; bias, as defined in Fig. 1.2. From (Heldt 2002).

1.5 Overview of the Text

The material presented in this text shows how to apply the principles and techniques of computing and numerical problem solving in a wide variety of problems that arise in biomedical engineering. The aim of this textbook is to provide the student with a *working knowledge* of Numerical Methods, i.e., to be able to read, understand and use Numerical Methods to solve biomedical engineering problems. A working knowledge has as its emphasis the understanding and application of the fundamentals. This approach provides the reader with exposure to a broad range of principles and techniques, but not theoretically rigorous derivations of the methods; the mathematical foundations are more appropriate for courses in numerical methods in a mathematics or computer science curriculum.

A second aim of the book is to give the reader examples of how to construct engineering models of biomedical systems. The *conservation laws first* theme from this chapter is reinforced in the models presented in the examples. Thus, the text is organized into four sections around the physical principles: fundamentals, using models of steady state behavior, using models of finite time behavior and using models of transient behavior. Throughout this text, examples from a variety of biomedical engineering specialties, including biomedical instrumentation, imaging, bioinformatics, biomechanics, and biomaterials, are used to reinforce the concepts.

1.5.1 Part I: Fundamentals

The first part of the text is an introduction to the fundamental principles of numerical methods. As programming is a necessary part of numerical methods, the examples, problems and applications in the text are given in MATLAB and basic terminology and principles of program development in the MATLAB language are reviewed. Since the reader may eventually implement a numerical method using another programming language, this section will help the reader relate implementation of common concepts of computer science: block structured design, data structures and analysis of algorithms. The emphasis is on design and the tradeoffs that a programmer makes when implementing an algorithm.

The introduction also includes a discussion of number representation and the effect that number representation has on the accuracy, precision and stability of the results of the computation. It is especially important in biomedical and healthcare applications that accuracy and stability be preserved and the system is as robust as possible. Whether or not MATLAB is used, keeping in mind concepts such as machine epsilon (the smallest number that can be represented by the computer) will help the programmer understand the importance of robust program and system design and controlling the propagation of error.

Lastly, this introduction includes a discussion of one of the most important concepts in numerical analysis, the role that a Taylor series approximation plays in mapping continuous models to their discrete analogs and methods for solving the discrete representation. The Taylor series plays an important role in deriving numerical algorithms and in characterizing the error introduced by the discrete approximation and also in the error propagated by performing a sequence of calculations.

1.5.2 Part II: Steady-state behavior (algebraic models)

Part II is an overview of techniques used to analyze systems that are in steady-state and whose models are formulated as algebraic equations that could be either linear or nonlinear. A single equation that is explicit in the unknown can easily be solved by methods from pre-college algebra; if the equation is implicit in the unknown, then root-finding techniques must be used. If the model is a set of simultaneous equations then numerical algebraic methods are used. Of course, the case of simultaneous, implicit equations is also treated. Each of these techniques is presented in this part of the text.

1.5.3 Part III: Dynamic behavior (differential equations)

Part III is of greatest interest to the biomedical engineer: modeling the transient behavior of dynamic systems and solving for the output of such systems. This section of the text includes methods for solving both ordinary and partial differential equations using numerical techniques as well as Simulink (a tutorial for Simulink is included in Appendix B).

Also in Part III, system behaviors during finite-time intervals, modeled by integral equations, are considered. In these cases, the solution of the model for an output parameter requires numerical integration and differentiation techniques. These methods are presented in this section, along with methods for improving the accuracy of the results. A recurring theme in this section is the trade-off that must be made between accuracy of the solution and the amount of computation that is performed.

1.5.4 Part IV: Modeling tools and applications

Part IV is an introduction to developing models of complex systems and to tools and techniques for analyzing complex behaviors. Examples of multicompartmental models of the circulatory, respiratory, and nervous systems are presented along with MATLAB implementations of the computational models. Tools and techniques for statistical and time series analysis are also presented with applications of these methods.

As the physical scale of problems continues to decrease, the need for mathematical models of biomedical systems continues to increase. The biomedical engineer who masters the material presented in this textbook will be well prepared to implement methods to solve mathematical models of biomedical systems for the steady-state, finite-time, or transient behavior of the system. He or she will be able to combine his or her analytical skills, computational skills, and mastery of the link between the two, numerical methods.

1.6 Lessons Learned in this Chapter

After studying this chapter, the biomedical engineering student will have learned the following:

- Mathematical models are tools that biomedical engineers use to predict the behavior of a system.
- Biomedical engineers model systems in one or more of three different states: steady-state behavior, behavior over a finite period of time, or transient behavior.
- Derivation of mathematical models begins with a conservation law.
- Numerical methods are the bridge between analytical formulation of the models (using algebra, calculus, or differential equations) and computer implementation.

1.7 Problems

1.1 List five applications of computers in biomedical engineering and briefly describe each application.

1.2 You have been assigned the task of designing a computerized patient-monitoring system for an intensive-care unit in a medium-sized community hospital. What parameters would you monitor? What role would you have the computer play?

1.3 Why is a computer-averaged, EEG-evoked response signal easier to analyze than a raw signal?

1.4 What advantage does a computer give in the automated clinical laboratory?

1.5 List the types of monitoring equipment normally found in the ICU/CCU of a hospital.

1.6 Are computer systems always applicable in biomedical equipment?

1.7 Describe three computer applications in medical research.

1.8 On a biomedical engineering examination, a student named George computes in feet and inches the maximum distance that a certain artificial heart design could pump blood upward against gravity. Unfortunately, in recording this distance on his examination paper, he reverses the numbers for feet and inches. As a result, his recorded answer is only 30% of the computed length, which was less than 10 feet with no fractional feet or inches. What length did George compute in feet and inches?

1.8 References

Enderle, J. D., Blanchard, S. M., and Bronzino, J. D. 2000. *Introduction to Biomedical Engineering*. San Diego, CA: Academic Press.

Fournier, R. L. 1999. *Basic Transport Phenomena in Biomedical Engineering*. Philadelphia, PA: Taylor & Francis.

Gevertz, J. L., Dunn S. M., Roth, C.M. 2005. Mathematical model of real-time PCR kinetics. *Biotechnol. Bioeng.*, in press.

Hallett. M. (2000). Transcranial Magnetic Stimulation and the Brain. *Nature*, **406:**147-150

Heldt, T., Shim, E. B., Kamm, R. D., Mark, R. G. 2002. Computational modeling of cardiovascular response to orthostatic stress. *J Appl Physiol.*, Mar:**92**(3): 1239-1254.

Melchior, F. M., Srinivasan, R.S., Charles, J.B. 1992. Mathematical modeling of human cardiovascular system for simulation of orthostatic response. *Am J Physiol.*, **262**(6 Pt 2):H1920-1933.

Norton, S. J. 2003. Can ultrasound be used to stimulate nerve tissue? *Biomed. Eng. Online*, **4**:2(1):6.

Nebeker, F. (2002). Golden accomplishments in biomedical engineering. *IEEE Engineering in Medicine and Biology Magazine*, **21**:17-47.

Pormann, J. B., Henriquez, C. S., Board, J. A., Rose, D. J., Harrild, D. M., and Henriquez, A. P. 2000. Computer Simulations of Cardiac Electrophysiology. Article No. 24, *Proceedings of the 2000 ACM/IEEE Conference on Supercomputing*. Dallas, TX.

Thompson, W. J. 2000. *Introduction to Transport Phenomena.* Upper Saddle River, NJ: Prentice Hall PTR.

Vierstraete, A. 1999. http://users.ugent.be/~avierstr/principles/pcr.html. Last viewed August 24, 2005.

Chapter 2
Introduction to Computing

2.1 Introduction

The computer is a ubiquitous tool for all engineering disciplines, and biomedical engineering is no exception. In biomedical engineering, computers are used to control instrumentation and data collection, analyze images, simulate models, and perform statistical analysis, among many other tasks. Biomedical engineers with good computing skills will be able to apply their expertise to any problem in this diverse field.

In this chapter, we will introduce the core areas of computing: programming languages and program design, data structures, and the analysis of algorithms. There are many MATLAB examples that illustrate how to develop a computer program using good program design principles.

The material in this chapter will enable the student to accomplish the following:

- Identify functions performed by computers in biomedical engineering applications
- Differentiate between imperative, functional, and object-oriented programming languages
- Identify the common data structures used in computer programming languages
- Describe how numbers are represented in computer programming languages
- Describe a computer algorithm and its role in biomedical engineering applications
- Identify a block-structured programming language and describe good programming practice
- Write example computer programs in MATLAB

2.2 The Role of Computers in Biomedical Engineering

The sphygmomanometer, a device for measuring blood pressure, is one example of biomedical instrumentation that is available to the general consumer because of the widespread use of computers in medical devices.

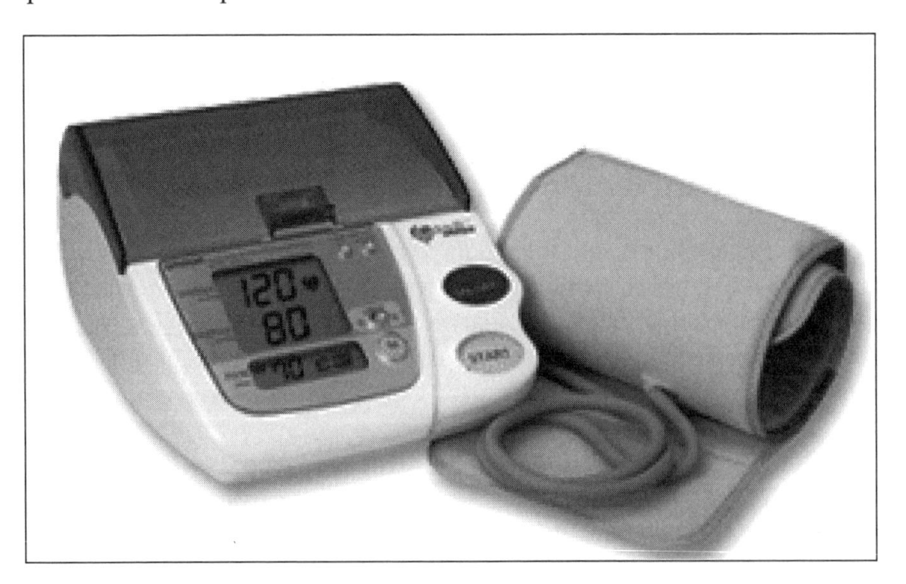

Figure 2.1 An automatic sphygmomanometer for home use. Image courtesy of Omron Healthcare.

The computer allows a consumer to operate the device by providing the:

1. primary user interface
2. primary control for the overall system
3. data storage for the system
4. primary signal processing functions for the system
5. safe and reliable operation of the overall system, including cuff inflation

The computer program in the blood pressure (BP) monitor must do all of these functions and execute them correctly each and every time the monitor is used. The biomedical engineer who writes the computer program must keep in mind the following criteria:

1. The program must be correct and operate the same way each and every time.
2. The users must have the confidence that all measurements are equally reliable.
3. The monitor must be easy to use.

The sphygmomanometer for home use is only one example of the impact that computers have had in biomedical engineering; other examples include medical imaging, especially computed tomography (CT) and magnetic resonance imaging (MRI); bioinformatics; hemodynamic and other simulation; and high-resolution microscopy.

Instrumentation such as the BP monitor, CT or MRI imager, and the microscope have a common structure. In all cases, the computer controls the electronic interface (the user interface and data collection), which is connected to the subsystems that sense the tissue properties and control the sensor.

All of these functions could be performed by a trained user, but then instruments such as the blood pressure monitor could not be put into the hands of the consumer for two reasons: First, the general consumer is untrained. Second, human performance changes over time. The computers embedded in the instruments have helped to standardize the operation, so that non-trained users can operate the BP monitor and other medical devices while being in better control of their healthcare.

Once a computer is given a standard set of instructions, the performance will be the same each time the device is used, unless there is a hardware failure. Some of the advantages of using computers are:

1. The actions directed by the computer and its program are *reproducible*: the operation will not change over time unless one or more parts (the electronics, actuator, or sensor) fail.

2. The precision and accuracy can be controlled: *precision* is the smallest possible resolvable event; it differs from *accuracy*, which is how close you can get to the truth (the real property). Precision and accuracy are dependent on both the sensors and the computation. The emphasis in this text will be on how we can best optimize the computer programs to maximize the precision and accuracy of the results.

Computers are not without their drawbacks. Some of the disadvantages of using computers instead of human control are:

1. Computers require precise and complete instructions. There is no room for ambiguity or assumptions.
2. All too often, program testing is not clear or complete. Testing a computer program, especially a large computer program, can be a full-time job by itself. You often have to worry about unanticipated input or conflicting input that is not expected. You may collect more data than you have accounted for. The number of possible combinations of inputs is exponential in the number of independent variables; for a moderate number of input variables, the number of possible inputs is too large to test.
3. There can be logic errors, or syntactic errors, that can later cause execution errors. In very large applications, like bioinformatics programs, special Web interfaces are built just to keep track of the bugs. A good example is http://www.biojava.org, the Web site for a set of Java-based tools for processing biomedical data, such as gene sequences.

Besides controlling or automating equipment, computers can make biomedical engineers more efficient by:

1. Providing design tools, including simulation (using tools such as Simulink, Pro-E, and ANSYS)
2. Keeping documentation (using tools such as MS Word or Excel)
3. Providing test procedures and results (using tools such as MATLAB and LabVIEW)

The emphasis in this book is to help the new biomedical engineer develop computer programs that control the precision and accuracy of the measurements and computations as an integral part of solutions to real-world bioengineering problems. First, the programmer must be familiar with the tools available in a programming environment such as MATLAB.

2.3 Programming Language Tools and Techniques

There are three classes of programming languages in use today: the imperative, or state transition class; the object-oriented class, and the functional class. In each class, there are a large number of programming languages. How can an engineer decide which class and which language to use?

The best examples of imperative or state-transition languages are C and Fortran. Programs written in these or other imperative languages are executed one statement at a time, in the order that the statements are written.

Object-oriented programming languages constitute the largest and most widely used class of languages today and include C++, Java, Python, and Smalltalk, the programming language that started the object-oriented revolution. Object-oriented programs execute much differently than imperative programs; the operations executed are dictated by the type of the data object and the desired action. The implementation of an action will depend on the type of the object (think about the difference between adding two integers and adding two vectors) although the *name* of the operation (e.g., *add*) may be the same in both cases. The examples in this chapter and throughout this book are written in MATLAB, an object-oriented programming language and environment. The user not familiar with MATLAB is strongly encouraged to read Appendix A, an introduction to the MATLAB programming language and environment.

Functional programming languages are written as a set of definitions of functions and a composition of these functions to be executed. The execution of this composition can be sequential, like the imperative languages, or parallel. Some common examples are Scheme, Sisal, or Caml. Functional languages often include object-oriented properties. This class of languages will not be considered here.

Both imperative and object-oriented languages have common programming language constructs that are designed to help the programmer develop readable and testable programs efficiently. These programming language constructs (statements) allow the programmer to manage the computation being performed and the flow of execution. Modern programming languages limit the control flow statements to one of three types: a sequence of statement; a statement for conditional execution; and a control structure for repeating a sequence of statement. These languages are called block-structured languages. All three of these statements have one path in and one path out; this standardizes program design and development.

2.3.1 Sequences of statements

A sequence of instructions is delimited by special symbols or keywords. These delimiters indicate the single beginning and end to the block.

Example 2.1 Programs that are sequences of statements.

Create a vector of the even whole numbers between 31 and 75.

Solution

The even whole numbers between 31 and 75 are the even numbers 32 to 74. There are two ways to create the vector: First, by typing the list of numbers or second, by using the MATLAB shorthand notation **first:increment:last**.

The solution using the first method is left to the reader. The solution using the second method is as follows:

1. The first number in the vector is 32.
2. The increment is 2, since the vector is to contain only the even numbers.
3. The last number in the vector is 74.

Therefore, the vector is specified as `[32:2:74]`. The MATLAB output is:

```
>> x = [32:2:74]

x =
   Columns 1 through 11
      32    34    36    38    40    42    44    46    48    50    52
   Columns 12 through 22
    54    56    58    60    62    64    66    68    70    72    74
```

2.3.2 Conditional execution

A conditional statement is required if there is a need for an execution path that depends on the values of the data input, or external input. These conditional statements also have delimiters: generically, `if…then` or `if…then…else`. There is still only one entrance to the block at the `if` keyword and one exit where the branch(es) return to the main line of the control flow.

2.3.2.1 If-then statements

The simplest form of conditional execution is the `if…then` statement, which in MATLAB has the form `if` *`expression statements`* `end`. If the *`expression`* is evaluated to be true, then the statement block is executed. The expression is always evaluated, and control flow always follows with the statement after the `end`.

2.3.2.2 If-then-else

The `if…then` statement provides the ability to program one type of conditional execution control flow. The `if…then…else` construct provides the ability to implement a control flow where one of two blocks, but not both, are executed.

In MATLAB, the syntax of the `if...then...else` statement is

```
if expression
   statement1
else
   statement2
end
```

The expression is always evaluated; if the expression is true, then the block `statement1` is executed; if the expression is evaluated to be false, then the block `statement2` is executed.

There is another variant of conditional execution in MATLAB, the `if...elseif...else...end` statement with the syntax

```
if expression1
   statement1
elseif expression2
   statement2
else
   statement3
end
```

The `elseif` clause allows for the evaluation of an alternate expression; if the first expression is false and the second expression is true, then the block `statement2` is executed. Here is an example:

Example 2.2 Simple control flow using `if...then...else`.

Write a MATLAB script to compute the following function:

$$f(x) = \begin{cases} -1 & x < 0 \\ 0 & x = 0 \\ 1 & x > 0 \end{cases}$$

Solution

In this example you are asked to create an m-file, a MATLAB script that computes the function f defined above. The value returned by the script is one of three values that depend on the input. The mathematics can be easily translated into MATLAB commands.

The valued to be returned by the m-file is -1 if the argument x is less than zero. The mathematical statement above is easily translated into the MATLAB statement

```
if x<0 f = -1;
```

where the variable `f` is the variable whose value is returned by the function.

If x is less than zero then the value to be returned has been determined (`f=-1`) and no more MATLAB statements are evaluated. If the value of `x` is not less than zero, then `f=-1` is not executed and the final value of `f` must still be determined.

If the value of `x` is not less than zero, then either it is equal to zero or it is greater than zero. According to the mathematical definition above, if `x` is equal to zero, then the value of `f` should be 0; If `x` is greater than zero, then the value of `f` should be 1.

The MATLAB script that implements this function will have three *expression-statement* pairs. However, the three statements should not be executed as a sequence; if the first assignment statement is executed, then the evaluation of f should be stopped. Can you tell why?

To execute only those statements that should be executed and no others, use the structure to test multiple conditions and evaluate only the statements that satisfy the conditions. The control structure to implement this design requires the `elseif` clause in the conditional statement. The final code is:

```
if x < 0
    f = -1;
elseif x == 0
    f = 0;
elseif x > 0
    f = 1;
end
```

As a followup to this problem, we leave it to the reader to show that this MATLAB code performs the same function as the built-in function `sign`.

2.3.2.3 Switch statement

In most block-structured programming languages, including MATLAB, there is a facility for conditionally executing a sequence of statements from an arbitrary number of alternatives. Of course, one could implement this control flow structure using a large `if...elseif...end` structure, but there is a more convenient shorthand notation in MATLAB, the `switch...case...otherwise...end` structure. The syntax of the MATLAB switch statement is

```
switch switch_expr
  case case_expr
    statement,...,statement
  case {case_expr1,case_expr2,case_expr3,...}
```

```
        statement,...,statement
   ...
     otherwise
        statement,...,statement
   end
```

Each alternative is called a case statement and has three parts: the `case` directive, one or more case expressions, and a sequence of one or more statements. When the `switch_expr` equals a `case_expr`, the corresponding sequence of statements is executed. If no `case_expr` equals the `switch_expr`, then the statements of the `otherwise` case are executed.
There is one important difference with the switch statement: The expressions must be either scalars or strings. MATLAB finds the right sequence of statements to execute by attempting to find the `case_expr` that matches the `switch_expr`.

Once the statements are executed, the control flow resumes with the statement that follows the switch statement, as with all other statements.

Example 2.3 Use of the `switch` statement.

Write a MATLAB script that reads a character variable (s or c) and an integer variable (1, 2, or 3) and plots a function based on the following chart:

	1	2	3
s	$\sin\theta$	$e^{-\theta}\sin^2\theta$	$\sin 3\theta / e^{-\theta}$
c	$\cos\theta$	$\cos 2\theta / e^{-2\theta}$	$e^{-\theta}\cos^3\theta$

You may assume that there are 100 points over the interval $0 \le \theta \le 2\pi$.

Solution

Instead of beginning by immediately starting to write MATLAB code for this problem, it is best to begin by becoming familiar with using the plot function. The function plot can take a number of different forms. The function call `plot(Y)` will plot the vector Y against its index (or independent variable). The function call `plot(X,Y)` will plot the values of the dependent variable in the vector Y against the values of the independent variable in the vector X. For example, the MATLAB commands result in Fig. 2.2.

```
>> x=1:100;
>> plot(sin(x/3));
```

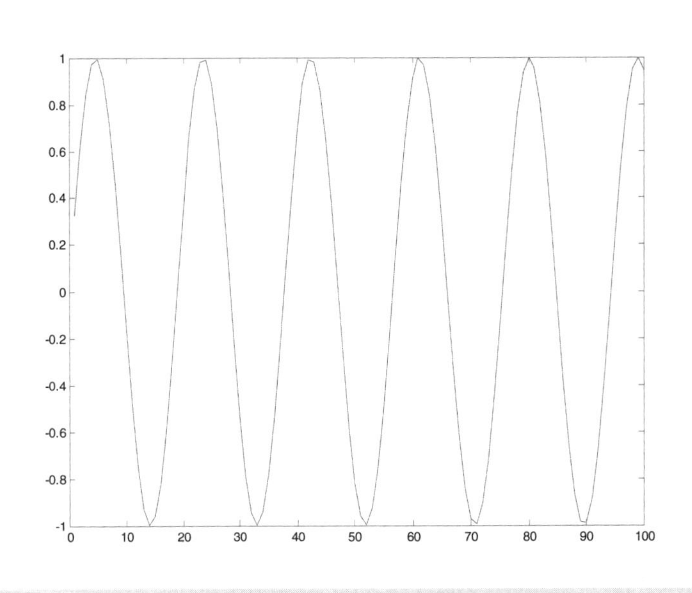

Figure 2.2 sin(x/3) curve.

Notice that the values of the **independent variable** x are plotted on the horizontal axis and the values of the **dependent variable** are plotted on the vertical axis. The independent variable for the given problem is the angle θ, in radians. Since the problem specifies using 100 values over the range, the vector

```
x=0:(2*pi)/100:2*pi;
```

specifies the values of the independent variable for this problem.

The solution to this problem is a MATLAB script that reads two inputs. These two inputs identify a row (indicated by either the character 's' representing the function sine or the character 'c' representing the function cosine) and column (indicated by one of the integer 1, 2, or 3) indices of the table above. The MATLAB script must evaluate and plot the function in the cell identified.

It is easiest to use `switch` statements, rather than a large number of `if-then-else` statements, to control the flow of execution of the program. An easy strategy will be to have one switch statement that selects between the two rows and in each statement block, have another switch statement that will select one of the columns. To begin,

```
switch letter
case 's'
a switch statement for row s goes here
case 'c'
a different switch statement for row c goes here
end
```

With this strategy, the switch statements for each row can be independently written and independently tested, keeping with our principles of good programming practice. The switch statement for the row *s*:

```
switch col
    case 1
        plot(theta,sin(theta));
    case 2
        plot(theta,exp(-theta).*(sin(theta)).^2);
    case 3
        plot(theta,sin(3*theta)./exp(-theta));
    end
```

can be inserted as the block following the case 's'. The complete MATLAB script for this exercise is:

```
twopi=2*pi;
theta=0:(twopi/100):twopi;
letter=input('Please enter either s or c: ');
col=input('Please enter either 1, 2 or 3: ');
figure(1);
switch letter
case 's'
    switch col
    case 1
        plot(theta,sin(theta));
    case 2
        plot(theta,exp(-theta).*(sin(theta)).^2);
    case 3
        plot(theta,sin(3*theta)./exp(-theta));
    end
case 'c'
    switch col
    case 1
        plot(theta,cos(theta));
    case 2
        plot(theta,cos(2*theta)./exp(-2*theta));
    case 3
        plot(theta,exp(-theta).*(cos(theta)).^3);
    end
end
```

A sample session using this script is below, and the results are shown in Fig. 2.3. Why does the `plot` function use the scale from 0 to 7 for the independent variable?

```
Please enter either s or c: 's'
Please enter either 1, 2 or 3: 2
```

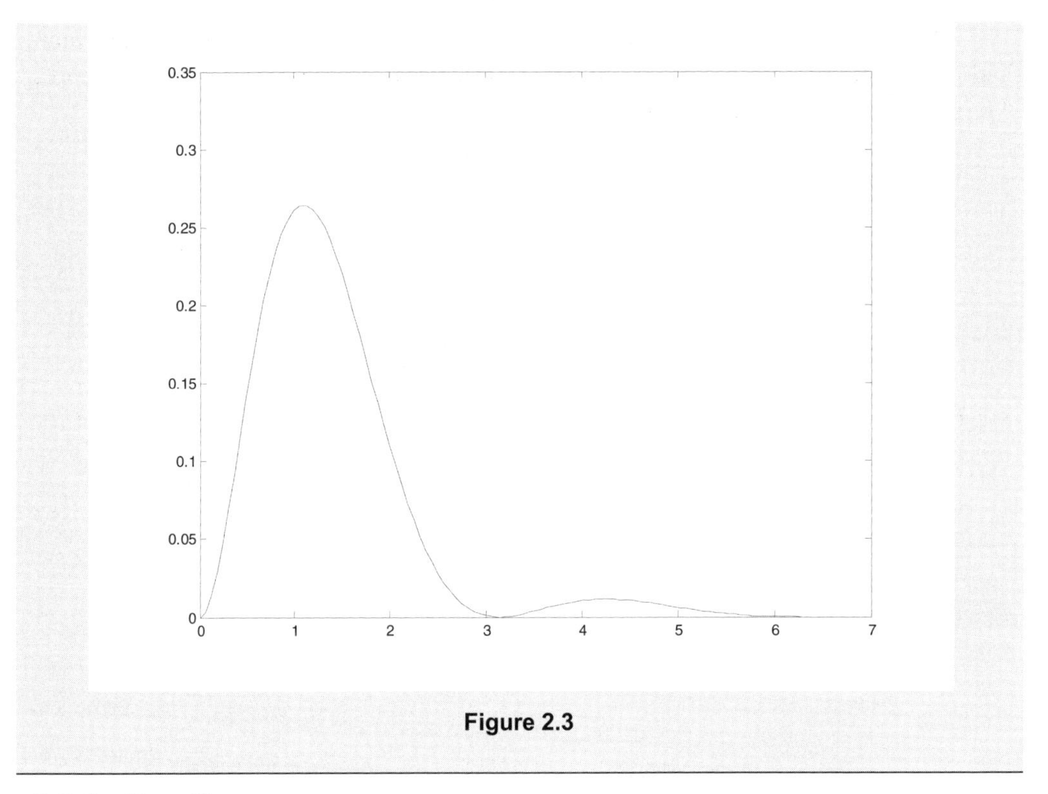

Figure 2.3

2.3.3 Iteration

2.3.3.1 While loops

Repetition is where a sequence of instructions, or block, is repeated until a condition is satisfied. The condition can be expressed in terms of state of the data or an external input. A generic repetition structure is a `while…do loop`.

Example 2.4 The use of `while` loops.

Write a script that computes the number of random numbers required to add up to 20 (or more). You are to use the random-number generator `rand` in this exercise.

Solution

The random number generator in MATLAB returns a pseudo-random number that is uniformly distributed between 0 and 1. We refer to it as a pseudo-random number because the sequence of numbers generated depend on the state of the computer random number generator at the time that the function call is made. For example, the following call to the random number generator

```
>> rand
ans =
     0.9501
```

returns the result shown. The random number returned can be equal to or greater than zero, but is strictly less than one.

The MATLAB script should stop executing once the sum of all the random numbers generated reaches 20. This means that the script will continue to **repeat** generating random numbers, adding each new one to the running sum, until the sum reaches at least 20. You also have to count the number of random numbers generated. Since it is not possible to predict exactly how many numbers need to be added, the program will have to repeat generating random numbers, until the condition is satisfied. The MATLAB `while...end` construct will be used to control the repetition.

The algorithm for solving this problem is:

1. The running sum of random numbers must start at 0 before the random numbers are generated. Set the count to be zero.
2. Check to see if the sum is greater than or equal to 20. If it is, then stop this script. If the sum is less than 20, then go to step 3.
3. Generate a random number and add it to the sum. Add 1 to the count of random numbers generated. Go to step 2 to check the sum.

The MATLAB script, shown below, is saved as `whileloop.m`

```
% whileloop.m
add=0;
count=0;
while add<20
    add=add+rand;
    count=count+1;
end
display(count);
display(add);
```

and is executed from the command window to obtain the output:

```
>> whileloop
count =
    43
add =
   20.0364
```

2.3.3.2 For loops

There are variations of iterative control structures, just as there were variations of the control structures for conditional execution. A *for loop* (`for...end`, in MATLAB) is a repetition structure where the block is repeated a fixed number of times, once for each value of a control variable. Typically, the execution of the block changes with the control variable—for example, accessing a different element of an array.

Example 2.5 Using `for...end` loops.

Given the vector x = [1 8 3 9 0 1], create a MATLAB script to add the values of the elements (Check your result with the `sum` command.)

Solution

The statement of this problem gives the clues that the script need only add up the values in the given vector, not an arbitrary vector. You know that the length of the vector is 6; one way to write the script would be the commands

```
x=[1 8 3 9 0 1];
add=0;
add=add+x(1);
add=add+x(2);
add=add+x(3);
add=add+x(4);
add=add+x(5);
add=add+x(6);
add
sum(x)
```

Notice that this sequence has a common structure: each element of the array is added to the sum. If the vector was longer, writing out this sequence of statements would become very tedious; since the statement are similar, the complete sequence can be replaced with a shorthand description that uses a *for loop*:

```
% forloop.m
x=[1 8 3 9 0 1];
add=0;
for i=1:length(x)
    add=add+x(i);
end
add
sum(x)
```

where the `for...end` loop is the shorthand notation for the six statements that add each element of x, in turn. The result of executing `forloop.m` is:

```
>> forloop
add =
    22
ans =
    22
```

Note that the equivalent of all the statements in the script forloop.m may be accomplished with a single command: `sum(x)`. In a small way, this demonstrates the power of MATLAB.

2.3.4 Encapsulation

The last key to writing well-structured MATLAB programs that are easy to read, understand and maintain is *encapsulation*. Encapsulation means to group together those instructions or statements that form one particular function. Encapsulation will improve readability and make the job of testing and debugging easier. It is hard to manage one large sequence of MATLAB statements.

Encapsulation is implemented in MATLAB by using *m-files*: text files that contain code in the MATLAB language. If large MATLAB programs are organized well, there will be one m-file for the main program and other m-files for the *functions* and *scripts* that are called by the main program or invoked at the MATLAB command prompt.

All modern programming languages have features for encapsulating sequences of instructions. There are typically two types of encapsulating programming constructs: *subroutines* and *functions*.

A *script* is a sequence of instructions that does not accept input arguments nor return output arguments. A script operates on or modifies existing data in the workspace. In other block structured programming languages, the organizational equivalent of MATLAB scripts will be referred to as *subroutines*.

A *function* is a sequence of instructions like a script, but the function may accept input arguments and return output arguments. The function may also have MATLAB variables that can only be accessed inside the function.

Like all block-structured programming tools, subroutines and functions each have only one control flow entry point and one exit.

Example 2.6 Using scripts and functions.

Write a *script* that asks for a temperature (in degrees Fahrenheit) and calls a *function* to compute the equivalent temperature in degrees Celsius. The script should keep running until no number is provided to convert (Note: The command `isempty` will be useful here).

Solution

The solution to this problem has two parts. The first is the function that coverts a temperature in Fahrenheit into its equivalent temperature in Celsius. The second part of the solution is the script that calls this function. The best strategy is to write and test the function first, and only after it is determined to be correct, write the script that calls the function.

The MATLAB function needs to compute the temperature in Celsius using the following formula:

$$C = \frac{5}{9}(F - 32)$$

The temperature in Fahrenheit is the argument to the function, which returns the temperature in degrees Celsius. The MATLAB function is:

```
% Far2Cel.m
function C=Far2Cel(F)
C=(5/9)*(F-32)
end
```

which should be tested before going on to complete the exercise. The script that uses this function should take as input from the keyboard a temperature in degrees Fahrenheit, display the temperature converted to degrees Celsius and then wait for the next input.

The description should suggest that there is a `while` loop, not a `for` loop in this script since the number of iterations is uncertain and cannot be computed *a priori*.

First, get the keyboard input of temperature in Fahrenheit and check to see if the input is empty. If the input is empty, stop execution; otherwise, convert the temperature to Celsius and then get another input temperature and once again check to see if the input is empty. Iterate until there is no more keyboard input. The script `F2C.m` is:

```
% F2C.m
F = input('The value of the temperature in degree Fahrenheit = ');
while ~isempty(F)
C = Far2Cel(F);
display(C);
F = input('The value of the temperature in degree Fahrenheit = ');
end
```

which is called from the MATLAB command line. In the sample execution below, note that the execution stops when there is no numeric input.

```
>> F2C
```

```
The value of the temperature in degree Fahrenheit = 32
C =
     0
The value of the temperature in degree Fahrenheit = 100
C =
   37.7778
The value of the temperature in degree Fahrenheit = 212
C =
   100
The value of the temperature in degree Fahrenheit = 77
C =
    25
The value of the temperature in degree Fahrenheit =
```

2.4 Fundamentals of Data Structures for MATLAB

MATLAB programs have two parts: a sequence of instructions for the computation to be performed and the representation of the data that are being operated on. The previous section is an overview of the basic programming language constructs that underlie all MATLAB programs. This section is an overview of the techniques available in MATLAB for representing data.

Although the emphasis of this text is on numerical methods and quantitative results, there is still the need to include other information, such as names of variables, the units of the results, and information on the interpretation. The format of the numerical output is important as well.

All modern programming languages include a facility for creating and using different data types. There are fundamental data types and mechanisms for forming aggregates of the data. This section is an overview of the six fundamental data types in MATLAB: `double`, `char`, `sparse`, `uint8`, `cell`, and `struct`.

2.4.1 Number representation

In calculus and courses on differential equations as well, the real numbers were used, whether this was made explicit or not. The real numbers have infinite precision, which is something that is not available in computer programs. Infinite precision would require infinite memory; no computer has infinite memory.

Thus, the real numbers have to be represented by an approximation — called the floating-point numbers. Floating-point numbers require only a finite amount of memory, but that means that there are real numbers that cannot be represented using a floating-point representation. This topic is treated in more detail in Chapter 3; floating-point representation has a large impact on the accuracy and precision of the computation and is an important topic to master before studying numerical methods in detail.

A number written in a MATLAB script or function is written using the conventional notation. MATLAB does not distinguish between an integer, real, or complex number; they are all stored as the same type of MATLAB variable.

A number may have a leading plus or minus sign and an optional decimal point. Numbers written in scientific notation use the letter `e` to specify the exponent (base 10). Imaginary numbers use either `i` or `j` as a suffix to indicate an imaginary component. Some examples of numbers that can be represented in MATLAB are:

Example 2.7 Number representation in MATLAB.

Positive integer	3
Negative integer	-45
Real number	0.00001
Scientific notation	2.71828e9
Complex number	3+5i
Imaginary number	-3.14159j

The amount of memory available to store each number is limited; there is an upper and lower limit to the real numbers that can be represented by the floating-point numbers in MATLAB. The MATLAB functions `realmin` and `realmax` return, respectively, the smallest and the largest real numbers that can be represented in MATLAB. The function `inf` returns the representation of infinity, and NaN returns a representation for Not-a-Number, the result of an undefined operation (such as 0.0/0.0) in MATLAB.

Complex numbers can be assigned directly to numeric variables or they can be created from two real components; the MATLAB command `c=complex(3,5)` creates a complex number `c` that is `3+5i`. The result of the command `c=complex(3,0)` is a complex number with an imaginary component 0 and is not a real number.

Example 2.8 Complex numbers.

Compute the inverse of the complex number 3+5i and verify the result by hand calculation.

Solution

Recall that the inverse of a complex number $a+bi$ can be computed by

$$(a+bi)^{-1} = \frac{(a-bi)}{(a+bi)(a-bi)}$$

The MATLAB function `inv()` computes the inverse of the argument; the command:

```
>> inv(3+5i)
ans =
    0.0882 - 0.1471i
```

can be verified by the calculation using the formula above, using the MATLAB function `conj()` to compute the complex conjugate of the argument:

```
>> x=3+5i;
>> conj(x)/(x*conj(x))
ans =
    0.0882 - 0.1471i
```

2.4.2 Arrays

An array is a data structure for representing a collection of objects that provides the ability to access each element of the collection in the same amount of time. You do not have to search all the way down a list, or through a matrix in order to get to a desired element.

All arrays in MATLAB are stored as column vectors, and each element can be accessed using only the single index into the column vector. However, if there is more than one index in the array reference, the location in the column vector is computed using the formula:

$$(j-1)*d+i$$

to reference the array element $A(i, j)$, where d is the length of one column. Subarrays of a multidimensional array can be referenced in their entirety as well.

Example 2.9 Indexing arrays in MATLAB.

Create a 5x5 matrix with elements $A(i,j)=1/(2^i+3^j)$. Print the entire matrix, the first row and the second column. Give two ways of displaying the element $A(3, 2)$.

Solution

Unlike many other programming languages, MATLAB does not have a dimension statement or method for declaring the space allocated for a variable; MATLAB automatically allocates storage for matrices and other arrays. For this problem, space for the matrix A is allocated at the time the elements are created. Since a formula is given for each element, the matrix can be created in a script that also includes the statements to print the elements (including the rows and columns). The size of the matrix is specified, so the elements can be computed using for loops, followed by the references to print the subelements. The MATLAB script is:

```
% ArraysRef.m
for i=1:5
    for j=1:5
        A(i,j)=1/(2^i+3^j);
    end
end
A
A(1,:)
A(:,2)
A(8)
A(3,2)
```

Notice that the first row is referred to as `A(1,:)`. The subscript 1 refers for the row and the second subscript is not a number, but rather the symbol ':' which refers to **all elements in this dimension**. The ':' in `A(:,2)` has the same meaning, but this time it refers to the row dimension and means all rows with the column being fixed at 2.

Remember that all MATLAB arrays are stored as column vectors, with the columns following each other in order. Therefore the element `A(3,2)` can be referred to as an element in a vector or as the element in the matrix (which is logically how we would expect to refer to it). The above script produces the following results when executed from the command window:

```
>> ArrayRefs
A =
    0.2000    0.0909    0.0345    0.0120    0.0041
    0.1429    0.0769    0.0323    0.0118    0.0040
    0.0909    0.0588    0.0286    0.0112    0.0040
    0.0526    0.0400    0.0233    0.0103    0.0039
    0.0286    0.0244    0.0169    0.0088    0.0036
ans =
    0.2000    0.0909    0.0345    0.0120    0.0041
ans =
    0.0909
    0.0769
    0.0588
    0.0400
    0.0244
ans =
    0.0588
ans =
    0.0588
```

Notice that the row is displayed as a row and the column is displayed as a column.

2.4.3 Characters and strings

Characters in MATLAB programs are represented the same way as in all other programming languages, using the ASCII character set. The numbers 32 to 127 are codes for digits, upper-case and lower-case letters. Characters with codes in the range 0 to 31 are special control codes, such as backspace. A string in MATLAB is simply an array of characters; and although mathematical operations cannot be applied to these arrays, there are operations that can be applied to strings: concatenation, character or string comparison or searching, and string conversion.

Example 2.10 Character strings as arrays.

Create two MATLAB variables with the strings 'Biomedical' and 'Engineering' respectively. Concatenate the two strings, print the combined string and then print the indices of all occurrences of the letter 'e'.

Solution

The two strings are first assigned to separate MATLAB variables and then concatenated using the MATLAB function `strcat`. The concatenated string is then search for all occurrences of the letter 'e' by first finding the length of the string and the comparing each character in the string (a character array) to the letter 'e'. A `for` loop can be used since the number of comparisons is fixed at the length of the string. The MATLAB code is:

```
% Strings.m
B='Biomedical';
E='Engineering';
BME=strcat(B,E)
L=length(BME);
for i=1:L
    if(BME(i)=='e')
        display(i);
    end
end
```

A sample execution of the script `Strings.m` is:

```
>> Strings
BME =
BiomedicalEngineering

i =
      5
```

```
i =
     16
i =
     17
```

First, notice that there is no space between the two words Biomedical and Engineering. Neither the variable `B` nor the variable `E` had a space at the end or beginning, respectively. Second, the expression in the `if` statement shows how a character string can be treated as an array. Each character in the string is referenced as an array element `BME(i)`, which is exactly how an element of a numeric array would be referenced. If the comparison of a character was to a number, then the array element `BME(i)` would be treated as its ASCII equivalent numeric code, instead of a character. Lastly, the capital E in Engineering was not detected, since the comparison was only to a lower case e.

2.4.4 Logical or Boolean data types

In MATLAB, there is not a distinct logical or Boolean data type; rather, a logical false (F) value is represented by a 0, and anything nonzero represents a logical true (T). There are three logical operators in MATLAB, the unary NOT operator (represented by ~), the binary AND (represented by &), and the inclusive OR (represented by |) operators. The exclusive OR operation is implemented by the MATLAB function `xor()` as in `xor(A,B)`.

Logical operators are most often used in expressions found in conditional or iterative execution statements. In MATLAB, though, there is another way to use logical operations, called *logical indexing*. Logical indexing is a technique for selecting elements from an array that satisfy specific criteria. The elements are selected by forming an *index array* whose elements are either 0 or 1. An index array can be constructed or can be created from another array by using the MATLAB function `logical()`, which takes as an argument a numeric array and returns an index array.

The index array is used as the subscript to the original array, and the resulting array is the set of elements that satisfied the original criteria.

Example 2.11 Logical indexing in MATLAB.

Given x = 1:10 and y = [3 1 5 6 8 2 9 4 7 0], execute and interpret the results of the following commands:

```
a. (x > 3) & (x < 8)
b. x(x > 5)
c. y(x <= 4)
d. x( (x < 2) | (x >= 8) )
```

```
e. y( (x < 2) | (x >= 8) )
f. x(y < 0)
```

Solution

First, create a MATLAB script `LogicalIndexing.m` with two statements to initialize the variables `x` and `y`, followed by the six expressions (a-f) above. Executing the script yields:

```
>> LogicalIndexing
x =
     1     2     3     4     5     6     7     8     9    10
y =
     3     1     5     6     8     2     9     4     7     0
ans =
     0     0     0     1     1     1     1     0     0     0
ans =
     6     7     8     9    10
ans =
     3     1     5     6
ans =
     1     8     9    10
ans =
     3     4     7     0
ans =
   Empty matrix: 1-by-0
>>
```

There are two terms in the expression in part a. The first term is `(x>3)`, which returns a logical array with the first three elements zero. The second term, `(x<8)`, returns a logical array with the last three terms set to zero. The result of the operation `(x > 3) & (x < 8)` is a logical array whose elements are the logical AND of the corresponding elements from each of `(x>3)` and `(x<8)`. The only terms whose elements are both true `(T)` are the fourth, fifth, sixth, and seventh elements, as shown above.

In part b, the expression `(x>5)` is a *logical index array*, with the first five elements set to zero since these elements are less than or equal to 5. Using this vector as an array of indices for the array `x` returns only those elements whose indices are 1; the elements whose indices are zero are not included in the result. Hence, result is [6 7 8 9 10]. The result of part c can be interpreted similarly.

Parts d and e are interpreted similar to parts b and c, except that the expression for the logical index array is similar to the expression in part a, with two terms. First, evaluate the index expression, the result of which will be similar to the result of part a. Next, use this array to index the arrays x and y respectively.

In part f, the index array is the array of elements of y that are less than zero. Since there are none, the index array is empty and consequently, no elements are selected from x, which results in the empty array.

Logical indexing is a powerful technique that can be used to efficiently evaluate complex functions. There are problems at the end of this chapter that rely on logical indexing.

2.4.5 Cells and cell arrays

In the discussion of MATLAB arrays so far, it has been implicitly assumed that the array is *homogeneous*, that is, each element in the array is the same type. There will be times, in complex applications, where data of different types will have to be grouped together. The only way to group dissimilar objects together is to use a data structure that can accommodate different data types.

A *cell array* is an array where each element, called a *cell*, can be an object of any of numbers, arrays, characters, or strings. A cell in a cell array can itself be a cell array. A cell array can be of any valid size or shape, including multidimensional structure arrays.

Array indices are referenced inside of parentheses as in the expression `A(1,2)`. Cell array indices are referenced using curly braces.

Example 2.12 Cell arrays and mixed data types.

One of the widely used gene expression data sets in bioinformatics is the data from *The Transcriptional Program in the Response of Human Fibroblasts to Serum* (Iyer et al., 1999). For each gene in the dataset, there are a number of attributes: the Gene cluster order, the Genbank Accession number, Clone IDs, Gene names, and expression changes over a 24 hour period. To see the data set, visit

http://gepas.bioinfo.cnio.es/data/fibroblasts/fibroblasts_ori.html.

Design a data structure to organize the 517 genes in the data set, including all of the attributes identified above.

Solution

The overarching organization of the data would be an array, since that permits accessing each set of the expression data in a constant time. However, the data is not homogenous, as there are both numbers and strings:

Numbers	Gene cluster order
	Clone ID

Expression data for 0, .25, .5, 1, 2, 4, 6, 8, 12, 16, 20 and 24 hours

Strings Gene name
Genbank Accession number

Each element in the array of genes would have all of these fields; the array can be neither a string array nor an array of numeric data. In this case, it is clear that we want an array of **cells**, where each cell contains these five items.

There are two possible ways to organize this cell array, and the choice depends in part on how the expression data is organized. Either the expression data can be organized as an array, or there can be one element in the cell array for each of the expression ratios.

The MATLAB code below illustrates the data structure that is based on the first design using a sample from the fibroblast data set.

Cluster order	4
Accession number	W88572
Clone ID	417426
Gene name	Homo sapiens protein 4.1-G
0HR 15MIN 30MIN	1 0.97 1
1HR 2HR 4HR	0.85 0.84 0.72
6HR 8HR 12HR	0.66 0.68 0.47
16HR 20HR 24HR	0.61 0.59 0.65

The cell array will be 517 rows by 5 columns. There is one column for each attribute. One of the columns is itself an array—the expression data for the gene. Each of the five attributes is associated with gene number 4, which is entered into the array in row 1. The syntax used in this example is one of two ways to index cell arrays:

```
Fibro{1,1} = 4;
Fibro{1,2}= 417426;
Fibro{1,3}= [1 0.97 1 0.85 0.84 0.72 0.66 0.68 0.47 0.61 0.59
0.65];
Fibro{1,4} = 'Homo sapiens protein 4.1-G';
Fibro{1,5} = 'W88572';
```

After entering this data at the MATLAB command line, the variable `Fibro` will appear in the workspace; the type is cell array, not a character or double precision numeric array.

Structures are multidimensional MATLAB arrays where elements are accessed by textual field designators, rather than integer indices. In the example above, it would be easier for the programmer to refer to the entries by symbolic name (such as 'name') rather than the index (in this case, 1).

Example 2.13 Structure arrays and mixed data types.

For the data set given in the previous example, design a data structure that uses MATLAB structures instead of cells.

Solution

The MATLAB code below is an example of how to enter the same gene expression data into a structure array instead of a cell array. Notice that the syntax reverts back to using parentheses to refer to an array element rather than the braces used to refer to a cell:

```
Fibro(1).number = 4;
Fibro(1).cloneid= 417426;
Fibro(1).expprofile= [1 0.97 1 0.85 0.84 0.72 0.66 0.68 0.47 0.61
0.59 0.65];
Fibro(1).name = 'Homo sapiens protein 4.1-G';
Fibro(1).accession = 'W88572';
```

Here the name `cloneid` refers to the gene Clone ID, `expprofile` refers to the expression profile, and `accession` refers to the accession number.

After entering this data at the MATLAB command line, the variable `Fibro` will appear in the workspace; the type is now structure array, not a character, double precision numeric, or cell array.

2.4.6 Data structures not explicitly found in MATLAB

Many common data structures are used in algorithms written in traditional imperative or object-oriented programming languages. These data structures—*Stacks, Queues, Linked Lists,* or *Trees*—are ways of organizing data sets to make particular operations more efficient. Whereas the array allows access to any element, each of these four data structures have controlled ways of accessing the data, but free the programmer from the constraint of a fixed amount of memory for the object. The first three items, Stacks, Queues, and Linked Lists, are methods for organizing lists of objects; a Tree is used to organize data hierarchically. Since the implementation of these data structures in MATLAB is beyond the scope of this book, please refer to one of the references listed at the end of this chapter for more information.

Example 2.14 Data structures in MATLAB: implementing a stack.

A stack is a data structure where data is added or removed from only one end. Unlike an array, there is not random access to the data, but only one datum can be accessed at a time. This special item is called the *top of the stack*. The functions `pop()`, to remove the top item, and `push()`, to add a new item to the top, are used to access the top of the stack.

Given a string that represents an amino acid sequence, show how a stack can be used to reverse the sequence.

Solution

The solution to this problem has two parts: First, to implement the stack and second, to use the stack to solve the problem, i.e., to reverse the amino acid sequence.

The stack itself will be represented as a MATLAB array. In this array, the first element is reserved to store the number of elements in the stack. Two MATLAB functions are required to implement the `push()` and `pop()` operations. The function `push` has two arguments: the first is the name of a stack and the second is the data to be placed on top of the stack. The function `pop` has one argument and returns one value: the argument is the name of a stack, and the value return is the data at the top. The MATLAB script `push.m` is:

```
% push.m
function push(A,x)
t=A(1);
A(t+1)=x;
A(1)=t+1;
end
```

Here the data `x` is being pushed onto the top of the stack `A`. First, we find the number of elements in the stack, which happens to be one less than the index of the current top element. Next, the data `x` is pushed onto the stack, in the next vacant element of the array. Finally, the number of data in the stack is increased by one. The MATLAB function `pop.m` is

```
% pop.m
function x=pop(A)
t=A(1);
x=A(t+1);
A(1)=t-1;
end
```

In this function, the number of data in the stack is retrieved and the top element is removed and stored in the variable `x`. The number of elements in the stack is decreased by one, since once the element is removed it is no longer part of the stack. Note that this function assumes that the stack has at least one element. The MATLAB function `reverse.m` to reverse the amino acid sequence is:

```
% reverse.m
function y=reverse(x)
l=length(x);
Stack=[];
Stack(1)=0;
for i=1:l
    c=x(i);
    push(Stack,c);
end
for i=1:l
    y(i)=pop(Stack)
end
```

First, the length of the sequence is determined and the stack is initialized. Then, each character (representing an amino acid) is pushed onto the stack. Once all the characters are on the stack, the characters are popped and put into the new character string, adding one character at a time. An example of using these functions is:

```
>> reverse('AACTGACT')
ans =
TCAGTCAA
```

The readers are encouraged to use paper and pencil to follow the sequence of operations in this example to convince themselves that the amino acid sequence is reversed.

2.4.7 Data type conversion

MATLAB, as all block-structured languages, provides the ability to convert objects in one data type to another. The entries in the Table 2.1 are the names of the MATLAB for converting the data type given by the row entry to the data type given by the column:

Table 2.1 MATLAB commands for data type conversion.

	String	Cell
Number	num2str	num2cell
Integer	int2str	
Matrix	mat2str	

The command `T = num2str(X)` converts the number `X` into a string representation `T` with about four digits and an exponent, if required. The command `I = int2str(N)` converts the integer `N` to a string representation `I`. Noninteger inputs are rounded before conversion. The input to either one of these commands can be a single number, a vector, or matrix of numbers. The command `C = num2cell(A)` converts a matrix `A` into a cell array by placing each element of `A` into a separate cell. The cell array will be the same size as the matrix `A`. Finally, the command `M = mat2str(A)` convert a 2-D matrix to a string `M` that may be used in `eval(M)` to reproduce the original matrix `A`.

Functions like `num2str` and `int2str` are useful for labeling and titling plots with numeric values.

Example 2.15 Data type conversion.

Adjust the parameters of the function $e^{-\theta}\sin^2\theta$ (shown in Example 2.3) to approximate an electrocardiogram (ECG) complex as closely as possible. Label the graph of each attempt.

Solution

In example 2.3, the function $e^{-\theta}\sin^2\theta$ appears to closely resemble the QRS complex and T wave of an ECG signal. In this problem, you are to vary the coefficients α and β in the expression $e^{-\alpha\theta}\sin^2\beta\theta$ to find a combination that models the QRST complex. It is helpful to plot the waveform and use a title that includes the values of α and β.

Here is an example where the `num2str` function can be used. The values of α and β are converted to strings and used in the string that is the title of the graph. The title is displayed using the `title()` function, which takes a single argument that is a string and prints the string above the graph. The function `findecg` below has two arguments that are values of α and β, computes a vector of function values and plots the vector, titling the graph with the expression so that one may find the expression that most closely models the QRST waveform.

```
% Example 2.15
function findecg(a,b)
twopi=2*pi;
twotheta-0:(twopi/100):twopi;
plot(twotheta,exp(-a*twotheta).*(sin(b*twotheta)).^2);
title(['Plot of e^{-', num2str(a), '\theta} * sin^2 ' , ...
    num2str(b) , '\theta']);
```

An example using this function is:

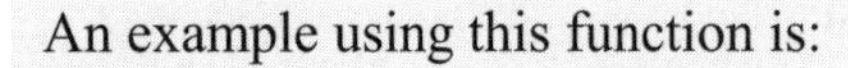

```
>> findecg(3,3)
```

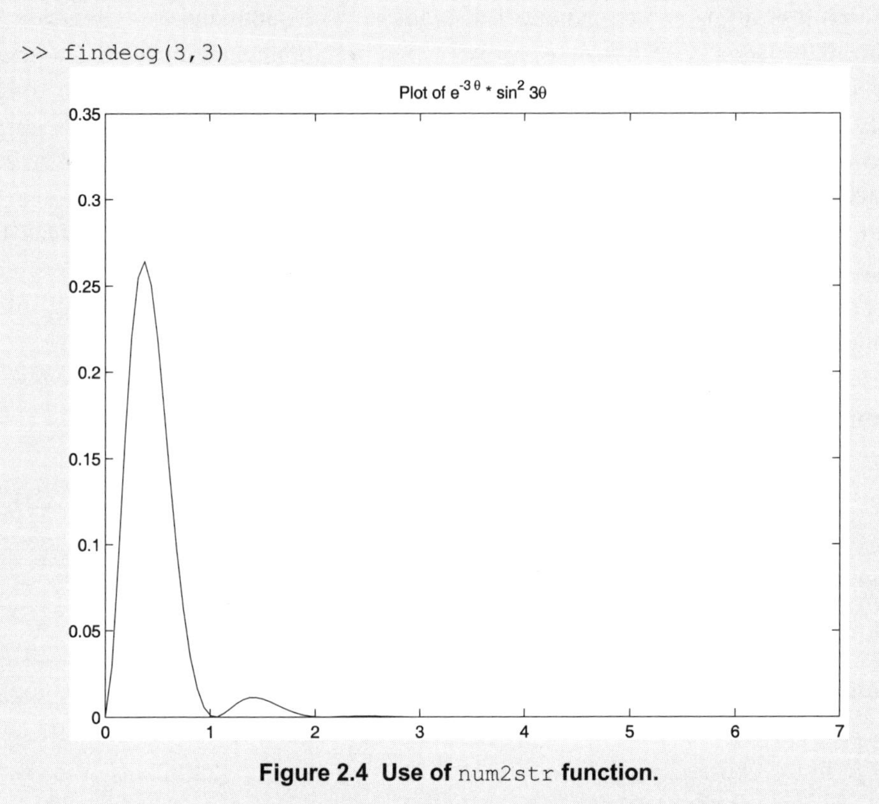

Figure 2.4 Use of `num2str` **function.**

Notice that the title of the graph in Fig. 2.4 includes the values of the independent variables that are passed to the function `findecg`. The function `title` has a single argument that is a string, which in this case is formed as a vector of substrings. The text substrings are delineated by single quotes, and there are two places where the function `num2str` is used to convert the parameter values to strings. The characters `'^'` in the strings are used to format superscripts and the special substring `'\theta'` is the character θ.

2.5 An Introduction to Object-Oriented Systems

The concept of *Object-Oriented Programming* (OOP) has revolutionized computer programming in the last 20 years. In OOP, programs are organized around *objects* rather than actions; you can also think of this as organizing programs around data rather than logic. Understanding the principles of OOP can help MATLAB programmers write correct programs and debug programs that have errors due to using incorrect operators and/or expressions.

The traditional view of a computer program is as a logical procedure that takes input data, processes the data, and produces output data. The programming challenge was to write the logic, not how to define the data. The principles of structured programming helped programmers by giving them a well-defined, concise set of rules to follow when organizing the logic of a program.

OOP languages are different. The programmer really cares about the data, that is, the objects that we want to manipulate, rather than the logic required to manipulate them. Examples of objects range from human beings (described by name, address, and so forth) to buildings and floors (whose properties can be described and managed), down to the little widgets on your computer desktop (such as buttons and scroll bars).

An object is a software element containing a collection of related data (in the form of *variables*) and *methods* (procedures) for operating on that data. An object is a specific instance of a class (a set). Objects are created by *instantiating*, creating an instance of an object, where an *instance* refers to an object that belongs to a particular class. For example, *California* is an instance of class *State*.

A *method* is a procedure contained in an object that is made available to other objects for the purpose of requesting services of that object.

A *message* is a signal from one object to another that requests the receiving object to carry out one of its methods. A message consists of three parts: the name of the receiver, the method it is to carry out; and any parameters the method may require to perform the action.

A *class* is a template that defines the variables and methods for a particular type of object. All objects of a particular class are identical in form and behavior but contain different data in their variables.

Encapsulation is a technique in which data is packaged together with its corresponding procedures. In OOP languages and systems, the mechanism for encapsulation is the object.

A tree structure representing the relationships among a set of classes is called the *class hierarchy*. A *superclass* is a class that is higher in the class hierarchy than another class; that is, a more general class. A *subclass* is a class that is a special case of another class.

A *class method* is a special kind of method that is invoked by sending a message to a class, rather than to one of its instances. Class methods usually perform tasks that can't or shouldn't be done at the instance level (e.g., creating and destroying instances of a class).

A class variable is a special kind of variable that stores its value in the class definition, rather than in the instances of that class. Class variables maintain information that is the same for all instances of the class.

Inheritance is a mechanism whereby classes can make use of methods and variables defined in classes above them in the class hierarchy.

An *instance method* is a normal method that is invoked by sending a message to an instance, rather than a class. This term is typically used to distinguish ordinary (instance) methods from the less common class methods.

An *instance variable* is a normal variable that stores a unique value in each instance of a class. This term is typically used to distinguish ordinary (instance) variables from the less common class variables.

For example, the class Mammal may have *subclasses* Human, Cat, and Dog. Dog, in turn, may have *subclasses* Beagle, Doberman, and Collie. The class Mammal is the *superclass* of Human. *Instances* of the class Beagle *inherit* operations (implemented as *messages* and *methods*) from the superclass Dog.

One of the first object-oriented computer languages was called Smalltalk. C++ and Java are the most popular object-oriented languages today. The Java programming language is designed especially for use in distributed applications. MATLAB is an example of an OOP language. A simple mathematical example to illustrate OOP principles is as follows:

Assume that you write the programming statement

```
z=x+y
```

in a computer program. If the programming language were a traditional block-structured language, then the statement would be compiled (i.e., translated) into a sequence of steps:

1. Load the contents of the variable `x`, convert the number to double precision, and store the number in register R1.
2. Load the contents of the variable `y`, convert the number to double precision, and store the number in register R2.
3. Add the contents of register R1 and R2; convert the double precision number to the type of the variable `z`.

If an OOP language is used, the variable `x` is taken to be an object that is sent the message '+' with an argument that is the variable `y`. A method, or operation, is executed that corresponds to the message '+'. The important difference is that *the method that is executed is dependent on the type of the object* `x`. If `x` is an integer, then the method performs an integer add; if `x` is a double precision number, then the method performs a double precision addition.

Note that the same symbol (or name) can be used for messages that invoke different methods. This is referred to as *operator overloading* and occurs quite frequently in OOP systems such as MATLAB. The example below shows operator overloading in MATLAB.

Example 2.16 Simple object-oriented programs that are sequences of statements.

Let x = [2 5 1 6]. Compute the result of each operation:

(a) Add 16 to each element
(b) Add 3 to just the odd-index elements
(c) Compute the square root of each element
(d) Compute the square of each element

Solution

The solutions to these four exercises rely on the object oriented properties of MATLAB. This means that while we know that we can add two numbers together, in MATLAB we can also add numbers to vectors, and the result will be defined by the operation of MATLAB.

(a) In this first part of the example, the object `x` is a vector of integers. The message sent to the object is addition, and the argument to this message is an integer object. The method that is invoked for the message "add an object to a vector" will first check to see if the argument is of a type that can be added to a vector. In this case, the argument is an integer, and the method, or rule, for adding an integer to a vector is clearly defined: The integer is added to each element of the vector.

First, create the vector and assign it to a variable. Next, if you add a constant (such as 16) to a vector, the result is defined by the mathematics. The interpretation of adding a number to a vector is that the number is added **to each element** of the vector.

```
x = [2 5 1 6]
x+16
```

Output of results

```
x =
     2     5     1     6
ans =
    18    21    17    22
```

(b) Since there are only four elements in the vector, the simplest way to add three to the odd elements is to enumerate the elements of the vector and to write a new expression for the first and third element of the vector:

```
[2+3 5 1+3 6]
```

Output of results

```
ans =
     5     5     4     6
```

The expression for each element of the vector does not have to be written out as shown if the same operation is being performed on each element of the vector. This example could also be done with logical array indexing (Sec. 2.4.4).

(c) In this part of the problem, you are asked to take the square root of each element in the vector; since MATLAB is object-oriented, the program only needs to take the square root of the vector.

```
sqrt(x)
```

Output of results

```
ans =
    1.4142    2.2361    1.0000    2.4495
```

In this part of the example, the vector `x` is the object, and the message is the MATLAB function `sqrt()`. The method invoked checks to see if the square root of the object can be computed; in this case the square root of the object can be computed with the results shown above.

(d) The general rule of OOP execution does not apply to squaring the elements of a vector. The MATLAB statement `x^2` is interpreted to mean sending the message raise to the power ('`^`') with the argument `2` to the object `x`.

If the message '`^`' is sent to a matrix (`x` is a 1x4 matrix), it is interpreted to mean that the operation of multiplying the matrix by itself should be performed. Therefore, the method invoked first checks to make sure that the matrix is square. Since `x` is a 1x4 matrix, it is not square, and the command will produce an error instead of squaring each element.

The message that must be specified is to square each element of the object. In MATLAB, an element-by-element operation is denoted by prefixing the operator symbol with the period. The symbol representing this message is '`.^`' as shown in the example below:

```
x.^2
```

Output of results

```
ans =
     4    25     1    36
```

2.6 Analyzing Algorithms and Programs

There is not a single correct way to write a MATLAB program. As an engineer, you will be faced with many choices for representing data and for implementing the algorithm. In many cases, the amount of data you will be working with is small, and the performance of the program (computation time and memory used) may not differ very much. When the amount of data is very large, such as in image reconstruction or multiple channels of long time series of EEG or ECG, the implementation becomes critical.

How does a biomedical engineer choose which method to use? The choice you will make will depend on a number of factors:

1. The accuracy and precision of the result
2. The amount of time the computation takes
3. The amount of memory space needed for the computation

This section is an introduction to the analytical tools that can help you decide how to represent data and to choose an implementation of an algorithm.

2.6.1 Polynomial complexity

Determining the amount of time or the amount of space required for a calculation is referred to as the *analysis of the algorithm*. Computing the amount of real time that a computer program takes is not a good measure of the time, since computing speed continues to increase as new processors are introduced.

If you do not know in advance how much data you will have, or if you need to write a program that will be general enough for a wide range of the amount of data, then you will need to analyze the algorithms in order to make an informed choice as to which algorithm to use.

Actual running time (clock time) depends on a number of factors, including the clock speed of the processor, the amount of memory, and whether or not other tasks are running. The only way to standardize how we talk about how "fast" a program is is to talk about the *asymptotic running time*, the amount of time required as a function of the number of data.

2.6.2 Operation counting

There are several general rules that one can use to determine the asymptotic (time) complexity of an algorithm. First, assume that our basic unit of time is a single addition, such as:

```
Z=x+y
```

The time for an assignment is less than the time for an addition, and so, it will be taken to be the same as the time of an addition. The time for a multiplication such as:

```
C=a*b
```

will take longer; the time for a multiplication is approximately the time for an addition multiplied by the length of the number.

Typically, either the time for an addition or the time for a multiplication will be the basic unit of measure for approximating or estimating asymptotic running time, depending on the problem to be solved. The more common operation performed will become the basic unit of measurement.

The following rules are used to compute the running time of an algorithm that is formed from a sequence of statements.

Rule 1: The time required to add (or subtract) two numbers is 1 unit of time, and the time required to multiply or divide two numbers is the length of the number (the number of bits used to represent the number) multiplied by the time for an addition.

Recall that multiplication is implemented as repeated addition. If the time required for a single addition is fixed, then the time required for a multiplication is just the number of times that terms are added, which is just the number of bits in the word.

Rule 2: The time for a sequence of statements is the sum of the times for the individual statements.

```
X = 10+Y
Z = X*3
Y = Z/X
```

The time required to compute `X`, `Y`, and `Z` is the sum of the time required for each of the statements. Remember that the time required for a multiplication operation is length of the number multiplied by the time required for an addition. Therefore, the time to compute `Z` or `Y` is much greater than the time required to compute `X`. The majority of the time required to compute `X`, `Y`, and `Z` is limited by the multiplication and the division. The amount of time required to compute `X` is insignificant by comparison.

Since the time to compute any addition is a constant, and therefore the time to compute a multiplication (or division) is constant, the time to compute `X`, `Y`, and `Z` is fixed, and can be expressed in terms of the number of operations. The total time therefore is a function only of the length of the sequence of statements.

Rule 3: The execution time for a loop is the number of iterations times the execution time of the body of the loop.

Example of a single loop

```
for j=1 to n
    sum = sum + a(i)*b(j)
end
```

Let w be the wordsize for your computer. The computing time of the body of the loop will take on the order of $w+2$ time; w is the time for the multiplication, and the remaining two units of time are for the addition and the assignment. Since this statement is repeated n times, the total execution time for this loop is $n*(w+2)$, or $nw + 2n$. Since it is safe to assume that w (the wordsize) is greater than 2, the term nw will grow faster that $2n$. We can safely estimate the running time as on the order of nw, which is written as $O(nw)$.

If there are nested loops in the program, then the number of iterations for each loop are multiplied, as in this example:

```
for i=1 to m
    for j=1 to n
            sum = sum + a(i)*b(j)
    end
end
```

in which the body of the inner loop takes $w+2$ operations. The inner loops takes $n*(w+2)$ operations, and the complete nested loop takes $m*n*(w+2)$ operations. Since we keep only the largest term in the expression, we write the asymptotic time as $O(mnw)$; the remaining terms in the expression will not grow as fast as mnw even as m, n, or w get larger.

Example 2.17 Analyzing execution time as a function of the amount of data.

You are asked to write a program to compute the energy or power in a single-channel ECG time series. Three samples of ECG data are used to test the MATLAB program: the first has 128 samples, the second has 4,096 samples, and the third has 65,536 samples.

The power can be computed in one of two ways. One way of computing the power is to take the sum of the squares of the voltage values; a second way is the sum square of the amplitudes of the frequency components. This second method is further complicated by the fact that the frequency spectra can be computed using either a DFT or FFT algorithm.

Choose the algorithm that has the minimum running time, as a function of the number of samples in the time series.

Solution

The energy in the ECG signal can be computed in three different ways:

1. The sum squared of the voltage values
2. The sum squared of the frequency components using the DFT
3. The sum squared of the frequency components using the FFT

These three are theoretically equivalent by Parseval's theorem (The energy, or information content, in the time domain signal is equal to the information content in the frequency domain signal). Aside from the accuracy of the result, the only difference between these three alternatives is the choice of implementation: method 1, 2, or 3.

Method 1: Let the ECG data be stored in the MATLAB array ecg. The program to compute the power is:

```
power=0;
for j=1 to length(ecg)
    power = power + ecg(i)*ecg(i)
end
```

If the length of the ECG array is n, and if a multiplication is the basic unit of time, then by Rule 3, the running time is $O(n)$.

Method 2: In MATLAB, the `fft` function computes the DFT if the length of the array is not a power of two. The program is:

```
freq=fft(ecg)
power=0;
for j=1 to length(freq)
    power = power + freq(i)*freq(i)
end
```

By Rule 2, the time for this method is the sum of the time for the DFT and the time to compute the energy. The MATLAB help for the DFT gives enough information to determine that the DFT takes $O(n^2)$ time, where n is the length of the ECG array. The total time is $O(n^2 + n)$; since n^2 grows much faster than n, the second term can be dropped. The running time is $O(n^2)$.

Method 3: In order to compute an FFT in MATLAB, the length of the array must be power of two. In this case, the computation time is $O(n \log_2 n)$. Since $\log_2 n$ is less than n, the time for an FFT is less than a DFT. The total computation time is $O(n \log_2 n + n)$; again the second term can be dropped and the running time is $O(n \log_2 n)$.

Therefore, the method that uses the least amount of computation time is Method 1, which computes the energy directly from the times series data. Fig. 2.5 shows the relationship between $O(n)$, $O(n \log_2 n)$, and $O(n^2)$ complexity.

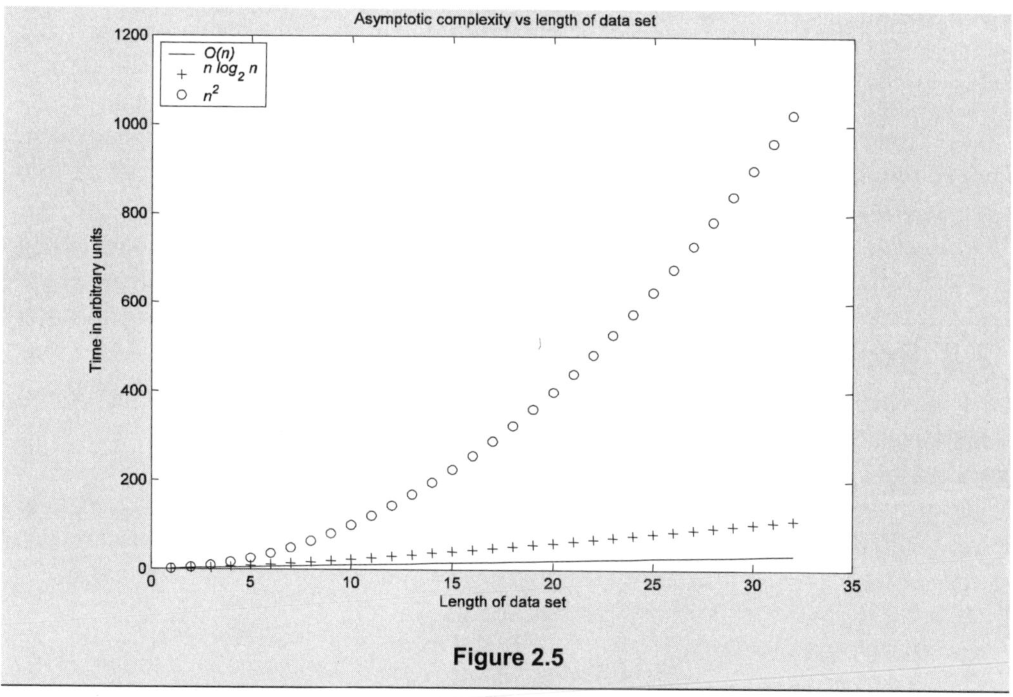

Figure 2.5

2.7 Lessons Learned in this Chapter

After studying this chapter, the student should have learned the following:

- Computers are used in biomedical engineering applications to perform repeated tasks without any degradation in performance and to provide an easy-to-use interface.
- Block-structured programming languages have a small set of control structures: sequences of statements, conditional execution, and iteration, each with one entry point and one exit point for the flow of computation. Using relatively few types of control structures gives programmers a mechanism for developing and debugging programs easily.
- In traditional imperative block-structured languages, the program is a sequence of actions.
- In object-oriented programming languages, the program is a specification of data objects and their interrelationships. MATLAB is an object-oriented programming language and system for performing matrix arithmetic easily.
- Data types and structures are methods for representing and organizing collections of data. The simple data types and structures in MATLAB are numbers and strings. Homogeneous collections are represented in arrays; heterogeneous (mixed) collections are represented in cell arrays and/or structure arrays.

- Simple tools for analyzing an algorithm or its implementation can help programmers predict how the execution time changes as the amount of data grows. There is always a tradeoff between the amount of space used for data and the amount of time the computation takes. A biomedical engineer has to use all of this information when deciding how to write a program.
- How to perform simple calculations in MATLAB.
- How to design and develop example MATLAB programs, including subroutines (MATLAB scripts) and functions.

2.8 Problems

2.1 Evaluate the following expressions, first by hand and then using MATLAB to check the answers:

a. 2/2*3
b. 6 – 2/5 + 7^2 – 1
c. 10/2\5 – 3 + 2*4
d. 3^2/4
e. 3^2^2
f. 2 + `round`(6/9 + 3*2)/2 – 3
g. 2 + `floor`(6/9 + 3*2)/2 – 3
h. 2 + `ceil`(6/9 + 3*2)/2 – 3

2.2 Create a vector, x, with the elements $x_n = (-1)^{n+1}/(2n-1)$. Add up the elements of the version of this vector that has 100 elements.

2.3 Plot the expression (determined in modeling the growth of the US population):

$$P(t) = \frac{197,273,000}{(1 + e^{-0.0313(t - 1913.25)})}$$

where t is the date, in years AD, using t = 1790 to 2000. What population is predicted in the year 2020?

2.4 Given the array `A = [2 7 9 7 ; 3 1 5 6 ; 8 1 2 5]`, explain the results of the following commands:

a. `A'`
b. `A(:,[1 4])`
c. `A([2 3],[3 1])`
d. `reshape(A,2,6)`
e. `A(:)`
f. `flipud(A)`
g. `fliplr(A)`

h. `[A A(end,:)]`
i. `A(1:3,:)`
j. `[A;A(1:2,:)]`
k. `sum(A)`
l. `sum(A')`
m. `sum(A,2)`
n. `[[A;sum(A)][sum(A,2);sum(A(:))]]`

2.5 Given `x = [3 15 9 12 -1 0 -12 9 6 1]`, provide the command(s) that will

a. Set the values of x that are positive to zero
b. Set values that are multiples of 3 to 3 (rem will help here)
c. Multiply the values of x that are even by 5
d. Extract the values of x that are greater than 10 into a vector called y
e. Set the values in x that are less than the mean to zero
f. Set the values in x that are above the mean to their difference from the mean

2.6 Write a MATLAB function to evaluate:

$$t(y) = \begin{cases} 200 & \text{when } y \text{ is below } 10{,}000 \\ 200 + 0.1(y - 10{,}000) & \text{when } y \text{ is between } 10{,}000 \text{ and } 20{,}000 \\ 1{,}200 + 0.15(y - 20{,}000) & \text{when } y \text{ is between } 20{,}000 \text{ and } 50{,}000 \\ 5{,}700 + 0.25(y - 50{,}000) & \text{when } y \text{ is above } 50{,}000 \end{cases}$$

With the test cases:

a. $y = 5{,}000 \quad t = 200$
b. $y = 17{,}000 \quad t = 900$
c. $y = 25{,}000 \quad t = 1{,}950$
d. $y = 75{,}000 \quad t = 11{,}950$

If you start each script with a request for input (using `input`), you'll be able to test that your code provides the correct results. Explain why the following if-block would **not** be a correct solution to this exercise:

```
if y < 10000
t = 200
elseif 10000 < y < 20000
t = 200 + 0.1*(y - 10000)
elseif 20000 < y < 50000
t = 1200 + 0.15*(y - 20000)
elseif y > 50000
t = 5700 + 0.25*(y - 50000)
end
```

2.7 Given $x = [8\ 2\ 2]$ and $y = [7\ 9\ 3]$, compute the following arrays

a. $a_{ij} = x_i y_j$

b. $b_{ij} = x_i / y_j$

c. $c_i = x_i y_i$ then add up the elements of c.

d. $d_{ij} = x_i / (2 + x_i + y_j)$

e. e_{ij} = reciprocal of the lesser of x_i and y_j

2.8 The Legendre polynomials $(P_n(x))$ are defined by the following recurrence relation:

$$(n+1)P_{n+1}(x) - (2n+1)P_n(x) + nP_{n-1}(x) = 0$$

with $P_0(x) = 1$, $P_1(x) = x$, and $P_2(x) = (3x^2 - 1)/2$. Compute the next three Legendre polynomials, and plot all six over the interval [-1,1].

2.9 Write a function that computes the cumulative product of the elements in a vector. The cumulative product of the jth element of the vector x, x_j, is defined by

$$p_j = (x_1)(x_2)\ldots(x_j)$$

for j = 1:length of the vector x. Create two different versions of this function:

a. One that uses two `for`-loops to explicitly carry out the calculations, element by element. An inner loop should accumulate the product and an outer loop should more through the elements of the vector p.

b. One that uses the built-in function `prod` to replace the inner loop.

In each case, you can check your results with the built-in function, `cumprod`.

2.10 The function plotted in exercises 2.3 and 2.15, with the arguments m='s' and n=2, comes close to approximating the shape of a QRS complex followed by a T wave. Write a MATLAB script that will generate a time series of three complexes with a constant heart rate.

Chapter 3
Concepts of Numerical Analysis

3.1 Scientific Computing

Scientific computing is the study of algorithms for solving mathematical problems that arise in various fields of science and engineering. A biomedical engineer starts with a continuous mathematical model to explain an observed phenomenon in biology or chemistry. The examples throughout this book show that these models are difficult or impossible to solve analytically, which is usually the case.

Numerical analysis is the mathematical theory that leads to algorithms for solving mathematical models approximately on a computer. Continuous functions are approximated by finite arrays of values, and algorithms that approximately solve the mathematical problem efficiently, accurately, and reliably are used to solve the problem.

There is a lot to consider when developing numerical methods and algorithms. The purpose of this textbook is to equip the biomedical engineer with the necessary tools and to show some examples of how numerical analysis and numerical methods are used in practice. The references at the end of this chapter will help the reader who needs some additional background information.

The material in this chapter will enable the student to accomplish the following:

- Appreciate the difference between absolute and relative errors
- Realize how Taylor series or Taylor formula is used to discretize a continuous function
- Understand how floating-point numbers, the representations of real numbers, are represented in MATLAB
- Be aware of how roundoff errors are propagated in numerical calculations
- Comprehend several techniques for avoiding cancellation errors

3.2 Numerical Algorithms and Errors

When using a computer approximation of mathematical models, error is always present in the calculation. The results will only be approximate, and the goal is to manage the error so that it is as small as possible. The biomedical engineer who is developing a numerical solution to a mathematical model has to be aware of relative and absolute error, the possible sources of error, and how to keep them small.

Given a quantity u and its approximation v, the absolute error in v is $|u - v|$. The relative error (assuming $u \neq 0$) is $|u - v|/|u|$. Relative error is usually a more meaningful measure than absolute error. This is especially true for errors in floating-point representation, which will be presented in Section 3.3. For example, the absolute and relative errors for several combinations of u and v are compared in Table 3.1:

Table 3.1 A comparison of absolute and relative errors.

u	v	Absolute Error	Relative Error
1	0.99	0.01	0.01
1	1.01	0.01	0.01
-1.5	-1.2	0.3	0.2
100	99.99	0.01	0.0001
100	99	1	0.01

When $u \approx 1$ there is not much difference between the absolute and relative errors, but when $|u| >> 1$, the relative error is a better reflection of the difference between u and its approximation v.

There are several types of error that may limit the accuracy of a numerical calculation. *Errors in the models* to be solved may be approximation errors when formulating the mathematical model. For example, stars or planets are often approximated by spheres when their properties are calculated or unimportant chemical reactions are often ignored in complex chemical modeling in order to arrive at a mathematical problem of a manageable size. There may also be *error in the input data.* Measurement systems or instrumentation are never perfectly accurate; even after careful calculation using a computer model, the solution will not quite match observations. This subject is treated in Chapter 9.

There is not much that can be done about either *model or measurement errors.* However, these sources of error need to be taken into consideration when determining the accuracy to which the problem should be solved.

Approximation errors arise when an approximate formula is used in place of the actual function to be evaluated. There are two types of approximation errors: *truncation errors* and *convergence errors.* Truncation errors arise from sampling a continuous process, such as interpolation, differentiation and integration. Convergence errors arise in iterative methods: For example, nonlinear problems must generally be solved approximately, by an iterative procedure that would converge to the exact solution in the limit (after infinitely many iterations), but the process is terminated after a finite number of iterations. Iterative methods often arise in linear algebra, where iterative procedures are terminated after a finite number of iterations before the exact solution is reached. These problems are treated in Chapter 4.

Roundoff errors are present in any computation that uses real numbers instead of integers, even when no approximation error is involved. Roundoff errors are present because real numbers cannot be represented exactly by computer; the representation of real numbers in MATLAB is discussed in Section 3.5. Roundoff errors are always propagated forward in the computation and could be greater than any model or approximation error in the algorithm. Managing error means to be aware of approximation error and roundoff error when selecting a particular numerical method.

3.3 Taylor Series

The key to connecting continuous and discrete versions of a formula is the *Taylor series* (and *Taylor formula*): Assume that $f(x)$ has $k + 1$ derivatives in an interval containing the points x_0 and $x_0 + h$. Then the infinite Taylor series:

$$f(x) = f(x_0) + (x - x_0) f'(x_0) + \frac{(x - x_0)^2}{2} f''(x_0) + \cdots + \frac{(x - x_0)^k}{(k)!} f^{(k)}(x_0) + \cdots \quad (3.1)$$

and the finite Taylor formula:

$$f(x) = f(x_0) + (x - x_0)f'(x_0) + \frac{(x - x_0)^2}{2} f''(x_0) + \cdots$$
$$+ \frac{(x - x_0)^k}{k!} f^{(k)}(x_0) + \frac{(x - x_0)^{k+1}}{(k+1)!} f^{(k+1)}(\xi) \tag{3.2}$$

are equivalent, where the last term

$$R^{n+1} = \frac{(x - x_0)^{n+1}}{(n+1)!} f^{n+1}(\xi) \tag{3.3}$$

is called the remainder and ξ is some point between x_0 and x.

The function $f(\ )$ is evaluated at the two points x_0 and $x = x_0 + h$, which are a distance, h apart. If the series in Eq. (3.1) is truncated after k terms, then it is equal to the Taylor formula with an error term:

$$\frac{h^{k+1}}{(k+1)!} f^{(k+1)}(\xi) \tag{3.4}$$

This is the truncation error due to approximating the continuous function by a finite number of terms.

Example 3.1 How truncation errors and roundoff errors arise.

Consider the problem of approximating the derivative $f'(x_0)$ of the function $f(x) = \sin(x)$, and let $x_0 = 0.5$. Thus, $f(x_0) = \sin(.5) \approx 0.479$. Now, consider evaluating $f(x)$ a point x near x_0, and suppose that $f'(x_0)$ is not directly available or cannot be computed. Therefore, $f'(x_0)$ has to be evaluated at the point x which is near x_0.

An algorithm can be constructed using the Taylor series and Taylor formula. For a small, positive value h, write:

$$f(x_0 + h) = f(x_0) + hf'(x_0) + \frac{h^2}{2} f''(x_0) + \frac{h^3}{6} f^{(3)}(x_0) + \frac{h^4}{24} f^{(4)}(\xi) + \cdots$$

and then:

$$f'(x_0) = \frac{f(x_0 + h) - f(x_0)}{h} - \left(\frac{h}{2} f''(x_0) + \frac{h^2}{6} f^{(3)}(x_0) + \frac{h^3}{24} f^{(4)}(\xi) + \cdots \right)$$

By the equivalence with the Taylor formula, the algorithm:

$$f'(x) \approx \frac{f(x_0+h)-f(x_0)}{h}$$

has the truncation error

$$\left| f'(x_0) - \frac{f(x_0+h)-f(x_0)}{h} \right| = \left| \left(\frac{h}{2} f''(x_0) + \frac{h^2}{6} f^{(3)}(x_0) + \frac{h^3}{24} f^{(4)}(\xi) + \cdots \right) \right|$$

The slope of the tangent at the point x_0 is approximated by the slope of the chord through neighboring points of $f(\,)$, as show in Fig. 3.1.

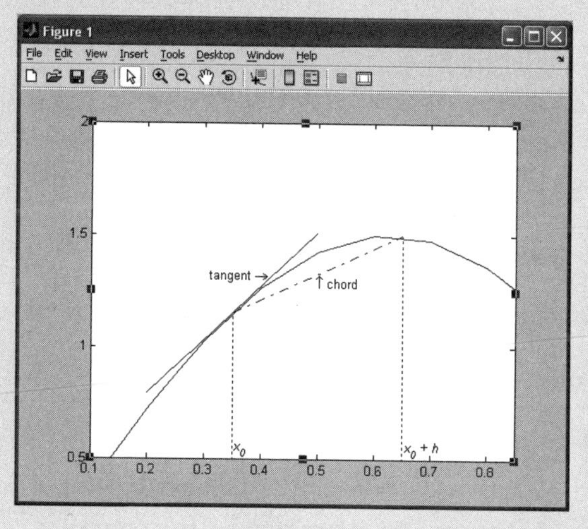

Figure 3.1 An instance of numerical differentiation: the tangent $f'(x_0)$ is approximated by the chord $(f(x_0+h)-f(x_0))/h$.

If $f''(x_0)$ is known and is nonzero, then, for small h, the truncation error can be estimated by:

$$\left| f'(x_0) - \frac{f(x_0+h)-f(x_0)}{h} \right| \approx \frac{h}{2} \left| f''(x_0) \right|.$$

Even without knowing $f''(x_0)$, provided that $f''(x_0) \neq 0$, this formula shows that the truncation error will decrease proportionally to h.

For $f(x) = \sin(x)$, the exact value of $f'(x_0) = \cos(0.5) = 0.877\ldots$. Using the Taylor formula approximation with $h = 0.1$, the result is:

$$f'(x_0) \approx (\sin(.6) - \sin(.5))/0.1 = 0.852\ldots$$

The absolute truncation error is approximately 0.025 and the relative error |0.877 – 0.852| / |0.877| is not very different.

Approximating $f'(x_0)$ using $h = 0.1$ is not very accurate. If the same Taylor series approximation is used with several smaller values of *h*, the errors are:

h	**Absolute truncation error**
0.1	2.541321e-002
0.01	2.411734e-003
0.001	2.398590e-004
0.0001	2.397274e-005
0.00001	2.397147e-006

The error decreases proportionally to *h*. Knowing that $f''(x) = -\sin(x)$ and that $\frac{1}{2}f''(x_0) \approx 0.240$, the sequence of errors suggests that the formula 0.24*h* is a good estimate for the error.

These calculations and those below were done in MATLAB. The results might suggest that a near-perfect accuracy can be achieved with an arbitrarily small *h*. Suppose that we want to have $\left|\cos(.5) - \frac{\sin(.5+h) - \sin(.5)}{h}\right| < 10^{-10}$. The results of Example 3.1 suggests that $h < 10^{-10}/0.25$ in the algorithm. Some values of the absolute truncation error for very small positive values of *h* are:

h	**Absolute truncation error**
1.0e-8	4.361050e-10
1.0e-9	5.594726e-8
1.0e-10	1.669696e-7
1.0e-11	7.938531e-6
1.0e-13	6.851746e-4
1.0e-15	8.173146e-2
1.0e-16	3.623578e-1

Fig. 3.2 is a log-log plot of the error as a function of step size h. As h decreases from right to left, at first the error decreases, but this trend changes and reverses directions. The MATLAB script that generates the plot in Fig. 3.2 is

```
x0 = .5;
f0 = sin(x0);
fp = cos(x0);
i = -20:0.5:0;
h = 10.^i;
err = abs (fp -(sin(x0+h) -f0)./h );
d_err = f0/2*h;
loglog (h,err,'-*');
hold on
loglog (h,d_err,'-.');
xlabel('Step size h')
ylabel('Absolute error')
```

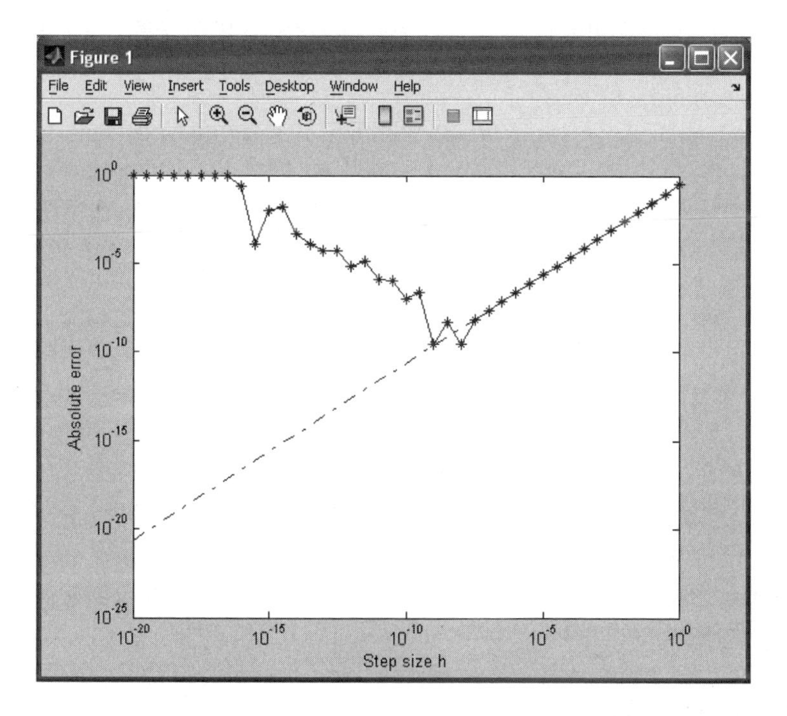

Figure 3.2: The combined effect of truncation and roundoff errors. The solid curve interpolates the computed values of $\left| f'(x_0) - \frac{f(x_0+h)-f(x_0)}{h} \right|$ **for** $f(x)=\sin(x)$, $x_0=0.5$. **Shown in dash-dot style is a straight line depicting the discretization error without roundoff error.**

The reason for the minimum at about $h = 10^{-8}$ is that the *total error* has both truncation and roundoff errors. The truncation error decreases as a function of h, and dominates the roundoff error when h is relatively large, but when h is less than 10^{-8}, the truncation error becomes very small and the roundoff error starts to dominate. Overall, the roundoff error increases as h decreases. This is one reason why it is very important to manage the error so that truncation error dominates when approximately solving problems involving numerical differentiation, such as differential equations. See Chapter 7 for more details on this important topic.

3.4 Keeping Errors Small

There are errors in each and every computed solution. Since the problem and the numerical algorithm both have errors, does the numerical solution of a similar problem (where similar means approximate data and approximate algorithms) differ by a little or a lot from the computed solution? If the difference is too great, the solution is meaningless.

This question is very important in medical imaging and leads to important theoretical concepts that will only be touched on here for reference. A problem in which the solution is sensitive to small error is said to be *ill-posed*, meaning that a small perturbation in the data can cause a large difference in the result. If the problem is ill-posed, then no algorithm may be found that would yield a solution insensitive to small errors. Some modification in the problem definition may be called for in such cases; medical imaging is a good example. Another example is the numerical differentiation in Example 3.1, which turns is ill-posed, noticeable when very high accuracy (i.e, a small value of h) is required.

In general, it is impossible to prevent accumulation of roundoff errors during a calculation. If ε_n is the relative error at the nth step of an algorithm, then $\varepsilon_n \approx c_0 n \varepsilon_n$ represents a linear propagation of the initial error, and $\varepsilon_n \approx c_1^n \varepsilon_0$ an exponential propagation of the initial error for some constants c_0 and $c_1 > 1$. Algorithms with exponential error growth must be avoided.

Example 3.2 An ill-posed problem.

Consider evaluating the integrals:

$$y_n = \int_0^1 \frac{x^n}{x+10} dx$$

for n=1, 2, …,25

From calculus:

$$y_n + 10y_{n-1} = \int_0^1 \frac{x^n + 10x^{n-1}}{x+10} dx = \int_0^1 x^{n-1} dx = \frac{1}{n}$$

and:

$$y_0 = \int_0^1 \frac{1}{x+10} dx = \ln(11) - \ln(10).$$

Thus, a numerical algorithm to compute the integrals is

Evaluate $y_0 = \ln(11) - \ln(10)$.

For n=1, …, 25 compute:

$$y_n = \frac{1}{n} - 10y_{n-1}$$

The sequence of 25 solutions is shown in the MATLAB figure in Fig. 3.3. The values would be exact if floating-point errors were not present; However, the roundoff error is multiplied by 10 at each step, so the error grows exponentially. The exact values all satisfy $0 < y_n < 1$ and the sequence shown below is meaningless.

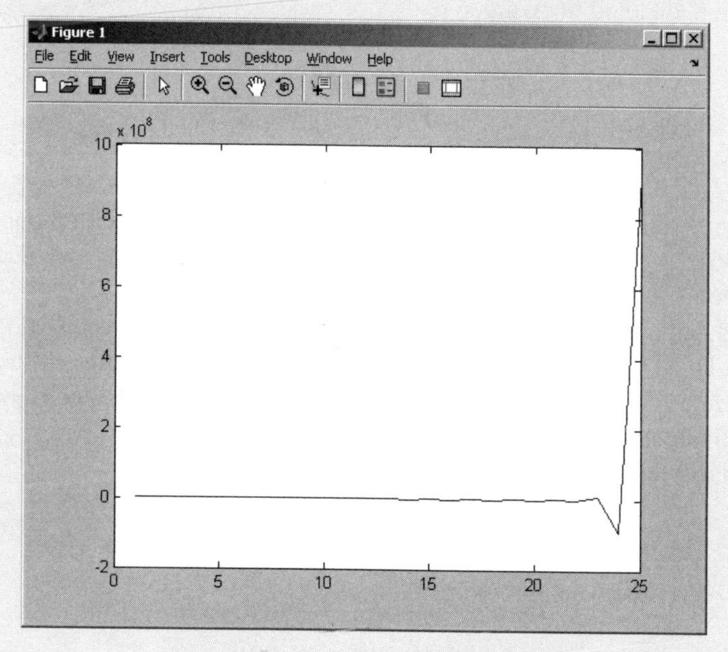

Fig. 3.3 Sequence of 25 solutions.

3.5 Floating-Point Representation in MATLAB

One of the many errors that can arise when calculating an approximate solution to a mathematical model is *roundoff error*: The error introduced by representing real numbers using finite precision.

Any real number can be accurately represented with an unlimited amount of pencil and paper. At worst, the representation is a repeating sequence, such as

$$\frac{8}{3} = 2.6666\ldots = \left(\frac{2}{10^1} + \frac{6}{10^2} + \frac{6}{10^3} + \frac{6}{10^4} + \frac{6}{10^5} + \cdots\right) \times 10^1$$

which is an example of an infinite series. However, computers have to use a finite amount of memory to represent real numbers. Therefore, only a finite number of digits may be used to represent any number, no matter what representation method is used. The infinite series representation of 8/3 can be truncated after 4 digits, yielding:

$$\frac{8}{3} = 2.6666\ldots = \left(\frac{2}{10^1} + \frac{6}{10^2} + \frac{6}{10^3} + \frac{6}{10^4}\right) \times 10^1 = 0.2666 \times 10^1$$

In general, there are t decimal digits (called the *precision*) in the floating-point representation of a real number. For each real number x, there is a floating-point representation, denoted fl(x), given by:

$$\begin{aligned} \mathrm{fl}(x) &= \pm 0.d_1 d_2 d_3 \ldots d_{t-1} d_t \times 10^e \\ &= \pm\left(\frac{d_1}{10^1} + \frac{d_2}{10^2} + \frac{d_3}{10^3} + \cdots + \frac{d_{t-1}}{10^{t-1}} + \frac{d_t}{10^t}\right) \times 10^e \end{aligned}$$

The sign, digits d_i, and exponent e are chosen so that fl(x) closely approximates the value of x. A floating-point representation is not unique; for instance, $0.2666 \times 10^1 = 0.02666 \times 10^2$.

The standard convention used is to normalize the representation by insisting that $d_1 \neq 0$. Thus, $1 \leq d_1 \leq 9$ and $0 \leq d_i \leq 9$ for $i = 2,\ldots, t$. The range of the exponents is also restricted: There are integers $U > 0$ and $L < 0$, such that all eligible exponents in a given floating-point system satisfy the inequality $L \leq e \leq U$. The largest number that is precisely representable in such a system is:

$$0.99\ldots99 \times 10^U < 10^U$$

and the smallest positive number is:

$$0.10\ldots00 \times 10^{L-1}$$

3.5.1 The IEEE 754 standard for floating-point representation

The common base for most computers today, following the IEEE standard set in 1985 (IEEE 754), is base 2. In this base each digit d_i may be only a 0 or 1; thus, the first (normalized) digit must be $d_1 = 1$ and need not be stored. The IEEE standard also shifts the exponent e, so that:

$$\mathrm{fl}(x) = \pm\left(1 + \frac{d_1}{2} + \frac{d_2}{4} + \frac{d_3}{8} + \cdots + \frac{d_t}{2^{t-1}}\right) \times 2^e$$

The standard floating-point word in the IEEE standard (which is used in MATLAB) requires 64 bits of storage and is called *double precision*. Of these 64 bits, 1 is allocated for the sign s (the number is negative if $s = 1$), 11 bits for the exponent, and 52 bits for the fraction. Since the fraction f contains 52 digits, the precision is $52 + 1 = 53$.

The exponent field needs to represent both positive and negative exponents. To do this, a *bias* is added to the actual exponent in order to get the stored exponent. For IEEE double-precision floating-point numbers, this value is 1,023. Thus, an exponent of zero means that 1,023 is stored in the exponent field. A stored value of 2,000 indicates an exponent of (2,000 – 1,023), or 977.

For double-precision words, the largest number representable is approximately 10^{308}, and the smallest positive number is approximately 2.2×10^{-308}.

Example 3.3 IEEE 754 floating-point representation.

The sequence of 64 binary digits:

0100000001111110100000000000000000000000000000000000000000000000

which is an IEEE 754 double precision floating-point number, is interpreted as follows:

- Since the first digit is 0, the number is positive
- The next 11 bits, 10000000111, represent the exponent e. These 11 bits represent the number 1,031; the exponent is $e = 1{,}031 - 1{,}023 = 8$
- The next 52 bits:
  1110100000000000000000000000000000000000000000000000
  represent the fractional portion, or mantissa, of the floating-point number. For this example, the mantissa is 0.90625.
- The floating-point number, in decimal, is

  $$\text{sign} \cdot (1 + \text{mantissa}) \times 2^{\text{exponent}}$$

  $$1.90625 \times 2^8 = 488$$

The exponents of 1,023 (all 0s) and +1,024 (all 1s) are reserved for special numbers. The number **zero** is not directly representable in the straight format, due to the assumption of a leading 1 (a true zero mantissa is needed to yield a value of zero). Zero is a special value, denoted with an exponent field of zero and a fraction field of zero. Note that -0 and +0 are distinct values, though they both compare as equal.

The values $+\infty$ and $-\infty$ are represented with an exponent of all 1s and a fraction of all 0s. The sign bit distinguishes between negative infinity and positive infinity. Being able to denote infinity as a specific value is useful, because it allows operations to continue past overflow situations.

The value **NaN** (*Not a Number*) is used to represent a value that does not represent a real number. NaNs are represented by a bit pattern with an exponent of all 1s and a non-zero fraction. NaN is used to detect problematic situations such as an attempt to divide 0 by 0, and handle the error without simply stopping.

An **overflow** occurs when a number is too large to fit into the floating-point system in use. An **underflow** occurs when the exponent is less than the smallest possible (-1,023 in IEEE 754 double precision). When overflow occurs during a calculation, this is generally fatal. But underflow is nonfatal: the system usually sets the number to 0 and continues. (MATLAB does this without warning messages).

3.5.2 Floating-point arithmetic, truncation, and rounding

There are two ways to store a real number using only t digits. **Truncating** ignores the digits $d_{t+1}, d_{t+2}, d_{t+3},\ldots$, and **rounding** represents a real number with t digits by adding 1 to d_t if $d_{t+1} \geq 5$. For example, with three digits,

x	Truncated	Rounded
5.672	5.67	5.67
-5.672	-5.67	-5.67
5.677	5.67	5.68
-5.677	-5.67	-5.68

The quantity 10^{1-t} is referred to as *eps*, the machine precision or machine epsilon. It is the smallest amount of (relative) error that is introduced for a single calculation. The machine epsilon, *eps*, is 2.2×10^{-16} for double-precision floating-point numbers in the IEEE standard. In MATLAB, typing *eps* displays approximately the value 2.2×10^{-16}. The quantity t-1 (for the rounding case) is often referred to as the number of significant digits.

Even if number representations were exact in the IEEE 754 system, arithmetic operations on numbers with exact representations introduce roundoff errors. The

IEEE standard requires operations to be performed with what is called *exact rounding*, in which the result of an arithmetic operation must be as though it is computed exactly and then rounded to the nearest floating-point number. With exact rounding, if fl(x) and fl(y) are floating-point representations, then:

$$\begin{aligned} \mathrm{fl}(\mathrm{fl}(x) \pm \mathrm{fl}(y)) &= (\mathrm{fl}(x) \pm \mathrm{fl}(y))(1+\varepsilon_1) \\ \mathrm{fl}(\mathrm{fl}(x) \times \mathrm{fl}(y)) &= (\mathrm{fl}(x) \times \mathrm{fl}(y))(1+\varepsilon_2) \\ \mathrm{fl}(\mathrm{fl}(x) \div \mathrm{fl}(y)) &= (\mathrm{fl}(x) \div \mathrm{fl}(y))(1+\varepsilon_3) \end{aligned}$$

where $\varepsilon_i \leq eps$. With exact rounding, relative errors remain small after each operation.

Example 3.4 Propagation of floating-point errors.

Let $x = 0.1103$ and let $y = 0.9963 \times 10^{-2}$; the exact value of the difference is $x - y =$ 0.100337. Compute the relative error if the result is rounded exactly or rounded to a precision of $t = 4$ digits.

The rounded difference to $t = 4$ digits is 0.1033. Therefore, the relative error with exact rounding is:

$$\frac{|0.100337 - 0.1003|}{|0.100337|} \approx 0.37 \times 10^{-3}$$

and the relative error if the floating-point numbers are rounded to 4 digits is:

$$\frac{|0.100337 - 0.1004|}{|0.100337|} \approx 0.63 \times 10^{-3}$$

In the latter case, the relative error has grown beyond the rounding unit of the floating-point system, $\frac{1}{2} \times 10^{-3}$.

Example 3.5 Machine precision in MATLAB.

In general, fl($1 + \alpha$) = 1 for any number α such that $\alpha \leq eps$. In MATLAB, the commands:

```
gamma = .5*2^(-100)
delta = (1+gamma)-1
```

produce the output:

```
gamma = 3.9443e-031
delta = 0
```

Explain why the error curve is flat for small values of h in Example 3.1.

Figure 3.2 shows that for small values of h, the roundoff error dominates. The MATLAB code in this example shows that there is a value of h (here `gamma`) that cannot be added or subtracted. Thus, the estimates of the derivative for this or smaller h will always be zero, and the error will be constant.

3.5.3 Roundoff error accumulation and cancellation error

Small errors are unavoidable in the many operations in a numerical algorithm. Besides knowing that each floating-point operation adds a small relative error, knowing how these errors accumulate is also important. However, there are a few rules to be aware of:

1. If x and y have markedly different magnitudes, then $x + y$ has a large absolute error.
2. If $|y| \ll 1$, then x/y has large relative and absolute errors. The same is true for xy if $|y| \gg 1$.
3. If $x = y$, then $x - y$ has a large relative error. This error is called a *cancellation error*.
4. Overflow can cause a loss of accuracy, but sometimes it can be avoided.

Example 3.6 Avoiding overflow.

Consider computing $c = \sqrt{a^2 + b^2}$ in a floating-point system with four digits of precision and two exponent digits, for $a = 10^{60}$ and $b = 1$. The correct result in this accuracy is $c = 10^{60}$. Show how rescaling ahead of time can avoid overflow results during the course of calculation.

Note that $c = s\sqrt{(a/s)^2 + (b/s)^2}$ for any $s \neq 0$. Using $s = a = 10^{60}$ gives an underflow when b/s is squared, which is set to zero. This yields the correct answer.

It is very important to be aware of cancellation error, because it often appears in practice. For example, in Example 3.1, the derivative of a smooth (differentiable) function $f(x)$ is approximated at a point $x = x_0$ by the difference of two neighboring values of f divided by the difference of the arguments:

$$f'(x_0) \approx \frac{f(x_0+h)-f(x_0)}{h}$$

For small h, there is a cancellation error in the numerator, which is then magnified by the denominator. Sometimes such errors can be avoided by a simple modification in the algorithm.

Example 3.7 Avoiding cancellation errors.

Suppose we wish to compute $y=\sqrt{x+1}-\sqrt{x}$ for $x = 100{,}000$ in a five-digit decimal arithmetic. The number 100,001 cannot be represented in this floating-point system exactly, and its representation in the system (regardless of whether truncating or rounding is used) is 100,000. In other words, for this value of x in this floating-point system, we have $x + 1 = x$. Blindly computing $\sqrt{x+1}-\sqrt{x}$ results in the value 0. Find a method for computing the correct result of the expression above.

The identity

$$\frac{\left(\sqrt{x+1}-\sqrt{x}\right)\left(\sqrt{x+1}+\sqrt{x}\right)}{\left(\sqrt{x+1}+\sqrt{x}\right)}=\frac{1}{\left(\sqrt{x+1}+\sqrt{x}\right)}$$

can be used to evaluate $\sqrt{x+1}-\sqrt{x}$. For x=100,000, the result is 0.15811×10^{-2}, which happens to be the correct value to five decimal digits.

One way of avoiding cancellation error is to use a Taylor series expansion.

Example 3.8 Using Taylor series expansions to avoid cancellation error.

Show how to compute $y=\sinh x=\left(e^{x}-e^{-x}\right)/2$ for a small value of $x>0, |x|\approx 0$ and minimize the cancellation error due to the subtraction.

The Taylor series expansion:

$$\sinh(x)=x+\frac{x^3}{6}+\frac{x^5}{120}+\cdots$$

can be used to compute the value of y and avoid the cancellation error, since there are only additions in the series. The truncated series $x+\frac{x^3}{6}$ is a reasonable approximation if x is small, since the truncation error in this approximation is $\approx \frac{x^5}{120}$ if $|x| \le 1$. Using the same five-digit decimal arithmetic as used in Example 3.7, sinh(0.1) = 0.10017 and sinh(0.01) = 0.01. These are "exact" values in this floating-point system. On the other hand, employing the formula without a discretization error that involves the exponential functions, we obtain 0.10018 and 0.010025 for these two values of x, respectively.

The effect of subtracting two nearly equal numbers in a floating-point system can be observed by doing the math: suppose $z = x - y$, where $x \approx y$. Then:

$$|z - \mathrm{fl}(z)| \le |x - \mathrm{fl}(x)| + |y - \mathrm{fl}(y)|$$

and the relative error is:

$$\frac{|z - \mathrm{fl}(z)|}{|z|} \le \frac{|x - \mathrm{fl}(x)| + |y - \mathrm{fl}(y)|}{|x - y|}$$

Since the denominator is very close to zero if $x \approx y$, the relative error could become very large.

When using rounding arithmetic, errors tend to change sign, whereas in truncation, errors do not. Statistically, there is a chance that occasional error cancellations will occur in a large-scale computation.

3.6 Lessons Learned in this Chapter

After studying this chapter, the student should have learned the following:

- The sources of error in scientific calculations are errors in the problem to be solved, errors in the input, errors due to approximating the continuous function(s), and round off errors in the calculations.
- The Taylor Series and Taylor Formula are the tools used to approximate continuous functions to develop numerical algorithms.
- Errors need to be kept small, so that the numerical solution is close to the exact solution.
- Absolute error is the combination of truncation error and roundoff error. When truncation error is small, roundoff error is large and vice versa. This is the most important tradeoff in numerical methods.
- Floating-point numbers are the computer representation of the real numbers and inherently have error, which can be due to either truncation or rounding.

- In MATLAB, floating-point numbers are represented using the IEEE 754 standard.
- Cancellation errors can often be avoided by rewriting the formula used in the calculation.

3.7 Problems

3.1 The function $f_1(x,\delta) = \cos(x+\delta) - \cos(x)$ can be transformed into another form, $f_2(x,\delta)$, using the trigonometric identity:

$$\cos(\phi) - \cos(\varphi) = -2\sin\left(\frac{\phi+\varphi}{2}\right)\sin\left(\frac{\phi-\varphi}{2}\right)$$

Thus, f_1 and f_2 have the same values, in exact arithmetic, for any given argument values x and δ.

Write a MATLAB script that will calculate $g_1(x,\delta) = f_1(x,\delta)/\delta + \sin(x)$ and $g_2(x,\delta) = f_2(x,\delta)/\delta + \sin(x)$ for x = 3 and $\delta = 1.0e-11$. Explain the difference in the results of the two calculations.

3.2 The function $f_1(x_0,h) = \sin(x_0+h) - \sin(x_0)$ can be transformed to another form, $f_2(x_0,h)$, using the trigonometric formula:

$$\sin(\phi) - \sin(\varphi) = 2\cos\left(\frac{\phi+\varphi}{2}\right)\sin\left(\frac{\phi-\varphi}{2}\right)$$

Thus, f_1 and f_2 have the same values, in exact arithmetic, for any pair of arguments x_0 and h.

Find a formula that avoids cancellation errors for computing the approximation $\dfrac{f(x_0+h)-f(x_0)}{h}$ to the derivative of $f(x) = \sin(x)$ at $x = x_0$. Write a MATLAB program that implements this formula and computes an approximation of $f'(.5)$, for $h = 1\times10^{-20},\ldots,1$. Compare your results to those in Example 3.1. Explain the difference between your results and the results reported in Example 3.1.

3.5 The number $\dfrac{8}{3} = 2.666\ldots$ has no exact representation in any decimal floating-point system with finite precision t. Is there a finite floating-point system in which this number does have an exact representation? If yes, then describe one such system. Answer this question for the irrational number π.

3.6 The roots of the quadratic equation $x^2 - 2bx + c = 0$ with $b^2 > c$ are given by:

$$x_{1,2} = b \pm \sqrt{b^2 - c}$$

Remember that $x_1 x_2 = c$. The MATLAB scripts calculate these roots in two different ways:

(a)

```
x1 = b + sqrt(b^2-c);
x2 = b -sqrt(b^2-c);
```

(b)

```
if b > 0
x1 = b + sqrt(b^2-c);
x2 =c / x1;
else
x2 = b - sqrt(b^2-c);
x1 =c / x2;
end
```

Which algorithm gives a more accurate result in general? Try to answer this question without any computing.

3.7 Write a MATLAB program that will:

a) Sum up $\frac{1}{n}$ for $n = 1,2,...,100{,}000$

b) Round each number $\frac{1}{n}$ to four decimal digits, and then sum them up in five-digit decimal arithmetic for $n = 1,2,...,100{,}000$.

c) Sum up the same rounded numbers (in four-digit decimal arithmetic) in reverse order, i.e., for n = 100,000,...,2, 1. Compare the three results and explain your observations.

3.8 In statistics (Chapter 9), one often needs to compute the quantities $\bar{x} = \frac{1}{n}\sum_{i=1}^{n} x_i$ and $s^2 = \frac{1}{n}\sum_{i=1}^{n}(x_i - \bar{x})^2$, where $x_1, x_2, \ldots, x_n$ are the given data. It is easy to see that s^2 can also be written as $s^2 = \frac{1}{n}\sum_{i=1}^{n} x_i^2 - \bar{x}^2$.

a) Which of the two methods to calculate s^2 requires fewer operations? Assume that $\bar{x}$ has already been calculated when counting the operations.
b) Which of the two methods is expected to give more accurate results for s^2 in general?
c) Write a MATLAB script to check whether your answer to the previous question was correct.

3.9 With the exact rounding of Section 3.5, each elementary operation has a relative error that is bounded in terms of the rounding unit, which is, in MATLAB, `eps`. Now consider exponentiation, which is implemented as $x^y = e^{y \ln x}$. Estimate the relative error in calculating x^y in floating point, assuming $\mathrm{fl}(x+y) = (x+y)(1+\varepsilon)$, $|\varepsilon| \le eps$, and that everything else is exact. Show that the bounds for elementary operations do not hold for exponentiation when x is very large.

3.8 References

Atkinson, K. A. 1989. *An Introduction to Numerical Analysis*. New York: John Wiley & Sons.

Burden , R. L. and Faires, J. D. 1993. *Numerical Analysis*. Boston, MA: PWS Kent.

Chapra, S. C. and Canale, R. P. 2006. *Numerical Methods for Engineers*. 5th ed. Boston, MA: McGraw Hill Book Company.

Hoffman, J. D. 2001. *Numerical Methods for Engineers and Scientists*. New York: Marcel Dekker.

Chapter 4
Linear Models of Biological Systems

4.1 Introduction

In this chapter, we discuss numerical methods for solving equations describing linear systems. Consider the set of simultaneous linear algebraic equations:

$$\begin{aligned}
a_{11}x_1 + a_{12}x_2 + \ldots\ldots &= b_1 \\
a_{21}x_1 + a_{22}x_2 + \ldots\ldots &= b_2 \\
\ldots\ldots\ldots\ldots\ldots\ldots\ldots\ldots\ldots & \\
a_{n1}x_1 + a_{n2}x_2 + \ldots\ldots &= b_n
\end{aligned} \tag{4.1}$$

This set may be represented in matrix notation by $\mathbf{Ax} = \mathbf{b}$, where $\mathbf{A}$ is the matrix of coefficients, $\mathbf{x}$ is the vector of unknowns, and $\mathbf{b}$ is the vector of constants, as defined by Eq.(4.2). If the matrix of coefficients, $\mathbf{A}$, is comprised of constants a_{ij}, which are not functions of $\mathbf{x}$, then the set of equations is *linear*. If $\mathbf{b} = 0$, then the set of equations is *homogeneous*.

$$\mathbf{A} = \begin{bmatrix} a_{11} & a_{12} & \cdots & a_{1n} \\ a_{21} & a_{22} & \cdots & a_{2n} \\ \cdots & \cdots & \cdots & \cdots \\ a_{n1} & a_{n2} & \cdots & a_{nn} \end{bmatrix}; \quad \mathbf{x} = \begin{bmatrix} x_1 \\ x_2 \\ . \\ . \\ x_n \end{bmatrix}; \quad \text{and } \mathbf{b} = \begin{bmatrix} b_1 \\ b_2 \\ . \\ . \\ b_n \end{bmatrix} \tag{4.2}$$

The material in this chapter will enable the student to accomplish the following:

- Learn the theoretical basis for numerical solution of simultaneous linear algebraic equations
- Develop MATLAB-based code for solving problems of $\mathbf{Ax} = \mathbf{b}$ type using two alternative approaches: noniterative methods (Gauss, Gauss-Jordan) and iterative methods (Jacobi, Gauss-Seidel)
- Formulate systems of linear equations $\mathbf{Ax} = \mathbf{b}$ for problems in biomedical engineering, and apply MATLAB-based codes to their efficient solution

4.2 Examples of Linear Biological Systems

Linear systems occur in several important problems of biomedical engineering. In this section, we will review three representative examples in (a) biomechanics, (b) biomedical imaging, and (c) metabolic engineering and cellular biotechnology.

4.2.1 Force balance in biomechanics

The field of biomechanics involves various complex nonlinear dynamics, yet linear problems are prevalent in the treatment of statics. The basis for linearity in statics arises from the linear decomposition of forces. Forces can be written in terms of scalar components and unit vectors. The 2-D vector $\mathbf{F}$ is composed of the $\mathbf{i}$ component, F_x, in the x direction and the $\mathbf{j}$ component, F_y, in the y direction, or:

$$\mathbf{F} = F_x\,\mathbf{i} + F_y\,\mathbf{j} \tag{4.3}$$

Vectors can be added by summing their components. Vectors can be multiplied in two ways: dot products (which physically means projection of one vector on to the other) or cross products (which yield a new vector that points along the axis of rotation).

In static equilibrium of biological systems, Newton's equations of motions can be applied to yield the following vector equations for force, **F**, and momentum, **M**:

$$\sum \mathbf{F} = 0 \tag{4.4}$$

and:

$$\sum \mathbf{M} = 0 \tag{4.5}$$

Typical problems in biomechanics require the drawing of a free-body diagram of the body segment(s) of interest with all externally applied loads and reaction forces at the supports. Consider a simple free-body diagram showing forces exerted at the shoulder due to the weight of an outstretched arm (Fig. 4.1).

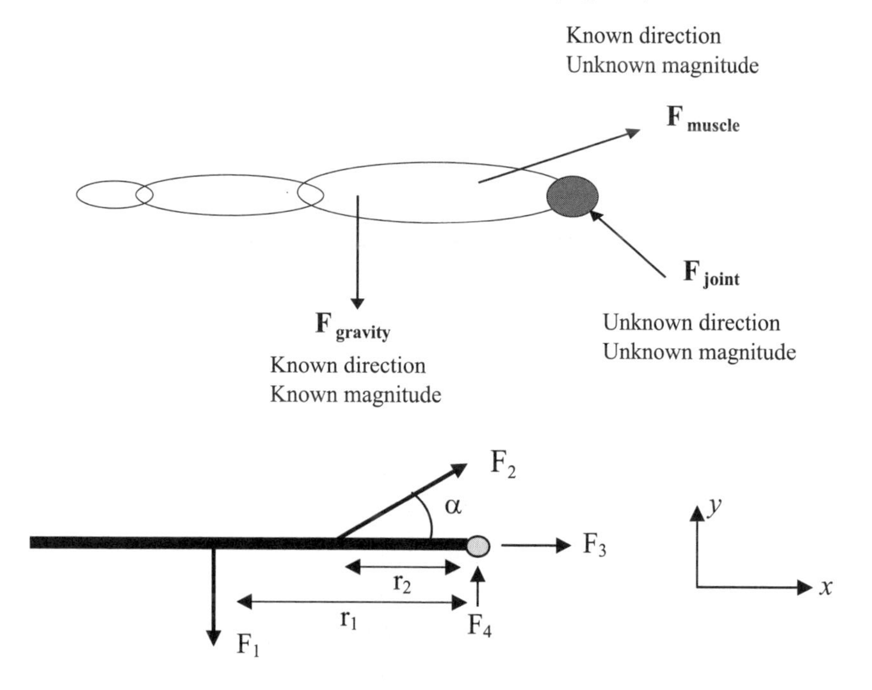

Figure 4.1 Free body diagram representation of forces acting on a hypothetical subject. The forces borne by the muscle and joint can be further resolved into their components, shown below. Using equilibrium conditions (for statics), the forces can be summed up along *x*- and *y*-directions, and the momentum is conserved along z. This problem leads to a linear algebraic system of equations that will be solved to quantify unknown forces experienced by the muscle and joint using the numerical techniques treated in this chapter. The F_{joint} can be resolved into two components, F_3 and F_4; the F_{muscle} represented by F_2, and $F_{gravity}$ by F_1, resulting in the lower free body diagram.

The static force equilibrium along x and y, and the moment balance along z, together yield:

$$\mathbf{F}_2 \cos\alpha + \mathbf{F}_3 = 0 \qquad (x\text{-direction}) \tag{4.6}$$

$$-\mathbf{F}_1 + \mathbf{F}_2 \sin\alpha + \mathbf{F}_4 = 0 \qquad (y\text{-direction}) \tag{4.7}$$

$$r_1\mathbf{F}_1 - r_2\mathbf{F}_2 \sin\alpha = 0 \qquad (z\text{-direction}) \tag{4.8}$$

The above equations necessitate the solution of simultaneous linear algebraic equations where the equations' matrix is a simple 3×3 system. For a more complicated biomechanical system, there can be many more equations that must be solved simultaneously. Methods to solve such systems numerically are the subject matter of this chapter.

4.2.2 Biomedical imaging and image processing

Another area of biomedical application wherein linear equations are found is the field of biomedical image processing, in particular the area of computer tomography for the reconstruction of three-dimensional images. An algebraic approach can be used for image reconstruction. In the figure shown below, for example, a square grid is superimposed on a hitherto unknown image $f(x, y)$.

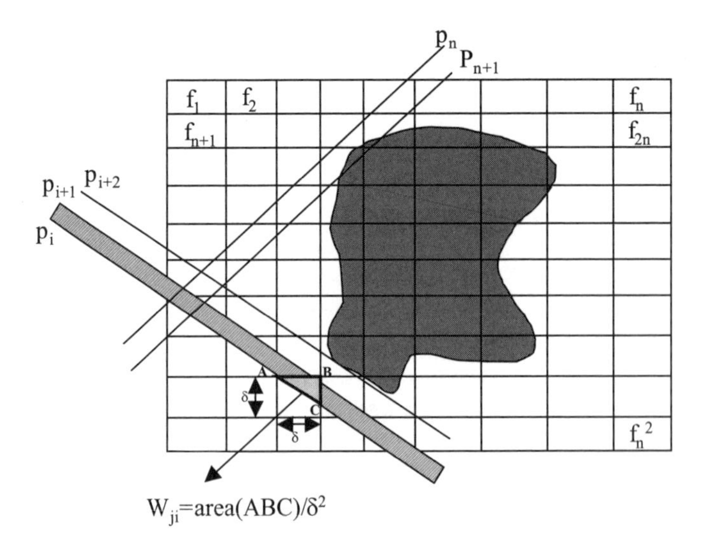

Figure 4.2 Algebraic reconstruction of a biomedical image. A square grid is superimposed over the unknown image and image values are assumed to be constant across each cell of the grid (Rosenfeld and Kak, 1982).

Each cell of the grid, j, is assumed to have a constant value for the function, f_j. A ray, i, of width τ approximately equal to the cell width, is drawn across the (x, y)-plane. The projected area of the ith ray is denoted by p_i. Then the contributions of function f_j to p_i can be expressed as an algebraic sum of M rays in all projections:

$$\sum_{j=1}^{N} w_{ij} f_j = p_i, \qquad i = 1, 2, \ldots, M \tag{4.9}$$

where w_{ij} is the weighting factor that represents the contribution of the jth cell to the ith ray integral. A simple problem would involve a small number of M and N. This system can be solved numerically by inverting the matrix. Methods for solving such problems are discussed further in this chapter. In reality, even for a 256×256 image, N can be larger than 50,000, and simplified procedures are necessary to reformulate the problem.

4.2.3 Metabolic engineering and cellular biotechnology

Consider the engineering design of microbial cell culture and growth in the field of biotechnology and metabolic engineering (Shuler and Kargi, 2002). Bioreactors are employed incorporating the use of biochemical substrates that provide essential chemical elements for the growth of the microorganisms. A typical stoichiometric representation of the aerobic growth of a microorganism with a defined chemical formula is shown below:

$$\underset{\text{(C-source)}}{C_2H_5OH} + \underset{\text{(O-source)}}{aO_2} + \underset{\text{(N-source)}}{bNH_3} \rightarrow \underset{\text{(Microorganism)}}{cCH_{1.7}N_{0.15}O_{0.4}} + dH_2O + eCO_2 \tag{4.10}$$

The ratio of moles of CO_2 produced per mole of O_2 consumed is called the respiratory quotient, RQ, which can be experimentally determined, and is therefore treated as a known quantity. To determine the theoretical yield of production (growth) of the microorganism, the stoichiometric coefficients, a, b, c, d, of Eq. (4.10) are required. One can easily formulate an elemental mass balance for each of the four key elements in Eq. (4.10), as shown below.

The mass balances for individual elements are:

$$\begin{aligned}
&\text{C (carbon)}: && 2 &&= c + (RQ)a \\
&\text{H (hydrogen)}: && 6 + 3b &&= 1.7c + 2d \\
&\text{O (oxygen)}: && 1 + 2a &&= 0.4c + d + 2(RQ)a \\
&\text{N (nitrogen)}: && b &&= 0.15c
\end{aligned} \tag{4.11}$$

In matrix notation:

$$\begin{bmatrix} RQ & 0 & 1 & 0 \\ 0 & -3 & 1.7 & 2 \\ -2 & 0 & 0.4 & 1 \\ 0 & -1 & 0.15 & 0 \end{bmatrix} \begin{bmatrix} a \\ b \\ c \\ d \end{bmatrix} = \begin{bmatrix} 2 \\ 6 \\ 1-2RQ \\ 0 \end{bmatrix} \tag{4.12}$$

Eq. (4.11) and its corresponding matrix form in Eq. (4.12) represent a set of four simultaneous linear algebraic equations involving four unknown metabolic stoichiometric coefficients. In the course of this chapter, the student will learn techniques to obtain numerical solutions to similar problems.

4.3 Simultaneous Linear Algebraic Equations

The most widely employed approach to solving simultaneous linear algebraic equations is the Gaussian elimination method. This is based on the principle of transforming a given "full" set of n algebraic linear equations to a triangular form of n equations, which can then be readily solved, as illustrated below.

4.3.1 Illustration of simple Gauss elimination for a 3×3 matrix

An example of a three-by-three system of equations is:

$$\begin{aligned} 2x + y - z &= 7 \\ 2x + 6y + 5z &= 0 \\ 3x + y + z &= 5 \end{aligned} \tag{4.13}$$

The Gaussian elimination consists of two simple steps:

Step 1

Use the first equation to eliminate x in the second and third equations:

- Multiply the first equation by –1 and add it to the second equation to get a new second equation with x variable eliminated.
- Also, multiply the first equation by –3/2 and add it to the third equation to get a new third equation with the x variable eliminated.

The resulting system is:

$$\begin{aligned} 2x + y - z &= 7 \\ 5y + 6z &= -7 \\ -\frac{1}{2}y + \frac{5}{2}z &= -\frac{11}{2} \end{aligned} \tag{4.14}$$

Step 2

Use the second equation to eliminate the y term in the third equation:

- Multiply the second equation by (1/10) and add it to the third equation to get a transformed third equation with the y variable eliminated.

The new set of equations looks like this:

$$\begin{aligned} 2x + y - z &= 7 \\ 5y + 6z &= -7 \\ \frac{31}{10}z &= -\frac{31}{5} \end{aligned} \tag{4.15}$$

This completes the "forward elimination" and leaves us with an upper triangular matrix system, represented below. The transformed matrix **A** is denoted by **U**. Thus the original equation **Ax** = **b** is transformed to:

$$\begin{bmatrix} 2 & 1 & -1 \\ 0 & 5 & 6 \\ 0 & 0 & \frac{31}{10} \end{bmatrix} \begin{bmatrix} x \\ y \\ z \end{bmatrix} = \begin{bmatrix} 7 \\ -7 \\ -\frac{31}{5} \end{bmatrix}$$

or:

$$\mathbf{Ux} = \bar{\mathbf{b}} \tag{4.16}$$

From Eqs. (4.15) or (4.16), we can now easily evaluate that $z = -1$. By substitution, $y = 1$ and $x = 2$.

4.3.2 Matrix notation of Gaussian elimination

The generic matrix form of linear algebraic equations is **Ax** = **b**. Gaussian elimination, shown above, can be readily applied to a system of linear algebraic equations including the vector of coefficient **b**. The net matrix including the coefficients is referred to as the augmented matrix.

In the example above from Eq. (4.10), the original augmented matrix is:

$$\left[\begin{array}{ccc|c} 2 & 1 & -1 & 7 \\ 2 & 6 & 5 & 0 \\ 3 & 1 & 1 & 5 \end{array}\right]$$

whereas the final augmented matrix is:

$$\left[\begin{array}{ccc|c} 2 & 1 & -1 & 7 \\ 0 & 5 & 6 & -7 \\ 0 & 0 & \frac{31}{10} & -\frac{31}{5} \end{array}\right] \tag{4.17}$$

Let us generalize the sequence of matrix operations performed under Gaussian elimination.

Declaration of the augmented Gaussian matrix

All the elements of the constant coefficients within the matrix **A** and the vector of constants **b** are included within the augmented matrix. This is done prior to any transformations, and therefore is referred to as the zeroth step ($k = 0$).

$$\begin{aligned} &\text{For} \quad i = 1,2,\ldots n && \text{(All rows)} \\ &a_{ij}^{(k=0)} = a_{ij} \qquad j = 1,2,\ldots,n && \\ &a_{ij}^{(k=0)} = c_{ij} \qquad j = n+1 && \text{(Augmented column)} \end{aligned} \tag{4.18}$$

Gaussian elimination

Let us review the matrix transformation one step at a time. We can generalize the division of each row by the diagonal coefficient, which is called the pivot element. At the first step (called $k = 1$), we get a modified row $R_2^{k=1}$ by subtracting a multiple of row R_1 (i=1) from the original row R_2. We can see that this multiple is the ratio of the first nonzero coefficient in the original row R_2 to the pivot element in R_1. The multiple is constant for the entire step. During the first step, the subtraction is carried out for all the column elements of the row R_2, and then for R_3, and so on. Thus, for the kth step, the $(k+1)^{st}$ row and all rows below this row are modified.

Result of the first step (k=1):

$$\begin{array}{c} \\ \\ R_1^{k=0} \rightarrow \\ R_2^{k=1} \rightarrow \\ R_3^{k=0} \rightarrow \end{array} \begin{array}{c} \begin{array}{cccc} \text{j=1} & \text{j=2} & \text{j=3} & \text{j=4} \\ \Downarrow & \Downarrow & \Downarrow & \Downarrow \end{array} \\ \left[\begin{array}{ccc|c} 2 & 1 & -1 & 7 \\ 2 \quad \frac{2}{2}\times 2 & 6-\frac{2}{2}\times 1 & 5-\frac{2}{2}\times(-1) & 0-\frac{2}{2}\times 7 \\ 3 & 1 & 1 & 5 \end{array}\right] \end{array} \tag{4.19}$$

Thus, after the first step (k = 1), each modified element of the second row (i = 2) can be generalized by:

$$a_{2j}^{k=1} = a_{2j}^{k=0} - \frac{a_{21}^{k=0}}{a_{11}^{k=0}} a_{1j}^{k=0}; \quad i = 2; j = 4,3,..,k; k = 1. \tag{4.20}$$

where the multiple was $\frac{2}{2}$.

Further, as shown below, after the first step (k = 1), each modified element of the third row (i = 3) is generalized by:

$$a_{3j}^{k=1} = a_{3j}^{k=0} - \frac{a_{31}^{k=0}}{a_{11}^{k=0}} a_{1j}^{k=0}; \quad i = 3; j = 4,3,..k; k = 1 \tag{4.21}$$

$$\begin{array}{c} \\ \\ R_1^{k=0} \rightarrow \\ R_2^{k=1} \rightarrow \\ R_3^{k=1} \rightarrow \end{array} \begin{array}{cccc} j=1 & j=2 & j=3 & j=4 \\ \Downarrow & \Downarrow & \Downarrow & \Downarrow \end{array}$$

$$\begin{array}{l} R_1^{k=0} \rightarrow \\ R_2^{k=1} \rightarrow \\ R_3^{k=1} \rightarrow \end{array} \left[\begin{array}{cccc} 2 & 1 & -1 & | \quad 7 \\ 0 & 5 & 6 & | \quad -7 \\ 3-\frac{3}{2}\times 2 & 1-\frac{3}{2}\times 1 & 1-\frac{3}{2}\times(-1) & | \quad 5-\frac{3}{2}\times 7 \end{array}\right]$$

where the multiple was $\frac{3}{2}$:

$$\begin{array}{l} R_1^{k=0} \rightarrow \\ R_2^{k=1} \rightarrow \\ R_3^{k=1} \rightarrow \end{array} \left[\begin{array}{ccc|c} 2 & 1 & -1 & 7 \\ 0 & 5 & 6 & -7 \\ 0 & -\frac{1}{2} & \frac{5}{2} & -\frac{11}{2} \end{array}\right] \tag{4.22}$$

Next, for k=2,

$$a_{3j}^{k=2} = a_{3j}^{k=1} - \frac{a_{32}^{k=1}}{a_{22}^{k=1}} a_{2j}^{k=1}; \quad i = 3; j = 4,3,..,k; k = 2, \tag{4.23}$$

and therefore the matrix transforms to:

$$\begin{array}{c} R_1^{k=0} \rightarrow \\ R_2^{k=1} \rightarrow \\ R^{k=2} \rightarrow \end{array} \left[\begin{array}{cccc} 2 & 1 & -1 & | & 7 \\ 0 & 5 & 6 & | & -7 \\ 0 & -\frac{1}{2}-\left(-\frac{1}{10}\right)\times 5 & \frac{5}{2}-\left(-\frac{1}{10}\right)\times 6 & | & -\frac{11}{2}-\left(-\frac{1}{10}\right)\times(-7) \end{array}\right]$$

which can be simplified to:

$$\begin{array}{c} R_1^{k=0} \rightarrow \\ R_2^{k=1} \rightarrow \\ R^{k=2} \rightarrow \end{array} \left[\begin{array}{ccccc} 2 & 1 & -1 & | & 7 \\ 0 & 5 & 6 & | & 7 \\ 0 & 0 & \frac{31}{10} & | & -\frac{62}{10} \end{array}\right] \tag{4.24}$$

The above steps can be combined into the following algorithm:

$$\begin{array}{ll} k = 1,2,\ldots,n-1 & \text{(Loop of incremental steps)} \\ \quad i = k+1, k+2, \ldots\ldots, n & \text{(Loop of rows)} \\ \quad\quad j = n+1, n, \ldots., k & \text{(Loop of columns for a given row)} \\ \quad\quad\quad a_{ij}^{(k)} = a_{ij}^{(k-1)} - \frac{a_{ik}^{(k-1)}}{a_{kk}^{(k-1)}} a_{kj}^{(k-1)} & \text{(Gauss transformation);} \quad a_{kk}^{(k-1)} \neq 0 \end{array} \tag{4.25}$$

Once the transformed augmented matrix is ready, a simple back substitution is executed, starting from the nth row and working backwards to the first row:

$$x_n = \frac{a_{n,n+1}}{a_{n,n}} \qquad \text{(Substitution for the } n\text{th row)}$$

Because, $a_{ii}x_i + \sum_{j=i+1}^{n} a_{ij}x_j = a_{i,n+1}$, we get the general back-substitution formula for all the rows above the nth row:

$$x_i = \frac{a_{i,n+1} - \sum_{j=i+1}^{n} a_{ij}x_j}{a_{ii}}, \quad i = n-1, n-2, \ldots, 1 \quad \text{(Loop of rows, working backwards)} \tag{4.26}$$

Discussion of the Gauss elimination approach

Although the Gauss elimination is an efficient method for solving simultaneous linear algebraic equations, there are two shortcomings associated with it.

The first issue arises from the nature of the matrix of coefficients of the equations. Particularly, when the pivot element of an encountered row is zero, and the entire pivot column below the pivot row is also zero, then the system of equations cannot be solved to yield a unique solution. There is either inconsistency or redundancy in the equations. If, however, a zero pivot element is encountered, but there is a nonzero pivot column below the pivot element, the straightforward Gauss method needs to be modified. This is formally achieved by interchanging the row whose pivot element is zero with a row that has the largest coefficient on the diagonal. This process is called (partial) pivoting, and the method, called *Gauss elimination with partial row pivoting*, is discussed in the next section. Partial pivoting reduces the possibility of division by zero, and it increases the accuracy of the Gauss elimination method by using the largest pivot element.

An additional approach exists for rearranging the matrix so as to identify the maximum possible pivot element. Here, in addition to rows, the columns are also searched, and columns (and their associated variables) are interchanged—pivoting performed with row and column interchange is called *complete pivoting*.

The second shortcoming is encountered when the matrix of coefficients is ill-conditioned. Typical ill-conditioned matrices may have an extremely small pivot element, in relation to the other pivot elements. During Gauss elimination, division by a very small coefficient can pose a problem because division by a number close to zero can lead to high sensitivity to coefficient rounding and to error propagation. An example is shown below:

$$\begin{aligned} 0.0001x_1 + x_2 &= 2 \\ x_1 + 2x_2 &= 5 \end{aligned} \tag{4.27}$$

If we were to use the basic Gauss elimination procedure, we would divide the first equation by pivot element of 0.0001 and subtract from the second equation, resulting in the following equations:

$$\begin{aligned} 0.0001x_1 + x_2 &= 2 \\ \left[2 - 1/0.0001\right]x_2 &= 5 - 2/0.0001 \end{aligned}$$

which, after rounding, gives an estimate of $x_2 = 2$, and therefore from the first equation, $x_1 = 0$. Clearly, this is not a good approximation to a solution for the second equation.

Instead, if we interchange the order of the equations by swapping the rows (partial row pivoting), we can work with a larger pivot element, as shown below

$$\begin{aligned} x_1 + 2x_2 &= 5 \\ 0.0001x_1 + x_2 &= 2 \end{aligned} \tag{4.28}$$

Now, based on Gauss elimination steps, we can multiply the first row by 0.0001 and subtract from the second equation:

$$\begin{aligned} x_1 + 2x_2 &= 5 \\ [1 - 2 \times 0.0001]x_2 &= 2 - 5 \times 0.0001 \end{aligned}$$

giving a value of $x_2 = 2$. Substituting this into the first equation gives $x_1 = 1$, which provides a significantly better approximation to a solution of the system.

The general algorithm for partial row pivoting is illustrated in the algorithm below.

Initialization steps

$a_{pj} = temp \quad p = k \quad |a_{kk}|$
For $i = 1, 2, \ldots n$ (All rows)
  $a_{ij}^{(k=0)} = a_{ij} \quad j = 1, 2, \ldots, n$
  $a_{ij}^{(k=0)} = c_{ij} \quad j = n + 1$ (Augmented column)
End

Partial row pivoting and forward elimination

For k = 1 to n - 1
  $Pivot = |a_{kk}|$ ($Pivot$ = Pivot element)
  $p = k$ (p = Pivot row)
  For $i = k + 1$ to n
    If $|a_{ik}| > Pivot$
      $Pivot = |a_{ik}|$ (Updating the pivot element)
      $p = i$ (Updating the pivot row)
    End
  End

If ($p>k$) (Interchange rows k and p)

For $j = 1, n+1$

$temp = a_{kj}$

$a_{kj} = a_{pj}$

$a_{pj} = temp$

End

End

For $i = k+1, k+2, \ldots\ldots, n$ (Loop of rows)

For $j = n+1, n, \ldots., k$ (Loop on columns for a given row)

$$a_{ij}^{(k)} = a_{ij}^{(k-1)} - \frac{a_{ik}^{(k-1)}}{a_{kk}^{(k-1)}} a_{kj}^{(k-1)} \quad \text{(Gauss transformation)} \tag{4.29}$$

End

End

End (End of loop on k, the counter for iteration step)

Back substitution

$x_n = \frac{a_{n,n+1}}{a_{n,n}}$ (Substitution for the nth row)

For $i = n-1, n-2, \ldots., 1$ (Loop of rows, working backwards)

$$x_i = \frac{a_{i,n+1} - \sum_{j=i+1}^{n} a_{ij} x_j}{a_{ii}} \tag{4.30}$$

End

Generally, a diagonally dominant matrix will work well with the Gauss elimination (i.e., $|a_{ii}| > \sum_{j \neq i} |a_{ij}|$)

Example 4.1 Solve a set of linear algebraic equations of the form Ax = b using the Gauss elimination method:

$$\begin{bmatrix} 2 & 1 & -1 \\ 2 & 6 & 5 \\ 3 & 1 & 1 \end{bmatrix} \begin{bmatrix} x_1 \\ x_2 \\ x_3 \end{bmatrix} = \begin{bmatrix} 7 \\ 0 \\ 5 \end{bmatrix}$$

Program

```
% example4_1.m - This program poses Ax = b in matrix form and
% solves for x using the Gauss elimination method. It calls the
% function GAUSS.M to transform the augmented matrix and do
% back-substitution.

clc; clear all;
% Input data
% Matrix of coefficients
A = ...
    [2, 1, -1
    2, 6, 5
    3, 1, 1];
% Vector of constants
b = [7; 0 ; 5];
% Solving the set of equations by Gauss elimination method
x = Gauss(A,b);
% Show the results
disp(' Results:')
fprintf(' x1 = %4.2f\n x2 = %4.2f\n x3 = %4.2f\n',x)
```

Function, developed by Constantinides and Mostoufi (1999), that performs the Gauss elimination method

```
function x = Gauss (A , c)
%GAUSS Solves a set of linear algebraic equations by the Gauss
%   elimination method.
%
%   GAUSS(A,C) finds unknowns of a set of linear algebraic
%   equations. A is the matrix of coefficients and C is the
%   vector of constants.
%
%   See also JORDAN, GSEIDEL

% (c) N. Mostoufi & A. Constantinides
% January 1, 1999

c = (c(:).')';     % Make sure it's a column vector
```

```
n = length(c);
[nr nc] = size(A);

% Check coefficient matrix and vector of constants
if nr ~= nc
   error('Coefficient matrix is not square.')
end
if nr ~= n
   error('Coefficient matrix and vector of constants do not have
the same length.')
end

% Check if the coefficient matrix is singular
if det(A) == 0
   fprintf('\n Rank = %7.3g\n',rank(A))
   error('The coefficient matrix is singular.')
end

unit = diag(ones( 1 , n));                 % Unit matrix
order = [1 : n];                            Order of unknowns
aug = [A c];                               % Augmented matrix

% Gauss elimination
for k = 1 : n - 1
   pivot = abs(aug(k , k));
   prow = k;
   pcol = k;

   % Locating the maximum pivot element
   for row = k : n
      for col = k : n
         if abs(aug(row , col)) > pivot
            pivot = abs(aug(row , col));
            prow = row;
            pcol = col;
         end
      end
   end

   % Interchanging the rows
   pr = unit;
   tmp = pr(k , :);
   pr(k , : ) = pr(prow , : );
   pr(prow , : ) = tmp;
   aug = pr * aug;

   % Interchanging the columns
   pc = unit;
   tmp = pc(k , : );
   pc(k , : ) = pc(pcol , : );
```

```
    pc(pcol , : ) = tmp;
    aug(1 : n , 1 : n) = aug(1 : n , 1 : n) * pc;
    order = order * pc;          % Keep track of the column
interchanges

    % Reducing the elements below diagonal to zero in the column k
    lk = unit;
    for m = k + 1 : n
       lk(m , k) = - aug(m , k) / aug(k , k);
    end
    aug = lk * aug;
end
x = zeros(n , 1);
% Back substitution
t(n) = aug(n , n + 1) / aug(n , n);
x(order(n)) = t(n);
for k = n  - 1 : -1 : 1
    t(k) = (aug(k,n+1) - sum(aug(k,k+1:n) .* t(k+1:n))) / aug(k,k);
    x (order(k)) = t(k);
end
```

Results

```
 x1 = 2.00
 x2 = 1.00
 x3 = -2.00
```

4.4 The Gauss-Jordan Reduction Method

Gauss-Jordan reduction is a variant of the Gauss elimination method. Here, the diagonal elements are normalized to 1, and all elements above and below the diagonal are rendered to 0, thus converting the matrix of coefficients to the identity matrix. By starting from the standard matrix form of:

$$\mathbf{A}\,\mathbf{x} = \mathbf{b} \tag{4.31}$$

the Gauss-Jordan reduction will result in:

$$\mathbf{I}\,\mathbf{x} = \tilde{\mathbf{b}} \tag{4.32}$$

where, notably, the solution vector will be the $\tilde{\mathbf{b}}$ vector.

Demonstration of Gauss-Jordan reduction

We will start with the same augmented matrix we considered previously and examine how the reduction of matrix coefficients leads to the identity matrix. For simplification, in this section we will not perform row or column interchanges related to pivoting. A combined Gauss-Jordan algorithm for reduction with pivoting will be treated in the following section. Starting with the augmented matrix:

$$\left[\begin{array}{ccc|c} 2 & 1 & -1 & 7 \\ 2 & 6 & 5 & 0 \\ 3 & 1 & 1 & 5 \end{array}\right]$$

Normalize the first row by dividing by 2:

$$\left[\begin{array}{ccc|c} 1 & 1/2 & -1/2 & 7/2 \\ 2 & 6 & 5 & 0 \\ 3 & 1 & 1 & 5 \end{array}\right]$$

Multiply the normalized first row by 2 and subtract it from the second row:

$$\left[\begin{array}{ccc|c} 1 & 1/2 & -1/2 & 7/2 \\ 0 & 5 & 6 & -7 \\ 3 & 1 & 1 & 5 \end{array}\right]$$

Multiply the normalized first row by 3 and subtract it from the third row:

$$\left[\begin{array}{ccc|c} 1 & 1/2 & -1/2 & 7/2 \\ 0 & 5 & 6 & -7 \\ 0 & -1/2 & 5/2 & -11/2 \end{array}\right]$$

Normalize the second row by dividing it by 5:

$$\left[\begin{array}{ccc|c} 1 & 1/2 & -1/2 & 7/2 \\ 0 & 1 & 6/5 & -7/5 \\ 0 & -1/2 & 5/2 & -11/2 \end{array}\right]$$

Multiply the normalized second row by 1/2 and subtract it from the first row:

$$\left[\begin{array}{ccc|c} 1 & 0 & -11/10 & 42/10 \\ 0 & 1 & 6/5 & -7/5 \\ 0 & -1/2 & 5/2 & -11/2 \end{array}\right]$$

Multiply the normalized second row by -1/2 and subtract it from the third row:

$$\begin{bmatrix} 1 & 0 & -11/10 & | & 42/10 \\ 0 & 1 & 6/5 & | & -7/5 \\ 0 & 0 & 31/10 & | & -62/10 \end{bmatrix}$$

Normalize the third row by dividing it by 31/10:

$$\begin{bmatrix} 1 & 0 & -11/10 & | & 42/10 \\ 0 & 1 & 6/5 & | & -7/5 \\ 0 & 0 & 1 & | & -2 \end{bmatrix}$$

Multiply the normalized third row by -11/10 and subtract it from the first row:

$$\begin{bmatrix} 1 & 0 & 0 & | & 2 \\ 0 & 1 & 6/5 & | & -7/5 \\ 0 & 0 & 1 & | & -2 \end{bmatrix}$$

Multiply the normalized third row by 6/5 and subtract it from the second row:

$$\begin{bmatrix} 1 & 0 & 0 & | & 2 \\ 0 & 1 & 0 & | & 1 \\ 0 & 0 & 1 & | & -2 \end{bmatrix}$$

Therefore, the answer to this problem is $x_1 = 2$, $x_2 = 1$, and $x_3 = -2$.

Example 4.2 Solve the Ax = b problem using the Gauss-Jordan reduction method.

$$\begin{bmatrix} 2 & 1 & -1 \\ 2 & 6 & 5 \\ 3 & 1 & 1 \end{bmatrix} \begin{bmatrix} x_1 \\ x_2 \\ x_3 \end{bmatrix} = \begin{bmatrix} 7 \\ 0 \\ 5 \end{bmatrix}$$

Program

```
% example4_2.m - This program solves the system Ax=b using the
% function JORDAN.M.

clc; clear all;

% Matrix of coefficients
A = [2, 1, -1; 2, 6, 5; 3, 1, 1];
```

```
% Vector of constants
b = [7; 0 ; 5];

% Solution
X = Jordan(A,b)
```

Function that performs the Gauss-Jordan reduction method

```
function x = Jordan (A , c)
%JORDAN Solves a set of linear algebraic equations by the
%   Gauss-Jordan method.
%
%   JORDAN(A,C) finds unknowns of a set of linear algebraic
%   equations. A is the matrix of coefficients and C is the
%   vector of constants.
%
%   See also GAUSS, GSEIDEL

% (c) N. Mostoufi & A. Constantinides
% January 1, 1999

c = (c(:).')';    % Make sure it's a column vector

n = length(c);
[nr nc] = size(A);

% Check coefficient matrix and vector of constants
if nr ~= nc
   error('Coefficient matrix is not square.')
end
if nr ~= n
   error('Coefficient matrix and vector of constants do not have
the same length.')
end

% Check if the coefficient matrix is singular
if det(A) == 0
   fprintf('\n Rank = %7.3g\n',rank(A))
   error('The coefficient matrix is singular.')
end

unit = diag(ones( 1 , n));    % Unit matrix
order = [1 : n];                      % Order of unknowns
aug = [A c];                          % Augmented matrix

% Gauss - Jordan algorithm
for k = 1 : n
   pivot = abs(aug(k , k));
   prow = k;
   pcol = k;
```

```
    % Locating the maximum pivot element
    for row = k : n
       for col = k : n
          if abs(aug(row , col)) > pivot
             pivot = abs(aug(row , col));
             prow = row;
             pcol = col;
          end
       end
    end

    % Interchanging the rows
    pr = unit;
    tmp = pr(k , :);
    pr(k , : ) = pr(prow , : );
    pr(prow , : ) = tmp;
    aug = pr * aug;

    % Interchanging the columns
    pc = unit;
    tmp = pc(k , : );
    pc(k , : ) = pc(pcol , : );
    pc(pcol , : ) = tmp;
    aug(1 : n , 1 : n) = aug(1 : n , 1 : n) * pc;
    order = order * pc; % Keep track of the column interchanges

    % Reducing the elements above and below diagonal to zero
    lk = unit;
    for m = 1 : n
       if m == k
          lk(m , k) = 1 / aug(k , k);
       else
          lk(m , k) = - aug(m , k) / aug(k , k);
       end
    end
    aug = lk * aug;
end
x = zeros(n , 1);
% Solution
for k = 1 : n
   x(order(k)) = aug(k , n + 1);
end
```

Results

```
X =
    2.0000
    1.0000
   -2.0000
```

4.5 Iterative Approach for Solution of Linear Systems

Iterative approaches can be used to solve simultaneous linear systems if the matrices are predominantly diagonal; that is, if the absolute value of each coefficient on the diagonal of the matrix is larger than the sum of the absolute values of the other coefficients in the same equation (same row of the matrix). Consider the system of equations solved earlier in this chapter:

$$\begin{aligned} 2x + y - z &= 7 \\ 2x + 6y + 5z &= 0 \\ 3x + y + z &= 5 \end{aligned} \tag{4.33}$$

Compare the absolute value of the diagonal elements of the matrix with the sum of the absolute values of the other coefficients:

$$|2| = |1| + |-1|.$$

and:

$$|6| < |2| + |5|$$

and:

$$|1| < |3| + |1|$$

thus, the above system of equations is not predominantly diagonal, and iterative methods discussed below cannot be used to solve this example.

4.5.1 The Jacobi method

The solution of **Ax** = **b** can be found through repeated trials, called iterations, by converting the problem to an equivalent system **x** = **Px** + **q** and generating a sequence of approximations: $x^{(1)}$, $x^{(2)}$,... . For each iteration, the relationship $x^{(k)} = \mathbf{P}\, x^{(k-1)} + \mathbf{q}$ should be obeyed. The basic idea is to solve the ith equation in the system for the ith variable. A simple three-by-three system, shown below:

$$\begin{aligned} a_{11}x_1 + a_{12}x_2 + a_{13}x_3 &= b_1 \\ a_{21}x_1 + a_{22}x_2 + a_{23}x_3 &= b_2 \\ a_{31}x_1 + a_{32}x_2 + a_{33}x_3 &= b_3 \end{aligned} \tag{4.34}$$

can be converted to the system **x** = **Px** + **q** where the matrix **P** has zeros on the diagonal, as shown below:

$$\begin{aligned}
x_1^{(k)} &= && -\frac{a_{12}}{a_{11}}x_2^{(k-1)} && -\frac{a_{13}}{a_{11}}x_3^{(k-1)} && +\frac{b_1}{a_{11}} \\
x_2^{(k)} &= -\frac{a_{21}}{a_{22}}x_1^{(k-1)} && && -\frac{a_{23}}{a_{22}}x_3^{(k-1)} && +\frac{b_2}{a_{22}} \\
x_3^{(k)} &= -\frac{a_{31}}{a_{33}}x_1^{(k-1)} && -\frac{a_{32}}{a_{33}}x_2^{(k-1)} && && +\frac{b_3}{a_{33}}
\end{aligned}$$

The Jacobi method starts with guess-estimates for x_1, x_2, x_3 for the right-hand side of the above equations. The updated vector **x** is computed on the left-hand side, and updated values of x_1, x_2, x_3 are iteratively substituted on the right-hand side to find the newer updated left-hand side, and so on. The iterations are stopped when the norm of the change in the solution vector **x** from one iteration to the next is sufficiently small or when the norm of the residual vector, $||\mathbf{Ax} - \mathbf{b}||$, is below a specified tolerance.

The effective algorithm for the Jacobi method is:

$$x_i^{(k+1)} = \frac{1}{A_{i,i}}\left(b_i - \sum_{\substack{j=1 \\ j\neq i}}^{N} A_{i,j} x_j^{(k)} \right) \tag{4.35}$$

Because the Jacobi method relies on guess-estimates of previous iterations, a simpler algorithm can be formulated to generalize the Jacobi method using matrix algebra. Consider again the problem:

$$\mathbf{A\,x} = \mathbf{b} \tag{4.36}$$

The matrix of coefficients **A** can be written in terms of a diagonal matrix, **D**, whose elements are the diagonal elements of **A**. Therefore, the matrix (**A** – **D**) will have its diagonal elements equal to zero. We use the following additive identity to formulate **A** in terms of the difference matrix (**A** – **D**) and **D**:

$$\mathbf{A} = (\mathbf{A} - \mathbf{D}) + \mathbf{D} \tag{4.37}$$

Then, substituting Eq. (4.37) into Eq. (4.36) yields:

$$\mathbf{D\,x} = \mathbf{b} - (\mathbf{A} - \mathbf{D})\mathbf{x} \tag{4.38}$$

from which the vector **x** can be evaluated:

$$\mathbf{x} = \mathbf{D}^{-1}\mathbf{b} - \mathbf{D}^{-1}(\mathbf{A} - \mathbf{D})\mathbf{x} \tag{4.39}$$

$$\mathbf{x} = \mathbf{D}^{-1}\mathbf{b} - (\mathbf{D}^{-1}\mathbf{A} - \mathbf{I})\mathbf{x} \tag{4.40}$$

Eq. 4.39 offers a convenient form for the Jacobi method using guess-estimates for **x** from one iteration to compute the estimate for **x** for the subsequent iteration. The working equation is shown below in (4.41), and is worked out in Example 4.3. This process is repeated until a desired tolerance is achieved.

$$\mathbf{x}^{(k)} = \mathbf{D}^{-1}\mathbf{b} - (\mathbf{D}^{-1}\mathbf{A} - \mathbf{I})\,\mathbf{x}^{(k-1)} \tag{4.41}$$

Example 4.3 Solution of Ax = b using an iterative Jacobi method.

$$\begin{bmatrix} 3 & 1 & -1 \\ 1 & 6 & 5 \\ 1 & 1 & 3 \end{bmatrix} \begin{bmatrix} x_1 \\ x_2 \\ x_3 \end{bmatrix} = \begin{bmatrix} 7 \\ 0 \\ 5 \end{bmatrix}$$

Program

```
% example4_3 - This program solves the system Ax=b using
% the function JACOBI.M. The Jacobi method is iterative, hence
% guess-estimates for x0 and a tolerance level should be provided.

clc; clear all;

n=input('Number of equations=');

% Matrix of coefficients (Input A based on your problem)
A = ...
[3, 1, -1
1, 6, 5
1, 1, 3];

% Vector of constants (Input b based on your problem formulation)
b = [7; 0 ; 5];
tol=input('Convergence criterion='); % Default tolerance is 1e-6
% Make sure you input a vector here
guess=input('Vector of initial guesses=');

% Solution
X = Jacobi(A,b,guess,tol,1)

for k=1:n
```

```
    fprintf('x(%2d)=%6.4g\n',k,X(k))
end
```

Function that performs the Jacobi method

```
function x = Jacobi(A, c, x0, tol, trace)
%JACOBI Solves a set of linear algebraic equations by the
%   Jacobi iterative method.
%
%   JACOBI(A,C,X0) finds unknowns of a set of linear algebraic
%   equations. A is the matrix of coefficients, C is the vector
%   of constants and X0 is the vector of initial guesses.
%
%   JACOBI(A,C,X0,TOL,TRACE) finds unknowns of a set of linear
%   algebraic equations and uses TOL as the convergence test.
%   A nonzero value for TRACE results in showing calculated
%   unknowns at the end of each iteration.
%
%   See also GAUSS, JORDAN

% (c) N. Mostoufi & A. Constantinides
% January 1, 1999

% Initialization
if nargin < 4 | isempty(tol)
   tol = 1e-6;
end
if nargin >= 4 & tol == 0
   tol = 1e-6;
end
if nargin < 5 | isempty(trace)
   trace = 0;
end
if trace
   fprintf('\n Initial guess :\n')
   fprintf('%8.6g  ',x0)
end

c = (c(:).')';            % Make sure it's a column vector
x0 = (x0(:).')';          % Make sure it's a column vector

n = length(c);
[nr nc] = size(A);

% Check coefficient matrix, vector of constants and
% vector of unknowns
if nr ~= nc
   error('Coefficient matrix is not square.')
end
if nr ~= n
```

```
   error('Coefficient matrix and vector of constants do not have
the same length.')
end
if length(x0) ~= n
   error('Vector of unknowns and vector of constants do not have
the same length.')
end

% Check if the coefficient matrix is singular
if det(A) == 0
   fprintf('\n Rank = %7.3g\n',rank(A))
   error('The coefficient matrix is singular.')
end

% Building modified coefficient matrix and modified
% vector of coefficients
D = diag(diag(A));             % The diagonal matrix
a0 = inv(D)*A - eye(n);        % Modified matrix of coefficients
c0 = inv(D)*c;                 % Modified vector of constants

x = x0;
x0 = x + 2 * tol;
iter = 0;

% Substitution procedure
while max(abs(x - x0)) >= tol
   x0 = x;
   x = c0 - a0 * x0;
   if trace
      iter = iter + 1;
      fprintf('\n Iteration no. %3d\n',iter)
      fprintf('%8.6g  ',x)
   end
end
```

Results

```
Number of equations = 3
Convergence criterion = 1e-2
Vector of initial guesses = [1;1;1]

 Initial Guess    :
        1         1         1
 Iteration no.   1
 2.33333        -1         1
 Iteration no.   2
       3  -1.22222   1.22222
 Iteration no.   3
 3.14815  -1.51852   1.07407
 Iteration no.   4
```

```
  3.19753  -1.41975   1.12346
  Iteration no.   5
  3.18107  -1.46914   1.07407
  Iteration no.   6
  3.18107  -1.42524   1.09602
  Iteration no.   7
  3.17375  -1.44353   1.08139
  Iteration no.   8
  3.17497  -1.43012   1.08993
  Iteration no.   9
  3.17335  -1.43743   1.08505
 x( 1) =  3.173
 x( 2) = -1.437
 x( 3) =  1.085>>
```

4.5.2 The Gauss-Seidel method

The Gauss-Seidel iteration uses the updated value of each variable of the solution vector *as soon as it is computed.* Thus, there is no need to maintain the old and new forms of the solution vector separately. For the three-by-three matrix problem illustrated above, the Gauss Seidel equations for each iteration become:

$$\begin{aligned} x_1^{(k)} &= \qquad\qquad\qquad -\frac{a_{12}}{a_{11}}x_2^{(k-1)} - \frac{a_{13}}{a_{11}}x_3^{(k-1)} + \frac{b_1}{a_{11}} \\ x_2^{(k)} &= -\frac{a_{21}}{a_{22}}x_1^{(k-1)} \qquad\qquad - \frac{a_{23}}{a_{22}}x_3^{(k-1)} + \frac{b_2}{a_{22}} \\ x_3^{(k)} &= -\frac{a_{31}}{a_{33}}x_1^{(k)} - \frac{a_{32}}{a_{33}}x_2^{(k)} \qquad\qquad + \frac{b_3}{a_{33}} \end{aligned} \tag{4.42}$$

Thus, the algorithm for the Gauss-Seidel method is:

$$x_i^{(k+1)} = \frac{1}{A_{i,i}}\left(b_i - \sum_{j=1}^{i-1} A_{i,j}x_j^{(k+1)} - \sum_{j=i+1}^{N} A_{i,j}x_j^{(k)} \right) \tag{4.43}$$

Another method, called LU-decomposition, is presented in Appendix C. Using LU-decomposition, the above equation can be recast into the following form:

$$(\mathbf{D} + \mathbf{L})\,\mathbf{x}^{(\mathbf{k+1})} = -\mathbf{U}\,\mathbf{x}^{(\mathbf{k})} + \mathbf{b} \tag{4.44}$$

where **D** is a diagonal matrix, **L** is a lower triangular matrix with zeros on the diagonal, and **U** is an upper triangular matrix with zeros on the diagonal. These matrices contain respectively, the diagonal, strictly lower triangular, and strictly upper triangular components of **A**.

Example 4.4 Solve a predominantly diagonal system of equations using the iterative Gauss-Seidel method.

$$\begin{bmatrix} 3 & 1 & -1 \\ 1 & 6 & 5 \\ 1 & 1 & 3 \end{bmatrix} \begin{bmatrix} x_1 \\ x_2 \\ x_3 \end{bmatrix} = \begin{bmatrix} 7 \\ 0 \\ 5 \end{bmatrix}$$

Program

```
% example4_4.m - This program solves the system Ax=b using the
% function GSEIDEL.M. The Gauss-Seidel method is iterative, hence
% guess-estimates for x0 and a tolerance level should be provided.

clc; clear all;
n=input('Number of equations = ');
% Matrix of coefficients (Input A based on your problem)
A = ...
    [3, 1, -1
    1, 6, 5
    1, 1, 3];
% Vector of constants (Input b based on your problem formulation)
b = [7; 0 ; 5];
tol=input('Convergence criterion = '); % Default tolerance is 1e-6
% Make sure you input a vector here
guess=input('Vector of initial guesses = ');
% Solution
X = GSeidel(A,b,guess,tol);
disp(' '); disp(' ')
for k=1:n
    fprintf('x(%2d) = %6.4g\n',k,X(k))
end

function x = GSeidel(A, b, x0, tol)
%   GSeidel Solves a set of linear algebraic equations by the
%   the Gauss Seidel iterative method.
%
%   GSeidel(A,b,X0, TOL) finds unknowns of a set of linear
%   algebraic equations. A is the matrix of coefficients,
%   b is the vector of constants, X0 is the vector of initial
%   guesses, and TOL is the convergence criterion.
% Initialization
fprintf('\n Initial guess :\n')
fprintf('%8.6g  ',x0)
```

```
b = (b(:).')';    % Define the vector of coefficients
x0 = (x0(:).')'; % Define the vector of guess estimates
d = diag(A);      % Extracting the diagonal elements of A
Q = A-diag(d);    % Defining the iteration matrix
n = length(b);
[N N] = size(A);
% Check if the coefficient matrix is singular
if det(A) == 0
    fprintf('\n Rank = %7.3g\n',rank(A))
    error('The coefficient matrix is singular.')
end
% Building modified coefficient matrix and modified
% vector of coefficients
x = x0;
x0 = x + 2 * tol;
iter = 0;
% Substitution procedure
while max(abs(x - x0)) >= tol
    x0 = x;
    for i=1:N
        x(i)=(b(i)-Q(i,:)*x)/d(i);
    end
    iter = iter + 1;
    fprintf('\n Iteration no. %3d\n',iter)
    fprintf('%8.6g  ',x)
end
```

Results

```
Number of equations = 3
Convergence criterion = 1e-2
Vector of initial guesses = [1;1;1]

 Initial guess :
       1         1         1
 Iteration no.   1
 2.33333  -1.22222    1.2963
 Iteration no.   2
 3.17284  -1.60905    1.1454
 Iteration no.   3
 3.25149  -1.49642   1.08164
 Iteration no.   4
 3.19269  -1.43348   1.08027
 Iteration no.   5
 3.17125  -1.42876   1.08584
```

```
 Iteration no.   6
 3.17153  -1.43345   1.08731

x( 1) =  3.172
x( 2) = -1.433
x( 3) =  1.087
```

In general, when the Jacobi method converges, the Gauss-Seidel method will also converge at an even faster rate, typically twice as fast. The general requirement for the application of Jacobi and Gauss-Seidel methods is that the matrix **A** be diagonally dominant. A sufficient but not necessary condition for the successive substitutions to converge is that the matrix **A** be symmetric and positive definite (see Appendix C).

Four numerical approximation methods were introduced in this chapter. Students are encouraged to apply these methods where appropriate, and use MATLAB's built-in routines to validate the answers obtained. The MATLAB routines should be used where possible as an efficient methodology for solution of matrices, especially nonsingular matrices of lower dimensions. For example, the command `x = A^-1*b` in the MATLAB command window can easily invert matrix **A** and perform the matrix multiplication between the inverse of **A** and the vector **b** to directly yield estimates for **x**.

Consider the problem from Examples 4.3 and 4.4 again.

```
>> A=[3,1,-1; 1,6,5; 1,1,3]
A =
      3     1    -1
      1     6     5
      1     1     3

>> b=[7; 0; 5]
ans =
      7
      0
      5
>> x=A^-1*b
x =
    3.1739
   -1.4348
    1.0870
```

4.6 Lessons Learned in this Chapter

After studying this chapter, the student should have learned the following:

- Linear algebraic systems in biomedical engineering can be formulated for numerical solution.
- Matrix algebra is a vital tool in the solution of systems of linear algebraic equations (refer to Appendix C).
- MATLAB commands for implementation of matrix operations exist.
- There are iterative and noniterative methods for solving **Ax** = **b** problems.
- Judicious choice can be made to identify the appropriate numerical solution strategy for linear algebraic equations.

4.7 Problems

4.1 Write a code to perform an LU decomposition of the coefficient matrix [**A**] (given below) using L-U decomposition with Gauss elimination. Your code should output [**L**] and [**U**] as well as verify that:

$$[\mathbf{A}]=[\mathbf{L}][\mathbf{U}]$$

Use the following MATLAB matrix function *only to check* your answer:

```
[L,U]=lu(A)
```

where `A = [1 3 6; 4 5 8; 1 2 3]`.

4.2 Linear equations in metabolic engineering

A microorganism is capable of using glucose, methanol, and hexadecane, singly. Its average cell composition (by weight) was analyzed to be 47% carbon, 6.5% hydrogen, 31% oxygen, 10% nitrogen, and the rest is ash. During active metabolism, the microorganism converts the substrate, oxygen, and ammonia into biomass, carbon dioxide, and water. A batch culture was carried out to estimate the cell yield based on substrate and oxygen (kg cells/kg substrate or oxygen). Air was supplied continuously and the exhaust gas was vented from the top of the bioreactor. A mass spectrometer was available for gas composition analysis. The water vapor was removed from both inlet and outlet gases before analysis. The inlet gas was ambient air (21% oxygen and 79% nitrogen). The composition of the off-gas on a volumetric basis is listed below. What are the estimated cell yield coefficients based on ammonia and oxygen for the consumption of glucose?

Substrate	% nitrogen	% carbon dioxide	% oxygen
glucose	78.8	10.2	11.0

4.3 Applications of spectroscopy for biological solutions

One of the most widely used applications of spectroscopy is for the quantitative determination of the concentration of biological molecules in solution. The absorbance of a solution, A_i, at wavelength i, is given by the sum of the product of $\varepsilon(k)$, the extinction coefficient/length of all components k (from 1 to N) obtained at the same wavelength and the concentrations of the components of the solution, $C(k)$ (Tinoco et al., 2002).

$$A_i = \left(\sum_{k=1}^{N} \varepsilon_i(k) C(k) \right)$$

Consider a hypothetic experiment when a protein solution with four amino acids, M, N, O, and P, was measured using four different wavelengths, and the absorbance values were recorded. The extinction coefficients for the four amino acids for the four wavelengths are tabulated below. Estimate the concentration of the four amino acids in M. Compare your answers and the efficiency of solution using the Gauss Jordan method and the iterative Jacobi method.

Wavelength	ε_M (M^{-1})	ε_N (M^{-1})	ε_O (M^{-1})	ε_P (M^{-1})	A_i
240	11300	8150	4500	4000	0.6320
250	5000	7500	3650	4200	0.5345
260	1900	3900	3000	4800	0.3310
280	1500	1400	2000	4850	0.1960

4.4 Three-phase liquid extraction of a biopharmaceutical drug

Consider the problem wherein a drug is being extracted from an organic phase into two aqueous phases. The volumetric flow rates of the two organic phases are denoted by H and K; and the volumetric flow rates of the two aqueous phases are denoted by L and M. Four different experiments are run by varying H, K, L, and M simultaneously, as tabulated below, in order to achieve the same exit concentration of the drug, y, w, x, and z, respectively. The total mass of solute entering each experiment is given by F. Estimate the exit concentrations of the drug in the four liquid streams.

Trial	H (L/min)	K (L/min)	L (L/min)	M (L/min)	F *(Total Mass in, g)*
1	100	125	125	62.5	6625
2	80	110	120	25	5290
3	140	80	120	100	7300
4	90	104.8	60	137.33	6539

The governing equation for mass distribution of the drug between the four liquid streams is shown below in terms of the total solute mass entering each experiment.

$$Hy + Kw + Lx + Mz = F$$

4.8 References

Constantinides, A. and Mostoufi, N. 1999. *Numerical Methods for Chemical Engineers with MATLAB Applications.* Upper Saddler River, NJ: Prentice Hall PTR.

Koehler, J. K. 1973. *Advanced Techniques in Biological Electron Microscopy.* New York: Springer Verlag.

Rosenfeld, A. and Kak, A. C. 1982. *Digital Picture Processing*, 2nd ed. New York: Academic Press.

Shuler, M. L. and Kargi, F. 2002. *Bioprocess Engineering: Basic Concepts*, 2nd ed. Englewood Cliffs, NJ: Prentice Hall.

Tinoco, I., Sauer, K., Wang, J. C., and Puglisi, J. D. 2002. *Physical Chemistry: Principles and Applications in Biological Sciences*, Upper Saddle River, NJ: Prentice Hall.

Chapter 5
Nonlinear Equations in Biomedical Engineering

5.1 Introduction

There are problems in biomedical engineering that require the solution of nonlinear equations. Most nonlinear equations cannot be explicitly solved using analytical methods, however. Unlike linear equations, whose roots can be found using analytical methods, *solvers* need to be formulated to determine the roots of nonlinear equations. This chapter introduces the student of biomedical engineering to the various approaches for identifying the roots of nonlinear equations. Nonlinear equations from varied core disciplines within Biomedical Engineering will be highlighted to illustrate the wide applications of simple numerical solution strategies.

The material in this chapter will enable the student to accomplish the following:

- Identify nonlinear equations within biomedical engineering models
- Be introduced to various techniques for finding roots of nonlinear equations (Successive Position; False Position; Newton Raphson)
- Review Newton's method for solution of simultaneous nonlinear equations
- Develop MATLAB code and implement it for the major nonlinear root-finding methods reviewed
- Identify methods for rapid convergence of root finding

5.2 General Form of Nonlinear Equations

Nonlinear equations can be readily distinguished from linear equations. For example, the following equations in x are clearly nonlinear, because x cannot be readily isolated to yield roots of this equation using analytical approaches.

$$\ln(x) + x = 4.6; \quad \frac{1}{\sqrt{x}} = \sin\left(\sqrt{x} + 0.01\right)$$

Any equation that contains such terms as $\log(x)$, square root of x, or $\sin(x)$ is nonlinear in nature. Generally, nonlinear equations are nonlinear functions of variables, or expressed in terms of higher orders of variables, for example, x^2, x^3, $\log(x)$, $\sin(x)$, or $\exp(x)$. Other common forms of nonlinear equations include polynomial equations that can be expressed as:

$$f(x) = a_n x^n + a_{n-1} x^{n-1} + \ldots\ldots + a_1 x + a_0 = 0$$

The roots of the equation are those values of the variable, x, where $f(\mathrm{x})$ equals zero. The number of roots of the equation depends on the order of $f(x)$. If $f(x)$ is a second-order, quadratic equation, it has two roots that can be determined analytically with ease. Third-order cubic equations, as well as higher order polynomials, have three or more roots. Such roots are generally more straightforward to determine using numerical methods to be discussed in the next section.

Beyond the number of roots, another important aspect of the root should be recognized: the nature of the roots, especially for higher order polynomial functions that appear in biomedical engineering problems. Roots of such functions can be (a) real and distinct; (b) real and repeated; (c) complex values; (d) a combination of any or all of (a) through (c). The real part of the roots can be positive, negative, or zero.

Consider, for example, two third-order polynomial equations. The first is:

$$x^3 - 7x + 6 = 0$$

the roots of which can be visualized by plotting this equation in MATLAB, as illustrated below using MATLAB commands and graphed in Fig. 5.1a.

```
x=linspace(-4,3,30);          % define the range of x
y=x.^3-7*x+6;                 % calculate the function
xx=linspace(0,0,30);          % define the y=0 line
plot(x,y, x,xx)               % plots the function y
xlabel('x'); ylabel('y');     % label the axes
```

Consider a second polynomial:

$$x^3 - 2x^2 + x - 2 = 0$$

the roots of which can be visualized by graphically plotting this equation in MATLAB, as shown in Fig. 5.1b.

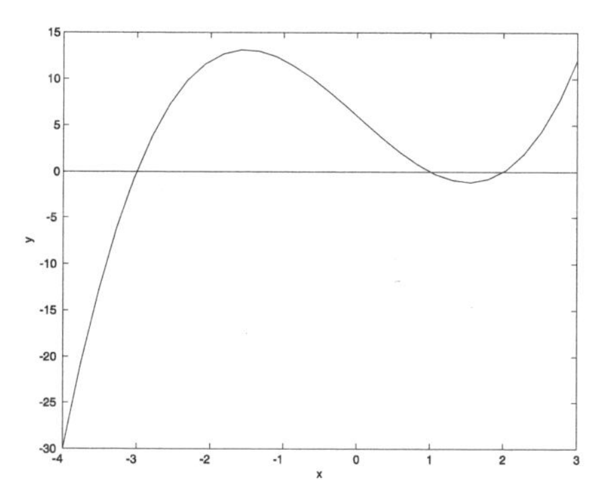

Figure 5.1a Function exhibits three real roots that can be identified where it intersects *y* = 0.

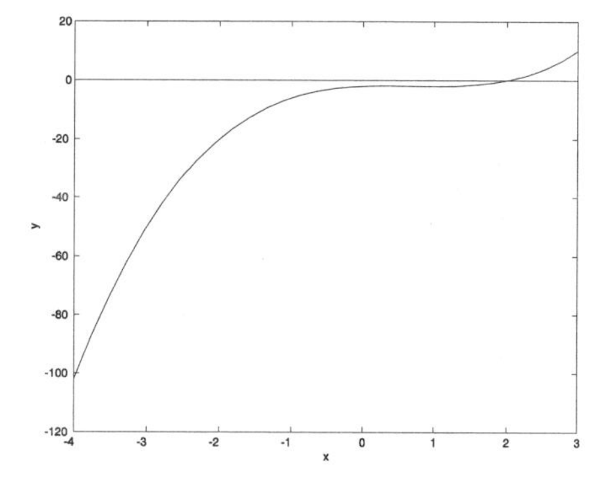

Figure 5.1b Function exhibits only one real root and two complementary complex roots that cannot be graphed.

The roots for the first equation are: 2, -3, and 1, and these are three real roots. The second equation has one real root, x=2, and two complementary complex roots, $x = -i$ and $x = i$. Note that only real roots can be found by the intersection of the graphical plot of the functions with $y = 0$.

Using the MATLAB function, `fzero`, the local roots of any one-dimensional equation can be found. To use this command, a separate function file should be created with the title `file_name.m`. The use of command `fzero('file_name',x_o)` identifies the local root where the function equals zero closest to the guess, x_o. This command has to be repeated for a different starting guess in search for every root. The solution to the second equation above is illustrated using the `fzero` command in MATLAB:

```
function y = function2(x) % a separate M-file called function2.m
y=x^3-2*x^2+x-2;

% the following commands are typed in the MATLAB workspace
>> x=linspace(-4,3,30);
>> x= fzero('function2',sqrt(-1))
x =0 + 1.0000i
>> x= fzero('function2',-sqrt(-1))
x =0 - 1.0000i
>> x= fzero('function2',4)
x =2
```

5.3 Examples of Nonlinear Equations in Biomedical Engineering

Several relevant examples of nonlinear equations in biomedical engineering are discussed below. These are classified according to the subareas of biomedical engineering in which they occur.

5.3.1 Molecular bioengineering

Enzymes are cell-derived proteins that act as catalysts during biochemical reactions. Enzymes typically act on substrates with high specificity. The consumption of a substrate source is described by the kinetic depletion of the substrate in terms of the enzyme activity. The kinetics of enzyme reactions can be formulated in terms of the Michaelis-Menten model, given as:

$$r_s = -\frac{ds}{dt} = \frac{V_{\max} s}{K_m + s}$$

where r_s is the reaction rate of the substrate in moles/volume/time; s is the substrate concentration in moles/volume; V_{max} is the maximal substrate consumption rate; and K_m is the concentration of the substrate that elicits half-maximal consumption rate. By rearranging the above equation and separating the variables, s, and time, t, one can integrate the equations to yield a nonlinear solution for the substrate concentration as a function of time:

$$K_m \ln\left(\frac{s_0}{s}\right) + (s_0 - s) = V_{\max} t$$

where s is the substrate concentration at any time t, and s_0 is the substrate concentration initially. To determine how the substrate depletes over time due to a given enzyme, the above equation has to be solved numerically. This solution will be illustrated using the Newton-Raphson method in this chapter. A typical curve for depletion of s is shown in Fig. 5.2. We will revisit this problem by solving an actual numerical example in Example 5.3, later in the chapter.

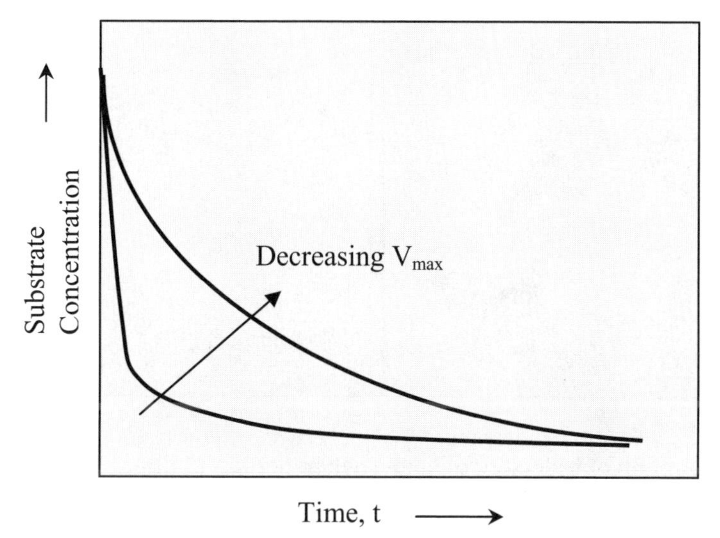

Figure 5.2 **The problem of solving for the concentration of a biochemical substrate being metabolized by an enzyme is nonlinear in nature. To solve for *s*, the substrate concentration at any time, *t*, will require a nonanalytical solution approach, discussed in this chapter.**

5.3.2 Cellular and tissue engineering

The movement of mammalian cells through an extracellular matrix of tissues is an important topic in cell and tissue engineering. Cell migration is necessary for successful repair of wounds, regeneration of tissues, and repopulation of implanted scaffolds for tissue-engineering therapies. For example, endothelial cells of microcirculatory blood vessel walls need to migrate out into tissue space to form new vascular networks, a process called angiogenesis. White blood cells need to migrate on vascular prosthetic materials to reach and eliminate any bacterial microorganisms, a process reminiscent of acute inflammatory reaction. The population migration of cells can be quantitatively described by the Dunn equation if the cell coordinates and trajectories are obtained through direct visualization by computer-assisted time-lapse digital microscopy:

$$< d^2 > = 2S^2[Pt - P^2(1 - e^{-t/P})]$$

The mean squared displacement of cells $<d^2>$ is a nonlinear function of elapsed time, t (Fig. 5.3). The independent parameters of cell movement are root-mean-square speed, S, and persistence time, P. These parameters are best determined by numerical regression of both estimates. If one of the estimates is known independently, the other can be found numerically.

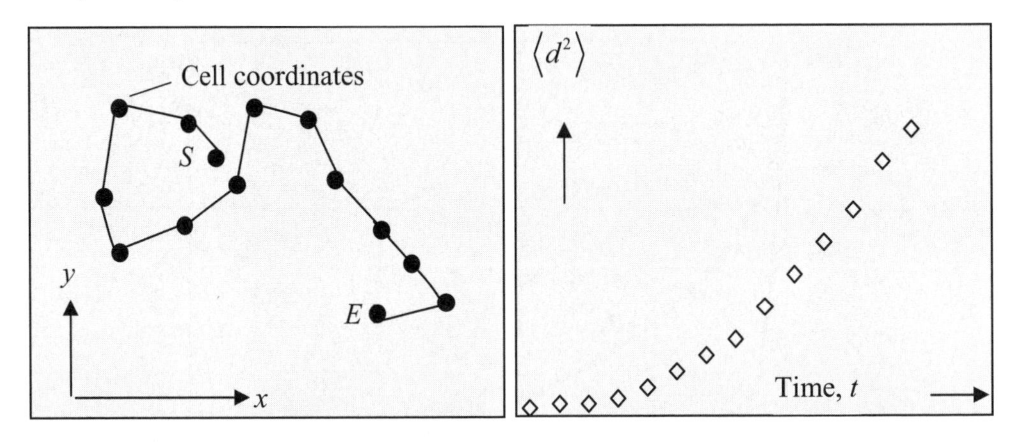

Figure 5.3 **Illustration of trajectory of single cell migration (*S* indicates starting position, *E* indicates end position) in two dimensions and population averaged mean square displacement, <d^2> versus time, *t*. Single cell speed and persistence time can be estimated through numerical solution of the Dunn equation using the iterative procedures discussed in this chapter.**

5.3.3 Bioheat transport: photothermal therapy

Light-induced tissue heating can be used for a variety of applications, including biostimulation of nerve growth and wound healing, sealing or welding blood vessels during surgeries, tissue necrosis used for the treatment of cancer or enlarged prostate tissues, and tissue vaporization due to ablation, which is employed in corrective eye surgery known as radial keratectomy (Enderle et al., 2005). The therapeutic use of a laser converts photonic energy to absorbed energy within the material phase of a tissue. Mathematical equations to describe laser irradiation with ablation result in partial differential equations in the temperature of the tissue surface as a function of time and space, graphed below. Here, θ is the dimensionless temperature, ξ is the dimensionless space (depth), and τ is dimensionless time prior to the onset of ablation.

By setting θ to 1 at $\xi = 0$, a transcendental algebraic equation is obtained in terms of the threshold ablation temperature, λ, and a dimensionless light absorption parameter, B, that must be solved numerically to yield the temperatures, λ, necessary to cause the onset of tissue ablation as a function of time, τ_{ab}.

$$B\lambda = \frac{2}{\sqrt{\pi}} B\sqrt{\tau_{ab}} + e^{B^2\tau_{ab}} erfc[B\sqrt{\tau_{ab}}] - 1$$

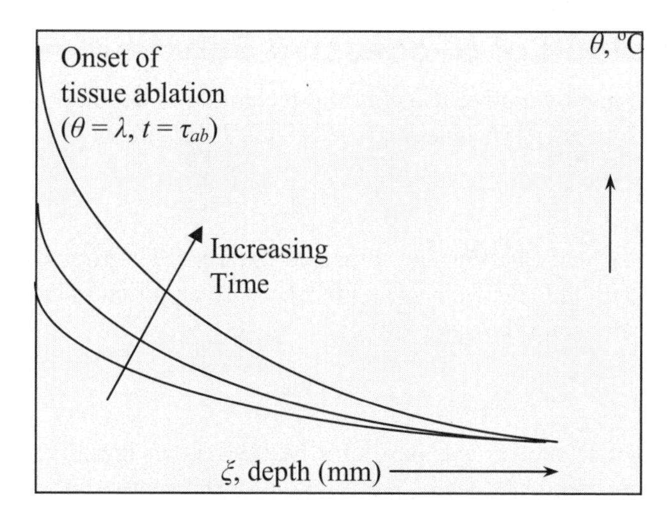

Figure 5.4 **Nondimensional surface temperature as a function of depth into the tissue at various times prior to the onset of tissue ablation. The interfacial temperatures can be obtained by numerical solution of nonlinear algebraic equation shown above, corresponding to the onset of ablation.**

5.3.4 Biomedical flow transport dynamics

Biofluid transport is relevant to a wide range of physiologic processes, including blood flow via convection, oxygen transport to peripheral tissues via concentration-driven diffusion, and ion transport into and out of cells due to ion pumps and electrochemical gradients. Biological fluids need to be frequently transported during therapeutical intervention, for example, during transfusion or perfusion prior to or during surgery. Imagine an organ perfusion system using flow of a physiologic fluid (density, $\rho = 1000$ kg/m^3; viscosity, $\mu = 0.001$ kg/m.s) through a catheter of 2.8 mm diameter (D) at a velocity of $u = 1.8$ m/sec. Typically, the use of catheters entails the use of laminar flow of fluids. In cases where high local velocities are reached within blood vessels, however, turbulent regimen may be attained. The ratio of inertial to viscous forces is a dimensionless group called the Reynold's number (Re), given by $Du\rho/\mu$. In instances when the value for this group is above the empirical threshold of 2000, the flow is considered turbulent. For such turbulent flows, the friction factor, b, is used to determine the change in the velocity (or energy losses) during the fluid flow in a channel or tubes. The empirical nonlinear equation that appears below is called the Colebrook Equation. Clearly, the solution for b will require the use of a nonanalytical method.

$$\frac{1}{\sqrt{b}} = 4.07\log\left(\mathrm{Re}\sqrt{b}\right) - 0.60$$

5.4 The Method of Successive Substitution

The simplest one-point iterative root-finding technique is developed by rearranging $f(x) = 0$, so that there is one instance of x on the left-hand side of the equation:

$$x = g(x) \tag{5.1}$$

The function $g(x)$ provides a formula to predict the root. The root is the intersection of the line $y = x$ with the curve $y = g(x)$. Starting with an initial value of x_1, as shown in Fig. 5.5a, we can obtain the value of x_2:

$$x_2 = g(x_1) \tag{5.2}$$

which is closer to the root x_l and may thus be used as an initial value for the next iteration. Therefore, the general iterative formula for this method is:

$$x_{n+1} = g(x_n) \tag{5.3}$$

which is known as the method of successive substitution, where the substitution is $x = g(x)$.

The condition for convergence by this method is that $|g'(x)| < 1$ for all x in the search interval. Fig. 5.5b shows the case when this condition is not satisfied and thus the method diverges. In a computer program, it is easier to determine this by comparing whether $|x_3 - x_2| < |x_2 - x_1|$ and, if so, the successive x_n values converge. The advantage of this method is that it can be started with only a single point, and without the need for calculating the derivative of the function.

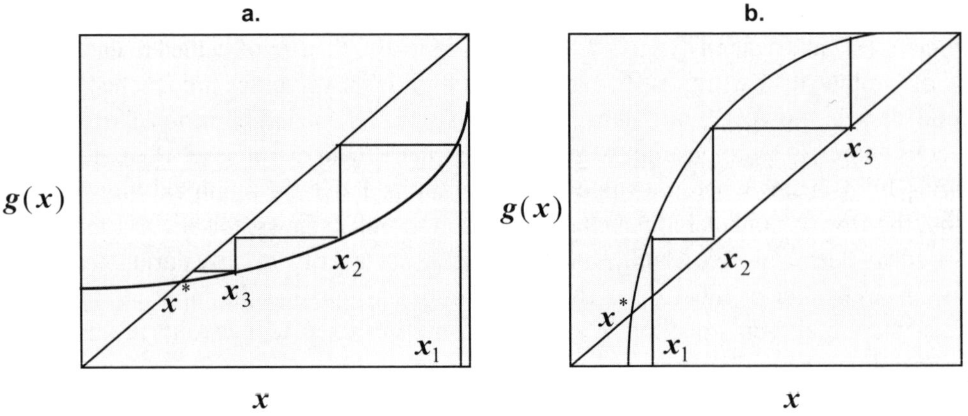

a. b.

$g(x)$ x^* x_3 x_2 x_1 x

$g(x)$ x^* x_1 x_2 x_3 x

Figure 5.5 Method of successive substitution: a. Convergence; b. Divergence.

5.5 The Method of False Position (Linear Interpolation)

This technique is root-finding based on linear interpolation between two points along the function. The points are chosen such that they straddle the root, i.e., they lie on either side of the root. For example, x_1 and x_2 in Fig. 5.6a are located on opposite sides of the root x^* of the nonlinear function $f(x)$. The interpolation is done by connecting the points $(x_1, f(x_1))$ and $(x_2, f(x_2))$ through a line segment called a chord, whose equation is:

$$y(x) = mx + c \tag{5.4}$$

Because this chord passes through the two points $(x_1, f(x_1))$ and $(x_2, f(x_2))$, its slope is:

$$m = \frac{f(x_2) - f(x_1)}{x_2 - x_1} \tag{5.5}$$

and its y intercept is:

$$c = f(x_1) - mx_1 \tag{5.6}$$

Eq. (5.4) then becomes:

$$y(x) = \left[\frac{f(x_2) - f(x_1)}{x_2 - x_1}\right]x + \left\{f(x_1) - \left[\frac{f(x_2) - f(x_1)}{x_2 - x_1}\right]x_1\right\} \tag{5.7}$$

The method locates the next best guess for the root at x_3 where $y(x_3) = 0$. By substituting $x = x_3$, $y(x_3) = 0$ in Eq. (5.7), we get:

$$x_3 = x_1 - \frac{f(x_1)(x_2 - x_1)}{f(x_2) - f(x_1)} \tag{5.8}$$

Note that for the shape of curve chosen in Fig. 5.6, x_3 is nearer to the root x^* than either x_1 or x_2. This may not be the case with all functions.

According to Fig. 5.6a, $f(x_3)$ has the same sign as $f(x_2)$; therefore, x_2 may be replaced by x_3. If the above operation is repeated, and the points $((x_1, f(x_1)))$ and $((x_3, f(x_3)))$ are connected with a new chord as shown in Fig. 5.6b, we obtain the value of x_4 by the second step of linear interpolation:

$$x_4 = x_1 - \frac{f(x_1)(x_3 - x_1)}{f(x_3) - f(x_1)} \tag{5.9}$$

which is nearer to the root than x_3. In general, consider x^+ to be the value at which $f(x^+) > 0$ and x^- to be the value at which $f(x^-)<0$. The next improved approximation of the root of the function may be calculated by successive application of the general formula:

$$x_n = x^+ - \frac{f(x^+)(x^+ - x^-)}{f(x^+) - f(x^-)} \tag{5.10}$$

For the next iteration (to find x_{n+1}), x^+ or x^- should be replaced by x_n according to the sign of $f(x_n)$.

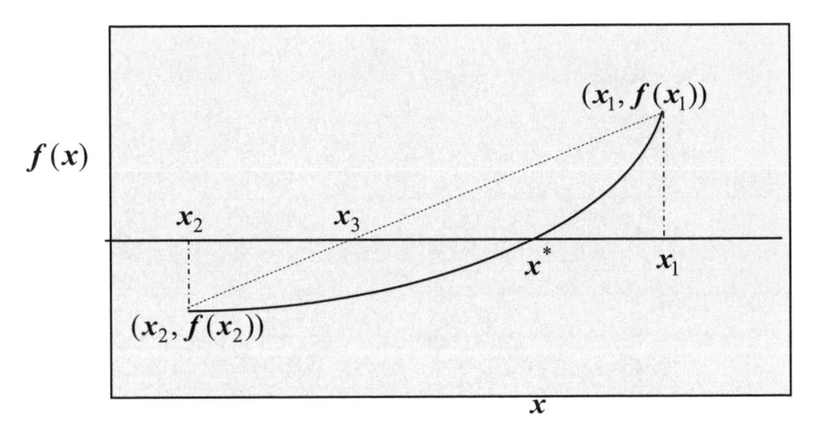

Fig. 5.6a The first step in linear interpolation.

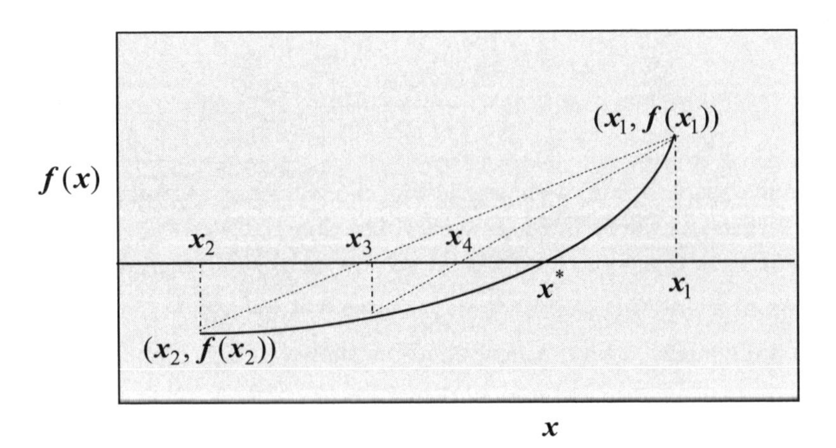

Fig. 5.6b The second step in linear interpolation.

This method is known by several names: method of chords, linear interpolation, false position. It is simple to calculate because it does not require evaluating derivatives of the function at each step. Thus, it is one of the least cumbersome methods for solving nonlinear equations. It has a major disadvantage, however. The accuracy and speed of convergence in finding the root are limited by the choice of x_1, which forms the pivot point for all subsequent iterations.

5.6 The Newton-Raphson Method

The major technique for locating roots of nonlinear equations is the *Newton-Raphson method.* The first step in using the Newton-Raphson method (abbreviated N-R method) is expressing the nonlinear equation as:

$$f(x) = 0 \tag{5.11}$$

For example, the Colebrook equation introduced in Section 5.3.4 can be readily expressed as:

$$\frac{1}{\sqrt{b}} - 4.07\log\left(\text{Re}\sqrt{b}\right) + 0.60 = 0$$

Once the nonlinear function is constructed, the Newton-Raphson treatment can be applied.

The student should first visually grasp the physical basis of the N-R method. At the outset, we do not know the value of x^* (that is, the root) at which the function will equal zero. Instead, we can begin to find some crude estimates for the root and refine them till we find a satisfactory value for the root. The N-R method allows us to formalize this process.

We express the function using the basic Taylor series expansion, shown below. The nonlinear function $f(x)$ can be expanded around its value at the arbitrary point, x_1, which is formally called the initial estimate of the root:

$$f(x) = f(x_1) + f'(x_1)(x - x_1) + \frac{f''(x_1)(x - x_1)^2}{2!} + ... \tag{5.12}$$

Because we seek the value of x that forces the function $f(x)$ to be zero, the left-hand side of Eq. (5.12) is set to zero, and the resulting equation is solved for x. However, the right-hand side is an infinite series. Therefore, a finite number of terms must be retained, and the series must be truncated. To do this, we simplify the sum to the first two terms, which contain the value of the function and its change (first derivative), and, in the process, linearize the function $f(x)$. This operation results in:

$$f(x) \approx f(x_1) + f'(x_1)(x - x_1) \tag{5.13}$$

Let the root of the function be $x = x^*$. At this value of the root, Eq. (5.13) becomes:

$$f(x^*) = f(x_1) + f'(x_1)(x^* - x_1) \tag{5.14}$$

Because we want to identify the x^* that forces $f(x)$ to equal zero, we can eliminate the function $f(x^*)$ and rearrange Eq. (5.14) to yield an expression for x^*:

$$x^* = x_1 - \frac{f(x_1)}{f'(x_1)} \tag{5.15}$$

Physically, Eq. (5.15) says that the value of x^* (the root at which the function $f(x) = 0$) can be calculated from the initial guess point x_1 by subtracting from this initial guess the term $f(x_1)/f'(x_1)$. Fig. 5.7a illustrates the significance of equation (5.15). Students should fully grasp the relevance of this figure. The value of x^* is obtained by moving from x_1 to x^* in the direction of the tangent $f'(x_1)$ of the function $f(x)$. The simple construction of a right-angle triangle with vertices at $(x_1,0)$, $(x^*,0)$, and $(x_1,f(x_1))$ can be used to verify that the slope of the hypotenuse, $f'(x_1) = f(x)/(x_1 - x^*)$.

In reality, however, we are not likely to reach x^* in one step, because we truncated the Taylor series at only two terms, and thus the value x^* obtained from (5.15) is not likely to satisfy $f(x) = 0$ (see Fig. 5.7a). Thus, at best, the first application of (5.15) can yield an estimate of the root shown by the arrow in Fig. 5.7a, which we can call x_2 as shown in the equation below and in Fig. 5.7b:

$$x_2 = x_1 - \frac{f(x_1)}{f'(x_1)} \tag{5.16}$$

In the N-R method, we reapply the Taylor series linearization at x_2 (shown in Fig. 5.7b) to obtain a yet better estimate of x^*, called x_3.

$$x_3 = x_2 - \frac{f(x_2)}{f'(x_2)} \tag{5.17}$$

This process is continued till the method converges on the root x^*. Repetitive application of the above steps yields the Newton-Raphson iterative formula:

$$x_{n+1} = x_n - \frac{f(x_n)}{f'(x_n)} \tag{5.18}$$

for the nth iteration.

Another common version of equation (5.18) designates the "*new*" and "*old*" values of estimates for the root. Students may find this version intuitively useful when encoding iterative steps with a numerical program—the x_{new} from the earlier iteration becomes x_{old} for the next iteration, and so on.

$$x_{new} = x_{old} - \frac{f(x_{old})}{f'(x_{old})} \tag{5.19}$$

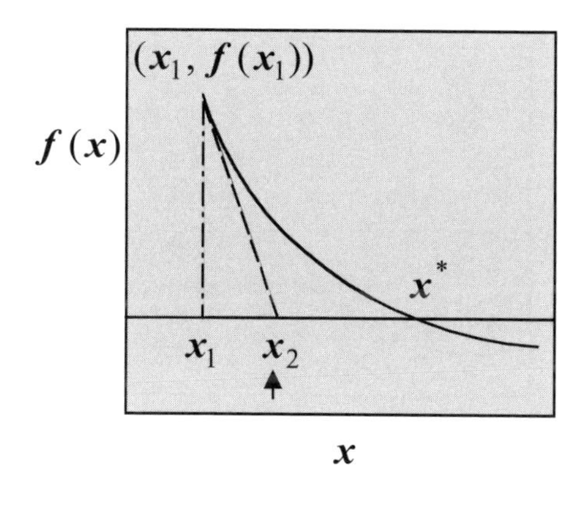

Figure 5.7a

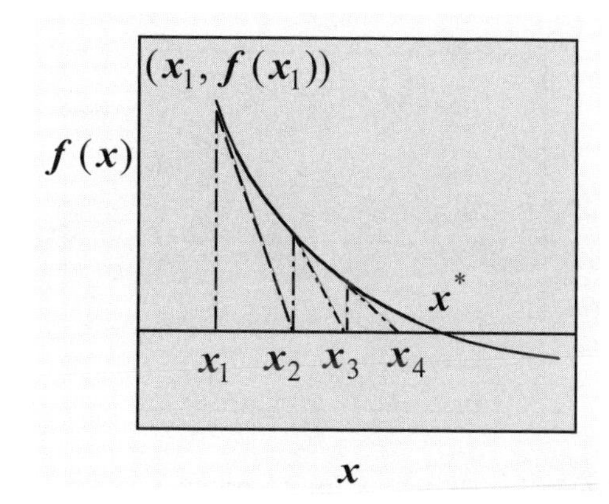

Figure 5.7b

In summary, the Newton-Raphson method uses the newly found position as the starting point for each subsequent iteration.

The above illustrations (Figs. 5.7a, and 5.7b), show how successive iterations converge toward a root in the search space. However, the shapes of nonlinear functions may vary drastically, and convergence is not always guaranteed. As a matter of fact, divergence is more likely to occur, as shown in Fig. 5.8a, unless extreme care is taken in the choice of the initial starting points to ensure convergence (Fig. 5.8b).

To investigate the convergence behavior of the Newton-Raphson method, one has to examine the term $-f(x_n)/f'(x_n)$ in Eq. (5.18). This is the error term, or correction term, applied to the previous estimate of the root at the nth iteration. A function with a strong vertical trajectory near the root will cause the first derivative of the function (the slope) and hence, the denominator of the error term, to be large; therefore, the convergence will be quite fast. If, however, $f(x)$ is nearly horizontal near the root, the convergence will be slow. Based on Eq. (5.18), at what point during the search would the N-R method fail to converge? Hint: when the ratio is indeterminate due to division by zero. Indeed, inflection points on the curve, within the region of search, are troublesome and may cause the search to diverge.

A condition for convergence of the Newton-Raphson method is: "If $f'(x)$ and $f''(x)$ do not change sign in the interval (x_1, x^*) (that is, the slope of $f(x)$ and slope of $f'(x)$ do not exhibit an inflection) and if $f'(x_1)$ and $f''(x_1)$ have the same sign, the iteration will always converge to x^*." These convergence criteria can be programmed as part of the MATLAB-based code that performs the Newton-Raphson search. The student should confirm that these criteria are not obeyed in Fig. 5.8a.

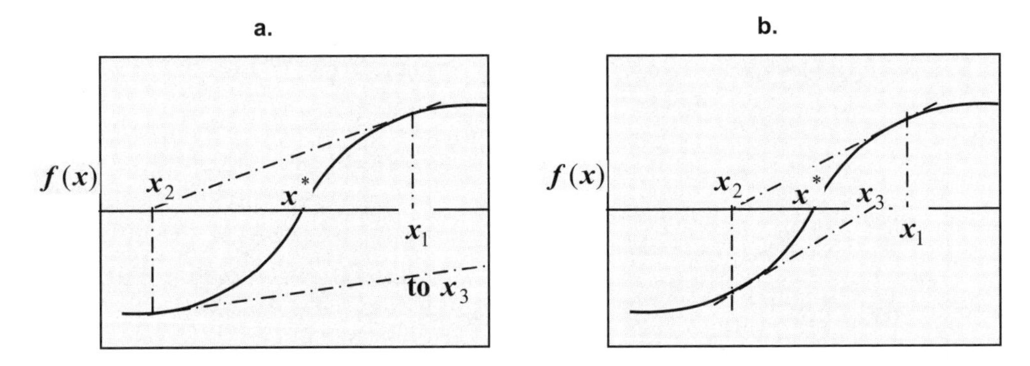

Figure 5.8 Choice of initial guesses affects convergence of Newton-Raphson method. a. Divergence; b. Convergence.

Example 5.1 Cardiovascular physiology—Solution of a simple polynomial using the Newton-Raphson method.

Statement of the problem

Assume that the left ventricular pressure (LVP) can be represented as a parabola with peak systolic pressure of 120 mmHg and the mean aortic pressure of 90 mmHg. Find the time at which the aortic valve opens and the time at which it closes, using the Newton-Raphson method. (Hint: opening and closing of the aortic valve occur approximately at the point when the mean pressure intercepts the LVP curve). Find these times if the peak pressure is varied from 120 mmHg to 100 mHg and 140 mmHg.

Solution

A parabola with peak pressure of 120 mmHg is drawn, intersecting the origin (0, 0) and end of a cycle (1, 0) as shown in Fig. 5.9. The governing equation for the parabola is:

$$(x-h)^2 = -4p(y-k)$$

where (h, k) are the coordinates of the vertex of the parabola, and p is the focus of the parabola. Starting values for (h, k) are given as (0.5, 120), assuming that the parabola is symmetric, and 0.5 is the midpoint of a cycle with unit time duration of 1. For k = 120 mmHg, since p is not provided, it is first determined from the intercept of the parabola with the x-axis (either (0, 0) or (1, 0)) to be 5.208 x 10^{-4}. Thus, for k = 120 mmHg, the locus of the parabola is as shown in Fig. 5.9.

$$(x-0.5)^2 = -4\cdot(5.208\times10^{-4})(y-120)$$

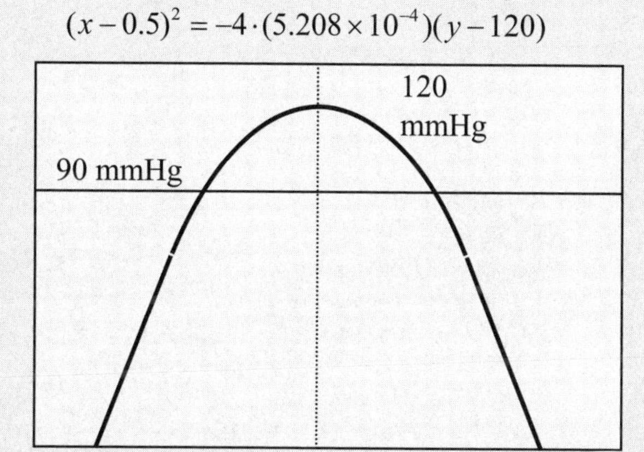

Figure 5.9 Plot of the parabolic equation.

The problem asks us to identify the roots of the parabola that intersect the y-axis at the mean aortic pressure of 90 mmHg. By substituting y = 90 in the parabola equation, we arrive at the following equation whose roots have to be identified using the Newton-Raphson method: $480x^2 - 480x + 90 = 0$

For other peak pressures, k, the equation can be generically shown to be:

$$x^2 - x + \frac{22.5}{k} = 0$$

Two different versions of solution strategies are given below. The first is a simplified code that incorporates the exact derivative of the function. To use this version, a file containing both the function and its derivative must be created in MATLAB and is designated as `parabola_and_derivative.m` file. Obviously, this file name can be changed for a given function at hand, and then appropriately referred to within the simplified Newton-Raphson program, called `simple_NR.m`.

The second version assumes that the exact derivative of the function is not known, and therefore approximates its value using a numerical approximation of the derivative, such as those in Chapter 6. This version utilizes the `NR.m` program.

Example 5_1a.m applies a simplified version of Newton-Raphson method that uses an analytical function for the derivative.

```
% example 5_1a.m
% This program calculates the roots of a parabolic
% equation using Newton-Raphson method for different
% peak pressures, denoted by k.

clc; clear all;

k = input(' Peak Pressure in mmHg, k = ');
x0 = input(' Starting value = ');

% simple_NR('F', X0, P1, P2, ..) is a simplified Newton-Raphson
% routine that finds the root of a nonlinear function
% contained in the MATLAB file called F with arguments P1, P2,…
% The file F contains both the function and its derivative.

xnew = simple_NR('parabola_and_derivative',x0,k);

fprintf('\n %3g %g %g\n',k,x0,xnew)
```

Function to be solved containing the analytical derivative

```
function [f,fprime] = parabola_and_derivative(x,k)
% alter the function below as necessary
f = x^2-x+22.5/k;
% modify the function derivative below as necessary
fprime = 2*x-1;
```

Function that applies the simplified version of Newton-Raphson method

```
function x = simple_NR(FunFcn,x0,varargin)
% simple_NR finds a zero of a defined function by the Newton-
% Raphson method.
%
% simple_NR('F',X0) finds a zero of the function described by the
% M-file F.M. that also contains the derivative of the function.
% This program differs from NR.m because it computes the
% derivative exactly, and does not numerically approximate the
% derivative. X0 is a starting guess.
%
% simple_NR('F',X0,P1,P2,...) allows for additional arguments
% which are passed to the function F(X,P1,P2,...).

% (c) S. Dunn, A. Constantinides, and P. Moghe
% August 1, 2003

% Initialization
```

```
tol = 1e-6;
iter = 0;
[fnk, fpr] = feval(FunFcn,x0,varargin{:});

header = ' Iteration        x            f(x)';
disp(header)
fprintf('%5.0d   %13.6g %13.6g  \n',iter, [x0 fnk])
x = x0;
x0 = x + 1;
itermax = 100;
% Main iteration loop
while abs(x - x0) > tol & iter <= itermax
    iter = iter + 1;
    x0 = x;
    x = x0 - fnk/fpr;
    [fnk,fpr] = feval(FunFcn,x,varargin{:});
    % Show the results of calculation
    fprintf('%5.0d   %13.6g   %13.6g \n',iter, [x fnk])
end
if iter >= itermax
    disp('Warning : Maximum iterations reached.')
end
```

Output (shown for only one value of peak pressure)

```
Peak Pressure in mmHg, k = 120
 Starting value = 0
 Iteration         x             f(x)
                    0          0.1875
    1          0.1875       0.0351562
    2         0.24375      0.00316406
    3        0.249924     3.81156e-05
    4            0.25     5.80764e-09
    5            0.25     1.11022e-16

 120 0 0.25
```

Example 5_1b.m solves the problem using the Newton-Raphson method, by approximating the derivative of the function numerically.

```
% example 5_1b.m
% This program calculates the roots of a parabolic
% equation using Newton-Raphson method for different
% peak pressures, denoted by k. Derivatives are numerically
% approximated. Use example5_1a.m as an alternative.

clc; clear all;

k = input(' Peak Pressure in mmHg, k = ');
```

```
x0 = input(' Starting value = ');
% NR2('F', X0, TOL, TRACE, P1, P2, ..) is a simplified
% Newton-Raphson routine that finds zero of function F with
% arguments P1, P2, etc. TOL is convergence tolerance.
% NR2 does not graphically show the path to convergence.
xnew = NR2('parabola',x0,[],0,k);
fprintf('%3g %g %g\n',k,x0,xnew)
```

Function that contains the equation to be solved

```
function y = parabola(x,k)
y = x^2-x+22.5/k;
```

Function that applies the Newton-Raphson method and approximates the derivative of the function numerically.

```
function x = NR2(FunFcn,x0,tol,trace,varargin)
%NR2 Finds a zero of a function by the Newton-Raphson method.
%
%   NR2('F',X0) finds a zero of the function described by the
%   Simplified Newton-Raphson Program -- M-file F.M. X0 is a
%   starting guess.
%
%   NR2('F',X0,TOL,TRACE) uses tolerance TOL for convergence test.
%   TRACE = 0 does not show the calculation steps
%   TRACE = 1 shows the calculation steps numerically
%
%   NR2('F',X0,TOL,TRACE,P1,P2,...) allows for additional
%   arguments which are passed to the function F(X,P1,P2,...).
%   Pass an empty matrix for TOL or TRACE to use the default
%   value.
%
%   See also FZERO, ROOTS, XGX, LI

% (c) S. Dunn, A. Constantinides, and P. Moghe
% August 1, 2003

% Initialization
if nargin < 3 | isempty(tol)
   tol = 1e-6;
end
if nargin < 4 | isempty(trace)
   trace = 0;
end
if tol == 0
   tol = 1e-6;
end
if (length(x0) > 1) | (~isfinite(x0))
   error('Second argument must be a finite scalar.')
```

```
end
iter = 0;
fnk = feval(FunFcn,x0,varargin{:});
if trace
  header = ' Iteration        x             f(x)';
  disp(header)
  fprintf('%5.0d   %13.6g %13.6g \n',iter, [x0 fnk])
end

x = x0;
x0 = x + 1;
itermax = 100;

% Main iteration loop
while abs(x - x0) > tol & iter <= itermax
   iter = iter + 1;
   x0 = x;

   % Set dx for differentiation
   if x ~= 0
      dx = x/100;
   else
      dx = 1/100;
   end

   % Differentiation
   a = x - dx;  fa = feval(FunFcn,a,varargin{:});
   b = x + dx;  fb = feval(FunFcn,b,varargin{:});
   df = (fb - fa)/(b - a);

   % Next approximation of the root
   if df == 0
      x = x0 + max(abs(dx),1.1*tol);
   else
      x = x0 - fnk/df;
   end

   fnk = feval(FunFcn,x,varargin{:});
   % Show the results of calculation
   if trace
      fprintf('%5.0d   %13.6g   %13.6g \n',iter, [x fnk])
   end
end

if iter >= itermax
   disp('Warning : Maximum iterations reached.')
end
```

Output of Example 5_1b.m

	Initial Guess, x(0) = 0	Initial Guess, x(0) = 1.0
Peak Pressure (mmHg)	Normalized time in cycle of opening of aortic valve	Normalized time in cycle of closing of aortic valve
100	0.34	0.66
120	0.25	0.75
140	0.2	0.79

Students should note that the roots of the quadratic parabolic equation above can be easily derived analytically as well, but the approaches shown here are useful if the pressure oscillations are described by more complex nonlinear dynamics.

Next, let us review the mechanics of solution methods through the example of a more complex nonlinear equation treated with three different routines. The first routine will involve the Newton-Raphson method for the solution of a nonlinear equation.

Example 5.2a Solution of the Colebrook equation—demonstration of the full implementation of Newton-Raphson method for a nonlinear equation.

Statement of the problem

Solve the Colebrook equation for the friction factor in a catheter tube. The Colebrook equation is given below in the Newton-Raphson format, i.e., $f(x)=0$.

$$\frac{1}{\sqrt{b}} - 4.07\log\left(\mathrm{Re}\sqrt{b}\right) + 0.06 = 0$$

Solution strategy

We will use Eq. (5.18) for the method of Newton-Raphson and repeat the procedure iteratively till the difference between two successive approximations of the root is less than a default value (say, 10^{-6}). We would like to show the convergence numerically and graphically, to gain insight into the convergence behavior.

Program description

Let us designate the Newton-Raphson code as NR.m. This code, which was developed by Constantinides and Mostoufi (1999), is used find the roots of nonlinear equations. The exact nonlinear function can be easily input as an argument in the program, so this MATLAB code should be readily applicable to other nonlinear equations.

```
% Example5_2a.m
% This program solves a nonlinear equation using the full
% Newton-Raphson method
% It calculates the friction factor from the Colebrook equation.

clc; clear all;

disp('Calculating the friction factor from Colebrook equation')

% Input
Re = input('\n Reynolds No.      = ');
fname = input('\n Function containing the Colebrook equation : ');
% Newton-Raphson
b0 = input(' Starting value of b = ');
f = NR(fname,b0,[],2,Re);
fprintf('\n Friction factor, b = %8.7f\n',b)
```

Function that contains the equation to be solved

```
function y = colebrook(b, Re)
% colebrook.m
% This function evaluates the value of Colebrook equation to be
% solved by the Newton-Raphson method.

y = 1/sqrt(b)-4.07*log(Re*sqrt(b))+0.6;
```

Function that applies the Newton-Raphson method, approximates the derivative of the function numerically, and plots the trajectory to the root.

```
function x = NR(FunFcn,x0,tol,trace,varargin)
%NR Finds a zero of a function by the Newton-Raphson method.
%
%   NR('F',X0) finds a zero of the function described by the
%   M-file F.M. X0 is a starting guess.
%
%   NR('F',X0,TOL,TRACE) uses tolerance TOL for convergence
%   test. TRACE=1 shows the calculation steps numerically and
%   TRACE=2 shows the calculation steps both numerically and
%   graphically.
%
%   NR('F',X0,TOL,TRACE,P1,P2,...) allows for additional
%   arguments which are passed to the function F(X,P1,P2,...).
%   Pass an empty matrix for TOL or TRACE to use the default
%   value.
%
%   See also FZERO, ROOTS, XGX, LI
```

```
% (c) N. Mostoufi & A. Constantinides
% January 1, 1999
% Initialization
if nargin < 3 | isempty(tol)
   tol = 1e-6;
end
if nargin < 4 | isempty(trace)
   trace = 0;
end
if tol == 0
   tol = 1e-6;
end
if (length(x0) > 1) | (~isfinite(x0))
   error('Second argument must be a finite scalar.')
end

iter = 0;
fnk = feval(FunFcn,x0,varargin{:});
if trace
  header = ' Iteration       x             f(x)';
  disp(header)
  fprintf('%5.0d   %13.6g %13.6g \n',iter, [x0 fnk])
  if trace == 2
     xpath = [x0 x0];
     ypath = [0 fnk];
  end
end

x = x0;
x0 = x + 1;
itermax = 100;

% Main iteration loop
while abs(x - x0) > tol & iter <= itermax
   iter = iter + 1;
   x0 = x;

   % Set dx for differentiation
   if x ~= 0
      dx = x/100;
   else
      dx = 1/100;
   end

   % Differentiation
   a = x - dx;  fa = feval(FunFcn,a,varargin{:});
   b = x + dx;  fb = feval(FunFcn,b,varargin{:});
   df = (fb - fa)/(b - a);

   % Next approximation of the root
```

```
   if df == 0
      x = x0 + max(abs(dx),1.1*tol);
   else
      x = x0 - fnk/df;
   end

   fnk = feval(FunFcn,x,varargin{:});
   % Show the results of calculation
   if trace
      fprintf('%5.0d   %13.6g %13.6g \n',iter, [x fnk])
      if trace == 2
         xpath = [xpath x x];
         ypath = [ypath 0 fnk];
      end
   end
end

if trace == 2
   % Plot the function and path to the root
   xmin = min(xpath);
   xmax = max(xpath);
   dx = xmax - xmin;
   xi = xmin - dx/10;
   xf = xmax + dx/10;
   yc = [];
   for xc = xi : (xf - xi)/99 : xf
      yc=[yc feval(FunFcn,xc,varargin{:})];
   end
   xc = linspace(xi,xf,100);
   ax = linspace(0,0,100);
   plot(xc,yc,xpath,ypath,xc,ax,xpath(1),ypath(2),'*',x,fnk,'o')
   axis([xi xf min(yc) max(yc)])
   xlabel('x')
   ylabel('f(x)')
   title('Newton-Raphson : The function and path to the root (* :
initial guess ; o : root)')
end

if iter >= itermax
   disp('Warning : Maximum iterations reached.')
end
if iter >= itermax
   disp('Warning : Maximum iterations reached.')
end
```

Input

```
Calculating the friction factor from Colebrook equation

 Reynolds No.      = 2e5
```

```
Function containing the Colebrook equation : 'colebrook'
Starting value of b = 0.0001
```

Output of results

```
Iteration          x              f(x)
               0.0001           69.6643
    1     0.000233872           33.3254
    2     0.000458271           13.2796
    3     0.000697928           3.56293
    4     0.000816553           0.38613
    5     0.000832694        0.00546019
    6     0.000832929      1.43852e-006

Friction factor, b = 0.0008329
```

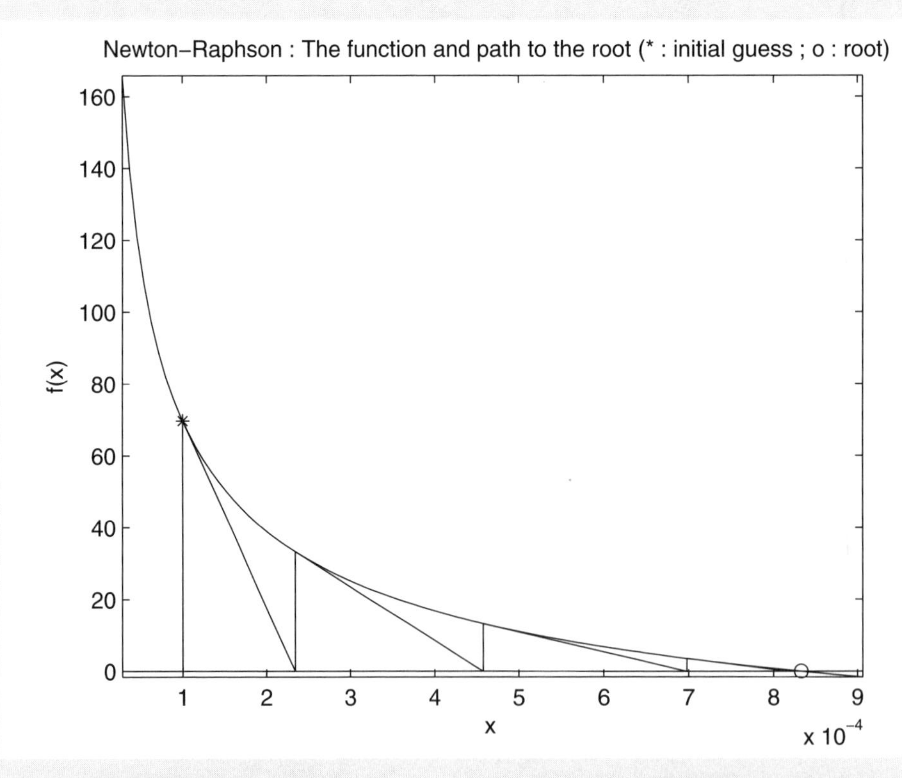

Figure 5.10 The path to the root using the Newton-Raphson method.

Discussion of results

Remarks on the Newton-Raphson solution: The Newton-Raphson treatment above rapidly converged to the root of the equation within six iterations. We suggest that students try the same approach using different parameter values (for example, the

Reynold's number can be varied from 5×10^3 to 1×10^4 to 5×10^4, etc.), or using different starting guesses for the friction factor.
For example, if we had assumed a different starting guess, keeping Reynold's number at 2×10^5, the Newton-Raphson search process would change significantly as shown below. The student should confirm this by working this out using MATLAB.

Starting Guess for x	**Convergence or Divergence**	**Remarks**
0.001	Convergence	4 iterations
0.0015	Convergence	5 iterations
0.0017	Convergence	6 iterations; imaginary roots also found
0.00175	Convergence	8 iterations; imaginary roots also found
0.0018	Divergence	

Next, we solve the problem in Example 5.2 using the two alternative methods of finding roots for nonlinear equations, the method of substitution, and the method of linear interpolation, respectively.

Example 5.2b Successive substitution method for solution of nonlinear equation.

Statement of the problem

Solve the Colebrook equation for the friction factor in a catheter tube. The Colebrook equation is given below in the format for the successive substitution method, i.e., $x = g(x)$.

$$b = \frac{1}{\left(4.07\log\left(\mathrm{Re}\sqrt{b}\right) - 0.60\right)^2}$$

Solution strategy

We will use Eq. (5.3) for the successive substitution method and repeat the procedure iteratively till the difference between two successive approximations of the root is less than a default value (say, 10^{-6}). We would like to show the convergence process numerically and graphically, to gain insight into the convergence behavior.

Program description

Let us designate the Successive Substitution code as XGX.m. This code is used to find the roots of nonlinear equations. The exact nonlinear function can be easily inputted as an argument in the program, so this MATLAB code should be readily applicable to other nonlinear equations.

```
% Example5_2b.m
% This program solves a nonlinear equation using the
% Successive Substitution method.
% It calculates the friction factor from the Colebrook equation.

clc; clear all;
disp('Calculating the friction factor from Colebrook equation')
% Input
Re = input('\n Reynolds No.        = ');
fname = input('\n Function containing the Colebrook equation : ');
% Successive Substitution method
b0 = input(' Starting value of b = ');
b = XGX(fname,b0,[],2,Re);
fprintf('\n Friction factor, b = %8.7f\n',f)
```

Function that contains the equation to be solved

```
function y = colebrookg(b, Re)
% colebrookg.m
% This function evaluates the value of Colebrook equation
% to be solved by the the XGX or successive substitution method.
y = (1/(4.07*log(Re*sqrt(b))-0.6))^2;
```

Function that applies the Successive Substitution method and plots the path to the root.

```
function x = XGX(FunFcn,x0,tol,trace,varargin)
%XGX Finds a zero of a function by x=g(x) method.
%
%   XGX('G',X0) finds the intersection of the curve y=g(x)
%   with the line y=x. The function g(x) is  described by the
%   M-file G.M. X0 is a starting guess.
%
%   XGX('G',X0,TOL,TRACE) uses tolerance TOL for convergence
%   test. TRACE=1 shows the calculation steps numerically and
%   TRACE=2 shows the calculation steps both numerically and
%   graphically.
%
%   XGX('G',X0,TOL,TRACE,P1,P2,...) allows for additional
%   arguments which are passed to the function G(X,P1,P2,...).
%   Pass an empty matrix for TOL or TRACE to use the default
%   value.
```

```
%
%   See also FZERO, ROOTS, NR, LI

% (c) N. Mostoufi & A. Constantinides
% January 1, 1999

% Initialization
if nargin < 3 | isempty(tol)
   tol = 1e-6;
end
if nargin < 4 | isempty(trace)
   trace = 0;
end
if tol == 0
   tol = 1e-6;
end
if (length(x0) > 1) | (~isfinite(x0))
   error('Second argument must be a finite scalar.')
end
if trace
  header = ' Iteration        x             g(x)';
  disp(header)
  if trace == 2
     xpath = [x0];
     ypath = [0];
  end
end

x = x0;
x0 = x + 1;
iter = 1;
itermax = 100;

% Main iteration loop
while abs(x - x0) > tol & iter <= itermax
   x0 = x;
   fnk = feval(FunFcn,x0,varargin{:});

   % Next approximation of the root
   x = fnk;

   % Show the results of calculation
   if trace
      fprintf('%5.0f   %13.6g %13.6g \n',iter, [x0 fnk])
      if trace == 2
         xpath = [xpath x0 x];
         ypath = [ypath fnk x];
```

```
         end
      end
      iter = iter + 1;
   end
   if trace == 2
      % Plot the function and path to the root
      xmin = min(xpath);
      xmax = max(xpath);
      dx = xmax - xmin;
      xi = xmin - dx/10;
      xf = xmax + dx/10;
      yc = [];
      for xc = xi : (xf - xi)/99 : xf
         yc=[yc feval(FunFcn,xc,varargin{:})];
      end
      xc = linspace(xi,xf,100);
      plot(xc,yc,xpath,ypath,xpath(2),ypath(2),'*', ...
         x,fnk,'o',[xi xf],[xi,xf],'--')
      axis([xi xf min(yc) max(yc)])
      xlabel('x', 'FontSize',12)
      ylabel('g(x) [-- : y=x]','FontSize',12)
      title('               x=g(x)  : The function and path to the root
 (* : initial guess ; o : root)','FontSize',12)
   end

   if iter >= itermax
      disp('Warning : Maximum iterations reached.')
   end
```

Solution

```
Calculating the friction factor from Colebrook equation

 Reynolds No.        = 2e5

 Function containing the Colebrook equation : 'colebrookg'
 Starting value of b= 0.0001
 Iteration         x              g(x)
     1          0.0001       0.00108666
     2       0.00108666      0.00080751
     3       0.00080751      0.00083597
     4       0.00083597     0.000832573
     5      0.000832573     0.000832971

 Friction factor, b = 0.0008330
```

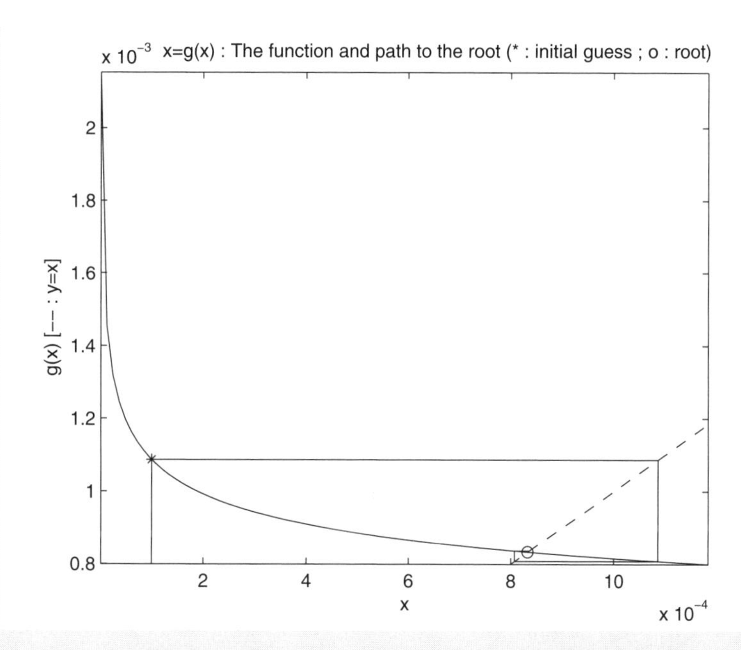

Figure 5.11 The path to the root using the Successive Substitution method.

Example 5.2c Linear interpolation method for solution of nonlinear equation.

Statement of the problem

Solve the Colebrook equation for the friction factor in a catheter tube. The Colebrook equation is given below in the $f(x)=0$ format.

$$\frac{1}{\sqrt{b}} - 4.07\log\left(\mathrm{Re}\sqrt{b}\right) + 0.06 = 0$$

Solution strategy

We will use Eq. (5.10) for the linear interpolation method and continue the procedure iteratively till the difference between two successive approximations of the root is less than a default value (say, 10^{-6}). We would like to show the convergence process numerically and graphically, to gain insight into the convergence behavior.

```
% Example5_2c.m
% This program solves a nonlinear equation using the
% Linear Interpolation method.
% It calculates the friction factor from the Colebrook equation.

clc; clear all;
```

```
disp('Calculating the friction factor from Colebrook equation')
% Input
Re = input('\n Reynolds No.        = ');
fname = input('\n Function containing the Colebrook equation : ');
% Linear Interpolation method
b1 = input(' First  starting value of b = ');
b2 = input(' Second starting value of b = ');
b = LI(fname,b1,b2,[],2,Re);
fprintf('\n Friction factor, b = %8.7f\n',b)
```

Function that contains the equation to be solved

The function `colebrook.m` is the same as the one used in Example 5.2a.

Function that applies the Linear Interpolation method and plots the path to the root. This code was developed by Constantinides and Mostoufi (1999).

```
function x = LI(FunFcn,x1,x2,tol,trace,varargin)
%LI Finds a zero of a function by the linear interpolation method.
%
%   LI('F',X1,X2) finds a zero of the function described by the
%   M-file F.M. X1 and X2 are starting points where the function
%   has different signs at these points.
%   LI('F',X1,X2,TOL,TRACE) uses tolerance TOL for convergence
%   test. TRACE=1 shows the calculation steps numerically and
%   TRACE=2 shows the calculation steps both numerically and
%   graphically.
%
%   LI('F',X1,X2,TOL,TRACE,P1,P2,...) allows for additional
%   arguments which are passed to the function F(X,P1,P2,...).
%   Pass an empty matrix for TOL or TRACE to use the default
%   value.
%
%   See also FZERO, ROOTS, XGX, NR

% (c) N. Mostoufi & A. Constantinides
% January 1, 1999

% Initialization
if nargin < 4 | isempty(tol)
   tol = 1e-6;
end
if nargin < 5 | isempty(trace)
   trace = 0;
end
if tol == 0
   tol = 1e-6;
end
if (length(x1) > 1) | (~isfinite(x1)) | (length(x2) > 1) | ...
```

```
        (~isfinite(x2))
      error('Second and third arguments must be finite scalars.')
   end
   if trace
     header = ' Iteration          x                f(x)';
     disp(header)
   end
   f1 = feval(FunFcn,x1,varargin{:});
   f2 = feval(FunFcn,x2,varargin{:});

   iter = 0;
   if trace
      % Display initial values
      fprintf('%5.0f    %13.6g %13.6g \n',iter, [x1 f1])
      fprintf('%5.0f    %13.6g %13.6g \n',iter, [x2 f2])
      if trace == 2
         xpath = [x1 x1 x2 x2];
         ypath = [0 f1 f2 0];
      end
   end

   if f1 < 0
      xm = x1;
      fm = f1;
      xp = x2;
      fp = f2;
   else
      xm = x2;
      fm = f2;
      xp = x1;
      fp = f1;
   end

   iter = iter + 1;
   itermax = 100;
   x = xp;
   x0 = xm;

   % Main iteration loop
   while abs(x - x0) > tol & iter <= itermax
      x0 = x;
      x = xp - fp * (xm - xp) / (fm - fp);
      fnk = feval(FunFcn,x,varargin{:});

      if fnk < 0
         xm = x;
         fm = fnk;
      else
         xp = x;
```

```
        fp = fnk;
    end

    % Show the results of calculation
    if trace
        fprintf('%5.0f   %13.6g %13.6g \n',iter, [x fnk])
        if trace == 2
            xpath = [xpath xm xm xp xp];
            ypath = [ypath 0 fm fp 0];
        end
    end
    iter = iter + 1;
end

if trace == 2
    % Plot the function and path to the root
    xmin = min(xpath);
    xmax = max(xpath);
    dx = xmax - xmin;
    xi = xmin - dx/10;
    xf = xmax + dx/10;
    yc = [];
    for xc = xi : (xf - xi)/99 : xf
        yc=[yc feval(FunFcn,xc,varargin{:})];
    end
    xc = linspace(xi,xf,100);
    ax = linspace(0,0,100);

plot(xc,yc,xpath,ypath,xc,ax,xpath(2:3),ypath(2:3),'*',x,fnk,'o')
    axis([xi xf min(yc) max(yc)])
    xlabel('x')
    ylabel('f(x)')
    title('Linear Interpolation : The function and path to the root
(* : initial guess ; o : root)')
end

if iter >= itermax
    disp('Warning : Maximum iterations reached.')
end
```

Solution

```
Calculating the friction factor from Colebrook equation

 Reynolds No.       = 2e5

 Function containing the Colebrook equation : 'colebrook'
 First  starting value b = 0.0003
 Second starting value b = 0.0015
```

```
Iteration        x              f(x)
   0          0.0003          25.1637
   0          0.0015         -10.0267
   1      0.00115809         -5.93487
   2      0.00099433         -3.29705
   3     0.000913895         -1.75924
   4     0.000873781        -0.917111
   5     0.000853604        -0.472089
   6     0.000843409        -0.241398
   7     0.000838246        -0.123012
   8     0.000835628       -0.0625739
   9     0.000834299       -0.0318015
  10     0.000833625       -0.0161549

Friction factor, b = 0.0008336
Friction factor, b = 0.0008360
```

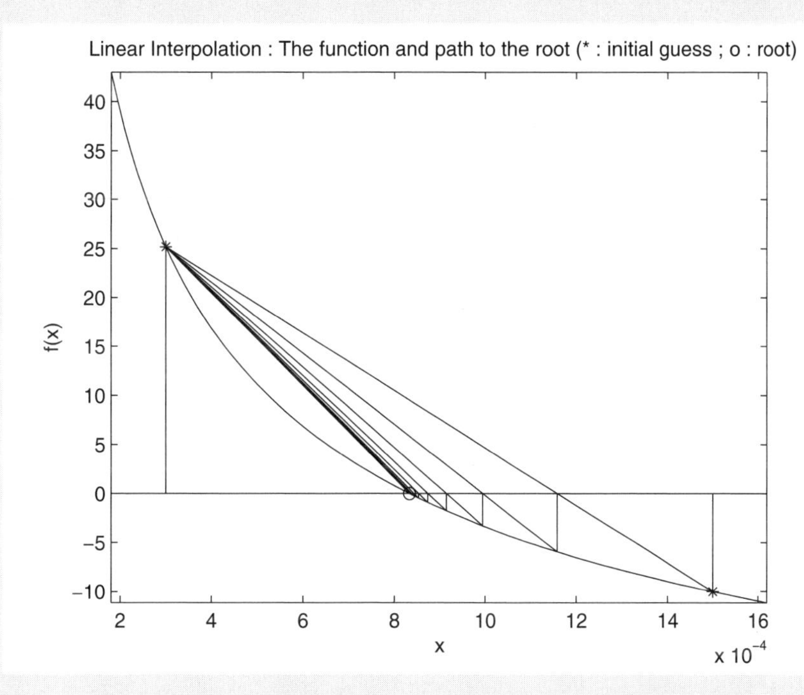

Figure 5.12 The path to the root using Linear Interpolation.

Discussion of results

Three alternative approaches were used to compute the solution to the Colebrook equation in Example 5.2a (Newton Raphson), 5.2b (successive substitution), and 5.2c (linear interpolation). In general, the Newton-Raphson method is the most commonly employed one among the three methods. However, its search process is sensitive to the choice of the initial guess for the root. The successive substitution method was able to identify the root above in the least number of steps. This method is the simplest of the three methods and does not require the computation of the derivative of the function. The linear interpolation method required the choice of two starting guesses for the extremities of the range of the search and required the highest number of iterations. The choice of the best method among the three alternatives presented is dependent on the nature of the function and the initial guesses. In general, students are urged to master the Newton-Raphson method.

Example 5.3 Molecular bioengineering problem involving the solution of a Michaelis-Menten kinetics equation using the Newton-Raphson method.

Statement of the problem

The analytical solution of the Michaelis-Menten model (Section 5.3.1) is given by:

$$K_m \ln\left(\frac{s_0}{s}\right) + (s_0 - s) = V_{\max} t$$

where K_m, the substrate concentration eliciting half-maximal substrate consumption rate, is equal to 0.5 mM; the maximal substrate consumption rate, V_{max} , is equal to 5.0 mM/min; and the initial substrate concentration, s_o, is 1.0 mM. Find the substrate concentration as a function of time (from 0 to 200 minutes) at 0.1 minute increments using the Newton-Raphson method.

```
% example5_3.m
% This program calculates the roots of the substrate
% consumption equation governed by Michealis-Menten kinetics

clc; clear all;

disp('Time      Concentration')
x0=0.95;
% initialize the counter and increment by unity during each
% iteration. We could use a while loop below;  a FOR loop is
% preferred because it initializes the counter and increments
% automatically at each iteration
for i=1:201
    t(i)=i-1;
    sub(i)= NR('enzymeconsump',x0,[],[],t,1.0,0.05,0.5);
    fprintf('%3g          %g \n',t(i),sub(i))
    x0=sub(i);
end
figure(1); plot(t,sub)
```

```
xlabel('Time, min'); ylabel('Concentration, mM')
title('Michaelis-Menten reaction')
```

Function that contains the Michaelis-Menten equation

```
function y = enzymeconsump(s,t,so,vmax,Km)
% enzymeconsump.m file
y = Km*log(so/s)+(so-s)-vmax*t;
```

Output (only graphical output shown below)

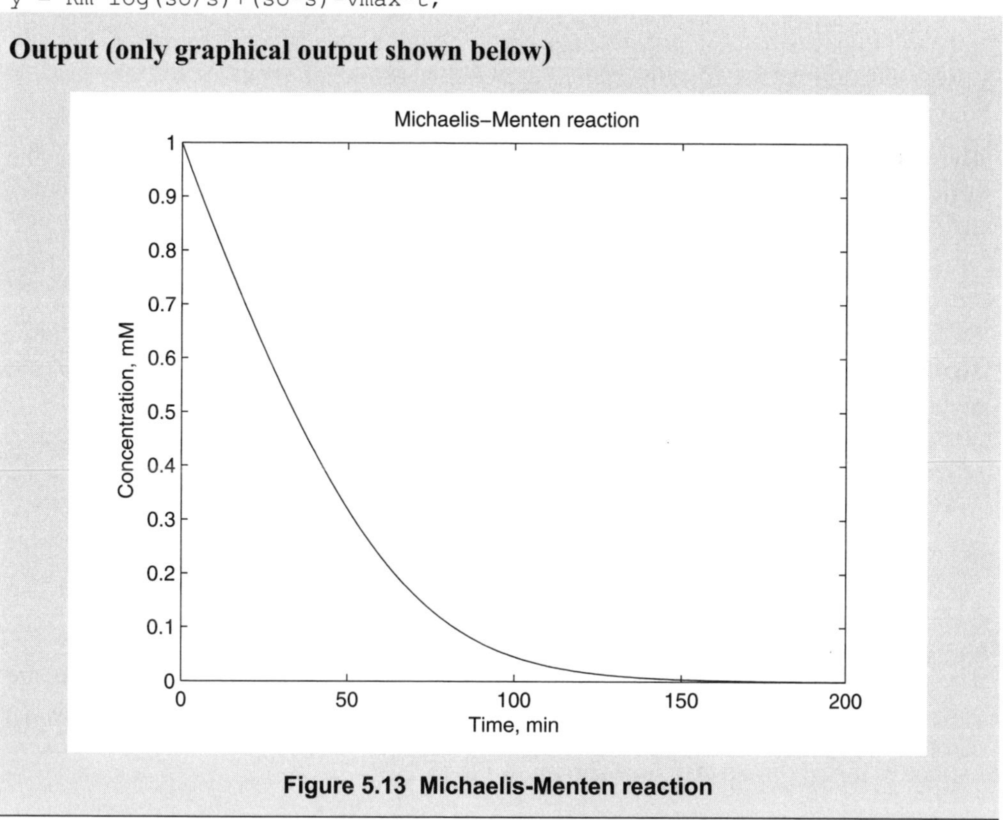

Figure 5.13 Michaelis-Menten reaction

5.7 Newton's Method for Simultaneous Nonlinear Equations

When the model involves two or more simultaneous nonlinear equations, the Newton-Raphson method can be extended to solve these equations simultaneously. Let us assume that the two unknown variables are x_1 and x_2, and two functions are given involving both x_1 and x_2:

$$\begin{aligned} f_1(x_1, x_2) &= 0 \\ f_2(x_1, x_2) &= 0 \end{aligned} \tag{5.20}$$

where f_1 and f_2 are nonlinear functions. We can expand both of these functions using two-dimensional Taylor series around initial estimates of $x_1^{(1)}$ and $x_2^{(1)}$,where the superscript (1) indicates the first iteration during the root-finding exercise.

$$\begin{aligned} f_1(x_1,x_2) &= f_1(x_1^{(1)},x_2^{(1)}) + \left.\frac{\partial f_1}{\partial x_1}\right|_{x^{(1)}} (x_1 - x_1^{(1)}) + \left.\frac{\partial f_1}{\partial x_2}\right|_{x^{(1)}} (x_2 - x_2^{(1)}) + \ldots. \\ f_2(x_1,x_2) &= f_2(x_1^{(1)},x_2^{(1)}) + \left.\frac{\partial f_2}{\partial x_1}\right|_{x^{(1)}} (x_1 - x_1^{(1)}) + \left.\frac{\partial f_2}{\partial x_2}\right|_{x^{(1)}} (x_2 - x_2^{(1)}) + \ldots \end{aligned} \tag{5.21}$$

The difference between the values of x at different iterations is defined as the correction variable:

$$\begin{aligned} \delta_1^{(1)} &= x_1 - x_1^{(1)} \\ \delta_2^{(1)} &= x_2 - x_2^{(1)} \end{aligned} \tag{5.22}$$

Setting the left-hand sides of Eq. (5.21) to zero and truncating the second-order and higher derivatives of the Taylor series results in the following equations:

$$\begin{aligned} \left.\frac{\partial f_1}{\partial x_1}\right|_{x^{(1)}} \delta_1^{(1)} + \left.\frac{\partial f_1}{\partial x_2}\right|_{x^{(1)}} \delta_2^{(1)} &= -f_1\left(x_1^{(1)},x_2^{(1)}\right) \\ \left.\frac{\partial f_2}{\partial x_1}\right|_{x^{(1)}} \delta_1^{(1)} + \left.\frac{\partial f_2}{\partial x_2}\right|_{x^{(1)}} \delta_2^{(1)} &= -f_2\left(x_1^{(1)},x_2^{(1)}\right) \end{aligned} \tag{5.23}$$

These are a set of simultaneous linear algebraic equations, where the unknowns are $\delta_1^{(1)}$ and $\delta_2^{(1)}$. This example, which has only two equations, can be solved explicitly for $\delta_2^{(1)}$ and $\delta_2^{(1)}$, as shown in Eq. (5.24):

$$\begin{aligned} \delta_1^{(1)} &= -\frac{\left[f_1 \frac{\partial f_2}{\partial x_2} - f_2 \frac{\partial f_1}{\partial x_2}\right]}{\left[\frac{\partial f_1}{\partial x_1}\frac{\partial f_2}{\partial x_2} - \frac{\partial f_1}{\partial x_2}\frac{\partial f_2}{\partial x_1}\right]} \\ \delta_2^{(1)} &= -\frac{\left[f_2 \frac{\partial f_1}{\partial x_1} - f_1 \frac{\partial f_2}{\partial x_1}\right]}{\left[\frac{\partial f_1}{\partial x_1}\frac{\partial f_2}{\partial x_2} - \frac{\partial f_1}{\partial x_2}\frac{\partial f_2}{\partial x_1}\right]} \end{aligned} \tag{5.24}$$

The new estimate for the solution can now be obtained from the previous estimate by adding it to the vector of correction variables:

$$x_i^{(n+1)} = x_i^{(n)} + \delta_i^{(n)} \tag{5.25}$$

For each iteration, the above equation is used to calculate the newest estimate. This process is repeated iteratively till all the correction variables reach a sufficiently small value (smaller than a prescribed tolerance).

For an example that involves more than two nonlinear simultaneous equations, a similar approach results in a series of equations that can be represented by the following matrix:

$$\begin{bmatrix} \frac{\partial f_1}{\partial x_1} & \cdots\cdots & \frac{\partial f_1}{\partial x_k} \\ & \cdots\cdots & \\ \frac{\partial f_k}{\partial x_1} & \cdots\cdots & \frac{\partial f_k}{\partial x_k} \end{bmatrix} \begin{bmatrix} \delta_1 \\ . \\ . \\ . \\ \delta_k \end{bmatrix} = - \begin{bmatrix} f_1 \\ . \\ . \\ . \\ f_k \end{bmatrix} \tag{5.26}$$

which can be represented in a matrix-vector notation to:

$$\mathbf{J\delta} = -\mathbf{f} \tag{5.27}$$

where $\mathbf{J}$ is the Jacobian matrix with partial derivatives, $\boldsymbol{\delta}$ is the correction vector, and $\mathbf{f}$ is the vector of functions. Readers should note that this is a set of linear algebraic equations. In order to find values of $\boldsymbol{\delta}$, the matrix of $\mathbf{J}$ will need to be first computed using numeric differentiation and then inverted. The inversion can be easily accomplished using the MATLAB command `inv`, provided the Jacobian matrix is nonsingular. A quick check for this is rendered if the determinant of the Jacobian is nonzero (using MATLAB command `det`).

Example 5.4 Determination of receptor occupancy during receptor-ligand dynamics using Newton's method for simultaneous nonlinear equations.

Formulation of the problem

Cell membrane receptors can bind to specific ligands, such as hormones and growth factors, and trigger intracellular signaling in mammalian cells (Lauffenburger and Linderman, 1993). Multimeric receptors can possess different units or epitopes that can bind to more than one ligand, thereby making the binding chemistry nonlinear. If we imagine a receptor that binds to three ligand molecules, and a second population of receptors that binds to two ligand molecules each, the following reversible binding reactions can be established:

Let C_A be the concentration of trimeric receptor.
Let C_B be the concentration of dimeric receptor.
Let C_L be the concentration of the ligand.
Let C_D be the concentration of all receptor-bound ligands.

$$\mathrm{A} + 3\mathrm{L} \Leftrightarrow \mathrm{D} \qquad K_{eq,1} = \frac{C_D}{C_A C_L^{\,3}}$$

$$\mathrm{B} + 2\mathrm{L} \Leftrightarrow \mathrm{D} \qquad K_{eq,2} = \frac{C_D}{C_B C_L^{\,2}}$$

The initial concentrations of A, B, and L are given as: $C_{A,0}$ = 5000/cell; $C_{B,0}$ = 10000/cell; $C_{L,0}$ = 10^{-7} M (1 M = 1 mole/liter = 6.023 x 10^{23} molecules/liter. For 1 cell equivalent volume of 4 x 10^{-9} cm^3, $C_{L,0}$ = 24000/cell). Find the fractional occupancy of the two receptors at equilibrium.

Analytical strategy for solution

Let the fractional occupancy of the two receptor populations, A and B, be x_1 and x_2, respectively. Then,

$$C_A = 5000(1 - x_1)$$
$$C_B = 10000(1 - x_2)$$
$$C_D = 5000x_1 + 10000x_2$$
$$C_L = 24000 - 5000x_1 - 10000x_2$$

Substituting in the above equations, we get:

$$K_{eq,1} = (5000x_1 + 10000x_2) / \{5000(1 - x_1)(24000 - 5000x_1 - 10000x_2)^3\}$$

$$K_{eq,2} = (5000x_1 + 10000x_2) / \{10000(1 - x_2)(24000 - 5000x_1 - 10000x_2)^2\}$$

Numerical solution

```
% example5_4.m
% Commands to solve the problem by calling on Newton's method

clc; clear all;
```

```
% Input two guesses for x1 and x2, respectively
x0=[.5  .2];
% Solution is provided by calling Newton.m and the prescribed
% function, with starting estimate vector x0
[x, iter]=Newton('receptor_ligand_func',x0)
```

Function to be solved

```
function f=receptor_ligand_func(x) % This is saved as a -M file.
x1=x(1); x2=x(2);
f(1)=(5000*x1+10000*x2)/(5000*(1-x1)*(24000-5000*x1-10000*x2)^3)  -
7.19e-14;
f(2)=(5000*x1+10000*x2)/(10000*(1-x2)*(24000-5000*x1-10000*x2)^2) -
6.0e-10;
f=f';
```

Function that applies Newton's method for simultaneous equations

```
function [xnew , iter] = Newton(fnctn , x0 , rho , tol , varargin)
%NEWTON   Solves a set of equations by Newton's method.
%
%   NEWTON('F',X0) finds a zero of the set of equations
%   described by the M-file F.M. X0 is a vector of starting
%   guesses.
%
%   NEWTON('F',X0,RHO,TOL) uses relaxation factor RHO and
%   tolerance TOL for convergence test.
%
%   NEWTON('F',X0,RHO,TOL,P1,P2,...) allows for additional
%   arguments which are passed to the function F(X,P1,P2,...).
%   Pass an empty matrix for RHO or TOL to use the default
%   value.

% (c) by N. Mostoufi & A. Constantinides
% January 1, 1999

% Initialization
if nargin < 4 | isempty(tol)
   tol = 1e-6;
end
if nargin < 3 | isempty(rho)
   rho = 1;
end

x0 = (x0(:).')'; % Make sure it's a column vector
nx = length(x0);
x = x0 * 1.1;
xnew = x0;
iter = 0;
```

```
maxiter = 100;

% Main iteration loop
while max(abs(x - xnew)) > tol & iter < maxiter
   iter = iter + 1;
   x = xnew;
   fnk = feval(fnctn,x,varargin{:});

   % Set dx for derivation
   for k = 1:nx
      if x(k) ~= 0
         dx(k) = x(k) / 100;
      else
         dx(k) = 1 / 100;
      end
   end

   % Calculation of the Jacobian matrix
   a = x;
   b = x;
   for k = 1 : nx
      a(k) = a(k) - dx(k);  fa = feval(fnctn,a,varargin{:});
      b(k) = b(k) + dx(k);  fb = feval(fnctn,b,varargin{:});
      jacob(:,k) = (fb - fa) / (b(k) - a(k));
      a(k) = a(k) + dx(k);
      b(k) = b(k) - dx(k);
   end

   % Next approximation of the roots
   if det(jacob) == 0
      xnew = x + max([abs(dx), 1.1*tol]);
   else
      xnew = x - rho * inv(jacob) * fnk;
   end
end

if iter >= maxiter
   disp('Warning : Maximum iterations reached.')
end
```

Solution

```
x =
    0.3009
    0.0994
iter =6
```

5.8 Lessons Learned in this Chapter

After studying this chapter, the student should have learned the following:

- The student should be able to identify nonlinear algebraic equations in biomedical engineering and formulate the problem for numerical solution.
- The key numerical tool learned in this chapter is the use of the Newton-Raphson method for solution of nonlinear algebraic equations. Two alternative strategies have also been treated: the method of successive substitution and the linear interpolation method.
- The importance of appropriate convergence to desired roots of such equation should be recognized from the Newton-Raphson solution. The starting guess is of paramount relevance, as Example 5.2a illustrated. This is because far from a root, where the higher order terms in the series are important, the Newton-Raphson formula can give grossly inaccurate corrections. In the search for a local maximum or minimum, if the first derivative nearly vanishes, then the Newton-Raphson sends its solution deviating, which points to the limitations of this approach. Students should carefully note the convergence criteria to be applied to problem solutions.
- Based on the three solved problems in Chapter 5, the student should be able to readily apply these MATLAB tools to solve nonlinear equations. A fourth problem illustrates how nonlinear equations can be simultaneously solved. To solve these problems, a master program or code is required, which refers to a separate function subprogram and a NR or Newton-based subroutine. Students should practice their skills at writing similar master programs for non-algebraic equations of varying complexity.

5.9 Problems

5.1 Magnetic resonance imaging (MRI) is a rapidly advancing area of biomedical imaging. MRI interrogates the nuclear environment of hydrogen protons in the soft tissues of the body, providing contrast imaging as well as functional imaging of tissues. A common coil used for MRI experiments is the solenoid. A solenoid coil can be used to couple energy into a sample and to detect the time-varying magnetic flux density during the acquisition of signals. The solenoid has length l and radius a. The long axis of the solenoid is along the y-axis. The solenoid produces a flux density at any observation point y, which depends on the number of turns in the coil, N, the amount of current, and μ_0, the permeability of free space, which is typically $4\pi \times 10^{-7}$ H/m. The inductance of a coiled solenoid is given approximately as:

$$L = \frac{\mu_0 N^2 \pi a^2}{l^2}\left[\left(l^2 + a^2\right)^{1/2} - a\right]$$

Compute, using the Newton-Raphson method in MATLAB, the radius of the solenoid coil, a, to provide inductance of 2.6 μH, assuming the following parameters:

N = 10 turns
l = 20 cm

5.2 When laser light is incident on tissue, the tissue acts as a transfer medium that can reflect, absorb, scatter, and transmit various portions of the incident wave of radiation. Diffusion equations can be constructed to describe the propagation of light intensity as a function of distance. Imagine an isotropic light source from a laser fiber within a tissue that can both absorb and scatter light. If the tissue acts as an infinite medium, under conditions of predominant scattering, the light fluence rate (in W/cm^2) at a large distance r from the fiber can be shown to be:

$$\phi(r) = \frac{\phi_o}{4\pi D}\frac{e^{-r/\delta}}{r}$$

where the penetration depth, $\delta = \sqrt{D/\mu_a}$; and μ_a is the tissue absorption coefficient. The anisotropy of the medium, g = 0.9; absorption coefficient, $\mu_a = 0.1$ cm^{-1}, scattering coefficient, $\mu_s = 100$ cm^{-1}. Find the distance away from the point source of laser where the light intensity drops to 10% of that of the source.

(Hint: $D = (1/3)\,\mu_{t'}$; $\mu_{s'} = \mu_S(1-g)$; $\mu_{t'} = \mu_a + \mu_{s'}$)

5.3 The following empirical relationship is available for the light propagation through biological tissues:

$$\frac{A-1}{A+1} = -1.440n_{rel}^{-2} + 0.710n_{rel}^{-1} + 0.688 + 0.0636n_{rel}$$

where A is the internal reflectance factor for a tissue, and n_{rel} is the ratio of the refractive indices of the tissue and the medium. When n_{rel} equals 1, A reduces to unity. Find the value of n_{rel} in order to give a reflectance factor of 4.0.

5.4 The governing equations for laser irradiation with ablation in one-dimension are differential equations.

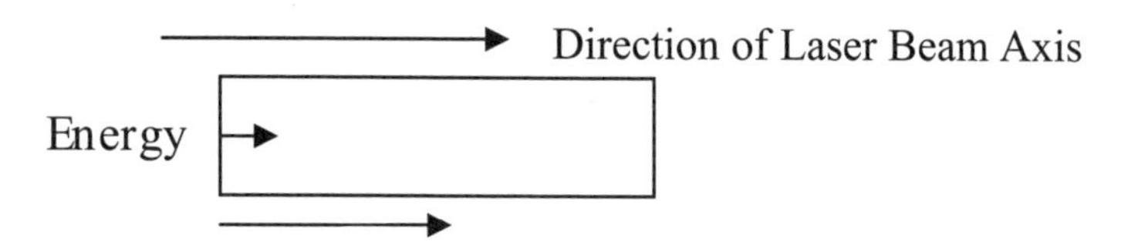

The solution to these equations is given below:

$$B\lambda = \frac{2}{\sqrt{\pi}} B\sqrt{\tau_{ab}} + e^{B^2\tau_{ab}} erfc\left[B\sqrt{\tau_{ab}} \right] - 1$$

where B is a dimensionless absorption parameter
λ is a dimensionless heating parameter
τ_{ab} is the onset of ablation (dimensionless time)
erfc is the complementary error function

Find the value of ($B\sqrt{\tau_{ab}}$) at which $(B\lambda) = 28.359$.

5.5 The mean squared displacement for cell migration in two-dimensional media is given by the Dunn equation (Dunn, 1983)

$$< d^2 >= 2S^2\left[Pt - P^2\left(1 - e^{-t/P}\right)\right]$$

where $<d^2>$ stands for mean squared displacement
S is the root mean squared cell speed
P is the directional persistence time (minutes)

White blood cells stimulated with chemoattractant factors migrated at 20 microns per minute on an expanded polytetrafluoroethylene, used as a vascular prosthetic biomaterial (Chang et al, 2000). What is the persistence time, P, necessary for a population of white blood cells to achieve a mean squared displacement of 4.3×10^{-3} cm^2 in 3 hours?

5.6 A major problem in the successful design of implantable tissues is the availability of oxygen for respiring tissues, which is determined by the spatial access of tissues to blood capillaries that bring oxygen carrying red blood cells. A classic model in this field is the Krogh cylinder (Fournier, 1999). Imagine a tissue space with cells surrounding a cylindrical capillary. Oxygen and other metabolites arriving into the capillary axially due to fresh flow of oxygenated blood will diffuse from the capillary radially toward the tissue, where they will be consumed by the cells. The solution of the Krogh cylinder problem yields an expression for the critical distance into the tissue, beyond which no more solute is available, denoted by r_{crit}:

$$y^2 \ln(y^2) - y^2 + 1 - \left[\frac{4D_T C_o}{R_o (r_c + t_m)^2}\right] + 4\frac{D_T}{Vr_c^2}\left[y^2 - 1\right]z + \frac{2D_T}{r_c K_o}\left[y^2 - 1\right] = 0$$

where $y = \dfrac{r_{crit}}{r_c + t_m}$.

Parameters given by Fournier (1999) are:

D_T = the metabolite tissue diffusivity = $8 \text{x} 10^{-6}$ cm^2/s
V = the blood plasma velocity = 0.005 cm/s
r_c = capillary radius = 0.0005 cm
t_m = capillary wall thickness = $5 \text{x} 10^{-5}$ cm
K_o = overall metabolite mass transfer rate = $5.75 \text{ x } 10^{-5}$ cm/s
C_o = 5 μmole/cm^3
R_o = 0.01 μmole/(cm^3 s)

Using the Newton-Raphson method, solve the above equation for r_{crit} as a function of z. Vary z from 0.001 to 0.1 cm in increments of 0.01 cm. Plot r_{crit} versus z.

5.7 Enzymes are biological catalysts that can be immobilized in porous matrices within reactors through which substrates can be diffused and reacted. Such enzyme-immobilized reactors have been conceptualized as extracorporeal devices for the selective elimination or transformation of specific components in blood, such as bilirubin in jaundice and the anticoagulant heparin (Lavin et al, 1985; Sung et al, 1986). The mass balance of substrates on an immobilized enzyme particle leads to an overall relationship between the efficiency of reaction, η, and the Thiele modulus, ϕ, which represents the ratio of substrate reaction rate to its diffusion rate. A large value for ϕ indicates that diffusion limits the overall substrate consumption rate, while a small value for ϕ indicates that reaction rate limits the overall consumption of the substrate. For a first-order reaction, this relationship becomes:

$$\eta = \frac{1}{\phi}\left(\frac{1}{\tanh 3\phi} - \frac{1}{3\phi}\right)$$

Determine and plot the values for the Thiele modulus when the efficiency of metabolite reaction varies from 30% to 60% at 5% increments.

5.10 References

Chang, C. C., Schloss, R. S., and Moghe, P. V. 2000. Quantitative Analysis of the Regulation of Leukocyte Chemosensory Migration by a Vascular Prosthetic Biomaterial. *J. Mater. Sci. Mater. Med.* **11**:337-344.

Constantinides, A. and Mostoufi, N. 1999. *Numerical Methods for Chemical Engineers with MATLAB Applications.* Upper Saddler River, NJ: Prentice Hall PTR.

Dunn, G. A. 1983. Characterising a kinesis response: time averaged measures of cell speed and directional persistence. *Agents and Actions* [Suppl.], **22**:14-33.

Enderle, J. D., Blanchard, S. M., and Bronzino, J. D. 2005. *Introduction to Biomedical Engineering.* 2nd Ed., San Diego, CA: Academic Press.

Fournier, R. L. 1999. *Basic Transport Phenomena in Biomedical Engineering.* Philadelphia, PA: Taylor & Francis.

Lauffenburger, D. A., and Linderman, J. J. 1993. *Receptors: Models for Binding, Trafficking, and Signaling.* New York: Oxford University Press.

Lavin, A., Sung, C., Klibanov, A.L., and Langer, R. 1985. A potential treatment for neonatal jaundice. *Science* **230**:543-545.

Sung, C., Lavin, A. , Klibanov, A. M., and Langer, R. 1986. An immobilized enzyme reactor for the detoxification of bilirubin. *Biotechnol. Bioeng.* **28**: 1531-1539.

Chapter 6
Finite Difference Methods, Interpolation and Integration

6.1 Introduction

The most commonly encountered mathematical models in engineering and science, including biomedical engineering, are in the form of differential equations. Systems that have one independent variable can be modeled by ordinary differential equations, whereas systems with two or more independent variables require partial differential equations. The great majority of differential equations, especially the nonlinear ones and those that involve large sets of simultaneous differential equations, do not have analytical solutions but require the application of numerical techniques for their solution.

Several numerical methods for differentiation, integration, and the solution of ordinary and partial differential equations are discussed in Chapters 6, 7, and 8 of this book. These methods are based on the concept of *finite differences*. Therefore, the purpose of thc early sections of this chapter is to develop the systematic terminology used in the *calculus of finite differences* and to derive the relationships between finite differences and differential operators, then to apply these in the development of differentiation and integration formulas, and in the numerical solution of ordinary and partial differential equations. The approach followed is essentially the same as that developed by Constantinides and Mostoufi (1999).

The calculus of finite differences may be characterized as a "two-way street" that enables the user to take a differential equation and integrate it numerically by calculating the values of the function at a discrete (finite) number of points. Or, conversely, if a set of finite values is available, such as experimental data, these may be differentiated, or integrated, using the calculus of finite differences. It should be pointed out, however, that numerical differentiation is inherently less accurate than numerical integration.

Another very useful application of the calculus of finite differences is in the derivation of interpolation/extrapolation formulas, the *interpolating polynomials*, which can be used to represent experimental data when the actual functionality of this data is not known. Interpolating polynomials, such as the Gregory-Newton interpolation formulas (Section 6.7), are the foundation of numerical integration methods, such as the Newton-Cotes formulas of integration (Section 6.1).

The material in this chapter will enable the student to accomplish the following:

- Understand the concepts of finite differences and apply these concepts to express derivatives in terms of finite differences
- Develop the tools needed for integration and differentiation of functions and data
- Develop the concepts of interpolating polynomials and apply these tools to the development of integration formulas

6.2 Symbolic Operators

In differential calculus, the definition of the derivative is given as

$$\left.\frac{df(x)}{dx}\right|_{x_0} = f'(x_0) = \lim_{x \to x_0} \frac{f(x) - f(x_0)}{x - x_0} \tag{6.1}$$

In the calculus of finite differences, the value of $(x - x_0)$ does not approach zero but remains a finite quantity. If we represent this quantity by h:

$$h = x - x_0 \tag{6.2}$$

then the derivative may be approximated by:

$$f'(x_0) \cong \frac{f(x_0+h)-f(x_0)}{h} \tag{6.3}$$

Under certain circumstances, there is a point, ξ, in the interval (a, b) for which the derivative can be calculated exactly from Eq. (6.3). This is given by the mean-value theorem of differential calculus:

Mean-value theorem: Let $f(x)$ be continuous in the range $a \le x \le b$ and differentiable in the range $a < x < b$; then there exists at least one ξ, $a < \xi < b$, for which:

$$f'(\xi) = \frac{f(b)-f(a)}{b-a} \tag{6.4}$$

This theorem forms the basis for both the differential calculus and the finite difference calculus.

A function $f(x)$, which is continuous and differentiable in the interval $[x_0, x]$, can be represented by a Taylor formula (see Chapter 3):

$$\begin{aligned} f(x) = f(x_0) + (x-x_0)f'(x_0) + \frac{(x-x_0)^2 f''(x_0)}{2!} + \frac{(x-x_0)^3 f'''(x_0)}{3!} \\ + \ldots + \frac{(x-x_0)^n f^{(n)}(x_0)}{n!} + R_n(x) \end{aligned} \tag{6.5}$$

where $R_n(x)$ is called the *remainder*. This term lumps together the remaining terms in the infinite series from $(n + 1)$ to infinity; it represents the *truncation* error, when the function is evaluated using the terms up to and including the nth-order term of the infinite series.

The mean-value theorem can be used to show that there exists a point ξ in the interval (x_0, x) so that the remainder term is given by:

$$R_n(x) = \frac{(x-x_0)^{n+1} f^{(n+1)}(\xi)}{(n+1)!} \tag{6.6}$$

The value of ξ is an unknown function of x; it is impossible to evaluate the remainder (truncation error term) exactly. However, the remainder,., a term of order $(n + 1)$, (because it is a function of $(x - x_0)^{n+1}$ and of the $(n + 1)^{th}$ derivative), can provide an upper bound on the error. For this reason, in our discussion of truncation errors we will always specify the order of the remainder term and will usually abbreviate it using the notation $O(h^{n+1})$, where h has been defined by Eq. (6.2).

The calculus of finite differences is used in conjunction with a series of discrete values, which can be either experimental data, such as:

$$y_{i-3} \qquad y_{i-2} \qquad y_{i-1} \qquad y_i \qquad y_{i+1} \qquad y_{i+2} \qquad y_{i+3}$$

or discrete values of a continuous function $y(x)$:

$$y(x-3h) \quad y(x-2h) \quad y(x-h) \quad y(x) \quad y(x+h) \quad y(x+2h) \quad y(x+3h)$$

or, equivalently, values of a function $f(x)$:

$$f(x-3h) \quad f(x-2h) \quad f(x-h) \quad f(x) \quad f(x+h) \quad f(x+2h) \quad f(x+3h)$$

In all the above cases, the values of the dependent variable y or f are those corresponding to *equally spaced* values of the independent variable x. This concept is demonstrated in Fig. 6.1 for a smooth function $y(x)$.

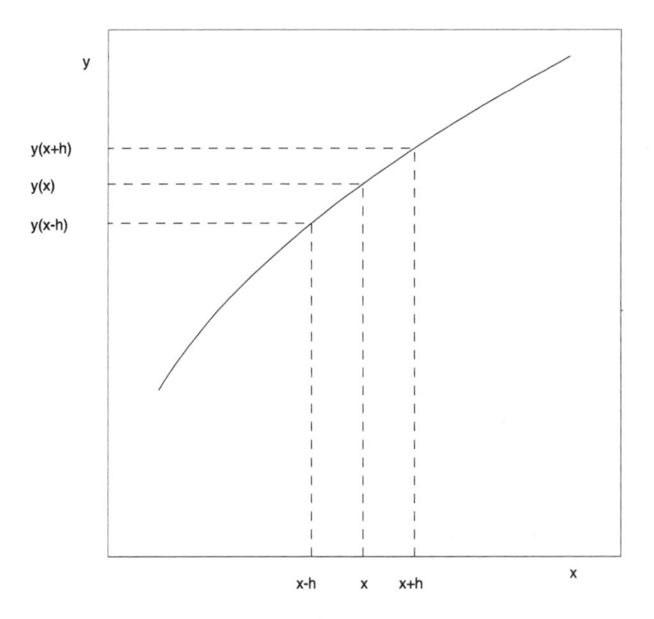

Figure 6.1 Values of function *y*(*x*) at equally spaced points of the independent variable *x*.

A set of *linear symbolic operators* drawn from differential calculus and from finite difference calculus will be defined in conjunction with the above series of discrete values. These definitions will then be used to derive the interrelationships between the operators. The linear symbolic operators are:

D = differential operator
∇ = backward difference operator
Δ = forward difference operator
δ = central difference operator

All these operators may be treated as algebraic variables, because they satisfy the distributive, commutative, and associative laws of algebra.

The first operator is well known from differential calculus. The *differential operator D*, when applied to the function $y(x)$, is

$$Dy(x) = \frac{dy(x)}{dx} = y'(x) \tag{6.7}$$

With these introductory concepts in mind, let us proceed to develop the backward, forward, and central difference operators.

6.3 Backward Finite Differences

Consider the set of values of y that are equally spaced in the x direction:

$$y_{i-3} \quad y_{i-2} \quad y_{i-1} \quad y_i \quad y_{i+1} \quad y_{i+2} \quad y_{i+3}$$

or the equivalent set:

$$y(x-3h) \quad y(x-2h) \quad y(x-h) \quad y(x) \quad y(x+h) \quad y(x+2h) \quad y(x+3h)$$

The *first backward difference* of y at i (or x) is defined as:

$$\nabla y_i = y_i - y_{i-1}$$
$$\text{or} \tag{6.8}$$
$$\nabla y(x) = y(x) - y(x-h)$$

The *second backward difference* of y at i is defined as:

$$\begin{aligned}\nabla^2 y_i &= \nabla(\nabla y_i) = \nabla(y_i - y_{i-1}) \\ &= \nabla y_i - \nabla y_{i-1} \\ &= (y_i - y_{i-1}) - (y_{i-1} - y_{i-2}) \\ &= y_i - 2y_{i-1} + y_{i-2}\end{aligned} \tag{6.9}$$

or equivalently for $y(x)$:

$$\nabla^2 y(x) = y(x) - 2y(x-h) + y(x-2h) \tag{6.9a}$$

The *third backward difference* of y at i is defined as:

$$\begin{aligned}\nabla^3 y_i &= \nabla\left(\nabla^2 y_i\right) = \nabla\left(y_i - 2y_{i-1} + y_{i-2}\right) \\ &= \nabla y_i - 2\nabla y_{i-1} + \nabla y_{i-2} \\ &= \left(y_i - y_{i-1}\right) - 2\left(y_{i-1} - y_{i-2}\right) + \left(y_{i-2} - y_{i-3}\right) \\ &= y_i - 3y_{i-1} + 3y_{i-2} - y_{i-3}\end{aligned} \tag{6.10}$$

Higher-order backward differences are derived similarly:

$$\nabla^4 y_i = y_i - 4y_{i-1} + 6y_{i-2} - 4y_{i-3} + y_{i-4} \tag{6.11}$$

$$\nabla^5 y_i = y_i - 5y_{i-1} + 10y_{i-2} - 10y_{i-3} + 5y_{i-4} - y_{i-5} \tag{6.12}$$

The coefficients of the terms in each of the above finite differences correspond to those of the binomial expansion $(a - b)^n$, where n is the order of the finite difference. Therefore, the general formula of the nth-order backward finite difference can be expressed as:

$$\nabla^n y_i = \sum_{m=0}^{n} (-1)^m \frac{n!}{(n-m)!m!} y_{i-m} \tag{6.13}$$

It should also be noted that the sum of the coefficients of the binomial expansion is always equal to zero. This can be used as a check to ensure that higher-order differences have been expanded correctly. It may be easier for the student to remember that the coefficients of the terms of the nth backward difference can be obtained from the nth row of Pascal's triangle, shown in Fig. 6.2.

Pascal's triangle	Backward finite difference
1 1	$\nabla y_i = y_i - y_{i-1}$
1 2 1	$\nabla^2 y_i = y_i - 2y_{i-1} + y_{i-2}$
1 3 3 1	$\nabla^3 y_i = y_i - 3y_{i-1} + 3y_{i-2} - y_{i-3}$
1 4 6 4 1	$\nabla^4 y_i = y_i - 4y_{i-1} + 6y_{i-2} - 4y_{i-3} + y_{i-4}$
1 5 10 10 5 1	$\nabla^5 y_i = y_i - 5y_{i-1} + 10y_{i-2} - 10y_{i-3} + 5y_{i-4} - y_{i-5}$
1 6 15 20 15 6 1	$\nabla^6 y_i = y_i - 6y_{i-1} + 15y_{i-2} - 20y_{i-3} + 15y_{i-4} - 6y_{i-5} + y_{i-6}$

Figure 6.2 Pascal's triangle and the corresponding backward finite differences.

The first and last elements in each row are unity, and the middle terms are obtained by adding the two adjacent terms in the row above it. The signs of the terms alternate between positive and negative.

Example 6.1 Express the first-order derivative in terms of backward finite differences with error of order *h*.

Solution

The general approach to expressing derivatives in terms of finite differences is to express function values in terms of Taylor expansions as presented in Chapter 3. We begin with the Taylor formula for y_{i-1} about y_i :

$$y_{i-1} = y_i + (x_{i-1} - x_i)y_i' + \frac{(x_{i-1} - x_i)^2}{2} y_i'' + \cdots + \frac{(x_{i-1} - x_i)^n}{n!} y_i^n + R^{n+1}$$

truncating after the squared term,

$$y_{i-1} = y_i + (x_{i-1} - x_i)y_i' + \frac{(x_{i-1} - x_i)^2}{2} y_i'' + R^3$$

substituting $(-h)$ for $(x_{i-1}-x_i)$ and rearranging

$$y_i' = \frac{y_i - y_{i-1}}{h} + \frac{h}{2} y_i'' + R^3$$

But the error is dominated by the term in y_i'', so the remainder is subsumed and

$$\frac{dy_i}{dx} = y_i' = \frac{1}{h}(y_i - y_{i-1}) + O(h)$$

The above equation, therefore, enables us to evaluate the first-order derivative of y at position i in terms of backward finite differences.

The term $O(h)$ is used to represent the order of the first term in the truncated portion of the series. When $h < 1.0$ and the function is smooth and continuous, the first term in the truncated portion of the series is the predominant term. It should be emphasized that for $h < 1.0$:

$$h > h^2 > h^3 > h^4 > \ldots > h^n$$

Therefore, formulas with higher-order error terms, $O(h^n)$, have smaller truncation errors, that is, they are more accurate approximations of derivatives.

On the other hand, when $h > 1.0$:

$$h < h^2 < h^3 < h^4 < \ldots < h^n$$

Therefore, formulas with higher-order error term have larger truncation errors and are less accurate approximations of derivatives.

It is obvious then, that the choice of step size h is very important in determining the accuracy and stability of numerical integration and differentiation. This concept will be discussed in more detail in Chapters 7 and 8.

Example 6.2 Express the first-order derivative in terms of backward finite differences with error of order h^2.

Solution

Start with the Taylor formula for y_{i-1} and y_{i-2} about y_i:

$$y_{i-1} = y_i + (x_{i-1} - x_i)y_i' + \frac{(x_{i-1} - x_i)^2}{2} y_i'' + \cdots + \frac{(x_{i-1} - x_i)^n}{n!} y_i^n + R^{n+1}$$

$$y_{i-2} = y_i + (x_{i-2} - x_i)y_i' + \frac{(x_{i-2} - x_i)^2}{2} y_i'' + \cdots + \frac{(x_{i-2} - x_i)^n}{n!} y_i^n + R^{n+1}$$

Substituting ($-h$) for (x_{i-1}-x_i) and ($-2h$) for (x_{i-2}-x_i) and truncating after the cubic term:

$$y_{i-1} = y_i - hy_i' + \frac{h^2}{2} y_i'' - \frac{h^3}{6} y_i''' + R^4$$

$$y_{i-2} = y_i - 2hy_i' + \frac{(2h)^2}{2} y_i'' - \frac{(2h)^3}{6} y_i''' + R^4$$

In order that the error be $O(h^2)$ and not $O(h)$, the term in y_i'' must cancel from the pair of equations. Multiplying the first equation by 4 and subtracting the second:

$$4y_{i-1} - y_{i-2} = 3y_i - 2hy_i' + O(h^3)$$

After rearranging:

$$\frac{dy_i}{dx} = y_i' = \frac{1}{2h}(3y_i - 4y_{i-1} + y_{i-2}) + O(h^2)$$

The above equation evaluates the first-order derivative of y at position i, in terms of backward finite differences, with error of order h^2.

The formulas for the first-order derivatives developed in the preceding two examples, together with those of the second-order, third-order, and fourth-order derivatives, are summarized in Table 6.1.

It can be concluded from these examples that any derivative can be expressed in terms of finite differences with any degree of accuracy desired. Higher accuracy, however, comes with the added cost of having to calculate a larger number of terms.

These formulas may be used to differentiate the function $y(x)$ given a set of values of this function at equally spaced intervals of x, such as a set of experiment data. Conversely, these same formulas will be used in the numerical integration of differential equations, as will be shown in Chapters 7 and 8.

Table 6.1 Derivatives in terms of backward finite differences

Error of order *h*

$$\frac{dy_i}{dx} = \frac{1}{h}\left(y_i - y_{i-1}\right) + O(h) \tag{6.14}$$

$$\frac{d^2y_i}{dx^2} = \frac{1}{h^2}\left(y_i - 2y_{i-1} + y_{i-2}\right) + O(h) \tag{6.15}$$

$$\frac{d^3y_i}{dx^3} = \frac{1}{h^3}\left(y_i - 3y_{i-1} + 3y_{i-2} - y_{i-3}\right) + O(h) \tag{6.16}$$

$$\frac{d^4y_i}{dx^4} = \frac{1}{h^4}\left(y_i - 4y_{i-1} + 6y_{i-2} - 4y_{i-3} + y_{i-4}\right) + O(h) \tag{6.17}$$

Error of order h^2

$$\frac{dy_i}{dx} = \frac{1}{2h}\left(3y_i - 4y_{i-1} + y_{i-2}\right) + O\left(h^2\right) \tag{6.18}$$

$$\frac{d^2y_i}{dx^2} = \frac{1}{h^2}\left(2y_i - 5y_{i-1} + 4y_{i-2} - y_{i-3}\right) + O\left(h^2\right) \tag{6.19}$$

$$\frac{d^3y_i}{dx^3} = \frac{1}{2h^3}\left(5y_i - 18y_{i-1} + 24y_{i-2} - 14y_{i-3} + 3y_{i-4}\right) + O\left(h^2\right) \tag{6.20}$$

$$\frac{d^4y_i}{dx^4} = \frac{1}{h^4}\left(3y_i - 14y_{i-1} + 26y_{i-2} - 24y_{i-3} + 11y_{i-4} - 2y_{i-5}\right) + O\left(h^2\right) \tag{6.21}$$

6.4 Forward Finite Differences

The derivation of forward finite differences follows a course parallel to that used in the development of backward differences. Only a summary of the forward differences relationships is given here. For the complete derivation of these results, the interested student is referred to Constantinides and Mostoufi (1999). The first few *forward differences* of y at i are defined below as:

First: $$\Delta y_i = y_{i+1} - y_i \tag{6.22}$$

Second: $$\Delta^2 y_i = y_{i+2} - 2y_{i+1} + y_i \tag{6.23}$$

Third: $$\Delta^3 y_i = y_{i+3} - 3y_{i+2} + 3y_{i+1} - y_i \tag{6.24}$$

Fourth: $$\Delta^4 y_i = y_{i+4} - 4y_{i+3} + 6y_{i+2} - 4y_{i+1} + y_i \tag{6.25}$$

Fifth: $$\Delta^5 y_i = y_{i+5} - 5y_{i+4} + 10y_{i+3} - 10y_{i+2} + 5y_{i+1} - y_i \tag{6.26}$$

In similarity to the backward finite differences, the forward finite differences also have coefficients that correspond to those of the binomial expansion $(a - b)^n$. The coefficients of the terms of the nth forward difference can be obtained easily from the nth row of Pascal's triangle, shown in Fig. 6.2.

In MATLAB, the function `diff(y)` returns forward finite differences of y. Values of nth-order forward finite difference may be obtained from `diff(y,n)`. For example:

```
>> y = [1 3 8 9 13 15]
y =
     1     3     8     9    13    15
>> diff(y)
ans =
     2     5     1     4     2
>> diff(y,2)
ans =
     3    -4     3    -2
>> diff(y,3)
ans =
    -7     7    -5
>> diff(y,4)
ans =
    14   -12
>> diff(y,5)
ans =
   -26
```

Example 6.3 Express the first-order derivative in terms of forward finite differences with error of order *h*.

Solution

Starting with the Taylor formula for y_{i+1} about y_i and truncating after the squared term:

$$y_{i+1} = y_i + (x_{i+1} - x_i)y'_i + \frac{(x_{i+1} - x_i)^2}{2} y''_i + R^3$$

Substituting h for $(x_{i+1} - x_i)$:

$$y_{i+1} = y_i + hy'_i + \frac{h^2}{2} y''_i + R^3$$

Rearranging:

$$y'_i = \frac{y_{i+1} - y_i}{h} + \frac{h}{2} y''_i + R^3$$

But the error is dominated by the term in y''_i,:

$$\frac{dy_i}{dx} = \frac{1}{h}(y_{i+1} - y_i) + O(h)$$

The above equation evaluates the first-order derivative of y at position i in terms of forward finite differences with error of order h.

Example 6.4 Express the second-order derivative in terms of forward finite differences with error of order *h*.

Solution

In order to develop a finite difference formula for a second derivative, the first derivative term cannot exist. This implies that we will need two equations, so that we can force the two first derivative terms to cancel.

In a fashion similar to Example 6.3:

$$y_{i+1} = y_i + (x_{i+1} - x_i)y_i' + \frac{(x_{i+1} - x_i)^2}{2} y_i'' + R^3$$

$$y_{i+2} = y_i + (x_{i+2} - x_i)y_i' + \frac{(x_{i+2} - x_i)^2}{2} y_i'' + R^3$$

Substituting h for $(x_{i+1}-x_i)$:

$$y_{i+1} = y_i + hy_i' + \frac{h^2}{2} y_i'' + R^3$$

$$y_{i+2} = y_i + 2hy_i' + \frac{4h^2}{2} y_i'' + R^3$$

Multiplying the first equation by 2 and subtracting it from the second:

$$y_{i+2} - 2y_{i+1} = -y_i + h^2 y_i'' + R^3$$

After rearranging:

$$\frac{d^2 y_i}{dx^2} = y_i'' = \frac{1}{h^2}(y_{i+2} - 2y_{i+1} + y_i) + O(h)$$

The above equation evaluates the second-order derivative of y at position i, in terms of forward finite differences, with error of order h.

The formulas developed in the preceding two examples for the first-order and second-order derivatives are summarized in Table 6.2, together with those of the third-order and fourth-order derivatives, including the corresponding set of equations with error of order h^2.

It should be pointed out that all the finite difference approximations of derivatives obtained in this section and the previous section have coefficients that add up to zero. This is a general rule of thumb that applies to all such combinations of finite differences.

By comparing Table 6.1 and Table 6.2, we conclude that derivatives can be expressed in their backward or forward finite differences, with formulas that are very similar to each other in the number of terms involved and in the order of truncation error. The choice between using forward or backward differences will depend on the geometry of the problem and its boundary conditions. This will be discussed further in Chapters 7 and 8.

Table 6.2 Derivatives in terms of forward finite differences

Error of order *h*

$$\frac{dy_i}{dx} = \frac{1}{h}(y_{i+1} - y_i) + O(h) \tag{6.27}$$

$$\frac{d^2 y_i}{dx^2} = \frac{1}{h^2}(y_{i+2} - 2y_{i+1} + y_i) + O(h) \tag{6.28}$$

$$\frac{d^3 y_i}{dx^3} = \frac{1}{h^3}(y_{i+3} - 3y_{i+2} + 3y_{i+1} - y_i) + O(h) \tag{6.29}$$

$$\frac{d^4 y_i}{dx^4} = \frac{1}{h^4}(y_{i+4} - 4y_{i+3} + 6y_{i+2} - 4y_{i+1} + y_i) + O(h) \tag{6.30}$$

Error of order h^2

$$\frac{dy_i}{dx} = \frac{1}{2h}(-y_{i+2} + 4y_{i+1} - 3y_i) + O(h^2) \tag{6.31}$$

$$\frac{d^2 y_i}{dx^2} = \frac{1}{h^2}(-y_{i+3} + 4y_{i+2} - 5y_{i+1} + 2y_i) + O(h^2) \tag{6.32}$$

$$\frac{d^3 y_i}{dx^3} = \frac{1}{2h^3}(-3y_{i+4} + 14y_{i+3} - 24y_{i+2} + 18y_{i+1} - 5y_i) + O(h^2) \tag{6.33}$$

$$\frac{d^4 y_i}{dx^4} = \frac{1}{h^4}(-2y_{i+5} + 11y_{i+4} - 24y_{i+3} + 26y_{i+2} - 14y_{i+1} + 3y_i) + O(h^2) \tag{6.34}$$

6.5 Central Finite Differences

As the name implies, central finite differences are *centered* at the pivot position and are evaluated using the values of the function to the right and to the left of the pivot position, but located only $h/2$ distance from it.

Consider the series of values used in the previous two sections, but with the additional value at the midpoints of the intervals:

$$y_{i-2} \quad y_{i-1½} \quad y_{i-1} \quad y_{i-½} \quad y_i \quad y_{i+½} \quad y_{i+1} \quad y_{i+1½} \quad y_{i+2}$$

Several *central difference* of y at i are listed below:

First: $$\delta y_i = y_{i+½} - y_{i-½} \tag{6.35}$$

Second: $$\delta^2 y_i = y_{i+1} - 2y_i + y_{i-1} \tag{6.36}$$

Third: $$\delta^3 y_i = y_{i+1½} - 3y_{i+½} + 3y_{i-½} - y_{i-1½} \tag{6.37}$$

Fourth: $$\delta^4 y_i = y_{i+2} - 4y_{i+1} + 6y_i - 4y_{i-1} + y_{i-2} \tag{6.38}$$

Fifth: $$\delta^5 y_i = y_{i+2½} - 5y_{i+1½} + 10y_{i+½} - 10y_{i-½} + 5y_{i-1½} - y_{i-2½} \tag{6.39}$$

In a manner similar with the other finite differences, the central finite differences also have coefficients that correspond to those of the binomial expansion $(a - b)^n$. The coefficients of the terms of the nth central difference can be obtained from the nth row of Pascal's triangle, shown on Fig. 6.2.

It should be noted that the *odd*-order central differences involve values of the function at the midpoint of the intervals, whereas the *even*-order central differences involve values at the full intervals only. To fully use odd- and even-order central differences, we need a set of values of the function y that includes twice as many points as that used in either backward or forward differences. This situation is rather uneconomical, especially in the case where these values must be obtained experimentally. This problem is alleviated by the use of the averager operator (see Constantinides and Mostoufi, 1999).

Example 6.5 Express the first-order derivative in terms of central finite differences with error of order h^2.

Solution

Starting with the Taylor formulas for y_{i+1} and y_{i-1}:

$$y_{i+1} = y_i + (x_{i+1} - x_i)y' + \frac{(x_{i+1} - x_i)^2}{2} y_i'' + \frac{(x_{i+1} - x_i)^3}{6} y''' + \cdots + \frac{(x_{i+1} - x_i)^n}{n!} y_i^n + R^{n+1}$$

$$y_{i-1} = y_i + (x_{i-1} - x_i)y_i' + \frac{(x_{i-1} - x_i)^2}{2} y_i'' + \frac{(x_{i-1} - x_i)^3}{6} y_i''' + \cdots + \frac{(x_{i-1} - x_i)^n}{n!} y_i^n + R^{n+1}$$

Subtracting the second formula from the first, taking into account that $h = (x_{i+1}\text{-}x_i)$ and $h = \text{-}(x_{i\text{-}1}\text{-}x_i)$, and truncating:

$$y_{i+1} - y_{i-1} = 2hy_i' + \frac{h^3}{3} y_i''' + R^5$$

The terms in even powers cancel out and only the odd powers remain (because of the change in sign). The term $\left(\frac{h^3}{3} y_i'''\right)$ dominates the remainder term, and the equation can be solved for y':

$$\frac{dy_i}{dx} = y_i' = \frac{1}{2h}\left(y_{i+1} - y_{i-1}\right) + O\left(h^2\right)$$

The above equation evaluates the first-order derivative of y at position i in terms of central finite differences. Comparing this equation with Eqs. (6.14) and (6.27) reveals that use of central differences increases the accuracy of the formulas for the same number of terms retained.

Example 6.6 Express the second-order derivative in terms of central finite differences with error of order h^2.

Solution

Starting with the Taylor formulas for y_{i+1} and y_{i-1}:

$$y_{i+1} = y_i + (x_{i+1} - x_i)y' + \frac{(x_{i+1} - x_i)^2}{2} y_i'' + \frac{(x_{i+1} - x_i)^3}{6} y''' + \cdots + \frac{(x_{i+1} - x_i)^n}{n!} y_i^n + R^{n+1}$$

$$y_{i-1} = y_i + (x_{i-1} - x_i)y_i' + \frac{(x_{i-1} - x_i)^2}{2} y_i'' + \frac{(x_{i-1} - x_i)^3}{6} y_i''' + \cdots + \frac{(x_{i-1} - x_i)^n}{n!} y_i^n + R^{n+1}$$

Add the two formulas together, taking into account that $h = (x_{i+1}-x_i)$ and $h = -(x_{i-1}-x_i)$, and truncating:

$$y_{i+1} + y_{i-1} = 2y_i + h^2 y_i'' + \frac{h^4}{12} y_i^{iv} + R^6$$

The terms in odd powers cancel out and only the even powers remain. Since $\left(\frac{h^4}{12} y_i^{iv}\right)$ dominates the remainder term, , and the equation can be solved for y':

$$\frac{d^2 y_i}{dx^2} = y_i'' = \frac{1}{h^2}(y_{i+1} - 2y_i + y_{i-1}) + O(h^2)$$

The above equation evaluates the second-order derivative of y at position i in terms of central finite differences, with error of order h^2. Comparing this equation with Eqs. (6.15) and (6.28) reveals that use of central differences increases the accuracy of the formulas for the same number of terms retained.

The formulas derived in Examples 6.5 and 6.6 for the first- and second-order derivatives are summarized in Table 6.3, along with those for the third- and fourth-order derivatives.

6.6 Interpolating Polynomials

Engineers and scientists often face the task of interpreting and correlating experimental observations, which are usually in the form of discrete data, and are called upon to either integrate or differentiate these data numerically or graphically. This task is facilitated by the use of interpolation/extrapolation formulas. The calculus of finite differences enables us to develop *interpolating polynomials* that can represent experimental data when the actual functionality of these data is not well known. But even more significantly, these polynomials can be used to approximate functions that are difficult to integrate or differentiate, thus making the task of numerical integration and differentiation somewhat easier to perform.

Table 6.3 Derivatives in terms of central finite differences

Error of order h^2

$$\frac{dy_i}{dx} = \frac{1}{2h}\left(y_{i+1} - y_{i-1}\right) + O\left(h^2\right) \quad (6.40)$$

$$\frac{d^2 y_i}{dx^2} = \frac{1}{h^2}\left(y_{i+1} - 2y_i + y_{i-1}\right) + O\left(h^2\right) \quad (6.41)$$

$$\frac{d^3 y_i}{dx^3} = \frac{1}{2h^3}\left(y_{i+2} - 2y_{i+1} + 2y_{i-1} - y_{i-2}\right) + O\left(h^2\right) \quad (6.42)$$

$$\frac{d^4 y_i}{dx^4} = \frac{1}{h^4}\left(y_{i+2} - 4y_{i+1} + 6y_i - 4y_{i-1} + y_{i-2}\right) + O\left(h^2\right) \quad (6.43)$$

Error of order h^4

$$\frac{dy_i}{dx} = \frac{1}{12h}\left(-y_{i+2} + 8y_{i+1} - 8y_{i-1} + y_{i-2}\right) + O\left(h^4\right) \quad (6.44)$$

$$\frac{d^2 y_i}{dx^2} = \frac{1}{12h^2}\left(-y_{i+2} + 16y_{i+1} - 30y_i + 16y_{i-1} - y_{i-2}\right) + O\left(h^4\right) \quad (6.45)$$

$$\frac{d^3 y_i}{dx^3} = \frac{1}{8h^3}\left(-y_{i+3} + 8y_{i+2} - 13y_{i+1} + 13y_{i-1} - 8y_{i-2} + y_{i-3}\right) + O\left(h^4\right) \quad (6.46)$$

$$\frac{d^4 y_i}{dx^4} = \frac{1}{6h^4}\left(-y_{i+3} + 12y_{i+2} - 39y_{i+1} + 56y_i - 39y_{i-1} + 12y_{i-2} - y_{i-3}\right) + O\left(h^4\right) \quad (6.47)$$

Let us assume that values of the function $f(x)$ are known at a set of $(n + 1)$ values of the independent variable x:

$$
\begin{array}{ll}
x_0 & f(x_0) \\
x_1 & f(x_1) \\
x_2 & f(x_2) \\
x_3 & f(x_3) \\
\cdot & \cdot \\
\cdot & \cdot \\
\cdot & \cdot \\
x_n & f(x_n)
\end{array}
$$

These values are called the *base points* of the function. They are shown graphically in Fig. 6.3a.

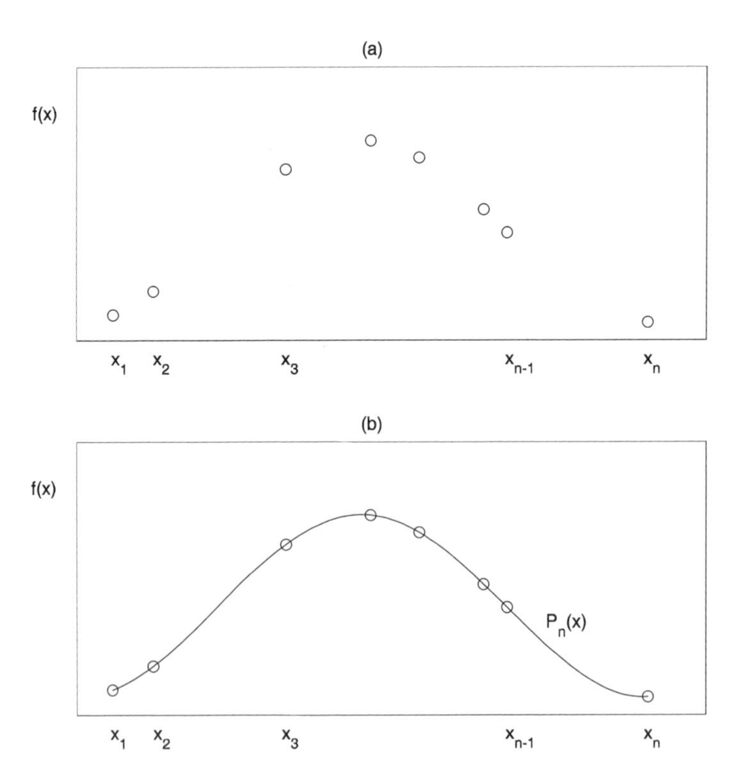

Figure 6.3 **a. Unequally spaced base points of the function *f*(x).**
b. Unequally spaced base points with interpolating polynomial.

The general objective in developing interpolating polynomials is to choose a polynomial of the form:

$$P_n(x) = a_0 + a_1 x + a_2 x^2 + a_3 x^3 + \ldots + a_n x^n \tag{6.48}$$

so that this equation fits exactly the base points of the function and connects these points with a smooth curve, as shown in Fig. 6.3b. This polynomial can then be used to approximate the function at any value of the independent variable x between the base points.

For the given set of $(n + 1)$ known base points, the polynomial must satisfy the equation:

$$P_n(x_i) = f(x_i) \qquad i = 0, 1, 2, \ldots, n \tag{6.49}$$

Substitution of the known values of $(x_i, f(x_i))$ in Eq. (6.48) yields a set of $(n + 1)$ simultaneous linear algebraic equations whose unknowns are the coefficients $a_0, \ldots, a_n$ of the polynomial equation. The solution of this set of linear algebraic equations is shown in more detail in Chapter 9 and is obtained using one of the algorithms discussed in Chapter 4. However, this solution results in an ill-conditioned linear system; therefore, other methods have been favored in the development of interpolating polynomials.

MATLAB has several functions for interpolation. The function `interp1`, when used in the form `yi = interp1(x,y,xi)`, takes the values of the independent variable `x` and the dependent variable `y` (base points) and does the one-dimensional interpolation based on `xi` to find `yi`. The default method of interpolation is linear. However, the user can choose the method of interpolation in the fourth input argument from `'nearest'` (nearest neighbor interpolation), `'linear'` (linear interpolation), `'spline'` (cubic spline interpolation), and `'cubic'` (cubic interpolation). If the vector of independent variable is not equally spaced, the function `interp1q` may be used instead. It is faster than `interp1` because it does not check the input arguments. MATLAB also has the function `spline` to perform one-dimensional interpolation by cubic splines, using the *not-a-knot* method. It can also return coefficients of piecewise polynomials, if required. The functions `interp2`, `interp3`, and `interpn` perform two-, three-, and n-dimensional interpolation, respectively.

6.7 Interpolation of Equally Spaced Points

In this section, we will develop the *Gregory-Newton formulas*, which are based on forward and backward differences for interpolating known or sampled data.

6.7.1 Gregory-Newton interpolation

First, we consider a set of known values of the function $f(x)$ at *equally spaced* values of x:

$$
\begin{array}{ll}
x - 3h & f(x - 3h) \\
x - 2h & f(x - 2h) \\
x - h & f(x - h) \\
x & f(x) \\
x + h & f(x + h) \\
x + 2h & f(x + 2h) \\
x + 3h & f(x + 3h)
\end{array}
$$

These points are represented graphically in Fig. 6.4 and are tabulated in Table 6.4 and Table 6.5. The first, second, and third backward differences of these base points are also tabulated in Table 6.4, and the corresponding forward differences in Table 6.5.

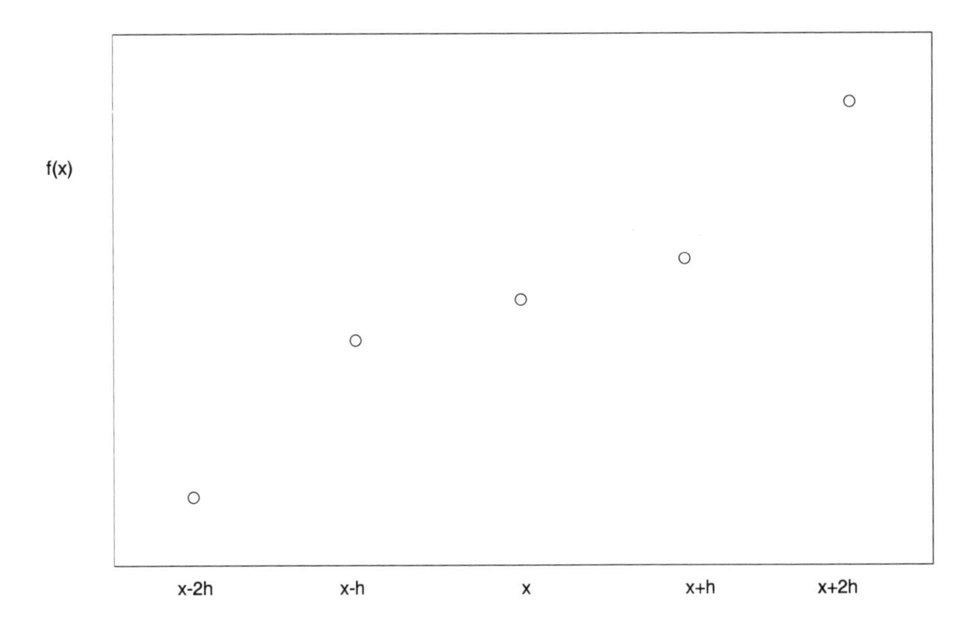

Figure 6.4 Equally spaced base points for interpolating polynomials.

Table 6.4 Backward difference table

i	x_i	$f(x_i)$	$\nabla f(x_i)$	$\nabla^2 f(x_i)$	$\nabla^3 f(x_i)$
-3	$x-3h$	$f(x-3h)$			
-2	$x-2h$	$f(x-2h)$	$f(x-2h)-f(x-3h)$		
-1	$x-h$	$f(x-h)$	$f(x-h)-f(x-2h)$	$f(x-h)-2f(x-2h)+f(x-3h)$	
0	x	$f(x)$	$f(x)-f(x-h)$	$f(x)-2f(x-h)+f(x-2h)$	$f(x)-3f(x-h)+3f(x-2h)-f(x-3h)$

Table 6.5 Forward difference table

i	x_i	$f(x_i)$	$\Delta f(x_i)$	$\Delta^2 f(x_i)$	$\Delta^3 f(x_i)$
0	x	$f(x)$	$f(x+h)-f(x)$	$f(x+2h)-2f(x+h)+f(x)$	$f(x+3h)-3f(x+2h)+3f(x+h)-f(x)$
1	$x+h$	$f(x+h)$	$f(x+2h)-f(x+h)$	$f(x+3h)-2f(x+2h)+f(x+h)$	
2	$x+2h$	$f(x+2h)$	$f(x+3h)-f(x+2h)$		
3	$x+3h$	$f(x+3h)$			

The *Gregory-Newton forward interpolation formula* can be derived using the definitions of the symbolic operators, given in Section 6.2, and the forward finite difference relations, given in Section 6.4. By definition,

$$\Delta f(x) = f(x+h) - f(x) \tag{6.50}$$

Therefore

$$f(x+h) = (1+\Delta) f(x) \tag{6.51}$$

Applying this to n steps, we obtain:

$$f(x+nh) = (1+\Delta)^n f(x) \tag{6.52}$$

The term $(1 + \Delta)^n$ can be expanded using the *binomial series*:

$$\begin{aligned}(1+\Delta)^n &= 1 + n\Delta + \frac{n(n-1)}{2!}\Delta^2 + \frac{n(n-1)(n-2)}{3!}\Delta^3 \\ &\quad + \frac{n(n-1)(n-2)(n-3)}{4!}\Delta^4 + \ldots\end{aligned} \tag{6.53}$$

Therefore, Eq. (6.52) becomes:

$$\begin{aligned}f(x+nh) &= f(x) + n\Delta f(x) + \frac{n(n-1)}{2!}\Delta^2 f(x) + \frac{n(n-1)(n-2)}{3!}\Delta^3 f(x) \\ &\quad + \frac{n(n-1)(n-2)(n-3)}{4!}\Delta^4 f(x) + \ldots\end{aligned} \tag{6.54}$$

When n is a positive integer, the binomial series has $(n + 1)$ terms; therefore, Eq. (6.54) is a polynomial of degree n. If $(n + 1)$ base-point values of the function f are known, this polynomial fits all $(n + 1)$ points exactly. Assume that these $(n + 1)$ base-points are $(x_0, f(x_0))$, $(x_1, f(x_1))$, . . . , $(x_n, f(x_n))$, where $(x_0, f(x_0))$ is the pivot point and x_i is defined as:

$$x_i = x_0 + ih \tag{6.55}$$

We can now designate the distance of the point of interest from the pivot point as $(x - x_0)$. The value of n is no longer an integer and is replaced by:

$$n = \frac{x - x_0}{h} \tag{6.56}$$

We write Eq. (6.54) at the origin (i.e., at $x = x_0$) and then substitute Eq. (6.56) to obtain:

$$\begin{aligned} f(x) = f(x_0) &+ \frac{(x-x_0)}{h}\Delta f(x_0) + \frac{(x-x_0)(x-x_1)}{2!h^2}\Delta^2 f(x_0) \\ &+ \frac{(x-x_0)(x-x_1)(x-x_2)}{3!h^3}\Delta^3 f(x_0) + \\ &+ \frac{(x-x_0)(x-x_1)(x-x_2)(x-x_3)}{4!h^4}\Delta^4 f(x_0) + \ldots \end{aligned} \tag{6.57}$$

This is the *Gregory-Newton forward interpolation formula.* The general formula of the above series is:

$$f(x) = f(x_0) + \sum_{k=1}^{n}\left(\prod_{m=0}^{k-1}(x-x_m)\right)\frac{\Delta^k f(x_0)}{k!h^k} \tag{6.58}$$

In a similar derivation, using backward differences, the *Gregory-Newton backward interpolation formula* is derived as:

$$\begin{aligned} f(x) = f(x_0) &+ \frac{(x-x_0)}{h}\nabla f(x_0) + \frac{(x-x_0)(x-x_{-1})}{2!h^2}\nabla^2 f(x_0) \\ &+ \frac{(x-x_0)(x-x_{-1})(x-x_{-2})}{3!h^3}\nabla^3 f(x_0) + \\ &+ \frac{(x-x_0)(x-x_{-1})(x-x_{-2})(x-x_{-3})}{4!h^4}\nabla^4 f(x_0) + \ldots \end{aligned} \tag{6.59}$$

The general formula of the above series is:

$$f(x) = f(x_0) + \sum_{k=1}^{n}\left(\prod_{m=0}^{k-1}(x-x_{-m})\right)\frac{\nabla^k f(x_0)}{k!h^k} \tag{6.60}$$

Recall that the binomial series (Eq. (6.53)) has a finite number of terms, $(n + 1)$, when n is a positive integer. However, in the Gregory-Newton interpolation formulas, n is not usually an integer; therefore, these polynomials have an infinite number of terms. It is known from algebra that if $|\Delta| \leq 1$, then the binomial series for $(1 + \Delta)^n$ converges to the value of $(1 + \Delta)^n$ as the number of terms becomes larger and larger. This implies that the finite differences must be small. This is true for a flat, smooth function, or, alternatively, if the known base points are close together; that is, if h is small. Of course, the number of terms that can be used in each formula depends on the highest order of finite differences that can be evaluated from the available known data. It is common sense that for evenly spaced data, the accuracy of interpolation is highest for a large number of data points that are closely spaced together.

For a given set of data points, the accuracy of interpolation can be further enhanced by choosing the pivot point as close to the point of interest as possible, so that $(x - x_0) < h$. If this is satisfied, then the interpolation formula should use as many terms as possible; that is, the number of finite differences in the equation should be maximized. The order of error of the formula applied in each case is equivalent to the order of the finite difference contained in the first truncated term of the series. Examination of Table 6.4 reveals that points at the bottom of the table have the largest number of backward differences, whereas Table 6.5 reveals that points at the top of the table have the largest available number of forward differences. Therefore, the forward formula should be used for interpolating between points near the top of the table, and the backward formula should be used for interpolation near the bottom of the table.

Example 6.7 Gregory-Newton method for interpolation of equally spaced data.

Statement of the problem

During the night and early morning, the body temperature of a hospital patient rose dramatically, for an as-yet-unknown reason, until the nurse detected this variation and administered medication. You need to interpret this change in temperature. As the first step, you must determine exactly when the patient's body reached its maximum temperature and what was the value of this temperature. The body temperature was recorded by a computer at one-hour intervals. These time-temperature data are listed in Table 6.6. Write a general MATLAB function for n^{th}-order one-dimensional interpolation by the Gregory-Newton forward interpolation formula to solve this problem.

Solution

The function uses the general formula of the Gregory-Newton forward interpolation (Eq. (6.58)) to perform the nth-order interpolation. The number of base points given to the function must be at least $(n + 1)$.

Program description

The MATLAB function `gregory_newton.m`, which was developed by Constantinides and Mostoufi (1999) to perform the Gregory-Newton forward interpolation, is used in this problem. The call to this function takes the general form `yi = gregory_newton(x,y,xi,n)`. The first and second input arguments are the coordinates of the base points. The third input argument is the vector of independent variables at which the interpolation of the dependent variable is required. The fourth input, n, is the order of interpolation. If no value is introduced to the function through the fourth argument, the function does linear interpolation. For performing second-order (or higher) interpolation, the value of n should be entered as the fourth input argument.

At the beginning, the function checks the inputs. The vectors of coordinates of base points have to be of the same size. The function also checks to see if the vector of independent variable is monotonic; otherwise, the function terminates calculations. The order of interpolation cannot be more than the intervals (number of base points minus one). In this case, the function displays a warning and continues with the maximum possible order of interpolation. The function then performs the interpolation according to Eq. (6.58).

The main program `example6_7.m` is written to solve the problem stated above. The program asks the user to input the vector of time (independent variable), vector of body temperature of the patient (dependent variable), and the order of interpolation. The program applies the function `gregory_newton.m` to interpolate the temperature between the recorded temperatures and find the maximum. The user can repeat the calculations with another order of interpolation.

Table 6.6 Patient's body temperature

Time (a.m.)	Temperature (°F)	Time (a.m.)	Temperature (°F)
1	98.9	7	104.0
2	99.5	8	104.1
3	99.9	9	102.5
4	101.3	10	101.2
5	101.6	11	100.5
6	102.5	12	100.2

Program

```
% example6_7.m - Interpolation of the time-temperature data
% given in Table 6.6 by Gregory-Newton forward interpolation
% formula to find the maximum temperature and the time this
```

```
% maximum occurred.

clc; clear all;

% Input data
time = input(' Vector of time = ');
temp = input(' Vector of temperature = ');
% Vector of time for interpolation
ti = linspace(min(time),max(time));
redo = 1;
while redo
   disp(' '); n = input(' Order of interpolation = ');
   te = gregory_newton(time,temp,ti,n);     % Interpolation
   [max_temp,k] = max(te);
   max_time = ti(k);
% Show the results
   fprintf('\n Maximum temperature of %4.1f F reached at
%4.2f.\n',max_temp,max_time)
% Show the results graphically
   figure(1); plot(time,temp,'o',ti,te)
   title('Patient''s Temperature Profile')
   xlabel('Time (a.m.)'); ylabel('Temperature (deg F)')
   axis([1 12 98 105])
   disp(' ')
   redo = input(' Repeat the calculation (1/0) : ');
end
```

Function that performs the Gregory-Newton interpolation

```
function yi = gregory_newton(x,y,xi,n)
%gregory_newton One dimensional interpolation.
%
%   YI = gregory_newton(X,Y,XI,N) applies the Nth-order
%   Gregory-Newton forward interpolation to find YI, the
%   values of the underlying function Y at the points in
%   the vector XI. The vector X specifies the points at
%   which the data Y is given.
%
%   YI = gregory_newton(X,Y,XI) is equivalent to the
%   linear interpolation.
%
%   See also INTERP1, NATURALSPLINE, Lagrange, SPLINE, INTERP1Q

% (c) by N. Mostoufi & A. Constantinides
% January 1, 1999

% Initialization
if nargin < 3
   error('Invalid number of inputs.')
end
% Check x for equal spacing and determining h
```

```
if min(diff(x)) ~= max(diff(x))
   error('Independent variable is not monotonic.')
else
   h = x(2) - x(1);
end

x = (x(:).')';     % Make sure it's a column vector
y = (y(:).')';     % Make sure it's a column vector
nx = length(x);
ny = length(y);
if nx ~= ny
   error('X and Y vectors are not the same size.');
end

% Check the order of interpolation
if nargin == 3 | n < 1
   n = 1;
end
n = floor(n);
if n >= nx
 fprintf('\nNot enough data points for %2d-order interpolation.',
n)
 fprintf('\n%2d-order interpolation will be performed instead.\n',
nx-1)
 n = nx - 1;
end

deltax(1,1:length(xi)) = ones(1,length(xi));
% Locating the required number of base points
for m = 1:length(xi)
   dx = xi(m) - x;
   % Locating xi
   [dxm , loc(m)] = min(abs(dx));
   % locating the first base point
   if dx(loc(m)) < 0
      loc(m) = loc(m) - 1;
   end
   if loc(m) < 1
      loc(m) = 1;
   end
   if loc(m)+n > nx
      loc(m) = nx - n;
   end
   deltax(2:n+1,m) = dx(loc(m):loc(m)+n-1);
   ytemp(1:n+1,m) = y(loc(m):loc(m)+n);
end

% Interpolation
yi = y(loc)';
for k = 1 : n
```

```
    yi = yi + prod(deltax(1:k+1,:)) .* diff(ytemp(1:k+1,:),k) /...
(gamma(k+1) * h^k);
end
```

Input and results

```
>>example6_7
 Vector of time = [1:12]
 Vector of temperature = [98.9, 99.5, 99.9, 101.3, 101.6, 102.5,
 104.0, 104.1, 102.5, 101.2, 100.5, 100.2]

 Order of interpolation = 2

 Maximum temperature of 104.3 F reached at 7.56.

 Repeat the calculation (1/0) : 0
```

Discussion of results

Graphical results are shown in Fig. 6.5. As can be seen from this plot and also from the numerical results, the patient's temperature reached the maximum of 104.3 °F at 7.56 hours (7:34 a.m.). The student can repeat the calculations using other values for order of interpolation to see the effect of fitting higher order polynomials.

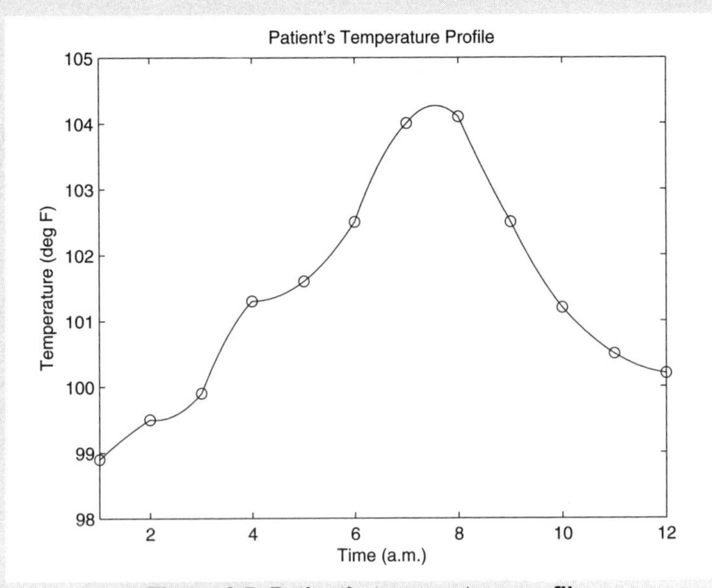

Figure 6.5 Patient's temperature profile.

6.8 Interpolation of Unequally Spaced Points

In this section, we will develop two interpolation methods for unequally spaced data: the *Lagrange polynomials* and *spline interpolation.*

6.8.1 Lagrange polynomials

Consider a set of unequally spaced base points, such as those shown in Fig. 6.3a. Define the polynomial:

$$P_n(x) = \sum_{k=0}^{n} p_k(x) f(x_k) \tag{6.61}$$

which is the sum of *weighted* values of the function at all $(n + 1)$ base points. The weights $p_k(x)$ are n^{th}-degree polynomial functions that correspond to each base point in a special way. Eq. (6.61) is actually a linear combination of nth-degree polynomials; therefore, $P_n(x)$ is also an nth-degree polynomial.

In order for the interpolating polynomial to fit the function exactly at all the base points, each particular weighting polynomial $p_k(x)$ must be chosen so that it has the value of unity when $x = x_k$, and the value of zero at all other base points, that is:

$$p_k(x_i) = \begin{cases} 0 & i \neq k \\ 1 & i = k \end{cases} \tag{6.62}$$

The *Lagrange polynomials*, which have the form:

$$p_k(x) = C_k \prod_{\substack{i=0 \\ i \neq k}}^{n} (x - x_i) \tag{6.63}$$

satisfy the first part of condition (6.62), because there will be a term $(x_i - x_i)$ in the product series of Eq. (6.63) whenever $x = x_i$. The constant C_k is evaluated to make the Lagrange polynomial satisfy the second part of condition (6.62):

$$C_k = \frac{1}{\prod_{\substack{i=0 \\ i \neq k}}^{n} (x_k - x_i)} \tag{6.64}$$

Combination of Eqs. (6.63) and (6.64) gives the Lagrange polynomials:

$$p_k(x) = \prod_{\substack{i=0 \\ i \neq k}}^{n} \left(\frac{x - x_i}{x_k - x_i} \right) \tag{6.65}$$

The interpolating polynomial $P_n(x)$ has a remainder term, which can be obtained from Eq. (6.6):

$$R_n(x) = \prod_{i=0}^{n}(x - x_i)\frac{f^{(n+1)}(\xi)}{(n+1)!} \qquad x_0 < \xi < x_n \tag{6.66}$$

6.8.2 Spline interpolation

When we deal with a large number of data points, high-degree interpolating polynomials are likely to fluctuate between base points instead of passing smoothly through them, as illustrated in Fig. 6.6a. Although the interpolating polynomial passes through all the base points, it is not able to predict the value of the function between these points satisfactorily. In order to avoid this undesirable behavior of the high-degree interpolating polynomial, a series of lower-degree polynomials may be used to connect a smaller number of base points. These sets of interpolating polynomials are called *spline functions*. Fig. 6.6b shows the result of such interpolation using third-degree (or cubic) splines for the same set of data as in Fig. 6.6a. Compared with the higher-order interpolation, third-degree splines provide a more acceptable approximation.

The most common spline used in engineering problems is the *cubic spline*. In this method, a cubic polynomial is used to approximate the curve between each two adjacent base points. Additional constraints are necessary to make the spline unique, because there would be an infinite number of third-degree polynomials passing through each pair of points. Therefore, it is set so all the polynomials should have equal first and second derivatives at the base points. These conditions imply that the slope and the curvature of the spline polynomials are continuous across the base points. The cubic spline of the interval $[x_{i-1}, x_i]$ has the following general form:

$$P_i(x) = a_i x^3 + b_i x^2 + c_i x + d_i \tag{6.67}$$

There are four unknown coefficients in Eq. (6.67) and n such polynomials for the whole range of data points $[x_0, x_n]$. Therefore, there are $4n$ unknown coefficients, and we need $4n$ equations to evaluate these coefficients. The required equations come from the following conditions:

a. Each spline passes through the base point at each end of its interval ($2n$ equations).
b. The first derivatives of the splines are continuous across the interior base points (n - 1 equations).
c. The second derivatives of the splines are continuous across the interior base points (n - 1 equations).

d. The second derivatives of the end splines are zero at the end base points (2 equations). This is called the *natural* condition. Another commonly used condition is to set the third derivative of the end splines equal to the third derivative of the neighboring splines. The latter is called *not-a-knot* condition.

Simultaneous solution of the above $4n$ linear algebraic equations results in the determination of all cubic interpolating polynomials. Chapter 9 gives further discussion and application of splines.

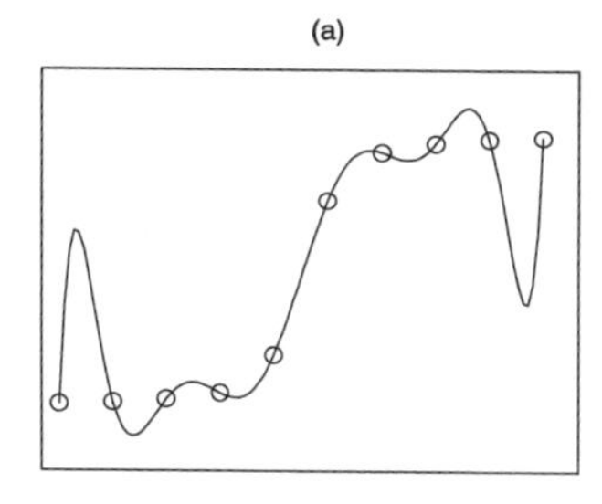

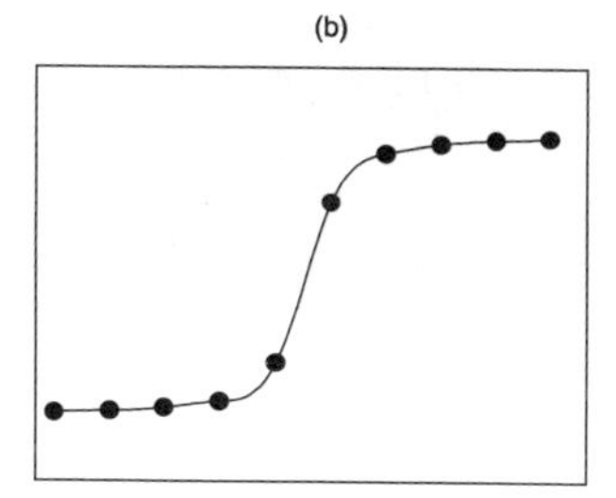

Figure 6.6 a. Fluctuation of high-degree interpolating polynomials between base points. b. Cubic spline interpolation.

6.9 Integration Formulas

In the following sections we develop the integration formulas. This operation is represented by:

$$I = \int_{x_0}^{x_n} f(x)\,dx \tag{6.68}$$

which is the integral of the function $y = f(x)$, or *integrand*, with respect to the independent variable, x, evaluated between the limits x_0 and x_n. If the function $f(x)$ is such that it can be integrated analytically, the numerical methods are not needed for this problem. However, in many cases, the function $f(x)$ is very complicated, or the function is only a set of tabulated values of x and y, such as experimental data. Under these circumstances, the integral in Eq. (6.68) must be calculated numerically. This operation is known as *numerical quadrature*.

It is known from differential calculus that the integral of a function $f(x)$ is equivalent to the area between the function and the x-axis enclosed within the limits of integration, as shown in Fig. 6.7a. Any portion of the area that is below the x-axis is counted as negative area (Fig. 6.7b). Therefore, one way of evaluating the integral:

$$\int_{x_0}^{x_n} y\,dx$$

is to plot the function graphically and then simply measure the area enclosed by the function. However, this is a very impractical and inaccurate way of evaluating integrals. A more accurate and systematic way of evaluating integrals is to perform the integration numerically. In the next section, we derive the Newton-Cotes integration formulas for equally spaced points.

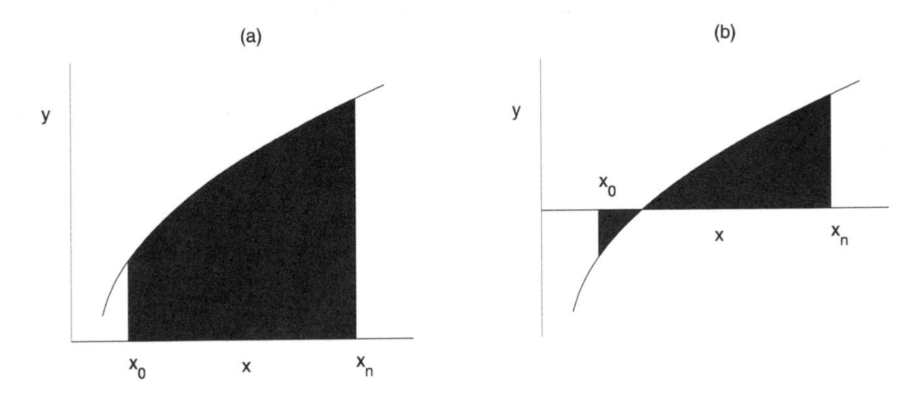

Figure 6.7 Graphical representation of the integral. a. Positive area only; b. Positive and negative areas.

6.10 The Newton-Cotes Formulas of Integration

This method is accomplished by first replacing the function $y = f(x)$ with a polynomial approximation, such as the Gregory-Newton forward interpolation formula (Eq. (6.57)). In practice, the interval $[x_0, x_n]$ is being divided into several segments, each of width h, and the Gregory-Newton forward interpolation formula becomes (note that $x_{i+1} = x_i + h$):

$$y = y_0 + \frac{(x - x_0)}{h}\Delta y_0 + \frac{(x - x_0)(x - x_1)}{2!h^2}\Delta^2 y_0 + \frac{(x - x_0)(x - x_1)(x - x_2)}{3!h^3}\Delta^3 y_0 + \ldots \tag{6.69}$$

Because this interpolation formula fits the function exactly at a finite number of $(n + 1)$ points, we divide the total interval of integration $[x_0, x_n]$ into n segments, each of width h. In the next step, by using Eq. (6.69), Eq. (6.68) can be integrated. The upper limits of integration can be chosen to include an increasing set of segments of integration, each of width h. In each case, we retain a number of finite differences in the finite series of Eq. (6.69) equal to the number of segments of integration. This operation yields the well-known *Newton-Cotes formulas of integration*. The first three of the Newton Cotes formulas are also known by the names *trapezoidal rule*, *Simpson's 1/3 rule*, and *Simpson's 3/8 rule*, respectively. These are developed in the next three sections.

6.10.1 The trapezoidal rule

In developing the first Newton-Cotes formula, we use one segment of width h and fit the polynomial through two points (x_0, y_0) and (x_1, y_1) (see Fig. 6.8). We retain the first two terms of the Gregory-Newton polynomial (up to and including the first forward finite difference) and group together the rest of the terms of the polynomial into the *remainder* term. This is tantamount to fitting a straight line between the two points. Thus, the integral equation becomes:

$$I_1 = \int_{x_0}^{x_1} \left[y_0 + \frac{(x - x_0)}{h} \Delta y_0 \right] dx + \int_{x_0}^{x_1} R_n(x)\,dx \tag{6.70}$$

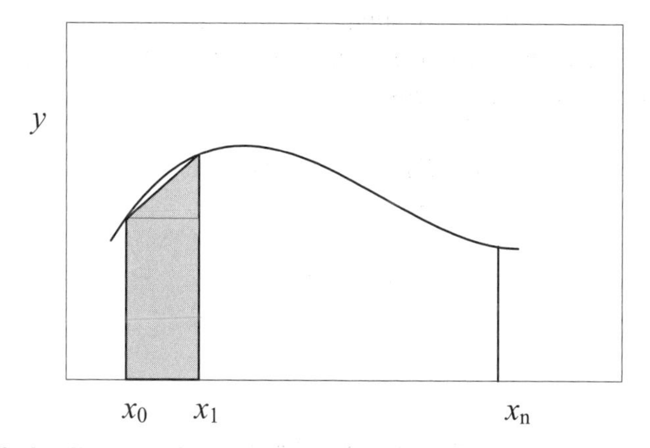

Figure 6.8 Application of the trapezoidal rule over one segment of integration.

The first integral on the right-hand side is integrated with respect to x, and the first forward difference is replaced with its definition of $\Delta y_0 = y_1 - y_0$, to obtain:

$$I_1 = \frac{h}{2}(y_0 + y_1) + \int_{x_0}^{x_1} R_n(x)\,dx \tag{6.71}$$

The remainder series can be replaced by one term evaluated at ξ_1 (for derivation see Constantinides and Mostoufi, 1999), therefore,

$$\int_{x_0}^{x_1} R_n(x)\,dx = -\frac{1}{12} h^3 D^2 f(\xi_1) \tag{6.72}$$

This is a term of order h^3 and is abbreviated by $O(h^3)$. Therefore, Eq. (6.71) can be written as:

$$I_1 = \frac{h}{2}(y_0 + y_1) + O(h^3) \tag{6.73}$$

This equation is known as the *trapezoidal rule*, because the term $(h/2)(y_0 + y_1)$ is the formula for calculating the area of a trapezoid. It was mentioned earlier that fitting a polynomial through only two points is equivalent to fitting a straight line through these points. This causes the shape of the integration segment to be a trapezoid, shown as the shaded area in Fig. 6.8. The area between $y = f(x)$ and the straight line that connects y_0 and y_1 represents the truncation error of the trapezoidal rule. If the function $f(x)$ is actually linear, then the trapezoidal rule calculates the integral exactly, because the second derivative is zero, causing the remainder term to vanish.

The trapezoidal rule in the form of Eq. (6.73) gives the integral of only one integration segment of width h. To obtain the total integral over n segments, Eq. (6.70) must be applied to each segment (with the appropriate limits of integration) to obtain the following series of equations:

$$I_1 = \frac{h}{2}(y_0 + y_1) + O(h^3) \tag{6.73}$$

$$I_2 = \frac{h}{2}(y_1 + y_2) + O(h^3) \tag{6.74}$$

$$\vdots$$

$$I_n = \frac{h}{2}(y_{n-1} + y_n) + O(h^3) \tag{6.75}$$

The total integral is approximated by the sum of the I_i. Addition of all these equations over the total interval gives the *composite trapezoidal rule*:

$$I = \frac{h}{2}\left(y_0 + 2\sum_{i=1}^{n-1} y_i + y_n\right) + nO(h^3) \tag{6.76}$$

For simplicity, the error term has been shown as $nO(h^3)$. This is only an approximation, because the remainder term includes terms in the second-order derivative of y evaluated at unknown values of ξ_i, each ξ_i being specific for that interval of integration. The absolute value of the error term cannot be calculated, but its relative magnitude can be measured by the order of the term. Because n is inversely proportional to h, as shown by Eq. (6.77),

$$n = \frac{x_n - x_0}{h} \tag{6.77}$$

the error term for the composite trapezoidal rule becomes:

$$nO\left(h^{3}\right)=\frac{x_{n}-x_{0}}{h}O\left(h^{3}\right)\cong O\left(h^{2}\right) \tag{6.78}$$

That is, the repeated application of the trapezoidal rule over multiple segments has lowered the accuracy of the method by approximately one order of magnitude. A more rigorous analysis of the truncation error is given in Appendix E.

6.10.2 Simpson's 1/3 rule

In the derivation of the second Newton-Cotes formula of integration, we use two segments of width h (see Fig. 6.9) and fit the polynomial through three points: (x_0, y_0), (x_1, y_1), and (x_2, y_2), equivalent to fitting a parabola through these points. We retain the first three terms of the Gregory-Newton polynomial (up to and including the second forward finite difference) and group together the rest of the terms of the polynomial into the remainder term. The integral equation becomes:

$$I_{1}=\int_{x_0}^{x_2}\left[y_{0}+\frac{\left(x-x_{0}\right)}{h}\Delta y_{0}+\frac{\left(x-x_{0}\right)\left(x-x_{1}\right)}{2!h^{2}}\Delta^{2}y_{0}\right]dx+\int_{x_0}^{x_2}R_{n}\left(x\right)dx \tag{6.79}$$

Integrating Eq. (6.79) and substituting the relevant finite difference relations simplifies this equation to:

$$I_{1}=\frac{h}{3}\left(y_{0}+4y_{1}+y_{2}\right)-\frac{1}{90}h^{5}D^{4}f\left(\xi_{1}\right) \tag{6.80}$$

The error term is of order h^5 and may be abbreviated by $O(h^5)$. We would have expected to obtain an error term of $O(h^4)$ because three terms were retained in the Gregory-Newton polynomial. However, the term containing h^4 in the remainder has a zero coefficient, thus giving this fortuitous result. The final form of the second Newton-Cotes formula, which is better known as *Simpson's 1/3 rule*, is:

$$I_{1}=\frac{h}{3}\left(y_{0}+4y_{1}+y_{2}\right)+O\left(h^{5}\right) \tag{6.81}$$

This equation calculates the integral over two segments of integration. Repeated application of Simpson's 1/3 rule over subsequent pairs of segments, and summation of all formulas over the total interval, gives the *composite Simpson's 1/3 rule*:

$$I=\frac{h}{3}\left(y_{0}+4\sum_{i=1}^{n/2}y_{2i-1}+2\sum_{i=1}^{n/2-1}y_{2i}+y_{n}\right)+O\left(h^{4}\right) \tag{6.82}$$

Simpson's 1/3 rule uses *pairs* of segments in an integration step; therefore, the total interval must be subdivided into an *even* number of segments. The first summation term in Eq. (6.82) sums up the odd-subscripted terms, and the second summation adds up the even-subscripted terms.

The accuracy of the composite Simpson's 1/3 rule was reduced by one order of magnitude to $O(h^4)$ for the same reason as in Section 6.10.1. Simpson's 1/3 rule is more accurate than the trapezoidal rule, but requires additional arithmetic operations.

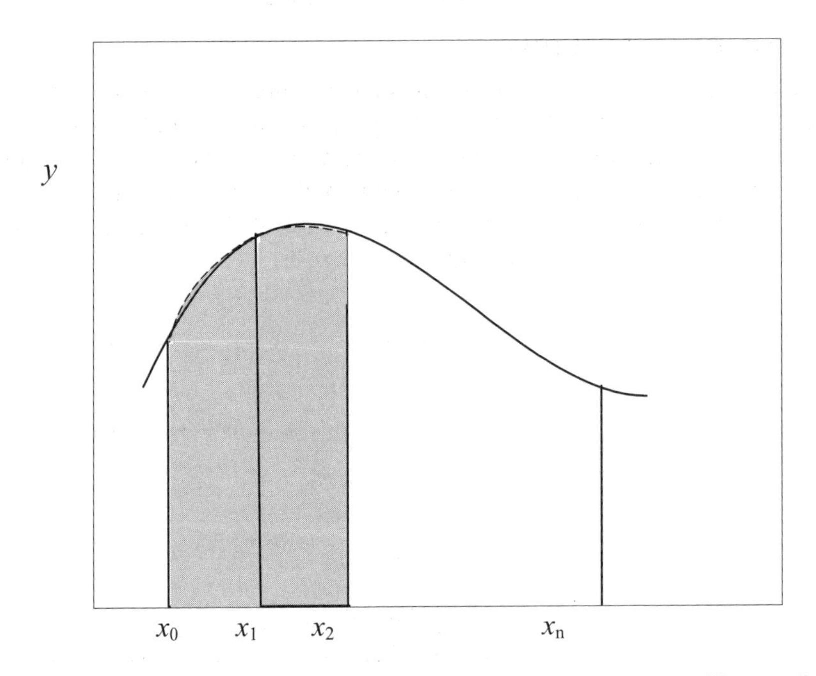

Figure 6.9 Application of Simpson's 1/3 rule over two segments of integration.

6.10.3 Simpson's 3/8 rule

In the derivation of the third Newton-Cotes formula of integration we use three segments of width h (see Fig. 6.10) and fit the polynomial through four points: (x_0, y_0), (x_1, y_1), (x_2, y_2), and (x_3, y_3). This is equivalent to fitting a cubic equation through the four points. We retain the first four terms of the Gregory-Newton polynomial (up to and including the third forward finite difference) and group together the rest of the terms of the polynomial into the remainder term. The integral equation becomes:

$$I_1 = \int_{x_0}^{x_3} \left[y_0 + \frac{(x-x_0)}{h}\Delta y_0 + \frac{(x-x_0)(x-x_1)}{2!h^2}\Delta^2 y_0 + \frac{(x-x_0)(x-x_1)(x-x_2)}{3!h^3}\Delta^3 y_0 \right] dx + \int_{x_0}^{x_3} R_n(x)\,dx \tag{6.83}$$

Integrating Eq. (6.83) and substituting the relevant finite difference relations simplifies the equation to:

$$I_1 = \frac{3h}{8}\left(y_0 + 3y_1 + 3y_2 + y_3\right) - \frac{3}{80}h^5 D^4 f\left(\xi_1\right) \tag{6.84}$$

The error term is of order h^5 and may be abbreviated by $O(h^5)$. The final form of this equation, which is better known as *Simpson's 3/8 rule*, is given by:

$$I_1 = \frac{3h}{8}\left(y_0 + 3y_1 + 3y_2 + y_3\right) + O\left(h^5\right) \tag{6.85}$$

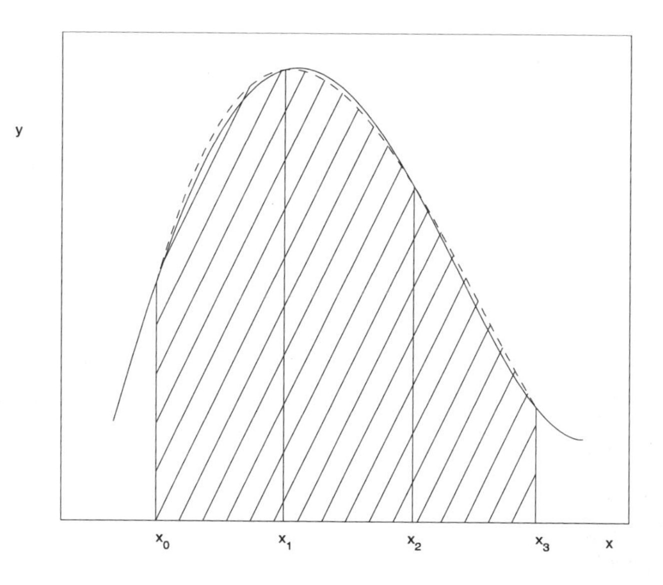

Figure 6.10 Application of Simpson's 3/8 rule over three segments of integration.

The *composite Simpson's 3/8 rule* is obtained by repeated application of Eq. (6.83) over triplets of segments and summation over the total interval of integration:

$$I = \frac{3h}{8}\left(y_0 + 3\sum_{i=1}^{n/3}\left(y_{3i-2} + y_{3i-1}\right) + 2\sum_{i=1}^{n/3-1} y_{3i} + y_n \right) + O\left(h^4\right) \tag{6.86}$$

Comparison of the error terms of Simpson's 1/3 rule and Simpson's 3/8 rule shows that they are both of the same order, with the latter being only slightly more accurate. For this reason, Simpson's 1/3 rule is usually preferred, because it achieves the same order of accuracy with three points, rather than the four points required by the 3/8 rule.

6.10.4 Summary of Newton-Cotes integration

The three Newton-Cotes formulas of integration derived in the previous sections are summarized in Table 6.7.

In the derivation of the Newton-Cotes formulas, the function $y = f(x)$ is approximated by the Gregory-Newton polynomial $P_n(x)$ of degree n with remainder $R_n(x)$. The evaluation of the integral is performed:

$$\int_a^b y dx = \int_a^b P_n(x) dx + \int_a^b R_n(x) dx \tag{6.87}$$

This results in a formula of the general form:

$$\int_a^b y dx = \sum_{i=0}^{n} w_i y_i + O\left[h^{n+2}, D^{n+1} f(\xi)\right] \tag{6.88}$$

where the x_i are $(n + 1)$ equally spaced base points in the interval $[a, b]$. The weights w_i are determined by fitting the $P_n(x)$ polynomial to $(n + 1)$ base points. The integral is exact, that is:

$$\int_a^b y dx = \sum_{i=0}^{n} w_i y_i \tag{6.89}$$

for any function $y = f(x)$ that is of polynomial form up to degree n, because the derivative $D^{n+1} f(\xi)$ is zero for polynomials of degree $\leq n$; thus, the error term $O[h^{n+2}, D^{n+1} f(\xi)]$ vanishes.

There are four functions in MATLAB (`trapz.m`, `cumtrapz`, `quad.m`, and `quad8.m`) that numerically evaluate the integral of a vector or a function using different Newton-Cotes formulas:

- The function `trapz(x,y)` calculates the integral of y (vector of data or function values) with respect to x (vector of variables) using the trapezoidal rule.
- The function `cumtrapz(x,y)` calculates the cumulative integral of y (vector of data or function values) with respect to x (vector of variables) using the trapezoidal rule.
- The function `quad('file_name',a,b)` evaluates the integral of the function represented in the m-file `file_name.m`, over the interval $[a, b]$ by Simpson's 1/3 rule.
- The function `quad8('filc_name',a,b)` evaluates the integral of the function introduced in the m-file `file_name.m` from a to b using 8-interval (9-point) Newton-Cotes formula.

Table 6.7 Summary of the Newton-Cotes numerical integration formulas

Formula	Integral	Error	Eq. #
Trapezoidal rule	$\int_{x_0}^{x_1} ydx = \frac{h}{2}(y_0 + y_1)$	$-\frac{1}{12}h^3D^2f(\xi)$	(6.90)
Simpson's 1/3 rule	$\int_{x_0}^{x_2} ydx = \frac{h}{3}(y_0 + 4y_1 + y_2)$	$-\frac{1}{90}h^5D^4f(\xi)$	(6.91)
Simpson's 3/8 rule	$\int_{x_0}^{x_3} ydx = \frac{3h}{8}(y_0 + 3y_1 + 3y_2 + y_3)$	$-\frac{3}{80}h^5D^4f(\xi)$	(6.92)
General quadrature formula	$\int_{x_0}^{x_n} ydx = \sum_{i=0}^{n} w_i y_i$	$O\left[h^{n+2}, D^{n+1}f(\xi)\right]$	(6.93)

Example 6.8 Integration formulas—trapezoidal and Simpson's 1/3 rules.

Statement of the problem

Two very important quantities in studying the growth of microorganisms in fermentation processes are the carbon dioxide evolution rate and the oxygen uptake rate. These are calculated from experimental analysis of the inlet and exit gases of the fermentor, and the flow rates, temperature, and pressure of these gases. The ratio of carbon dioxide evolution rate to oxygen uptake rate yields the respiratory quotient, which is a good barometer of the metabolic activity of the microorganism. In addition, the above rates can be integrated to obtain the total amounts of carbon dioxide produced and oxygen consumed during the fermentation. These total amounts form the basis of the material balancing techniques used in modeling of fermentation processes. Table 6.8 shows a set of rates calculated from the fermentation of *Penicillium chrysogenum*, which produces penicillin antibiotics.

Write a general MATLAB function for integrating experimental data using Simpson's 1/3 rule, and use it to calculate the total amounts of carbon dioxide produced and oxygen consumed during this ten-hour period of fermentation. Compare the results of this function and those obtained by using the existing MATLAB function `trapz` (trapezoidal rule).

Table 6.8 Fermentation data

Time of fermentation (h)	Carbon dioxide evolution rate (g/h)	Oxygen uptake rate (g/h)
140	15.72	15.49
141	15.53	16.16
142	15.19	15.35
143	16.56	15.13
144	16.21	14.20
145	17.39	14.23
146	17.36	14.29
147	17.42	12.74
148	17.60	14.74
149	17.75	13.68
150	18.95	14.51

Method of solution

In this problem, the carbon dioxide evolution rate data and the oxygen uptake rate data are integrated separately. There are 11 data points (10 intervals) for each rate; therefore, we can use either the trapezoidal rule or Simpson's 1/3 rule for this integration. We first use Simpson's 1/3 rule and then repeat using the trapezoidal rule, as the problem specifies.

Program description

The MATLAB function `simpson.m`, which was developed by Constantinides and Mostoufi (1999), first tests the input arguments, which are the vector of independent variable (x) and the vector of function values (y). These two vectors should be of the same length. Elements of vector x have to be equally spaced values. Also, the number of elements of these vectors (n) should be odd (even number of intervals). If the vectors contain an even number of elements (odd number of intervals), the function calculates the value of the integral up to the point (n - 1) and adds the value of the integral, approximated by the trapezoidal rule, for the last interval. The user should pay special attention to this case because the truncation errors for Simpson's 1/3 rule and trapezoidal rule are not of the same order. After checking the above conditions, the function calculates the value of the integral based on Eq. (6.82). If necessary, the function adds the value of the integral for the last segment according to Eq. (6.75).

The main program, `example6_8.m`, asks the user to input the data from the keyboard, calls the functions `trapz.m` and `simpson.m` for integration, and displays the results.

Program

```
% example6_8.m - Calculates carbon dioxide evolved and
% oxygen uptaken in a fermentation process using TRAPZ
% (trapezoidal rule) and SIMPSON (Simpson's 1/3 rule)
functions.

clc; clear all

% Input data
t = input(' Vector of time = ');
r_CO2 = input(' Carbon dioxide evolution rate (g/h) = ');
r_O2 = input(' Oxygen uptake rate (g/h) = ');

% Integration
m1CO2 = trapz(t,r_CO2);
m2CO2 = Simpson(t,r_CO2);
m1O2 = trapz(t,r_O2);
m2O2 = Simpson(t,r_O2);

% Output
fprintf('\n Total carbon dioxide evolution  = %9.4f (evaluated
by the trapezoidal rule)',m1CO2)
fprintf('\n Total carbon dioxide evolution  = %9.4f (evaluated
by the Simpson 1/3 rule)',m2CO2)
fprintf('\n Total oxygen uptake             = %9.4f (evaluated
by the trapezoidal rule)',m1O2)
fprintf('\n Total oxygen uptake             = %9.4f (evaluated
by the Simpson 1/3 rule)\n',m2O2)
```

Function that evaluates Simpson's 1/3 rule

```
function Q = Simpson(x , y)
%SIMPSON Numerical evaluation of integral by Simpson's 1/3
rule.
%
```

```
%   SIMPSON(X,Y) numerically evaluates the integral of the
%   vector of function values Y with respect to X by
%   Simpson's 1/3 rule. X is the vector of equally spaced
%   independent variable. Length of Y has to be odd (even
%   number of intervals). If length of Y is even, the function
%   calculates the integral for [LENGTH(Y)-1] points by
%   Simpson's 1/3 rule and adds to it the value of the
%   integral for the last interval by trapezoidal rule.
%   See also TRAPZ , QUAD , QUAD8, GAUSSLEGENDRE

% (c) N. Mostoufi & A. Constantinides
% January 1, 1999

points = length(x);
if length(y) ~= points
   error('x and y are not of the same length')
   break
end

dx = diff(x);
maxi = max([min(abs(x))/1000 , 1e-10]);
if max(dx)-min(dx) > maxi
   error('X is not equally spaced.')
   break
end

h = dx(1);
if mod(points,2) == 0
   warning('Odd number of intervals; Trapezoidal rule will be
used for the last interval.')
   n = points - 1;
else
   n = points;
end
% Integration
y1 = y(2 : 2 : n - 1);
y2 = y(3 : 2 : n - 2);
Q = (y(1) + 4 * sum(y1) + 2 * sum(y2) + y(n)) * h /3;

if n ~= points
   Q = Q + (y(points) + y(n)) * h / 2;
end
```

Input and results

```
>>example6_8

Vector of time = [140:150]
Carbon dioxide evolution rate (g/h) = [15.72, 15.53, 15.19,
16.56, 16.21, 17.39, 17.36, 17.42, 17.60, 17.75, 18.95]
```

```
Oxygen uptake rate (g/h) = [15.49, 16.16, 15.35, 15.13, 14.20,
14.23, 14.29, 12.74, 14.74, 13.68, 14.51]

Total carbon dioxide evolution  =  168.3450 (evaluated by the
trapezoidal rule)
Total carbon dioxide evolution  =  168.6633 (evaluated by the
Simpson 1/3 rule)
Total oxygen uptake             =  145.5200 (evaluated by the
trapezoidal rule)
Total oxygen uptake             =  144.9733 (evaluated by the
Simpson 1/3 rule)
```

Discussion of results

The integration of the experimental data, using both Simpson's 1/3 rule and the trapezoidal rule, yield the total amounts of carbon dioxide and oxygen shown in Table 6.9.

Table 6.9 Comparison of results

	Simpson's 1/3	Trapezoidal
Total CO_2 (g)	168.6633	168.3450
Total O_2 (g)	144.9733	145.5200

6.11 Lessons Learned in this Chapter

After studying this chapter, the student should have learned the following:

- Finite differences are vital tools in the development of differentiation and integration methods.
- Derivatives may be expressed in terms of finite differences with any degree of accuracy desired.
- Both data and functions may be integrated or differentiated using the methods developed in this chapter.
- Interpolating polynomials are used in the development of integration formulas.
- The Newton-Cotes formulas of integration may be used to integrate experimental data.
- Splines are commonly used in the interpolation of unequally spaced data.

6.12 Problems

6.1 Derive the equation that expresses the third-order derivative of y in terms of backward finite differences, with: (a) Error of order h; (b) Error of order h^2.

6.2 Derive the equations for the first-, and second-order derivatives of y in terms of forward finite differences with error of order h^3.

6.3 Derive the Gregory-Newton backward interpolation formula.

6.4 Using the experimental data in Table 6.10:

(a) Develop the forward difference table. Verify your results by using the MATLAB function `diff`.

(b) Develop the backward difference table.

(c) Apply the Gregory-Newton interpolation formulas to evaluate the function at $x = 10, 50, 90, 130, 170$, and 190.

Table 6.10 Data of penicillin fermentation

Time (h)	Penicillin concentration (units/mL)	Time (h)	Penicillin concentration (units/mL)
0	0	120	9430
20	106	140	10950
40	1600	160	10280
60	3000	180	9620
80	5810	200	9400
100	8600		

6.5 Write a MATLAB function that uses the Gregory-Newton backward interpolation formula to evaluate the function $f(x)$ from a set of $(n + 1)$ equally spaced input values. Write the function in a general fashion so that n can be any positive integer. Also, write a MATLAB script that reads the data and shows how this MATLAB function fits the data. Use the experimental data of Table 6.10 to verify the program, and evaluate the function at $x = 10, 50, 90,130, 170$, and 190.

6.6 With the set of unequally spaced data points in Table 6.11, use Lagrange polynomials and spline interpolation to evaluate the function at $x = 2, 4, 5, 8, 9$, and 11.

Table 6.11 Unequally spaced data

x	$f(x)$	x	$f(x)$
1	7.0	10	8.2
3	3.5	12	9.0
6	3.2	13	9.2
7	3.9		

6.7 In studying the mixing characteristics of chemical reactors, a sharp pulse of a non-reacting tracer is injected into the reactor at time $t = 0$. The concentration of material in the effluent from the reactor is measured as a function of time $c(t)$. The residence time distribution (RTD) function for the reactor is defined as:

$$E(t) = \frac{c(t)}{\int_0^\infty c(t)\,dt}$$

and the cumulative distribution function is defined as:

$$F(t) = \int_0^t E(t)\,dt$$

The mean residence time of the reactor is calculated from:

$$t_m = \frac{V}{q} = \int_0^\infty tE(t)\,dt$$

where V is the volume of the reactor and q is the flow rate. The variance of the RTD function is defined by:

$$\sigma^2 = \int_0^\infty (t - t_m)E(t)\,dt$$

The exit concentration data shown in Table 6.12 were obtained from a tracer experiment studying the mixing characteristics of a continuous flow reactor. Calculate the RTD function, cumulative distribution function, mean residence time, and the variance of the RTD function of this reactor.

Table 6.12 Exit concentration data of chemical reactor

Time (s)	$c(t)$ (mg/L)	Time (s)	$c(t)$ (mg/L)
0	0	5	5
1	2	6	2
2	4	7	1
3	7	8	0
4	6		

6.13 References

Chapra, S. C., and Canale, R. P. 2006. *Numerical Methods for Engineers*, 4th ed. New York: McGraw-Hill Book Company.

Constantinides, A., and Mostoufi, N. 1999. *Numerical Methods for Chemical Engineers with MATLAB Applications*. Upper Saddle River, NJ: Prentice Hall PTR.

Hanselman, D., and Littlefield, B. 2005. *Mastering MATLAB 7*. Upper Saddle River, NJ: Prentice Hall PTR.

Chapter 7
Dynamic Systems: Ordinary Differential Equations

7.1 Introduction

Mathematical modeling of physiological systems will often result in ordinary or partial differential equations. The fundamental reason underlying this is that biosystems are dynamic in nature. Their behavior constantly evolves with time or varies with respect to position in space. When the transient behavior of these systems disappears, the models become algebraic equations of the type solved in Chapters 4 and 5 of this book.

In this chapter we will consider the numerical solution of ordinary differential equations (ODEs). These are the models that arise from the study of the dynamics of physiological systems that have one independent variable. The latter may be either the space variable x or the time variable t, depending on the geometry of the system and its boundary conditions. Ordinary differential equations may arise from modeling the metabolic pathways of living cells; the complex interactions of pharmacokinetics; the kinetics of the oxygen/hemoglobin system; the transfer of nutrients across cells; the dynamics of membrane and nerve cell potentials; the transformation and replication of stem cells; the mechanism of migration and binding of tissue cells; or the dynamics of interacting populations of bacteria and the human species.

The material in this chapter will enable the student to accomplish the following:

- Model the dynamics of physiological systems using ordinary differential equations
- Obtain numerical solutions of the differential equations, plot the numerical results, and interpret the dynamic behavior of the biosystems under a variety of conditions
- Appreciate the accuracy and stability of the models and the numerical solutions obtained from these models

Several examples of interest to biomedical engineers are discussed briefly in the rest of this section. The numerical solution of some of these cases will be presented later in the chapter as worked examples.

7.1.1 Pharmacokinetics: the dynamics of drug absorption

Pharmacokinetics is the study of the processes that affect drug distribution and the rate of change of drug concentrations within the body (Fournier, 1999). Drugs can enter the body through the gastrointestinal tract, referred to as the enteral route, or through a variety of other pathways that include intravenous injection, inhalation, subcutaneous penetration, and so on. These are referred to as parenteral routes.

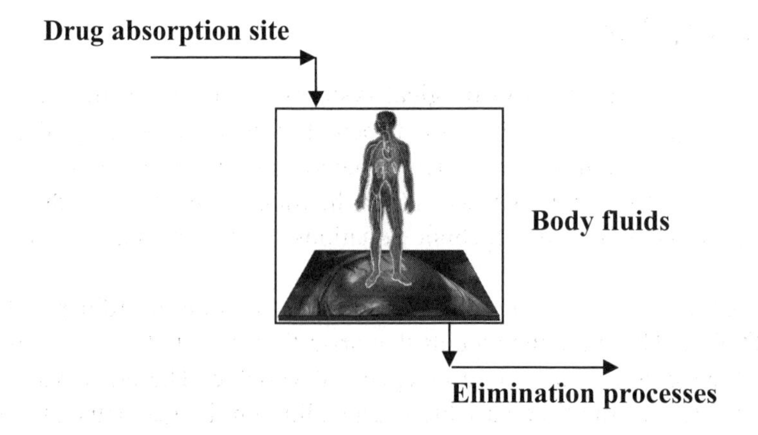

Figure 7.1 Simplified drug absorption model.

The drug distribution throughout the body is affected by several factors, such as blood perfusion rate, capillary permeability, drug biological affinity, the metabolism of the drug, and renal excretion. The drug is eliminated from the body by enzymatic reactions in the liver and by excretion into the urine stream via the kidneys. A simplified model for drug absorption and elimination is shown in Fig. 7.1. This model treats all body fluids as a single-compartment unit. A mathematical simulation of this model results in a set of linear ordinary differential equations. Methods for the solution of such a set are developed in Section 7.5 of this chapter, and are demonstrated in Example 7.3.

7.1.2 Tissue engineering: cell differentiation, cell adhesion and migration dynamics

Cell differentiation is a critical dynamic process that underlies the progressive specialization of the various embryonic and progenitor cells to multifunctional tissues in the body. For example, embryonic stem cells in a growing fetus replicate and differentiate to develop into specialized types of cells, such as bone cells, skin cells, liver cells, muscle cells, and so on. The differentiation process involves a series of changes in cell phenotype and morphology that typically become more pronounced and easier to observe directly at the later stages of the process (Palsson and Bhatia, 2004). This process begins with the stem cells' commitment to differentiation, followed by a coordinated series of gene-expression events, causing the cell to differentiate to a new state. A series of such progressive states leads to fully mature specialized cells. These mature cells perform their intended function in the body and eventually die, or undergo change to another type of cell through a process called *transdifferentiation*. The progressive series of events that converts a stem cell to a fully mature specialized cell may be modeled as a multicompartment model. The unsteady state balances on these compartments result in a set of simultaneous ordinary differential equations. The solution of such a set of equations is demonstrated in Example 7.6, which presents and discusses stem cell differentiation.

An important aspect of tissue engineering is the proper design and manufacture of porous matrices that imitate the properties of the epidermis and may be used as prosthetic scaffolding to promote dermal regeneration, thus enhancing the healing process of wounded or burned skin. A cellular dynamic process, relevant to wound repair and tissue regeneration, is cell migration (Lauffenburger and Horowitz, 1996). Cell migration is necessary for cells to repopulate a healing wound and an implanted scaffold for tissue regeneration, and during embryogenesis for cell sorting and organ development. Cell migration is also relevant to cancer and tumor metastasis.

Cellular migration is a coordinated process that results from the interaction of specific cell surface receptors with ligands, which are typically biomolecules of an extracellular matrix (Fig. 7.2). Quantitative descriptions of the cell migration process involve establishing relationships between the cell motility response (e.g., cell speed, cell directional persistence, population cell motility) and the various attributes of the ligands. A number of ligand properties, such as ligand surface concentration, degree of receptor occupancy, and ligand affinity, affect the activation of cell motility. An interesting mode of complex cell migration has been quantitatively analyzed by Moghe and coworkers (Tjia and Moghe, 2002a, 2002b). This migration involves cellular internalization (*endocytosis* or *phagocytosis*, depending on the nature of ligand carriers) of the ligands after receptor-ligand binding. The dynamics of cell-ligand interactions have been modeled from a kinetic-mechanistic point of view (Tjia and Moghe, 2002c) using diffusion-reaction descriptions and equations similar to those in the traditional Michaelis-Menten kinetics. A model of cell migration is presented and solved in Example 7.7.

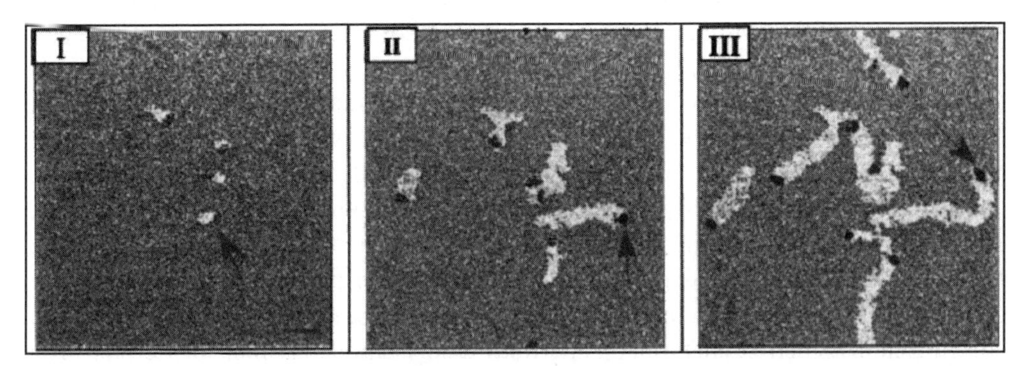

Figure 7.2 **The migration of keratinocytes is enhanced by the presence of ligand-bound microcarriers. (I), (II), and (III) show the progressive clearance of cellular tracks (from Tjia and Moghe, 2002c) .**

7.1.3 Metabolic Engineering: Glycolysis pathways of living cells

Living cells break down glucose to produce carbon dioxide and water in a complex process called *glycolysis* that involves several enzyme-catalyzed reactions. This process generates chemical energy, which is in turn used in the biological synthesis of other compounds, such as proteins. The energy produced in glycolysis is stored by the cell in the form of adenosine triphosphate (ATP). The net effect of this pathway is:

$$C_6H_{12}O_6 + 6O_2 \longrightarrow 6CO_2 + 6H_2O + \text{energy}$$

Many of the chemical reactions in the glycolysis pathway are catalyzed by enzymes, such as the reaction shown here:

$$S + E \underset{k_{-1}}{\overset{k_1}{\rightleftarrows}} ES \xrightarrow{k_2} P + E$$

An enzyme, E, catalyzes the conversion of a substrate, S, to form a product, P, via the formation of an intermediate complex, $[ES]$. The dynamic behavior of enzymatic reactions is modeled by ordinary differential equations. . Methods of solution for sets of ordinary differential equations are developed in Section 7.4 of this chapter and are applied to obtain the solution of an enzyme catalysis problem in Example 7.2. The steady state analysis of such reactions simplifies the model to algebraic equations, the solution to which may be obtained by the methods discussed in Chapters 4 and 5 of this book

7.1.4 Transport of molecules across biological membranes

The transport of molecules across biological membranes is vital to the physiology of living cells. The supply of nutrients to the cell for growth and reproduction, and the transfer of waste products from cell to the extracellular medium, is a complex process that is facilitated by many mechanisms (Fig. 7.3).

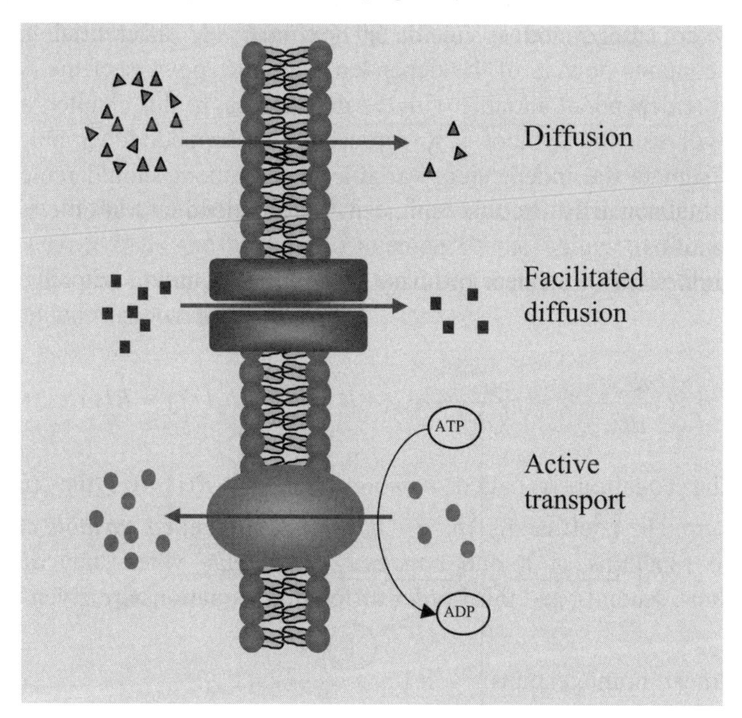

Figure 7.3 Diffusion across biological membranes

There is passive transport of molecules due to the combined effects of concentration gradients and electrical potential differences that exist across the cell membrane. Neutral molecules diffuse from regions of high concentration to regions of low concentration. In addition, charged molecules move along a voltage gradient that normally exists across a cell membrane, such as in neural cells and axons. Carrier-mediated transport and active transport are additional mechanisms that facilitate the movement of molecules across cell boundaries. The transport mechanism of molecules may be modeled using ordinary and partial differential equations. In this chapter, we will discuss dynamic transport systems of one independent variable that may be modeled by ordinary differential equations. In Example 7.5, we solve the Hodgkin-Huxley model that simulates the dynamics of membrane and nerve cell potentials. In Chapter 8, we will examine transport systems of two or more independent variables that result in partial differential equations.

7.2 Classification of Ordinary Differential Equations

Ordinary differential equations are classified according to their *order*, *linearity*, *homogeneity*, and *boundary conditions*. The order of a differential equation is the order of the highest derivative present in that equation. Ordinary differential equations may be categorized as *linear* or *nonlinear*. A differential equation is nonlinear if it contains powers of the dependent variable, powers of the derivatives, or products of the dependent variable with the derivatives. In this chapter, as much as possible, we will use the symbol y to represent the dependent variable, and the symbol t to designate the independent variable. The student should remember that either t or x, is customarily used to represent the independent variable in ordinary differential equations.

The general form of a linear ordinary differential equation of order n may be written as:

$$b_n(t)\frac{d^n y}{dt^n} + b_{n-1}(t)\frac{d^{n-1}y}{dt^{n-1}} + \ldots + b_1(t)\frac{dy}{dt} + b_0(t)y = R(t) \tag{7.1}$$

If $R(t)=0$, the equation is called *homogeneous*. If $R(t)\neq 0$, the equation is *nonhomogeneous*. The coefficients $\{b_i \mid i = n, \ldots, 0\}$ are called *variable coefficients* when they are functions of t, and *constant coefficients* when they are scalars. Examples of first-, second-, and third-order differential equations are given below:

First-order, linear, homogeneous:
$$\frac{dy}{dt} + y = 0 \tag{7.2}$$

First-order, linear, nonhomogeneous:
$$\frac{dy}{dt} + y = kt \tag{7.3}$$

First-order, nonlinear, nonhomogeneous:
$$\frac{dy}{dt} + y^2 = kt \tag{7.4}$$

Second-order, linear, nonhomogeneous: $$\frac{d^2y}{dt^2}+\frac{dy}{dt}+y=e^t \tag{7.5}$$

Second-order, nonlinear, nonhomogeneous: $$y\frac{d^2y}{dt^2}+\frac{dy}{dt}+y=\cos(t) \tag{7.6}$$

Third-order, linear, homogeneous: $$\frac{d^3y}{dt^3}+a\frac{d^2y}{dt^2}+b\frac{dy}{dt}+y=0 \tag{7.7}$$

Third-order, nonlinear, nonhomogeneous: $$\frac{d^3y}{dt^3}+a\left(\frac{d^2y}{dt^2}\right)^2+\frac{dy}{dt}+y=\sin(t) \tag{7.8}$$

Eqs. (7.4), (7.6), and (7.8) are nonlinear because they contain the terms y^2, $y(d^2y/dt^2)$, and $(d^2y/dt^2)^2$, respectively, whereas Eqs. (7.2), (7.3), (7.5), and (7.7) are linear.

To obtain a unique solution of a nth-order differential equation, or of a set of n simultaneous first-order differential equations, it is necessary to specify n values of the dependent variables (or their derivatives) at specific values of the independent variable. These are the initial or boundary conditions of the problem.

Ordinary differential equations may be classified as *initial-value* problems or *boundary-value* problems. In initial-value problems, the values of the dependent variables and/or their derivatives are *all* known at the initial value of the independent variable. A problem whose dependent variables, and/or their derivatives, are all known at the final value of the independent variable (rather than the initial value) is identical to the initial-value problem, because only the direction of integration must be reversed. Therefore, the term initial-value problem refers to either case. In boundary-value problems, the dependent variables and/or their derivatives are known at more than one point of the independent variable. If some of the dependent variables (or their derivatives) are specified at the initial value of the independent variable, and the remaining variables (or their derivatives) are specified at the final value of the independent variable, then this is a *two-point boundary-value* problem.

The methods for solution of initial-value problems are developed in Section 7.4. The methods for solution of boundary-value problems will not be covered in this book. The interested student is referred to Constantinides and Mostoufi (1999) and Kubícek and Hlavàcek (1975).

7.3 Transformation to Canonical Form

A large number of methods for integrating ordinary differential equations require that the system consists of a set of n simultaneous first-order ordinary differential equations of the form:

$$\begin{aligned} \frac{dy_1}{dt} &= f_1(t, y_1, y_2, \ldots, y_n) \qquad & y_1(t_0) &= y_{1,0} \\ \frac{dy_2}{dt} &= f_2(t, y_1, y_2, \ldots, y_n) \qquad & y_2(t_0) &= y_{2,0} \\ &\vdots \\ \frac{dy_n}{dt} &= f_n(t, y_1, y_2, \ldots, y_n) \qquad & y_n(t_0) &= y_{n,0} \end{aligned} \tag{7.9}$$

This is called the *canonical* form of the equations. When the initial conditions are given at a common point, t_0, then the set of equations (7.9) has solutions of the form:

$$\begin{aligned} y_1 &= F_1(t) \\ y_2 &= F_2(t) \\ &\vdots \\ y_n &= F_n(t) \end{aligned} \tag{7.10}$$

The above problem can be condensed into matrix notation, where the system equations are represented by:

$$\frac{d\mathbf{y}}{dt} = \mathbf{f}(t, \mathbf{y}) \tag{7.11}$$

the vector of initial conditions is:

$$\mathbf{y}(t_0) = \mathbf{y_0} \tag{7.12}$$

and the vector of solutions is:

$$\mathbf{y} = \mathbf{F}(t) \tag{7.13}$$

Differential equations of higher order, or systems containing equations of mixed order, can be transformed to the canonical form by a series of substitutions.

For example, consider the nth-order differential equation:

$$\frac{d^n z}{dt^n} = G\left(z,\ \frac{dz}{dt},\ \frac{d^2 z}{dt^2},\ \dots\ \frac{d^n z}{dt^n},\ t \right) \tag{7.14}$$

The following transformations:

$$\begin{aligned} z &= y_1 \\ \frac{dz}{dt} &= \frac{dy_1}{dt} = y_2 \\ \frac{d^2 z}{dt^2} &= \frac{dy_2}{dt} = y_3 \\ &\vdots \\ \frac{d^{n-1} z}{dt^{n-1}} &= \frac{dy_{n-1}}{dt} = y_n \\ \frac{d^n z}{dt^n} &= \frac{dy_n}{dt} \end{aligned} \tag{7.15}$$

when substituted into the nth-order equation (7.14), give the equivalent set of n first-order equations of canonical form:

$$\begin{aligned} \frac{dy_1}{dt} &= y_2 \\ \frac{dy_2}{dt} &= y_3 \\ &\vdots \\ \frac{dy_n}{dt} &= G(y_1,\ y_2,\ y_3, \dots, y_n,\ t) \end{aligned} \tag{7.16}$$

If the right-hand side of the differential equations is not a function of the independent variable, then the above set of equations may be written as:

$$\frac{d\mathbf{y}}{dt} = \mathbf{f}(\mathbf{y}) \tag{7.17}$$

If the functions $\mathbf{f}(\mathbf{y})$ are linear in terms of $\mathbf{y}$, then the equations can be written in matrix form:

$$\mathbf{y}' = \mathbf{A}\mathbf{y} \tag{7.18}$$

as in Example 7.1 (*a*) and (*b*). The methods for solution of nonlinear sets are discussed in Section 7.4. Solutions for linear sets of ordinary differential equations are developed in Section 7.5.

The next example demonstrates the technique for converting higher-order linear and nonlinear differential equations to canonical form.

Example 7.1 Transformation of ordinary differential equations into their canonical form.

Statement of the problem

Apply the transformations defined by Eqs. (7.15) and (7.16) to the following ordinary differential equations:

(a) $$\frac{d^4z}{dt^4}+5\frac{d^3z}{dt^3}-2\frac{d^2z}{dt^2}-6\frac{dz}{dt}+3z=0 \qquad \text{(Linear, homogeneous)}$$

With initial conditions

$$\text{at } t=0,\quad \left.\frac{d^3z}{dt^3}\right|_0=2,\quad \left.\frac{d^2z}{dt^2}\right|_0=1.5,\quad \left.\frac{dz}{dt}\right|_0=1,\quad z|_0=0.5$$

(b) $$\frac{d^4z}{dt^4}+5\frac{d^3z}{dt^3}-2\frac{d^2z}{dt^2}-6\frac{dz}{dt}+3z=e^{-t} \qquad \text{(Linear, nonhomogeneous)}$$

With initial conditions

$$\text{at } t=0,\quad \left.\frac{d^3z}{dt^3}\right|_0=2,\quad \left.\frac{d^2z}{dt^2}\right|_0=1.5,\quad \left.\frac{dz}{dt}\right|_0=1,\quad z|_0=0.5$$

(c) $$\frac{d^3z}{dt^3}+z^2\frac{d^2z}{dt^2}-\left(\frac{dz}{dt}\right)^3-2z=0 \qquad \text{(Nonlinear, homogeneous)}$$

With boundary conditions

$$\text{at t}=0,\quad \left.\frac{d^2z}{dt^2}\right|_0=1,\quad \left.\frac{dz}{dt}\right|_0=2,\quad z|_0=3$$

Solution

(a) Apply the transformation according to Eqs. (7.15) to obtain the following four equations:

$$\frac{dy_1}{dt} = y_2 \qquad y_1(0) = 0.5$$

$$\frac{dy_2}{dt} = y_3 \qquad y_2(0) = 1$$

$$\frac{dy_3}{dt} = y_4 \qquad y_3(0) = 1.5$$

$$\frac{dy_4}{dt} = -3y_1 + 6y_2 + 2y_3 - 5y_4 \qquad y_4(0) = 2$$

This is a set of linear ordinary differential equations that can be represented in matrix form by Eq. (7.18), where matrix **A** is given by:

$$\mathbf{A} = \begin{bmatrix} 0 & 1 & 0 & 0 \\ 0 & 0 & 1 & 0 \\ 0 & 0 & 0 & 1 \\ -3 & 6 & 2 & -5 \end{bmatrix}$$

The method for obtaining the solution of sets of linear ordinary differential equations is discussed in Section 7.5.

(b) The presence of the term e^{-t} on the right-hand side of this equation makes it nonhomogeneous. The left-hand side is identical to that of Eq. (a), so that the transformations of Eq. (a) are applicable. An additional transformation is needed to replace the e^{-t} term. This transformation is:

$$y_5 = e^{-t}$$

$$\frac{dy_5}{dt} = -e^{-t} = -y_5$$

Make the substitutions into Eq. (b) to obtain the following set of five linear ordinary differential equations:

$$\frac{dy_1}{dt} = y_2 \qquad y_1(0) = 0.5$$

$$\frac{dy_2}{dt} = y_3 \qquad y_2(0) = 1$$

$$\frac{dy_3}{dt} = y_4 \qquad y_3(0) = 1.5$$

$$\frac{dy_4}{dt} = -3y_1 + 6y_2 + 2y_3 - 5y_4 + y_5 \qquad y_4(0) = 2$$

$$\frac{dy_5}{dt} = -y_5 \qquad y_5(0) = 1$$

The above set is linear, therefore it condenses into the matrix form of Eq. (7.18), with the matrix $\boldsymbol{A}$ given by:

$$\mathbf{A} = \begin{bmatrix} 0 & 1 & 0 & 0 & 0 \\ 0 & 0 & 1 & 0 & 0 \\ 0 & 0 & 0 & 1 & 0 \\ -3 & 6 & 2 & -5 & 1 \\ 0 & 0 & 0 & 0 & -1 \end{bmatrix}$$

(c) This problem is nonlinear, however, similar transformations may be applied:

$$z = y_1$$

$$\frac{dz}{dt} = \frac{dy_1}{dt} = y_2$$

$$\frac{d^2z}{dt^2} = \frac{dy_2}{dt} = y_3$$

$$\frac{d^3z}{dt^3} = \frac{dy_3}{dt}$$

Make the substitutions into Eq. (c) to obtain the set:

$$\frac{dy_1}{dt} = y_2 \qquad y_1(0) = 3$$

$$\frac{dy_2}{dt} = y_3 \qquad y_2(0) = 2$$

$$\frac{dy_3}{dt} = 2y_1 + y_2^3 - y_1^2 y_3 \qquad y_3(0) = 1$$

As expected, this is a set of *nonlinear* differential equations, which cannot be expressed in matrix form. The methods for solution of nonlinear differential equations are developed in Section 7.4.

7.4 Nonlinear Ordinary Differential Equations

In this section, we develop numerical solutions for a set of ordinary differential equations in their canonical form:

$$\frac{d\mathbf{y}}{dt} = \mathbf{f}(t, \mathbf{y}) \tag{7.11}$$

with the vector of initial conditions given by:

$$\mathbf{y}(t_0) = \mathbf{y_0} \tag{7.12}$$

In order to be able to illustrate these methods graphically, we treat $\boldsymbol{y}$ as a single variable rather than as a vector of variables. The formulas developed for the solution of a single differential equation are readily expandable to those for a set of differential equations, which must be solved *simultaneously*. This concept is demonstrated in Section 7.4.3.

We begin the development of these methods by first rearranging Eq. (7.11) and integrating both sides between the limits of $t_i \leq t \leq t_{i+1}$ and $y_i \leq y \leq y_{i+1}$:

$$\int_{y_i}^{y_{i+1}} dy = \int_{t_i}^{t_{i+1}} f(t, y)\, dt \tag{7.19}$$

The left side integrates readily to obtain:

$$y_{i+1} - y_i = \int_{t_i}^{t_{i+1}} f(t, y)\, dt \tag{7.20}$$

One method for integrating Eq. (7.20) is to take the left-hand side of this equation and use Taylor series for its approximation. This technique works directly with the tangential trajectories of the dependent variable y rather than with the areas under the function $f(t, y)$. This is the procedure used in Sections 7.4.1 and 7.4.2.

7.4.1 The Euler and modified Euler methods

One of the earliest techniques developed for the solution of ordinary differential equations is the *Euler method*. This is simply obtained from Eq. (6.27), which was derived in Example 6.3 by expanding y_{i+1} about y_i in Taylor series:

$$\frac{dy_i}{dx} = \frac{1}{h}(y_{i+1} - y_i) + O(h) \tag{6.27}$$

This equation is rearranged to give a *forward marching* formula for evaluating y:

$$y_{i+1} = y_i + h\frac{dy_i}{dt} + O\left(h^2\right) \tag{7.21}$$

Eq (7.21) is the *explicit Euler formula* for integrating differential equations. The term *explicit* refers to the fact that only one unknown value, y_{i+1}, is on the left-hand side of the equation and may be evaluated, in terms of known values, on the right-hand side of the equation. The derivative is replaced by its equivalent y'_i or $f(t_i, y_i)$ to give the more commonly used form of the explicit Euler method:

$$y_{i+1} = y_i + hf\left(t_i, y_i\right) + O\left(h^2\right) \tag{7.22}$$

From here on, the terms y'_i and $f(t_i, y_i)$ will be used interchangeably. The student should remember that these are equal to each other because of the differential equation (7.11).

The Euler method, Eq. (7.22), simply states that the next value of y is obtained from the previous value by moving a step of width h in the tangential direction of y. This is demonstrated graphically in Fig. 7.4a. This Euler formula is rather inaccurate because it has a truncation error of only $O(h^2)$. If h is large the trajectory of y can quickly deviate from its true value, as demonstrated in Fig. 7.4b.

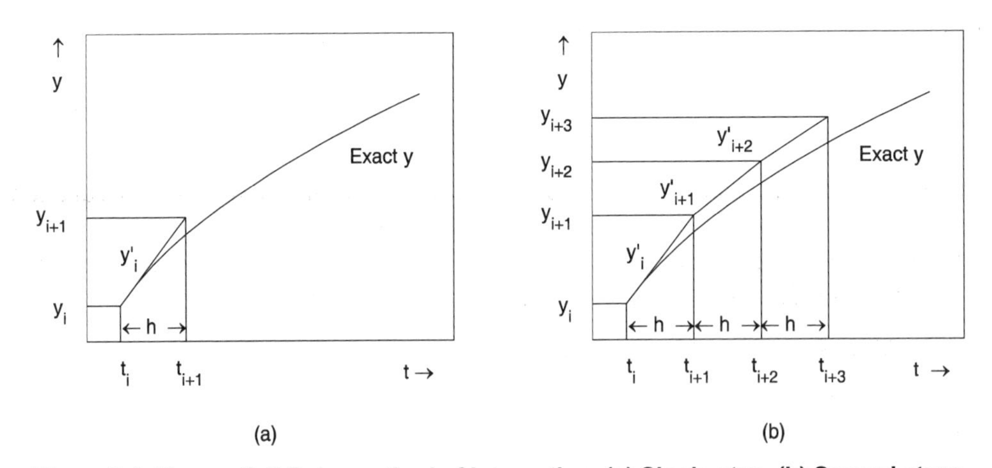

Figure 7.4 The explicit Euler method of integration. (a) Single step. (b) Several steps.

The accuracy of the Euler method can be improved by utilizing a combination of forward and backward Taylor series. To do this we start with Eq. (6.14), that was derived in Example 6.1 using backward Taylor series and backward finite differences:

$$\frac{dy_i}{dx} = \frac{1}{h}(y_i - y_{i-1}) + O(h) \tag{6.14}$$

We rewrite Eq. (6.14) for the derivative at $(i + 1)$:

$$\frac{dy_{i+1}}{dx} = \frac{1}{h}(y_{i+1} - y_i) + O(h) \tag{7.23}$$

and rearrange to solve for y_{i+1}, also replace the derivative by its equivalent $f(t_{i+1}, y_{i+1})$:

$$y_{i+1} = y_i + hf(t_{i+1}, y_{i+1}) + O(h^2) \tag{7.24}$$

This is called the *implicit Euler formula* (or backward Euler), because it involves the calculation of function f at an unknown value of y_{i+1}. Eq. (7.24) can be viewed as taking a step forward from position i to $(i + 1)$ in a gradient direction that must be evaluated at $(i + 1)$.

Implicit equations cannot be solved individually but must be set up as sets of simultaneous algebraic equations. When these sets are linear, the problem can be solved by the application of the Gauss elimination methods developed in Chapter 4. If the set consists of nonlinear equations, the problem is much more difficult and must be solved using Newton's method for simultaneous nonlinear algebraic equations developed in Chapter 5.

In the case of the Euler methods, the problem can be simplified by first applying the explicit method to *predict* a value y_{i+1}:

$$(y_{i+1})_{\text{Predicted}} = y_i + hf(t_i, y_i) + O(h^2) \tag{7.25}$$

and then using this predicted value in the implicit Euler formula (7.24) to get a *corrected* value:

$$(y_{i+1})_{\text{Corrected}} = y_i + hf\left(t_{i+1}, (y_{i+1})_{\text{Predicted}}\right) + O(h^2) \tag{7.26}$$

This combination of steps is known as the *Euler predictor-corrector* (or *modified Euler*) method. Correction by Eq. (7.26) may be applied more than once until the corrected value converges, that is, the difference between the two consecutive corrected values becomes less than the convergence criterion. However, not much more accuracy is achieved after the second application of the corrector.

The explicit, as well as the implicit, forms of the Euler methods have error of order (h^2). However, when used in combination as predictor-corrector, their accuracy is enhanced, yielding an error of order (h^3). This conclusion can be reached by adding

$y_{i+1} = y_i + \Delta y_i$ to $y_{i+1} = y_i + \nabla y_{i+1}$ and rearranging:

$$y_{i+1} = y_i + \frac{1}{2}\left(\Delta y_i + \nabla y_{i+1}\right) \tag{7.27}$$

and utilizing Taylor series expansions to obtain:

$$y_{i+1} = y_i + \frac{h}{2}\left[f\left(t_i, y_i\right) + f\left(t_{i+1}, y_{i+1}\right)\right] + O\left(h^3\right) \tag{7.28}$$

The terms of order (h^2) cancel out because they have opposite sign, thus giving a formula of higher accuracy.

It can be seen by writing Eq. (7.28) in the form:

$$y_{i+1} = y_i + \frac{h}{2} f\left(t_i, y_i\right) + \frac{h}{2} f\left(t_{i+1}, y_{i+1}\right) + O\left(h^3\right) \tag{7.29}$$

that this method uses the weighted trajectories of the function y evaluated at two positions that are located one full step of width h apart and weighted equally. In this form, Eq. (7.29) is also known as the *Crank-Nicolson* method.

Eq. (7.29) can be written in a more general form as:

$$y_{i+1} = y_i + w_1 k_1 + w_2 k_2 \tag{7.30}$$

where, in this case:

$$k_1 = hf\left(t_i, y_i\right) \tag{7.31}$$

$$k_2 = hf\left(t_i + c_2 h,\ y_i + a_{21} k_1\right) \tag{7.32}$$

The choice of the weighting factors, w_1 and w_2, and the positions i and $(i + 1)$ at which to evaluate the trajectories is dictated by the accuracy required of the integration formula, that is, by the number of terms retained in the infinite series expansion.

This concept forms the basis for a whole series of integration formulas, with increasingly higher accuracies, for ordinary differential equations. These are discussed in the following section.

7.4.2 The Runge-Kutta methods

The most widely used methods of integration for ordinary differential equations are the series of methods called Runge-Kutta second-, third-, fourth-, and fifth-order, plus a number of other techniques that are variations on the Runge-Kutta theme. These methods are based on the concept of weighted trajectories formulated at the end of Section 7.4.1. In a more general fashion, the forward marching integration formula for the differential equation (7.11) is given by the recurrence equation:

$$y_{i+1} = y_i + w_1 k_1 + w_2 k_2 + w_3 k_3 + \ldots + w_m k_m \tag{7.33}$$

where each of the trajectories k_i are evaluated by:

$$\begin{aligned} k_1 &= hf\left(t_i, y_i\right) \\ k_2 &= hf\left(t_i + c_2 h, y_i + a_{21} k_1\right) \\ k_3 &= hf\left(x_i + c_3 h, y_i + a_{31} k_1 + a_{32} k_2\right) \\ &\vdots \\ k_m &= hf\left(x_i + c_m h, y_i + a_{m1} k_1 + a_{m2} k_2 + \ldots + a_{m,m-1} k_{m-1}\right) \end{aligned} \tag{7.34}$$

These equations can be written in a compact form as:

$$y_{i+1} = y_i + \sum_{j=1}^{m} w_j k_j \tag{7.35}$$

$$k_j = hf\left(x_i + c_j h, y_i + \sum_{l=1}^{j-1} a_{jl} k_l\right) \tag{7.36}$$

where $c_1 = 0$ and $a_{1j} = 0$. The value of m, which determines the complexity and accuracy of the method, is set when $(m + 1)$ terms are retained in the infinite series expansion of y_{i+1}.

The procedure for deriving the Runge-Kutta methods is outside the scope of the book. The interested student is referred to Constantinides and Mostoufi (1999) for a detailed derivation of the second-order Runge-Kutta method.

Several Runge-Kutta formulas are listed in Table 7.1. The fourth-order Runge-Kutta, which has an error of $O(h^5)$, is probably the most widely used numerical integration method for ordinary differential equations. Implicit Runge-Kutta methods that offer wider regions of stability than the explicit methods have been developed and are thoroughly discussed by Hairer (Hairer et al., 1980, Hairer and Wanner, 1991a, 1991b). These methods, such as Radau5, which uses an implicit fifth-order Runge-Kutta method with step size control, are recommended for the solution of stiff differential equations. Discussion of these methods is outside the scope of this book. The interested user may read the aforementioned references for more details.

Table 7.1 Summary of the Runge-Kutta integration formulas

Second-order Runge-Kutta method (same as Crank-Nicolson method)

$$\begin{aligned} y_{i+1} &= y_i + \frac{1}{2}\left(k_1 + k_2\right) + O\left(h^3\right) \\ k_1 &= hf\left(t_i, y_i\right) \\ k_2 &= hf\left(t_i + h, y_i + k_1\right) \end{aligned} \tag{7.37}$$

Third-order Runge-Kutta method

$$\begin{aligned} y_{i+1} &= y_i + \frac{1}{6}\left(k_1 + 4k_2 + k_3\right) + O\left(h^4\right) \\ k_1 &= hf\left(t_i, y_i\right) \\ k_2 &= hf\left(t_i + \frac{h}{2}, y_i + \frac{k_1}{2}\right) \\ k_3 &= hf\left(t_i + h, y_i + 2k_2 - k_1\right) \end{aligned} \tag{7.38}$$

Fourth-order Runge-Kutta method

$$\begin{aligned} y_{i+1} &= y_i + \frac{1}{6}\left(k_1 + 2k_2 + 2k_3 + k_4\right) + O\left(h^5\right) \\ k_1 &= hf\left(t_i, y_i\right) \\ k_2 &= hf\left(t_i + \frac{h}{2}, y_i + \frac{k_1}{2}\right) \\ k_3 &= hf\left(t_i + \frac{h}{2}, y_i + \frac{k_2}{2}\right) \\ k_4 &= hf\left(t_i + h, y_i + k_3\right) \end{aligned} \tag{7.39}$$

7.4.3 Simultaneous differential equations

It was mentioned at the beginning of Section 7.4 that the methods for the solution of a single differential equation are readily adaptable for solving sets of simultaneous differential equations. To illustrate this, we use the set of n simultaneous ordinary differential equations in their canonical form:

$$\begin{aligned} \frac{dy_1}{dt} &= f_1(t, y_1, y_2, \ldots, y_n) \\ \frac{dy_2}{dt} &= f_2(t, y_1, y_2, \ldots, y_n) \\ &\vdots \\ \frac{dy_n}{dt} &= f_n(t, y_1, y_2, \ldots, y_n) \end{aligned} \tag{7.40}$$

and expand, for example, the fourth-order Runge-Kutta formulas to:

$$\begin{aligned} y_{i+1,j} &= y_i + \frac{1}{6}\left(k_{1j} + 2k_{2j} + 2k_{3j} + k_{4j}\right) + O\left(h^5\right) && j = 1, 2, \ldots, n \\ k_{1j} &= hf_j\left(t_i, y_{i1}, y_{i2}, \ldots, y_{in}\right) && j = 1, 2, \ldots, n \\ k_{2j} &= hf_j\left(t_i + \frac{h}{2}, y_{i1} + \frac{k_{11}}{2}, y_{i2} + \frac{k_{12}}{2}, \ldots, y_{in} + \frac{k_{1n}}{2}\right) && j = 1, 2, \ldots, n \\ k_{3j} &= hf_j\left(t_i + \frac{h}{2}, y_{i1} + \frac{k_{21}}{2}, y_{i2} + \frac{k_{22}}{2}, \ldots, y_{in} + \frac{k_{2n}}{2}\right) && j = 1, 2, \ldots, n \\ k_{4j} &= hf_j\left(t_i + h, y_{i1} + k_{31}, y_{i2} + k_{32}, \ldots, y_{in} + k_{3n}\right) && j = 1, 2, \ldots, n \end{aligned} \tag{7.41}$$

This method is programmable using nested loops. In MATLAB, the values of k and y_i can be put into vectors, thus easily evaluating Eq. (7.41) in matrix form.

7.4.4 MATLAB functions for nonlinear equations

There are several functions in MATLAB that may be used for the integration of sets of ordinary differential equations of the form of (7.11). These solvers, along with their method of solution, are listed in Table 7.2.

Any one of the following statements may be used to call an ODE solver:

```
[T, Y] = solver(@name_func, tspan, y0)
[T, Y] = solver(@name_func, tspan, y0, options)
[T, Y] = solver(@name_func, tspan, y0, options, p1, p2,...)
```

where "`solver`" is one of `ode23`, `ode45`, `ode113`, `ode15s`, `ode23s`, `ode23t`, or `ode23tb`.

Table 7.2 Ordinary differential equation (ODE) solvers in MATLAB

Solver	Method of solution
`ode23`	Runge-Kutta lower-order (second-order, three stages)
`ode45`	Runge-Kutta higher-order (fourth-order, five stages)
`ode113`	Adams-Bashforth-Moulton of varying order (1-13)
`ode15s`	Implicit, multistep of varying order (1-5), for stiff differential equations
`ode23s`	Modified Rosenbrock of second-order, for stiff differential equations
`ode23t`	Implementation of the trapezoidal rule using a "free" interpolant, for moderately stiff differential equations
`ode23tb`	Implementation of an implicit Runge-Kutta formula with a first stage that is a trapezoidal rule step and a second stage that is a backward differentiation formula of second-order, for stiff differential equations

The arguments that are passed to the solver are:

`name_func`: The name of the *m*-file containing the function that evaluates the right-hand side of the differential equations. Function `name_func(t, y)` must return a column vector corresponding to $\mathbf{f}(t, \mathbf{y})$ of the differential equations.

`tspan`: A vector specifying the interval of integration, `[t0,tf]`. To obtain solutions at specific points of `t` (all increasing or all decreasing), use `tspan=[t0,t1,...,tf]`; to obtain solutions at equally spaced intervals, specify `tspan = [t0:delt:tf]`, where `delt` is the user's choice of spacing between points where output will be given.

`y0`: The vector containing the initial conditions of the differential equations.

`options`: Optional integration argument created using the `odeset` function. For details on this function give the command `help odeset` in the MATALB Command Window.

p1, p2, ...: Optional parameters that the solver passes to `name_func` and all the functions specified in options.

[T, Y]: The solver returns the values of independent and dependent variables in the vectors `T, Y,` respectively. The vector of independent variable is not equally spaced, because the integrating solver controls the step size unless the user has specified the `tspan`, as described above.

For example:

```
[T,Y] = ode45(@test1_func,[0:10],[1,0],[],0.1, 0.02, 0.1)

function dydt = test1_func(t, y, p1, p2, p3)
dydt = [p1*y(1)-p2*y(2)^2; p3*exp(y(1))];
```

This function should return the value(s) of the derivative(s) as a column vector. The first input to this function has to be the independent variable, t, even if it is not explicitly used in the definition of the derivative. The second input argument to the function is the vector of dependent variables, y. The additional parameters, `p1, p2, p3`, are the last three values in the `ode45` call, (..., 0.1, 0.02, 0.1), which get passed on to the `test1_func` function.

An alternate way of using these functions is:

```
[T,Y] = ode45('test2_func',[0:10],[1,0],[],0.1, 0.02, 0.1)

function dydt = test2_func(t, y, flag, p1, p2, p3)
dydt = [p1*y(1)-p2*y(2)^2; p3*exp(y(1))];
```

It should be noted that in this case the third input to *test2_func* has to be an empty variable, *flag*, and the additional parameters are introduced starting with the fourth argument.

Example 7.2 Solution of enzyme catalysis reactions.

Statement of the problem

An enzyme, E, catalyzes the conversion of a substrate, S, to form a product, P, via the formation of an intermediate complex, ES, as shown below:

$$S + E \underset{k_{-1}}{\overset{k_1}{\rightleftarrows}} ES \xrightarrow{k_2} P + E$$

Apply the law of mass action to this simple enzymatic reaction to obtain the differential equations that describe the dynamics of the reaction. Use the following values of initial conditions and rate constants to integrate the differential equations and plot the time profiles for all variables in the model:

Initial Conditions: $[S]_0 = 1.0\ \mu\text{M}$ $[E]_0 = 0.1\ \mu\text{M}$ $[ES]_0 = 0$ $[P]_0 = 0$

Constants: $k_1 = 0.1\ (\mu\text{M})^{-1}\text{s}^{-1}$ $k_{-1} = 0.1\ \text{s}^{-1}$ $k_2 = 0.3\ \text{s}^{-1}$

Determine the time (in seconds) it takes for the reaction to reach 99.9% conversion of the substrate.

Solution

The law of mass action states that the rate of molecular collision of two species in a dilute gas or solution is proportional to the product of the two concentrations. Based on this, the model equations are:

$$\frac{d[S]}{dt} = -k_1[S][E] + k_{-1}[ES] \qquad [S]_0 = 1.0$$

$$\frac{d[E]}{dt} = -k_1[S][E] + k_{-1}[ES] + k_2[ES] \qquad [E]_0 = 0.1$$

$$\frac{d[ES]}{dt} = k_1[S][E] - k_{-1}[ES] - k_2[ES] \qquad [ES]_0 = 0$$

$$\frac{d[P]}{dt} = k_2[ES] \qquad [P]_0 = 0$$

We integrate the equations for the period 0 to 1000 seconds using the program listed below as `example7_2.m` and the function `enzyme_kinetics_equations.m`.

Program

```
% example7_2.m - Integration of simple enzyme kinetics model
% using MATLAB function ode45.m to integrate the differential
% equations that are contained in the file:
% enzyme_kinetics_equations.m

clc; clear all;
% Set the initial conditions, constants, & time span
yzero=[1, 0.1, 0, 0];
k1=0.1; k_1=0.1; k2=0.3;
tspan=[0 1000];

% Integrate the equations
[t,y]=ode45(@enzyme_kinetics_equations,tspan,yzero,[],k1,k_1,k2);
n=length(t);
```

```
% Print out the results
n=length(y);
for i=1:n
    if y(i,1)<=0.001*yzero(1)
    fprintf('Reaction is 99.9 percent complete at time = %4.0f
seconds',t(i));
    break
    end
end
% Plot concentration profiles
clf; figure(1); plot(t,y(:,1),'-',t,y(:,4),'-.')
title('Concentration Profiles of Substrate and Product',
'FontSize',12)
xlabel('Time, s','FontSize',12);
ylabel('Concentration, \muM', 'FontSize',12);
legend('S','P');
figure(2); plot(t,y(:,2),'-',t,y(:,3),'-.')
title('Concentration Profiles of Enzyme and Complex',
'FontSize',12)
xlabel('Time, s','FontSize',12);
ylabel('Concentration, \muM', 'FontSize',12);
legend('E','ES')
```

Function that contains equations (`enzyme_kinetics_equations.m`)

```
function dy=enzyme_kinetics_equations(t,y,k1,k_1,k2)
% enzyme_kinetics_equations.m
% Contains the equations for example7_2

% Variables
S=y(1); E=y(2); ES=y(3);
% Equations
dy=[-k1*S*E+k_1*ES
    -k1*S*E+k_1*ES+k2*ES
    k1*S*E-k_1*ES-k2*ES
    k2*ES];
```

Results

The plots (Fig. 7.5a and 7.5b) show that the enzyme complex, $[ES]$, forms quickly within the first few seconds of the reaction. The substrate gets converted steadily to product. The program determines that the reaction reaches 99.9% conversion at 960 seconds. By this time, the enzyme complex disappears and the enzyme returns back to its original free state.

```
Reaction is 99.9 percent complete at time = 960 seconds
```

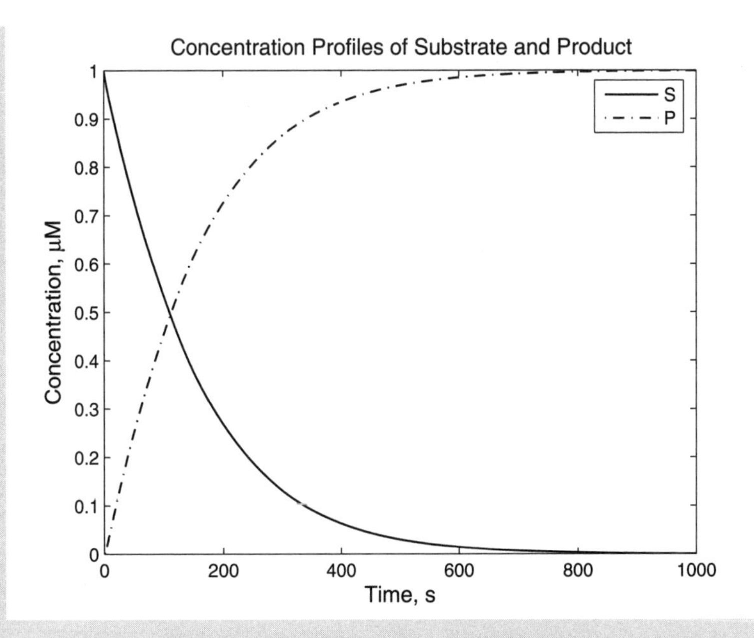

Figure 7.5a Concentration profiles of substrate and product.

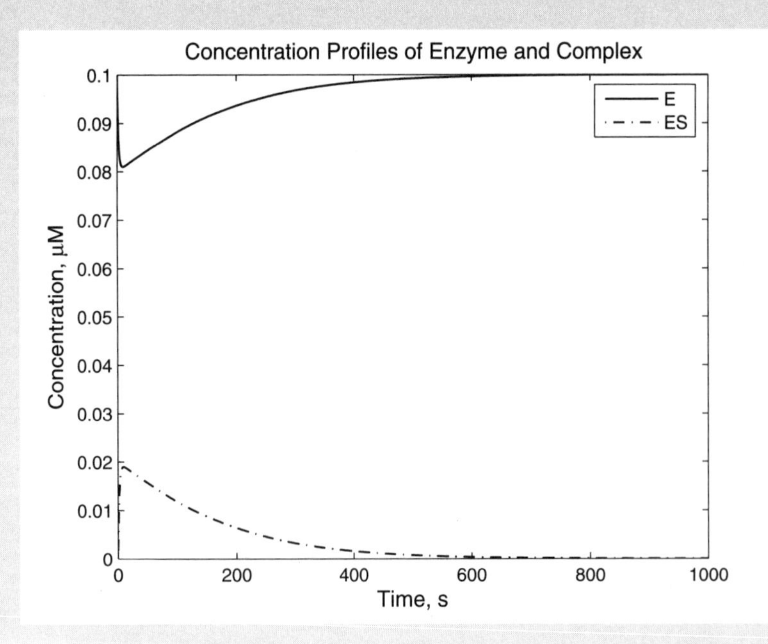

Figure 7.5b Concentration profiles of enzyme and complex.

7.5 Linear Ordinary Differential Equations

The analysis of many bioengineering systems yields mathematical models that are sets of *linear* ordinary differential equations with constant coefficients and can be reduced to the form:

$$\mathbf{y}' = \mathbf{A}\mathbf{y} \tag{7.18}$$

with given initial conditions:

$$\mathbf{y}(0) = \mathbf{y_0} \tag{7.42}$$

7.5.1 Method using eigenvalues and eigenvectors

Sets of linear ordinary differential equations with constant coefficients have closed-form solutions that can be readily obtained from the eigenvalues and eigenvectors of matrix **A**. In order to develop this solution, let us first consider a single linear differential equation of the type:

$$\frac{dy}{dt} = ay \tag{7.43}$$

with the given initial condition:

$$y(0) = y_0 \tag{7.44}$$

Eq. (7.43) is essentially the scalar form of the matrix set of Eq. (7.18). The solution of the scalar equation can be obtained by separating the variables and integrating both sides of the equation:

$$\begin{aligned} \int_{y_0}^{y} \frac{dy}{y} &= \int_0^t a\,dt \\ \ln\frac{y}{y_0} &= at \\ y &= e^{at} y_0 \end{aligned} \tag{7.45}$$

In an analogous fashion, the matrix set can be integrated to obtain the solution:

$$\mathbf{y} = \mathbf{e}^{\mathbf{A}t}\mathbf{y_0} \tag{7.46}$$

In this case, **y** and $\mathbf{y_0}$ are *vectors* of the dependent variables and the initial conditions, respectively. The term $\mathbf{e}^{\mathbf{A}t}$ is the matrix exponential function, which can be obtained from Eq. (7.47):

$$\mathbf{e}^{\mathbf{A}t} = \mathbf{I} + \mathbf{A}t + \frac{\mathbf{A}^2 t^2}{2!} + \frac{\mathbf{A}^3 t^3}{3!} + \frac{\mathbf{A}^4 t^4}{4!} + \dots \tag{7.47}$$

It can be demonstrated that Eq. (7.46) is a solution of Eq. (7.18) by differentiating the former:

$$\begin{aligned}
\frac{d\mathbf{y}}{dt} &= \frac{d}{dt}\left(\mathbf{e}^{\mathbf{A}t}\right)\mathbf{y}_0 \\
&= \frac{d}{dt}\left(\mathbf{I} + \mathbf{A}t + \frac{\mathbf{A}^2 t^2}{2!} + \frac{\mathbf{A}^3 t^3}{3!} + \frac{\mathbf{A}^4 t^4}{4!} + \dots\right)\mathbf{y}_0 \\
&= \left(\mathbf{A} + \mathbf{A}^2 t + \frac{\mathbf{A}^3 t^2}{2!} + \frac{\mathbf{A}^4 t^3}{3!} + \dots\right)\mathbf{y}_0 \\
&= \mathbf{A}\left(\mathbf{I} + \mathbf{A}t + \frac{\mathbf{A}^2 t^2}{2!} + \frac{\mathbf{A}^3 t^3}{3!} + \dots\right)\mathbf{y}_0 \\
&= \mathbf{A}\left(\mathbf{e}^{\mathbf{A}t}\right)\mathbf{y}_0 \\
&= \mathbf{A}y
\end{aligned} \tag{7.48}$$

The solution of the set of linear ordinary differential equations is very cumbersome to evaluate in the form of Eq. (7.47) because it requires the evaluation of the infinite series of the exponential term $\mathbf{e}^{\mathbf{At}}$. However, this solution can be modified by further algebraic manipulation to express it in terms of the eigenvalues and eigenvectors of the matrix **A**. In Chapter 4, we showed that a nonsingular matrix **A** of order n has n eigenvectors and n nonzero eigenvalues, whose definitions are given by:

$$\begin{aligned}
\mathbf{A}\mathbf{x}_1 &= \lambda_1 \mathbf{x}_1 \\
\mathbf{A}\mathbf{x}_2 &= \lambda_2 \mathbf{x}_2 \\
&\vdots \\
\mathbf{A}\mathbf{x}_n &= \lambda_n \mathbf{x}_n
\end{aligned} \tag{7.49}$$

All the above eigenvectors and eigenvalues can be represented in a more compact form, as follows:

$$\mathbf{AX} = \mathbf{X\Lambda} \tag{7.50}$$

where the columns of matrix **X** are the individual eigenvectors:

$$\mathbf{X} = \left[\mathbf{x}_1, \mathbf{x}_2, \mathbf{x}_3, \dots, \mathbf{x}_n\right] \tag{7.51}$$

and $\mathbf{\Lambda}$ is a diagonal matrix with the eigenvalues of **A** on its diagonal:

$$\mathbf{\Lambda} = \begin{bmatrix} \lambda_1 & 0 & 0 & \dots & 0 \\ 0 & \lambda_2 & 0 & \dots & 0 \\ 0 & 0 & \lambda_3 & \dots & 0 \\ \dots & \dots & \dots & \dots & \dots \\ 0 & 0 & 0 & \dots & \lambda_n \end{bmatrix} \tag{7.52}$$

Through a series of matrix operations, Eqs. (7.47) and (7.50) can be combined to express the matrix exponential as follows:

$$\mathbf{e}^{\mathbf{A}t} = \mathbf{X}\mathbf{e}^{\mathbf{\Lambda}t}\mathbf{X}^{-1} \tag{7.53}$$

For a complete derivation of this equation, see Constantinides and Mostoufi (1999).

The solution of the linear differential equations can now be expressed in terms of eigenvalues and eigenvectors by combining Eqs. (7.46) and (7.53):

$$\mathbf{y} = \left[\mathbf{X}\mathbf{e}^{\mathbf{\Lambda}t}\mathbf{X}^{-1}\right]\mathbf{y}_0 \tag{7.54}$$

This method will always work, provided that we can find n linearly independent eigenvectors of the $(n \times n)$ matrix **A**. This is equivalent to saying that matrix **X** must be nonsingular so that its inverse may be calculated. The eigenvalues and eigenvectors of matrix **A** can be calculated using the techniques developed in Appendix C or simply by applying the built-in MATLAB functions described below.

7.5.2 MATLAB functions for linear equations

MATLAB has several functions that may be used to calculate matrix exponentials and eigenvalues/eigenvectors:

`expm(A)`: Calculates the matrix exponential of **A** using a scaling and squaring algorithm with a Pade approximation (Burden et al., 1981).

`expm2(A)`: Calculates the matrix exponential of **A** via Taylor series. As a practical numerical method, this is often slow and inaccurate.

`expm3(A)`: Calculates the matrix exponential of **A** via eigenvalues and eigenvectors. The accuracy of this method is determined by the condition of the eigenvector matrix.

`eig(A)`: Calculates the eigenvalues of matrix **A**.

`[X, LAMBDA] = eig(A)`: Produces a diagonal matrix LAMBDA of eigenvalues, as in Eq. (7.52), and a full matrix X whose columns are the corresponding eigenvectors, as in Eq. (7.51), so that Eq. (7.50) is satisfied, that is, A*X = X*LAMBDA.

Eq. (7.54) may be evaluated using some of the above MATLAB functions as follows:

```
syms t
A = [enter the elements of matrix A]
y0 = [enter the elements of vector y0]
[X, LAMBDA] = eig(A)
y = X*expm(LAMBDA*t)*X^-1*y0
```

The use of these functions is demonstrated in Example 7.3.

Example 7.3 The dynamics of drug absorption.

Statement of the problem

The drug absorption mechanism in the body may be modeled, in its simplest form, as a three-step process, shown diagrammatically in Figure 7.6.

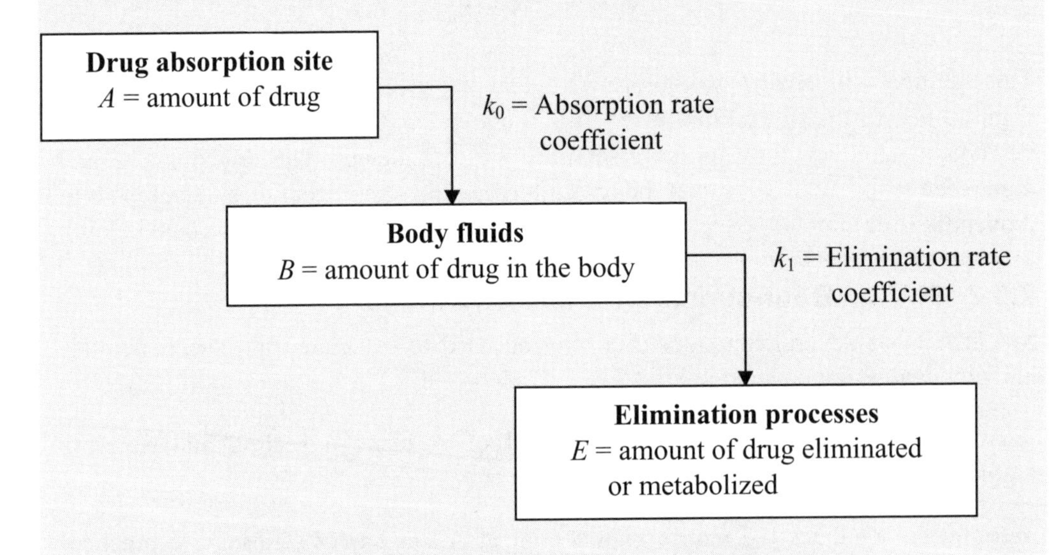

Figure 7.6 Drug absorption and elimination mechanism.

All body fluids are treated as a single unit. Unsteady-state mass balances around each of the three steps yield three linear ordinary differential equations. The equation that describes the rate of change of the amount of drug at the absorption site is:

$$\frac{dA}{dt} = -k_0 A, \qquad A(0) = A_0 \tag{7.55}$$

The rate of change of the amount of drug in the body is described by:

$$\frac{dB}{dt} = k_0 A - k_1 B, \qquad B(0) = 0 \tag{7.56}$$

and the rate of change of the amount of drug eliminated is measured by:

$$\frac{dE}{dt} = k_1 B, \qquad E(0) = 0 \tag{7.57}$$

Equations (7.55), (7.56), and (7.57) constitute a set of simultaneous first-order linear ordinary differential equations whose solutions, $A(t)$, $B(t)$, and $E(t)$, correspond to the concentrations of drug being delivered, being in the body, and being eliminated, respectively. It has been determined that values of $k_0 = 0.01\ \text{min}^{-1}$ and $k_1 = 0.035\ \text{min}^{-1}$ are reasonable values for this system. Use the analytical and numerical solution of these equations to calculate the time, , t_{max}, at which the concentration of drug in the body reaches its maximum value, $B_{max} = B(t_{max})$, and plot the profiles for all three concentrations as functions of time.

Solution

(a) The analytical solutions to the differential equations may be obtained with the Symbolic Math Toolbox command of MATLAB `dsolve` as follows:

```
>> [A,B,E]=dsolve('DA=-k0*A','DB=k0*A-k1*B','DE=k1*B', 'A(0)=A0',
'B(0)=0', 'E(0)=0');
>> A=simplify(A)
A =
A0*exp(-k0*t)
>> B=simplify(B)
B =
k0*A0*(-exp(-k1*t)+exp(-k0*t))/(-k0+k1)
>> E=simplify(E)
E =
-A0*(exp(-k0*t)*k1-k1+k0-exp(-k1*t)*k0)/(-k0+k1)
```

From this output, we conclude that the analytical solutions for A, B, and E are:

$$A(t) = A_0 e^{-k_0 t}$$

$$B(t) = \frac{k_0 A_0}{k_1 - k_0}(e^{-k_0 t} - e^{-k_1 t})$$

$$E(t) = \frac{-A_0(k_1 e^{-k_0 t} - k_0 e^{-k_1 t}) + A_0(k_1 - k_0)}{(k_1 - k_0)}$$

The law of conservation of mass predicts that:

$$A(t) + B(t) + E(t) = A_0 + B_0 + E_0$$

This is easily verified by the MATLAB command (remember that B_0 and E_0 are equal to zero in this problem):

```
>> simplify(A+B+E)
ans =
A0
```

The value of t_{max} is obtained by taking the derivative of $B(t)$, equating it to zero, and solving for t, using the values $k_0 = 0.01$ and $k_1 = 0.035$:

```
>> dB = diff(B)
dB =
k0*A0*(k1*exp(-k1*t)-k0*exp(-k0*t))/(-k0+k1)
>> tmax = solve(dB,'t')
tmax =
log(k1/k0)/(-k0+k1)
>> k0 = 0.01; k1 = 0.035;
>> eval(tmax)
ans =
   50.1105
```

This predicts that the maximum concentration of the drug in the body is reached at approximately 50 minutes after injection.

(b) This problem will now be solved using the eigenvalue-eigenvector method of Eq. (7.54) and the matrix exponential method of Eq. (7.46). The following MATLAB script was written for this purpose. This program is called `example7_3b.m` and is included in the biosystems software that accompanies this book:

Program

```
% example7_3b.m - Solution of the drug absorption problem,
% both symbolically and numerically, using the eigenvalue-
% eigenvector method and the matrix exponential method.

clc; clear all;
syms c t
% Constants
k0=0.01; k1=0.035;
disp('Initial concentrations:')
c0=[1; 0; 0]
disp(' '); disp('Matrix of coefficients:')
K=[-k0 0 0; k0 -k1 0; 0 k1 0]
```

```
% Eigenvalue-eigenvector method
[X,lambda]=eig(K);
disp(' '), disp('Eigenvectors (each column of matrix X):'),   X
disp(' ')
disp('Eigenvalues (on the diagonal of matrix lambda):'),   lambda
disp(' '), disp('Inverse of X:'),   X^-1
disp(' ');
disp('Concentrations using eigenvalue-eigenvector method:')
c=X*expm(lambda*t)*X^-1*c0

% Evaluate concentration profiles
t=[0:100]; c=eval(c);

% Find the maximum concentration and time of drug in the body
[Cmax,tm]=max(c(2,:));
fprintf('\nMaximum concentration in the body = %6.4f at tmax =
%4.2f min.\n',Cmax, tm-1)

% Plot the results
clf; figure(1); h=plot(t,c(1,:), t,c(2,:),':',t,c(3,:),'--');
title('Eigenvalue-Eigenvector Solution')
ylabel('Concentration'); xlabel('Time, min');
legend('C_A','C_B','C_C')

% Matrix exponential method
disp(' '); disp('Concentrations using matrix exponential method:')
syms t
c=expm(K*t)*c0
t=[0:100]; c=eval(c);

% Plot the results
figure(2); h=plot(t,c(1,:), t,c(2,:),':',t,c(3,:),'--');
title('Matrix Exponential Solution')
xlabel('Time, min'); ylabel('Concentration');
legend('C_A','C_B','C_C')
```

Output of results

```
Initial concentrations:
c0 =
     1
     0
     0

Matrix of coefficients:
K =
   -0.0100         0         0
    0.0100   -0.0350         0
         0    0.0350         0
```

```
Eigenvectors (each column of matrix X):
X =
         0         0    0.5661
         0    0.7071    0.2265
    1.0000   -0.7071   -0.7926

Eigenvalues (on the diagonal of matrix lambda):
lambda =
         0         0         0
         0   -0.0350         0
         0         0   -0.0100

Inverse of X:
ans =
    1.0000    1.0000    1.0000
   -0.5657    1.4142         0
    1.7664         0         0

Concentrations using eigenvalue-eigenvector method:
c =
                                exp(-1/100*t)
   -2/5*exp(-7/200*t)+2/5*exp(-1/100*t)
  1+2/5*exp(-7/200*t)-7/5*exp(-1/100*t)

Maximum concentration in the body = 0.1731 at tmax = 50.00 min.

Concentrations using matrix exponential method:
c =
                                exp(-1/100*t)
   -2/5*exp(-7/200*t)+2/5*exp(-1/100*t)
  1+2/5*exp(-7/200*t)-7/5*exp(-1/100*t)
```

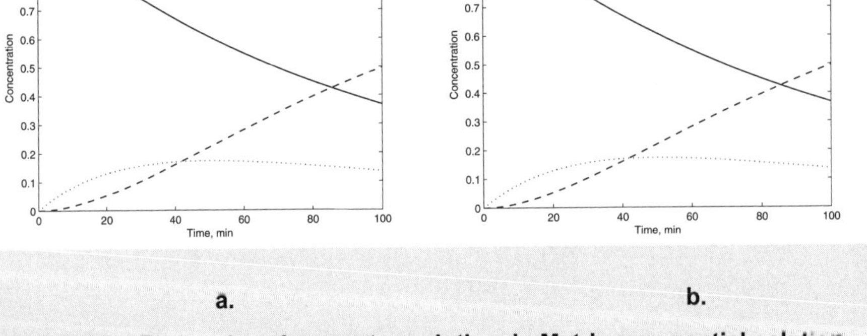

a. b.

Figure 7.7a. Eigenvalue-eigenvector solution, b. Matrix exponential solution.

Discussion of results

As expected, the results from the two methods are identical, and they also confirm the results of the analytical method. The values of t_{max} and B_{max} are 50 min and 0.1731, respectively.

7.6 Steady-State Solutions and Stability Analysis

Before we attempt to obtain the numerical solution of a set of differential equations, let us examine the steady state solution of the problem. The steady state is reached when variations with respect to time become zero. To accomplish this mathematically, we force the time-derivatives to become zero and solve the resulting algebraic equations. It is likely that the set of equations will have multiple steady states, including the trivial case, where all variables are zero. We demonstrate these concepts by analyzing a set of two simultaneous nonlinear ordinary differential equations of the form:

$$\begin{aligned} \frac{dN_1}{dt} &= f_1(N_1, N_2) \\ \frac{dN_2}{dt} &= f_2(N_1, N_2) \end{aligned} \tag{7.58}$$

At steady state the derivatives are set to zero to obtain:

$$f_1\left(N_1^*, N_2^*\right) = 0 \qquad f_2\left(N_1^*, N_2^*\right) = 0 \tag{7.59}$$

where N_1^* and N_2^* are the steady-state values of the dependent variables. We also define the small deviations (perturbations), $\bar{N}_1$ and $\bar{N}_2$, away from the steady state, so that:

$$N_1 = N_1^* + \bar{N}_1 \qquad N_2 = N_2^* + \bar{N}_2 \tag{7.60}$$

By direct substitution of Eqs. (7.60) into Eqs. (7.58), we obtain:

$$\begin{aligned} \frac{d\left(N_1^* + \bar{N}_1\right)}{dt} &= f_1\left(N_1^* + \bar{N}_1, N_2^* + \bar{N}_2\right) \\ \frac{d\left(N_2^* + \bar{N}_2\right)}{dt} &= f_2\left(N_1^* + \bar{N}_1, N_2^* + \bar{N}_2\right) \end{aligned} \tag{7.61}$$

The left-hand sides are expanded into the corresponding two derivatives, and the right-hand sides into Taylor series:

$$
\begin{aligned}
\frac{dN_1^*}{dt}+\frac{d\bar{N}_1}{dt} &= f_1\left(N_1^*,N_2^*\right)+\left(\frac{\partial f_1}{\partial N_1}\right)^*\bar{N}_1+\left(\frac{\partial f_1}{\partial N_2}\right)^*\bar{N}_2+\text{higher order terms}\\
\frac{dN_2^*}{dt}+\frac{d\bar{N}_2}{dt} &= f_2\left(N_1^*,N_2^*\right)+\left(\frac{\partial f_2}{\partial N_1}\right)^*\bar{N}_1+\left(\frac{\partial f_2}{\partial N_2}\right)^*\bar{N}_2+\text{higher order terms}
\end{aligned}
\qquad (7.62)
$$

We apply the condition of steady state (time-derivatives and functions at steady state are zero), and assume that the perturbations around the steady state are small. The latter assumption enables us to drop the higher-order terms that involve $\bar{N}_1^2, \bar{N}_2^2, \bar{N}_1^3, \bar{N}_2^3$, etc., thus essentially linearizing the equations that describe the perturbation around the steady state. Eqs. (7.62) simplify to:

$$
\begin{aligned}
\frac{d\bar{N}_1}{dt} &= \left(\frac{\partial f_1}{\partial N_1}\right)^*\bar{N}_1+\left(\frac{\partial f_1}{\partial N_2}\right)^*\bar{N}_2\\
\frac{d\bar{N}_2}{dt} &= \left(\frac{\partial f_2}{\partial N_1}\right)^*\bar{N}_1+\left(\frac{\partial f_2}{\partial N_2}\right)^*\bar{N}_2
\end{aligned}
\qquad (7.63)
$$

The matrix of partial derivatives is the *Jacobian* of the original set of differential equations evaluated near the neighborhood of the steady state:

$$
\mathbf{J}^* = \begin{bmatrix} \left(\dfrac{\partial f_1}{\partial N_1}\right)^* & \left(\dfrac{\partial f_1}{\partial N_2}\right)^* \\ \left(\dfrac{\partial f_2}{\partial N_1}\right)^* & \left(\dfrac{\partial f_2}{\partial N_2}\right)^* \end{bmatrix} \qquad (7.64)
$$

It should be obvious that Eq. (7.63) is a set of simultaneous linear ordinary differential equations of the form:

$$
\bar{\mathbf{N}}' = \mathbf{J}^*\bar{\mathbf{N}} \qquad (7.65)
$$

It was demonstrated in Section 7.5 that the solution of a set of linear ordinary differential equations of the form of Eq. (7.18) can be obtained from the eigenvalues of the matrix **A**. Similarly, the solution of Eq. (7.65) will depend on the eigenvalues of the Jacobian matrix $\mathbf{J}^*$. The eigenvalues could be real positive, real negative, and/or complex with positive or negative real parts.

Let us show the eigenvalues in their most general form:

$$\lambda_k = a_k \pm b_k i \qquad k = 1, 2, \ldots, n \tag{7.66}$$

where a_k are the real parts, b_k are the coefficients of the imaginary part of the eigenvalues, and $i = \sqrt{-1}$, remembering that complex eigenvalues appear as conjugate pairs. We now summarize all possible cases and their stability analysis in Table 7.3, and show time profiles and phase plots of (N_1 vs. N_2) for the corresponding cases in Fig. 7.8. Negative eigenvalues result in stable solutions (Cases 1 & 2), while positive eigenvalues cause instability (Cases 3 & 4). The presence of complex eigenvalues introduces oscillatory behavior in the solutions (Cases 2, 4, & 6). If both positive and negative real values exist, the solution is a metastable saddle point (Case 5). Finally, if the eigenvalues are complex and the real parts are zero, the results are neutrally stable oscillatory (Case 6).

Similar analysis applies to sets of equations that contain n dependent variables (where $n > 2$). In that case, the Jacobian is of size ($n \times n$), and phase plots of pairs of variables are constructed. Three-dimensional phase plots may also be constructed, if their use is deemed instructive.

Table 7.3 Stability Analysis Based on the Eigenvalues of the Jacobian Matrix.

Case	a_k	b_k	Stability analysis
1	All negative	Zero	Stable, nonoscillatory
2	All negative	Nonzero	Stable, oscillatory
3	All positive	Zero	Unstable, nonoscillatory
4	All positive	Nonzero	Unstable, oscillatory
5	Positive and negative	Zero	Metastable, saddle point
6	Zero	Nonzero	Neutrally stable, oscillatory

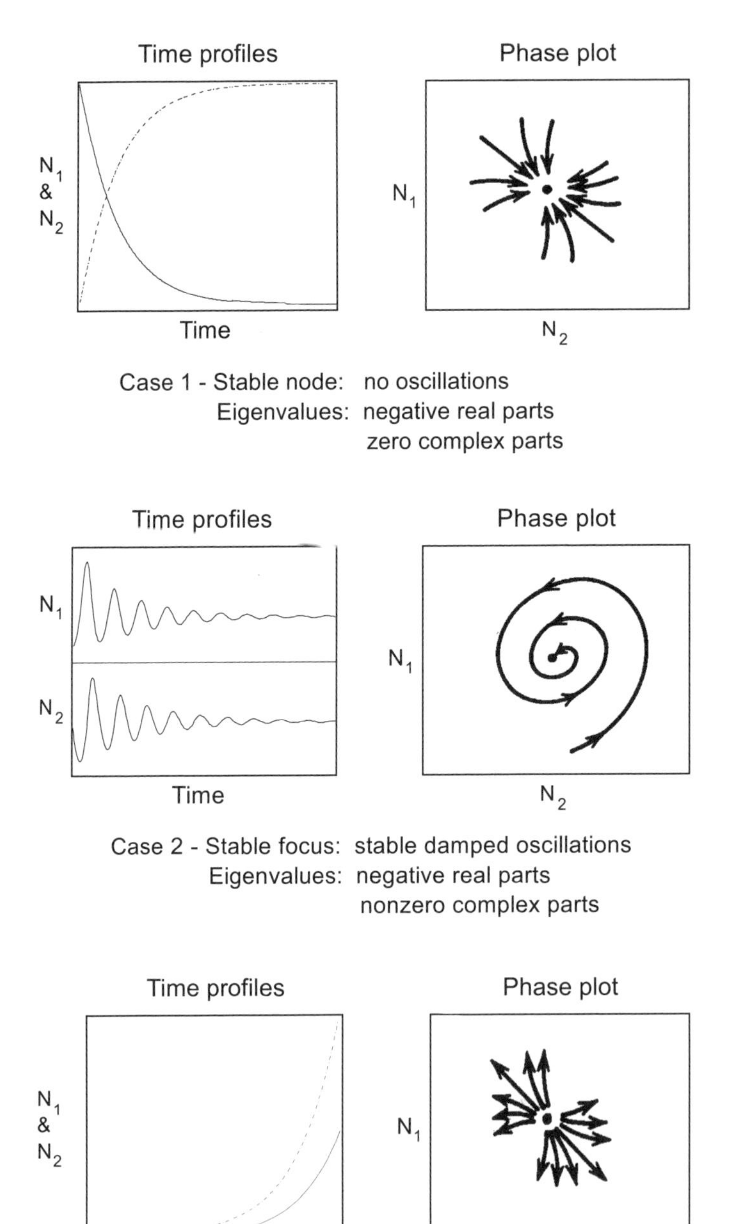

Figure 7.8 Time profiles and phase plots for stability analysis.

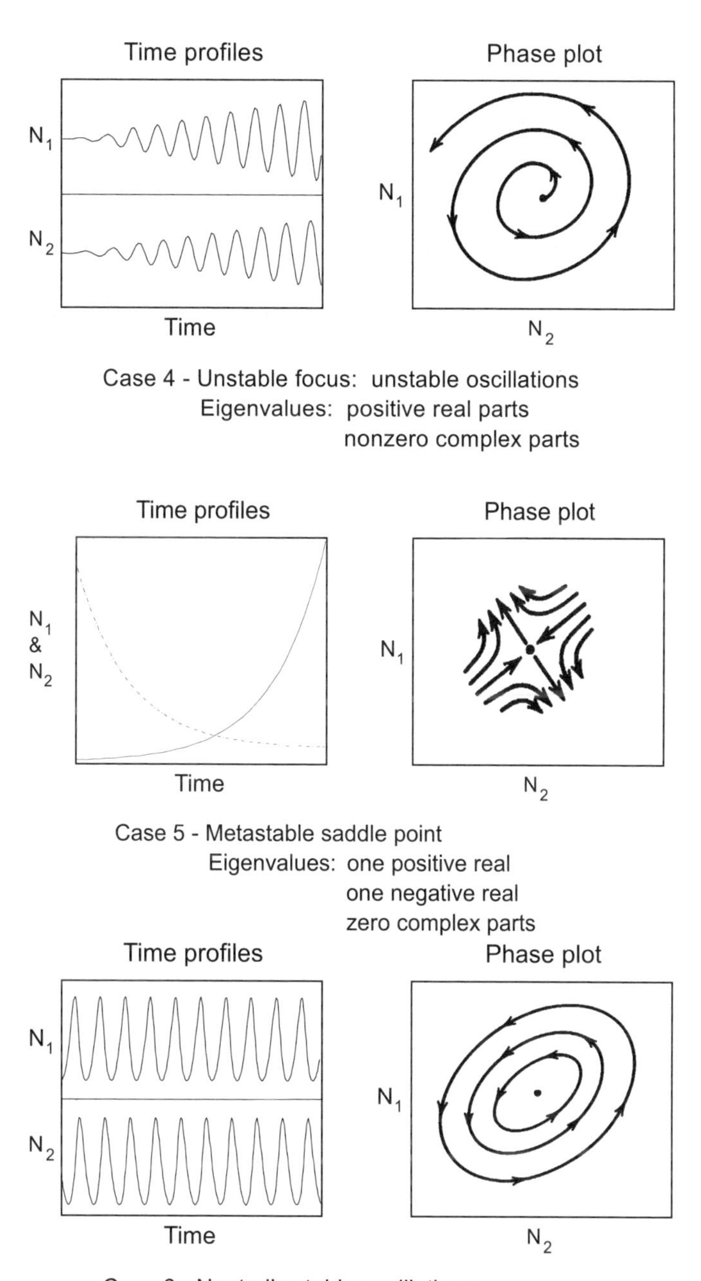

Case 6 - Neutrally stable oscillations
Eigenvalues: zero real parts
nonzero complex parts

Figure 7.8 (cont.) Time profiles and phase plots for stability analysis.

7.7 Numerical Stability and Error Propagation

Topics of paramount importance in the numerical integration of differential equations are the error propagation, stability, and convergence of these solutions. Two types of stability considerations enter in the solution of ordinary differential equations: inherent stability (or instability) and numerical stability (or instability). Inherent stability is determined by the mathematical formulation of the problem and is dependent on the eigenvalues of the Jacobian matrix of the differential equations, as was shown in Section 7.6. On the other hand, numerical stability is a function of the error propagation in the numerical integration method. The behavior of error propagation depends on the values of the characteristic roots of the difference equations that yield the numerical solution. In this section, we concern ourselves with numerical stability considerations as they apply to the numerical integration of ordinary differential equations.

There are three types of errors present in the application of numerical integration methods. These are the *truncation error*, the *roundoff error*, and the *propagation error*. The truncation error is a function of the number of terms that are retained in the approximation of the solution from the infinite series expansion. The truncation error may be reduced by retaining a larger number of terms in the series or by reducing the step size of integration h. The plethora of available numerical methods of integration of ordinary differential equations provides a choice of increasingly higher accuracy (lower truncation error), at an escalating cost in the number of arithmetic operations to be performed, and with the concomitant accumulation of roundoff errors.

Computers carry numbers using a finite number of significant figures, as was discussed in Chapter 3. A roundoff error is introduced in the calculation when the computer rounds up or down (or just chops) the number to n significant figures. Roundoff errors may be reduced significantly by the use of double precision. However, even a very small roundoff error may affect the accuracy of the solution, especially in numerical integration methods that march forward (or backward) for hundreds or thousands of steps, each step being performed using rounded numbers.

The truncation and roundoff errors in numerical integration accumulate and propagate, creating the propagation error, which, in some cases, may grow in exponential or oscillatory pattern, thus causing the calculated solution to deviate drastically from the correct solution.

Fig. 7.9 illustrates the propagation of error in a numerical integration method. Starting with a known initial condition y_0, the method calculates the value y_1, which contains the truncation error for this step and a small roundoff error introduced by the computer. The error has been magnified on the figure in order to illustrate it more clearly. The next step starts with y_1 as the initial point and calculates y_2. But because y_1 already contains truncation and roundoff errors, the value obtained for y_2 contains these errors propagated, in addition to the new truncation and roundoff errors from the second step. The same process occurs in subsequent steps.

Error propagation in numerical integration methods is a complex operation that depends on several factors. Roundoff error, which contributes to propagation error, is entirely determined by the accuracy of the computer being used. The truncation error is fixed by the choice of method being applied, by the step size of integration, and by the values of the derivatives of the functions being integrated. For these reasons, it is necessary to examine the error propagation and stability of each method individually and in connection with the differential equations to be integrated. Some techniques work well with one class of differential equations but fail with others.

More detailed discussions of stability of numerical integration methods, and other advanced topics, such as step size control and integration of stiff differential equations, are included in Appendix E of this book.

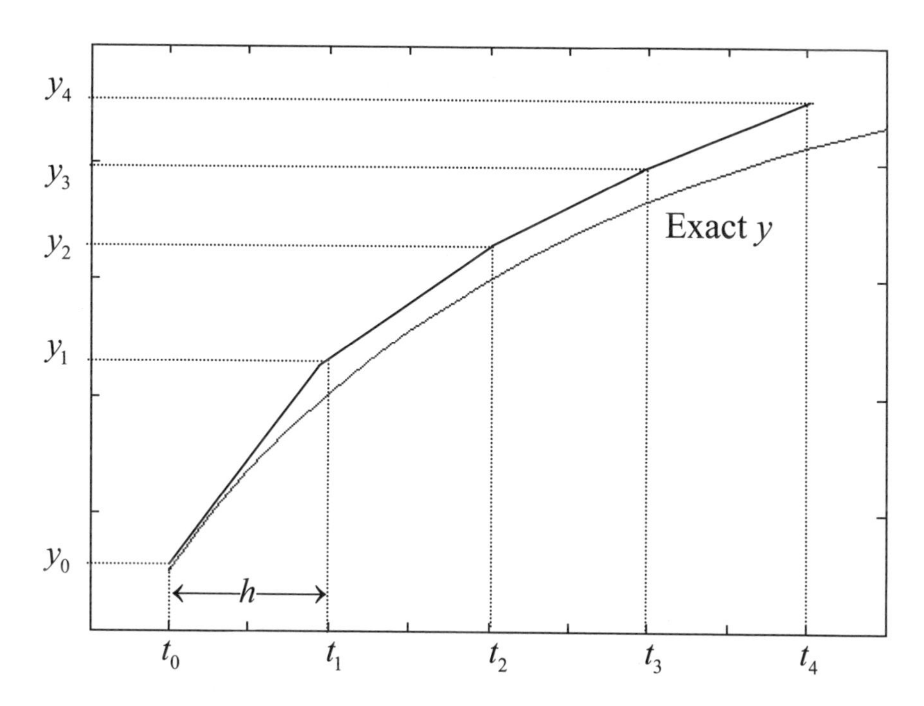

Figure 7.9 **Error propagation in numerical integration methods. The error has been magnified in order to illustrate it more clearly.**

7.8 Advanced Examples

Example 7.4 Metabolic engineering: Modeling the glycolysis pathways of living cells.

Statement of the problem

An important step in the glycolytic pathway is the phosphorylation of fructose 6-phosphate to fructose 1,6-biphosphate. This reaction is catalyzed by the enzyme phosphofructokinase. This enzyme is an example of an allosteric enzyme that is inhibited by ATP and stimulated by adenosine diphosphate (ADP) or by adenosine-monophosphate (AMP). The enzyme becomes active when it combines with γ molecules of ADP:

$$\text{Enzyme} + \gamma\text{ADP} \underset{k_{-3}}{\overset{k_3}{\rightleftarrows}} \text{Enzyme-ADP}^{\gamma}$$

The active complex catalyzes the reaction of fructose 6-phosphate to fructose 1,6-bi-phosphate, and in this process it converts one molecule of ATP to one molecule of ADP, as follows:

$$\begin{array}{c} \text{Fructose 6-phosphate} \\ \Updownarrow \\ \text{ATP} + \text{Enzyme-ADP}^{\gamma} \underset{k_{-1}}{\overset{k_1}{\rightleftarrows}} \text{ATP-Enzyme-ADP}^{\gamma} \xrightarrow{k_2} \text{Enzyme-ADP}^{\gamma} + \text{ADP} \\ \Updownarrow \\ \text{Fructose 1,6-biphosphate} \end{array}$$

This is the Sel'kov model as discussed by Keener and Sneyd (1998). Since the net result of this reaction is the formation of an additional ADP molecule that may further activate the enzyme, this reaction has a positive feedback effect on itself. Assuming that there is a steady supply of the ATP available to this reaction at the rate of v_1, and an irreversible flow of ADP away from the reaction at the rate of v_2, the steps of the reaction that involve the consumption of ATP and the formation of ADP, via the formation of enzyme complexes, may be shown schematically as:

$$\xrightarrow{v_1} \text{S}_1$$

$$\gamma\text{S}_2 + \text{E} \underset{k_{-3}}{\overset{k_3}{\rightleftarrows}} \text{ES}_2^{\gamma}$$

$$\text{S}_1 + \text{ES}_2^{\gamma} \underset{k_{-1}}{\overset{k_1}{\rightleftarrows}} \text{S}_1\text{ES}_2^{\gamma} \xrightarrow{k_2} \text{ES}_2^{\gamma} + \text{S}_2$$

$$\text{S}_2 \xrightarrow{v_2}$$

where S_1 represents the ATP molecule, S_2 stands for the ADP molecule, and E represents the enzyme phosphofructokinase.

Keener and Sneyd applied the law of mass action to this reaction scheme to obtain the following set of ordinary differential equations that describe the dynamics of the reactions:

$$
\begin{aligned}
\frac{ds_1}{dt} &= v_1 - k_1 s_1 x_1 + k_{-1} x_2 \\
\frac{ds_2}{dt} &= k_2 x_2 - k_3 s_2^{\gamma} e + k_{-3} x_1 - v_2 s_2 \\
\frac{dx_1}{dt} &= -k_1 s_1 x_1 + (k_{-1} + k_2) x_2 + k_3 s_2^{\gamma} e - k_{-3} x_1 \\
\frac{dx_2}{dt} &= k_1 s_1 x_1 - (k_{-1} + k_2) x_2 \\
\frac{de}{dt} &= -\frac{dx_1}{dt} - \frac{dx_2}{dt}
\end{aligned}
\tag{7.67}
$$

where $s_1 = [S_1] = [ATP]$, $s_2 = [S_2] = [ADP]$, $e = [E]$, $x_1 = [ES_2^{\gamma}]$, $x_2 = [S_1ES_2^{\gamma}]$. Square brackets are used to denote concentration of the particular compound in the cell. The last equation that describes the rate of change (de/dt) of the free enzyme is obtained from the balance equation for the total enzyme in the cell (e_0), assuming that the total amount of enzyme remains constant:

$$e + x_1 + x_2 = e_0 \tag{7.68}$$

The above equations are a set of simultaneous first-order nonlinear ordinary differential equations. Methods of solution for such a set were developed in Section 7.4, and are applied here to obtain the solution of the glycolysis problem in this example.

Solution

(a) It is well known in the literature that the rate of glycolysis is oscillatory. To show this, integrate the above set of differential equations with the following initial conditions and constants:

Initial Conditions: $s_1(0) = 1.0$ $s_2(0) = 0.2$ $x_1(0) = 0$ $x_2(0) = 0$ $e_0(0) = 1.4$

Constants: $\gamma = 2.0$ $v_1 = 0.003$ $v_2 = 2.5 * v_1$ $k_1 = 0.1$
$k_{-1} = 0.2$ $k_2 = 0.1$ $k_3 = 0.2$ $k_{-3} = 0.2$

Note: The constants contain units of time (seconds) and concentrations (nM) as needed for unit consistency of the equations.

Plot the concentration profiles of all five dependent variables and discuss the results. Plot the phase plot of s_1 and s_2, and discuss what this phase plot demonstrates.

(b) Perform a stability analysis of these equations by examining the eigenvalues of the Jacobian matrix evaluated around the steady state. How do the eigenvalues predict the oscillatory behavior of the concentration vs. time profiles?

(a) **Integration of equations**

The program `example7_4a.m`, listed below, integrates the differential equations using *ode45* and plots the results.

Program

```
% example7_4a.m - Integration of the glycolysis model
% using the MATLAB function ode45.m to integrate the
% differential equations that are contained in the file:
% glycolysis_equations.m

clc; clear all;

% Set the initial conditions & time span
yzero=[1, .2, 0, 0, 1.4];
tspan=[0 3000];

% Integrate the equations
[t,y]=ode45(@glycolysis_equations,tspan,yzero);
n=length(t);

% Plot concentration profiles
clf; figure(1); plot(t,y)
title('Concentration Profiles of Glycolysis')
xlabel('Time, s'); ylabel('Concentration')
text(530,1.35,'ATP (s_1)'); text(900,0.65,'ADP (s_2)')
text(1600,0.25,'Enzyme-ADP complex (x_1)')
text(1600,0.09,'ATP-Enzyme-ADP complex (x_2)')
text(1600,1.28,'free enzyme (e)')

% Plot phase diagrams
figure(2); plot(y(:,1),y(:,2))
title('Phase Plot of Glycolysis')
xlabel('ATP (s_1)'); ylabel('ADP (s_2)')
```

Function that contains equations (`glycolysis_equations.m`)

```
function dy=glycolysis_equations(t,y)
% glycolysis_equations.m
% Contains the glycolysis model for example7_4a
% Constants
gamma=2; neu1=0.003; neu2=2.5*neu1;
k1=0.1; k_1=2*k1; k2=0.1; k3=0.2; k_3=k3;
s1=y(1); s2=y(2); x1=y(3); x2=y(4); e =y(5);
% Equations
dy=[neu1-k1*s1*x1+k_1*x2
    k2*x2-k3*s2^gamma*e+k_3*x1-neu2*s2
    -k1*s1*x1+(k_1+k2)*x2+k3*s2^gamma*e-k_3*x1
    k1*s1*x1-(k_1+k2)*x2
    -(-k1*s1*x1+(k_1+k2)*x2+k3*s2^gamma*e-k_3*x1)-(k1*s1*x1-
(k_1+k2)*x2)];
```

Results of integration

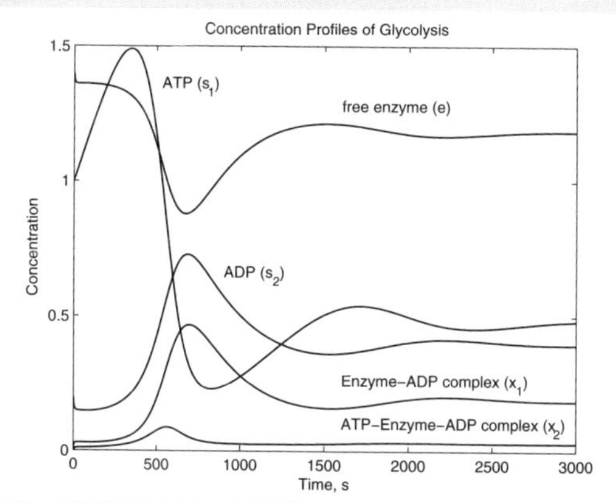

Figure 7.10a Concentration profiles of glycolysis.

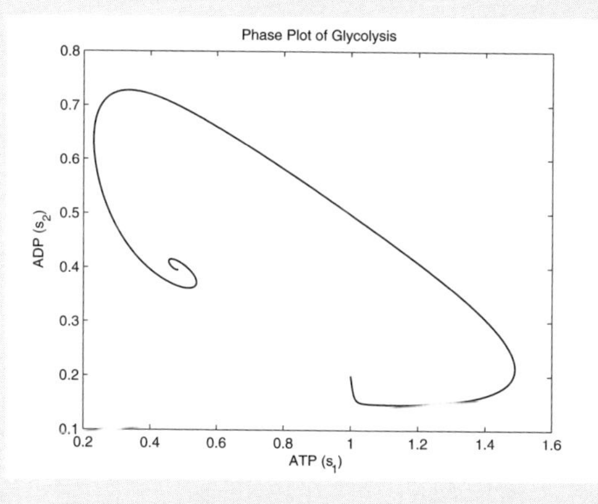

Figure 7.10b Phase plot of glycolysis.

The concentration profiles (Fig. 7.10a) of the glycolysis system of equations indicate that the above set of constants and initial conditions represent a case that is oscillatory at first, but approaches steady state within 3,000 seconds (50 minutes). The phase plot of ADP vs. ATP (Fig. 7.10b) exhibits a stable focus of the type shown on Fig. 7.8 Case (2).

(b) **Steady-state analysis of glycolysis equations**

The program `example7_4b.m`, listed below, performs the stability analysis of Eqs. (7.67) by first evaluating the Jacobian matrix of the differential equations using the MATLAB command `jacobian(dy,v)`, where `dy` is the vector of derivatives and `v` is the vector of variables. Next, it calculates the steady-state solution of the differential equations by setting the derivatives equal to zero and solving for the unknown variables using the MATLAB command *solve*. Finally, the stability of the steady state is examined by obtaining the eigenvalues of the Jacobian matrix around the steady state, using the command `eig`.

Program

```
% example7_4b.m - Steady state analysis of the glycolysis model
% using MATLAB functions jacobian.m and eig.m

clc; clear all;

% Set the constants
e0=1.4; gamma=2; neu1=0.003; neu2=2.5*neu1;
k1=0.1; k_1=2*k1; k2=0.1; k3=0.2; k_3=k3;

% Evaluate the Jacobian matrix
syms s1 s2 x1 x2 e
disp('Steady State Analysis of the Glycolysis Equations:')
v=[s1, s2, x1, x2, e];
dy=[neu1-k1*s1*x1+k_1*x2;
    k2*x2-k3*s2^gamma*e+k_3*x1-neu2*s2;
    -k1*s1*x1+(k_1+k2)*x2+k3*s2^gamma*e-k_3*x1;
    k1*s1*x1-(k_1+k2)*x2;
    -(-k1*s1*x1+(k_1+k2)*x2+k3*s2^gamma*e-k_3*x1)-(k1*s1*x1-
(k_1+k2)*x2)];
J=jacobian(dy,v);
disp('The Jacobian matrix is:'), J
% Evaluate the steady state solution
[SteadyState]=solve('neu1-k1*s1*x1+k_1*x2=0',...
    'k2*x2-k3*s2^gamma*e+k_3*x1-neu2*s2=0',...
    '-k1*s1*x1+(k_1+k2)*x2+k3*s2^gamma*e-k_3*x1=0',...
    'k1*s1*x1-(k_1+k2)*x2=0',...
```

```
    'e+x1+x2=e0', 's1,s2,x1,x2,e');
disp(' '), disp('The steady state values of each variable are:')
disp('s1'),disp(SteadyState.s1),disp(' ')
disp('s2'),disp(SteadyState.s2),disp(' ')
disp('x1'),disp(SteadyState.x1),disp(' ')
disp('x2'),disp(SteadyState.x2),disp(' ')
disp('e '),disp(SteadyState.e), disp(' ')
n=length(SteadyState.s1);
disp('Value of each variable at the steady state(s):')
disp('           s1          s2          x1          x2          e')
for i=1:n;
    s1=eval(SteadyState.s1); s2=eval(SteadyState.s2);
    x1=eval(SteadyState.x1); x2=eval(SteadyState.x2);
    e =eval(SteadyState.e);
    fprintf(' %2i   %9.4f  %9.4f  %9.4f  %9.4f  %9.4f   \n',...
       i, s1(i), s2(i), x1(i), x2(i), e);
end
for i=1:n
    s1=eval(SteadyState.s1); s2=eval(SteadyState.s2);
    x1=eval(SteadyState.x1); x2=eval(SteadyState.x2);
    e =eval(SteadyState.e);
   fprintf('\nSteady state %2i \n',i)
   disp(' '); disp('Jacobian matrix at steady state:'), eval(J)
   disp(' '); disp('Eigenvalues of Jacobian at steady state:');
eig(eval(J))
end
```

Results of steady state analysis

```
Steady State Analysis of the Glycolysis Equations:
The Jacobian matrix is:
J =
[  -1/10*x1,                  0,       -1/10*s1,      1/5,          0]
[         0,  -2/5*s2*e-3/400,              1/5,     1/10,   -1/5*s2^2]
[  -1/10*x1,          2/5*s2*e,   -1/10*s1-1/5,     3/10,    1/5*s2^2]
[   1/10*x1,                  0,        1/10*s1,    -3/10,          0]
[         0,         -2/5*s2*e,              1/5,       0,   -1/5*s2^2]


The steady state values of each variable are:

s1
neu1*(k_1+k2)*(k3*exp(log(neu1/neu2)*gamma)+k_3)/k1/exp(log(neu1/
   neu2)*gamma)/k3/(-neu1+e0*k2)

s2
neu1/neu2
x1
```

```
exp(log(neu1/neu2)*gamma)*k3*(-neu1+e0*k2)/k2/
   (k3*exp(log(neu1/neu2)*gamma)+k_3)

x2
neu1/k2

e
k_3*(-neu1+e0*k2)/k2/(k3*exp(log(neu1/neu2)*gamma)+k_3)

Value of each variable at the steady state(s):
            s1          s2          x1          x2          e
  1      0.4763      0.4000      0.1890      0.0300      1.1810

Steady state  1

Jacobian matrix at steady state:
ans =
   -0.0189         0   -0.0476    0.2000         0
         0   -0.1965    0.2000    0.1000   -0.0320
   -0.0189    0.1890   -0.2476    0.3000    0.0320
    0.0189         0    0.0476   -0.3000         0
         0   -0.1890    0.2000         0   -0.0320

Eigenvalues of Jacobian at steady state:
ans =
  -0.4859
  -0.3060
  -0.0015 + 0.0044i
  -0.0015 - 0.0044i
  -0.0000
```

For this system of equations and constants, the steady state analysis shows that one steady state exists at which the concentrations of the main components are:

[ATP] = 0.4763   [ADP] = 0.4000
[Enzyme-ADP complex] = 0.1890
[ATP-Enzyme-ADP complex] = 0.0300
[free Enzyme] = 1.1810

The eigenvalues of the Jacobian matrix of this system are: two real negative, two complex with negative real parts, and one zero eigenvalue. Such a combination of eigenvalues predicts an oscillatory behavior with damped oscillations approaching a steady state. The zero eigenvalue is a direct consequence of the conservation of mass principle applied to the enzyme. These results confirm the evolution of the system shown by the concentration profiles.

Example 7.5 The dynamics of membrane and nerve cell potentials.

Formulation of the problem

The activation and inactivation of the potassium/sodium channels and the role they play in the generation of nerve action potential formed the basis of the Nobel Prize winning work of Hodgkin and Huxley in the 1940s and 1950s (Hodgkin and Huxley, 1952). They studied the effect of the application of voltage potentials on the Na+ and K+ channels on the giant squid axon and developed mathematical models that describe the dynamics of the processes.

Numerous papers and books have been written on the Hodgkin-Huxley model. A very concise description of this model is that of Keener and Sneyd (1998). They begin by showing that the cell membrane can be modeled as a capacitor in parallel with an ionic current, and since there can be no buildup of charge on either side of the membrane, the sum of the ionic and capacitive currents must be zero, resulting in the equation:

$$C_m \frac{dV}{dt} + I_{ion} = 0 \tag{7.69}$$

where V denotes the internal minus the external potential. In the giant squid axon, and in many nerve cells, the principal ionic currents are the sodium current, I_{Na}, and the potassium current, I_K. Other currents that are present, such as the chloride current, are lumped together into one current called the *leakage current*, I_L. The ionic currents for sodium and potassium ions can be modeled by the current-voltage relationships:

$$I_{Na} = g_{Na}(V - V_{Na}) \tag{7.70}$$

$$I_K = g_K(V - V_K) \tag{7.71}$$

and the leakage current may be shown as:

$$I_L = g_L(V - V_L) \tag{7.72}$$

where g_{Na} and g_K are the membrane conductances for sodium and potassium ions, respectively, and g_L is a combined conductance for leakage current. V_{Na} and V_K are the equilibrium membrane potentials due to concentration differences of the two ions, sodium and potassium, and V_L is the potential at which the leakage current due to chloride and other ions is zero.

The sodium and potassium potentials are calculated from the Nernst equation:

$$V_{\mathrm{Na}} = \frac{RT}{zF} \ln\left(\frac{\left[\mathrm{Na}^+\right]_e}{\left[\mathrm{Na}^+\right]_i} \right) \tag{7.73}$$

$$V_{\mathrm{K}} = \frac{RT}{zF} \ln\left(\frac{\left[\mathrm{K}^+\right]_e}{\left[\mathrm{K}^+\right]_i} \right) \tag{7.74}$$

Ionic channels open and close in response to a voltage. This behavior of ionic channels in response to changes in membrane potential is the basis for electrical excitability, and is of fundamental significance to neurophysiology. According to Keener and Sneyd (1998), the current flow through a population of channels is the product of two terms:

$$I = \eta(V,t)\,\phi(V) \tag{7.75}$$

where $\eta(V,t)$ is the proportion of open channels in a population, and $\phi(V)$ is the *I-V* curve of a single channel. The simplest model for the K+ channel assumes that the channel can exist either in the closed state in the proportion of $(1-\eta)$, or in the open state in the proportion of η:

$$\overbrace{(1-\eta)}^{Closed} \underset{\beta(V)}{\overset{\alpha(V)}{\rightleftarrows}} \overbrace{(\eta)}^{Open} \tag{7.76}$$

Then the rate of change of the open channels may be modeled by the differential equation:

$$\frac{d\eta}{dt} = \alpha(V)(1-\eta) - \beta(V)\eta \tag{7.77}$$

It is sometimes instructive to write Eq. (7.77) in the form:

$$\tau_\eta(V)\frac{d\eta}{dt} = \eta_\infty(V) - \eta \tag{7.78}$$

where $\eta_\infty(V)$ is the steady state value of η, that may be obtained from Eq. (7.77) as:

$$\eta_\infty(V) = \frac{\alpha}{\alpha+\beta} \tag{7.79}$$

and τ_η is the time constant of approach to steady state:

$$\tau_\eta = \frac{1}{\alpha + \beta} \tag{7.80}$$

Hodgkin and Huxley used a *voltage clamp* in their studies of the giant squid axon. The user of a voltage clamp fixes the membrane potential by applying a rapid step from one voltage to another and then measures the current that must be applied, I_{app}, to hold the voltage constant. Based on their experimental data, Hodgkin and Huxley modified the potassium conductance, g_K, in order to obtain sigmoidal increase and exponential decrease:

$$g_{\text{K}} = \overline{g}_K n^4 \tag{7.81}$$

They also modified the sodium conductance, g_{Na}, to account for two processes at work, one that turns on the sodium current and one that turns it off:

$$g_{\text{Na}} = \overline{g}_{Na} m^3 h \tag{7.82}$$

Keener and Sneyd (1998) interpret the potassium mechanism to be equivalent to having four "n" gates per potassium channel, all of which must be open for potassium to flow. They also elucidate the mechanism of the Na^+ channel as consisting of three "m" gates and one "h" gate, each of which can be either closed or open. Combining equations (7.69)-(7.72), (7.77), (7.81), and (7.82) results in the complete Hodgkin-Huxley model:

$$\begin{aligned}
C_m \frac{dv}{dt} &= -\overline{g}_K n^4 (v - v_{\text{K}}) - \overline{g}_{Na} m^3 h (v - v_{\text{Na}}) - \overline{g}_L (v - v_{\text{L}}) + I_{\text{app}} \\
\frac{dn}{dt} &= \alpha_n (1 - n) - \beta_n n \\
\frac{dm}{dt} &= \alpha_m (1 - m) - \beta_m m \\
\frac{dh}{dt} &= \alpha_h (1 - h) - \beta_h h
\end{aligned} \tag{7.83}$$

The potential, v, is the deviation from rest potential ($v = V - V_{eq}$) measured in units of mV; current density I is in units of μA/cm^2; conductances are in units of mS/cm^2; and capacitance C_m is in μF/cm^2. The rate constants of α and β are, in units of (ms)$^{-1}$:

$$
\begin{aligned}
\alpha_n &= 0.01\frac{10-v}{e^{\left(\frac{10-v}{10}\right)}-1} & \beta_n &= 0.125e^{\left(\frac{-v}{80}\right)} \\
\alpha_m &= 0.1\frac{25-v}{e^{\left(\frac{25-v}{10}\right)}-1} & \beta_m &= 4e^{\left(\frac{-v}{18}\right)} \\
\alpha_h &= 0.07e^{\left(\frac{-v}{20}\right)} & \beta_h &= \frac{1}{e^{\left(\frac{30-v}{10}\right)}+1}
\end{aligned}
\tag{7.84}
$$

The steady state values of the gating variables and the time constants are:

$$
\begin{aligned}
n_\infty &= \frac{\alpha_n}{\alpha_n+\beta_n} & m_\infty &= \frac{\alpha_m}{\alpha_m+\beta_m} & h_\infty &= \frac{\alpha_h}{\alpha_h+\beta_h} \\
\tau_n &= \frac{1}{\alpha_n+\beta_n} & \tau_m &= \frac{1}{\alpha_m+\beta_m} & \tau_h &= \frac{1}{\alpha_h+\beta_h}
\end{aligned}
\tag{7.85}
$$

The constants and initial conditions for this simulation are:

$$
\begin{aligned}
&\bar{g}_{\mathrm{K}} = 36\ \mathrm{mS/cm^2} && \bar{g}_{\mathrm{Na}} = 120\ \mathrm{mS/cm^2} && \bar{g}_L = 0.3\ \mathrm{mS/cm^2} \\
&v_{\mathrm{K}} = \text{-}12\ \mathrm{mV} && v_{\mathrm{Na}} = 115\ \mathrm{mV} && v_{\mathrm{L}} = 10.6\ \mathrm{mV} \\
&v(0) = 8\ \mathrm{mV} && n(0) = 0.3177 && m(0) = 0.0529 \\
&h(0) = 0.5961
\end{aligned}
$$

The initial conditions for the four variables (v, n, m, and h) in Eq. (7.83) are chosen based on the following statement made by Hodgkin and Huxley:

> "By a membrane action potential is meant one in which the membrane potential is uniform, at each instant, over the whole of the length of the fibre considered. There is no current along the axis of the cylinder and the net membrane current must therefore always be zero, except during the stimulus. If the stimulus is a short shock at t = 0, the form of the action potential should be given by solving Eq. (7.83) with I = 0 and the initial conditions that $V = V_0$ and m, n, and h have their resting steady state values, when t = 0."

(a) First verify the values of the initial conditions, $n(0)$, $m(0)$, and $h(0)$; they must be the resting steady state values of these variables (when v = 0). Integrate the differential equations for the time span of 0 to 20 ms, using an initial voltage of 8 mV. There is no current along the axis of the cylinder, and the net membrane current must always be zero; therefore, use a current density of 0 μA/cm^2 and a membrane capacitance of 1 μF/cm^2. Plot the time profiles of the potential, v, the gating variables, n, m, and h, and the conductances, g_K and g_{Na} (Eqs. (7.81) and (7.82)).

(b) Calculate and plot the steady-state values of the time constants and the gating variables (Eqs. (7.85)) as functions of the potential in the range of voltages from –100 mV to +100 mV.

Solution

(a) Integration of equations

The program `example7_5.m`, listed below, first calculates the initial conditions of the gating variables using Eqs. (7.85), then integrates the differential equations that are contained in the function `hodgkin_huxley_equations.m` using the MATLAB function `ode45.m`. The program also uses the function `rate_constants.m` to calculate the values of α and β. The same program also calculates the steady-state values of the time constants and the gating variables, and plots the results.

Program

```
% example7_5.m - Simulation of the Hodgkin-Huxley model
% using MATLAB function ode45.m to integrate the differential
% equations that are contained in the file:
% hodgkin_huxley_equations.m

clc; clear all;
warning off MATLAB:divideByZero

% Evaluate the initial conditions for gating variables
v=0;
[alpha_n,beta_n,alpha_m,beta_m,alpha_h,beta_h]=rate_constants(v);
tau_n=1./(alpha_n+beta_n);
n_ss=alpha_n.*tau_n;
tau_m=1./(alpha_m+beta_m);
m_ss=alpha_m.*tau_m;
tau_h=1./(alpha_h+beta_h);
h_ss=alpha_h.*tau_h;
fprintf('\n The following initial conditions of the gating
variables are used:')
fprintf('\n n_ss= %5.4g \n m_ss= %5.4g \n h_ss= %5.4g ',
n_ss,m_ss,h_ss)
fprintf('\n They are the resting steady state values of these
variables (when v=0).')
% Integrate the equations
yzero=[8,n_ss,m_ss,h_ss]; tspan=[0,20];
[t,y]=ode45(@hodgkin_huxley_equations,tspan,yzero);
% Evaluate the conductances
ggK=36; ggNa=120;
gK=ggK*y(:,2).^4; gNa=ggNa*y(:,3).^3.*y(:,4);
```

```
% Plot the results
clf; figure(1); plot(t,y(:,1),'k');
title('Time Profile of Membrane Potential in Nerve Cells')
xlabel('Time (ms)'); ylabel('Potential (mV)')
figure(2); plot(t,y(:,2:4));
title('Time Profiles of Gating Variables')
xlabel('Time (ms)'); ylabel('Gating variables')
text(7,0.6,'\leftarrow n(t)'); text(4.5,0.9,'\leftarrow m(t)');
text(7,0.25,'\leftarrow h(t)')
figure(3); plot(t,gK,t,gNa);
title('Time Profiles of Conductances')
xlabel('Time (ms)'); ylabel('Conductances')
text(7,6,'g _K'); text(3.6,25,'g _{Na}');

% Evaluate the rate constants
v=[-100:1:100];
[alpha_n,beta_n,alpha_m,beta_m,alpha_h,beta_h]=rate_constants(v);

% Evaluating time constants and gating variables at steady state
tau_n=1./(alpha_n+beta_n);
n_ss=alpha_n.*tau_n;
tau_m=1./(alpha_m+beta_m);
m_ss=alpha_m.*tau_m;
tau_h=1./(alpha_h+beta_h);
h_ss=alpha_h.*tau_h;

% Plot the time constants
figure(4); plot(v,tau_n,v,tau_m,v,tau_h)
axis([-100 100 0 10])
title('Time Constants as Functions of Potential')
xlabel('Potential (mV)'); ylabel('Time constants (ms)')

text(-75,4,'\tau _n'); text(0,0.8,'\tau _m');text(15,8,'\tau _h');

% Plot the gating variables at steady state
figure(5); plot(v,n_ss,v,m_ss,v,h_ss)
axis([-100 100 0 1])
title('Gating Variables at Steady State as Functions of
Potential')
xlabel('Potential (mV)'); ylabel('Gating variables at steady
state')
text(-35,0.1,'n_\infty'); text(25,0.4,'m_\infty');
text(-20,0.8,'h_\infty');
```

Function that contains equations (`hodgkin_huxley_equations.m`)

```
function dy=hodgkin_huxley_equations(t,y)
% hodgkin_huxley_equations.m
% Contains the Hodgkin-Huxley model for example7_5

% Constants
ggK=36; ggNa=120; ggL=0.3;
vK=-12; vNa=115; vL=10.6;
```

```
Iapp=0;  Cm=1;
% Equations
v=y(1); n=y(2); m=y(3); h=y(4);
[alpha_n,beta_n,alpha_m,beta_m,alpha_h,beta_h]=rate_constants(v);

dy=[(-ggK*n^4*(v-vK)-ggNa*m^3*h*(v-vNa)-ggL*(v-vL)+Iapp)/Cm
    alpha_n*(1-n)-beta_n*n
    alpha_m*(1-m)-beta_m*m
    alpha_h*(1-h)-beta_h*h];
```

Function that calculates the rate constants (`rate_constants.m`)

```
function [alpha_n,beta_n,alpha_m,beta_m,alpha_h,beta_h] =
 rate_constants(v)
% rate_constants.m
% Calculates the rate constants for the Hodgkin-Huxley model

alpha_n=0.01*(10-v)./(exp((10-v)/10)-1);
beta_n=0.125*exp(-v/80);
alpha_m=0.1*(25-v)./(exp((25-v)/10)-1);
beta_m=4*exp(-v/18);
alpha_h=0.07*exp(-v/20);
beta_h=1./(exp((30-v)/10)+1);
```

Results

```
    The following initial conditions of the gating variables are used:
    n_ss= 0.3177
    m_ss= 0.05293
    h_ss= 0.5961
They are the resting steady state values of these variables (when
v=0).
```

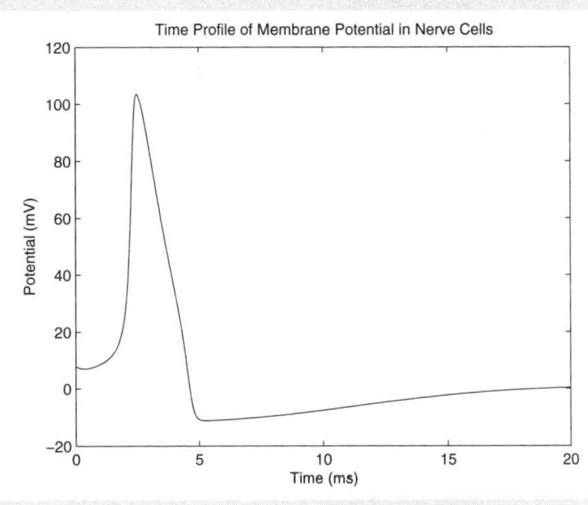

Figure 7.11a Time profile of membrane potential in nerve cells.

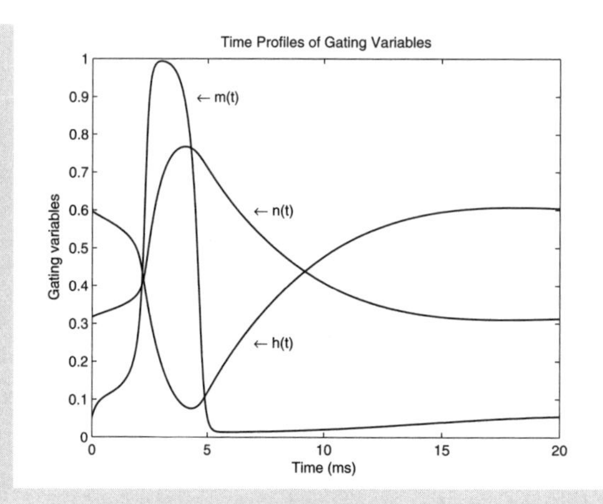

Figure 7.11b Time profiles of gating variables.

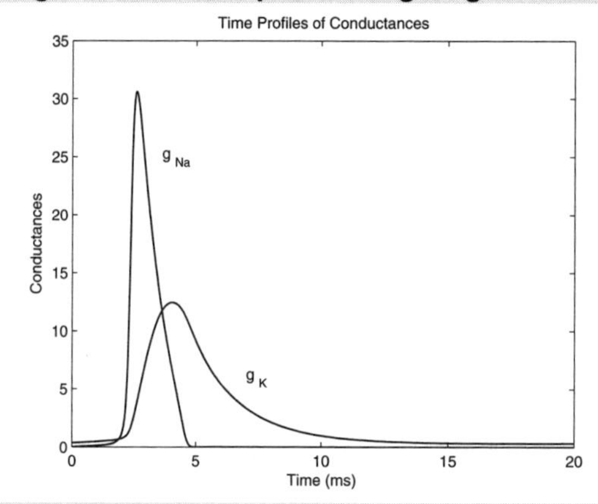

Figure 7.11c Time profiles of conductances.

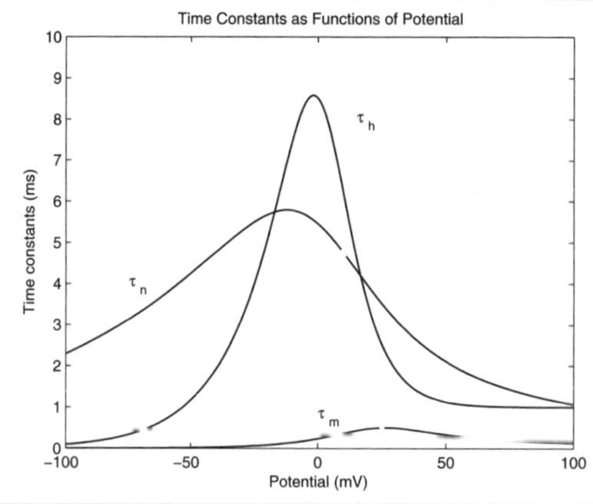

Figure 7.11d Time constants as functions of potential.

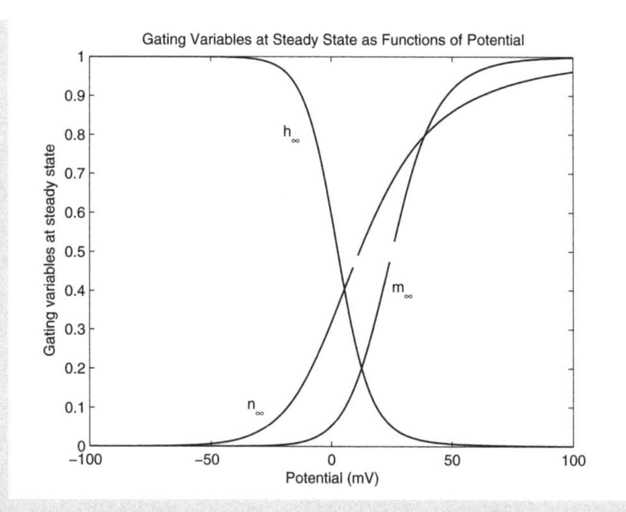

Figure 7.11e Gating variables at steady state as functions of potential.

Discussion of results

The application of a stimulus to the cell, in the form of a voltage of 8 mV at t = 0, raises the membrane potential above the threshold value and causes the generation of a self-propagating action potential, as shown in Fig. 7.11a. The membrane potential rises rapidly to over 100 mV and then drops back to its resting potential, all in a matter of less than 20 milliseconds. This action is explained as follows: the sodium gates have a much smaller time constant, τ_m, (Fig. 7.11d), therefore $m(t)$ responds faster, i.e., the sodium channels open faster, allowing the flow of Na^+ into the cell, thus making the potential more positive. As the potential rises, the value of h_∞ goes to zero (Fig. 7.11e), thus causing the sodium current to inactivate because its conductance, g_{Na}, goes to zero (Fig. 7.11c). This mechanism, however, has a higher time constant, thus it is slower to show its effect. The voltage-gated potassium channels also open when the membrane potential becomes more positive than during the resting state; however, unlike the sodium channels, they open more slowly and become fully opened only after the sodium channels have closed. The potassium channels then remain open until the membrane potential has returned to near its resting value.

The student is encouraged to work out Problem 7.1 (at end of this chapter), which applies a constant current of 10 μA/cm^2, and to observe and interpret the results.

Example 7.6 The dynamics of stem cell differentiation.

Formulation of the problem

Stem cells in a growing fetus replicate and differentiate to develop into specialized types of cells, such as bone cells, skin cells, liver cells, muscle cells, and so on. In an adult human body, the bone marrow contains stem cells, such as *hematopoietic* cells that generate red blood cells, and *mesenchymal* cells that produce connective tissue cells. The differentiation process involves a series of changes in cell phenotype and morphology that typically become more pronounced and easier to observe directly at the latter stages of the process (Palsson and Bhatia, 2004). This process begins with the stem cell's commitment to differentiation, followed by a coordinated series of gene-expression events, causing the cell to differentiate to a new state. A series of such progressive states leads to fully mature specialized cells. These mature cells perform their intended function in the body and eventually die, or undergo change to another type of cell through a process called *transdifferentiation*. The progressive series of events that converts a stem cell to a fully mature specialized cell may be depicted schematically as shown in Figure 7.12.

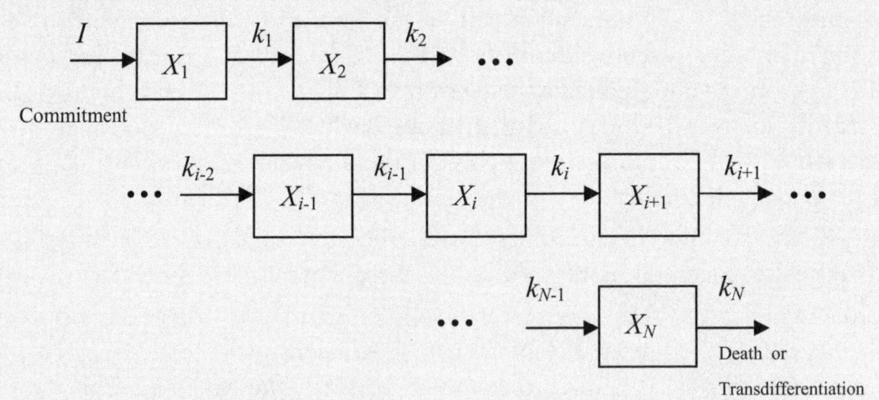

Figure 7.12 Mechanism of stem cell replication and differentiation.

where X_i = number of cells in stage i of differentiation (cells)
I = number of cells entering the differentiation process (cells/day)
k_i = the transition rate of cells from stage i to stage i+1 (1/day)
N = the total number of stages of differentiation (may be as high as 16 to 18)

The final stage of the process, N, may be considered as the intended goal of the differentiation, i.e., the state of specialized mature cells. This last stage may have a zero transition rate constant. That is, if k_N is equal to zero, then cells do not die or transdifferentiate.

The dynamics of the differentiation process may be easily simulated using a multicompartment model. Assuming that each stage is homogeneous in its cellular content, an unsteady-state balance on each compartment yields the following set of ordinary differential equations:

$$\frac{dX_1}{dt} = I - k_1 X_1$$

$$\frac{dX_2}{dt} = k_1 X_1 - k_2 X_2$$

$$\vdots$$

$$\frac{dX_i}{dt} = k_{i-1} X_{i-1} - k_i X_i$$

$$\vdots$$

$$\frac{dX_N}{dt} = k_{N-1} X_{N-1} - k_N X_N$$

The above model reflects the transition of cells from one stage of differentiation to the next, with no cell division and no self-renewal. These two concepts are explored in Problems 7.9 and 7.10 at the end of this chapter.

Using the above differential equations, simulate numerically the following stem cell differentiation cases:

(a) Stem cells commit to the differentiation process at a continuous rate of $I = 5{,}000$ cells/day. Assume that these cells undergo 10 stages of differentiation ($N = 10$). No death occurs at the last step in the process ($k_N = 0$). Integrate the differential equations and trace the path of these cells through the 10 stages of differentiation, using the following initial conditions and constants:

$$I = 5000 \text{ cells/day}$$

$$X_i(0) = 0 \text{ cells, for } i = 1,\ldots,N$$

$$k_i = 2.2 \text{ day}^{-1}, \text{ for } i = 1,\ldots,(N-1)$$

$$k_N = 0 \quad \text{no death or transdifferentiation}$$

Examine and discuss the time profiles. Does this case reach a steady state?

(b) There are no new cells entering the process, i.e., $I = 0$, but the initial number of cells in the first stage of differentiation, $X_1(0)$, is 5,000. Assume that these cells undergo the same number of stages of differentiation as in case (a). No death occurs at the last stage of the process. Integrate the differential equations and trace the path of these cells through the 10 stages of differentiation, using the following initial conditions and constants:

$$I = 0 \text{ cells/day}$$
$$X_1(0) = 5000 \text{ cells}, \quad X_i(0) = 0 \text{ cells, for } i = 2,\ldots,N$$
$$k_i = 2.2 \text{ day}^{-1}, \text{ for } i = 1,\ldots,(N-1)$$
$$k_N = 0 \quad \text{no death or transdifferentiation}$$

Examine and discuss the time profiles. How many days does it take for the completion of this process?

(c) This is the same as case (a), except for the occurrence of death at the completion of stage 10. Examine the time profiles and predict the steady-state behavior of this system using the following initial conditions and constants:

$$I = 5000 \text{ cells/day}$$
$$X_i(0) = 0 \text{ cells, for } i = 1,\ldots,N$$
$$k_i = 2.2 \text{ day}^{-1}, \text{ for } i = 1,\ldots,N, \text{ with death}$$

Solution

(c) The MATLAB program and function that solve all three cases are listed below:

Program

```
% example7_6.m - Solution of the stem cell differentiation model
% using MATLAB function ode45.m to integrate the differential
% equations that are contained in the file:
% cell_differentiation_equations.m

clc; clear all;

% Set the number of stages & time span
N=10; tzero=0; tmax=10; tspan=[tzero:0.1:tmax];
% Case (a): With continuous input; no death
I=5000;                     % Input
Xzero=zeros(N,1);           % Initial conditions
k=2.2*ones(N-1,1); k(N)=0;  % Transiton rate constants, no death
% Integrate the equations
[t,X]=ode45('cell_differentiation_equations',tspan,Xzero,[],N,I,k);
% Pseudo steady state values for stages 1 to N-1
SS=I/k(1); X_last=X(length(X),N);
disp('Case (a)')
fprintf('The pseudo steady state number of cells in stages %1d
to%2d = %4.0f',1,N-1,SS)
fprintf('\nThe number of cells in stage %2d, at %2d days = %4.0f
```

```
\n',N,tmax,X_last)
% Plot concentration profiles
clf; figure(1); subplot(2,1,1), plot(t,X(:,1:1:N-1))
title(['Figure 7.13 (a):  Continuous input (I = ',num2str(I),...
        '); no death (k(1:', num2str(N-1),') = ',num2str(k(1)),...
        ', k(', num2str(N),') = ',num2str(k(N)),') '])
text(0.4,SS,'i = 1'); text(0.45*tmax,SS/2,['i = ', num2str(N-1)]);
xlabel('Time, days'); ylabel('Number of cells');
subplot(2,1,2), plot(t,X(:,N)/1000)
axis([tzero, tmax, 0, 1.1*X_last/1000])
text(tmax/2,X_last/2000,['i = ', num2str(N)]);
xlabel('Time, days'); ylabel('Number of cells (thousands)');
% Case (b): With no new input; no death
I=0;                                % Input
Xzero=zeros(N,1); Xzero(1)=5000; % Initial conditions
% Integrate the equations
[t,X]=ode45('cell_differentiation_equations',tspan,Xzero,[],N,I,k);
% Steady state values for stages 1 to N-1
SS=I/k(1);
X_last=X(length(X),N);
disp('Case (b)')
fprintf('The steady state number of cells in stages %1d to%2d =
%4.0f',1,N-1,SS)
fprintf('\nThe final number of cells in stage %2d at %2d days =
%4.0f \n',N,tmax,X_last)
% Plot concentration profiles
figure(2); plot(t,X(:,1:1:N))
title(['Figure 7.14 (b):  No new input (I = ', num2str(I),...
        '); no death (k(1:', num2str(N-1),') = ', num2str(k(1)),...
        ', k(', num2str(N),') = ', num2str(k(N)),')'])
text(0.4,0.8*X(1),'i = 1');
text(0.45*tmax,X(1)/2,['i = ',num2str(N)]);
xlabel('Time, days'); ylabel('Number of cells');
% Case (c): With continuous input; with death (or transdiff.)
I=5000;             % Input
Xzero=zeros(N,1);   % Initial conditions
k(N)=k(1);          % reset the death rate constant
% Transiton rate constants, with death
% Integrate the equations
[t,X]=ode45('cell_differentiation_equations',tspan,Xzero,[],N,I,k);
% Pseudo steady state values for stages 1 to N-1
SS=I/k(1);
X_last=X(length(X),N);
disp('Case (c)')
fprintf('The steady state number of cells in all stages =
%4.0f',SS)
% Plot concentration profiles
figure(3); plot(t,X(:,1:1:N))
axis([tzero, tmax, 0, 1.1*I/k(1)])
title(['Figure 7.15 (c):  Continuous input (I = ',...
```

```
        num2str(I), '); with death (k(1:', num2str(N),...
        ') = ',num2str(k(1)),')'])
text(0.4,SS,'i = 1'); text(0.5*tmax,SS/2,['i = ', num2str(N)]);
xlabel('Time, days'); ylabel('Number of cells');
```

Function that contains the equations (`cell_differentiation_equations.m`)

```
function dX=cell_differentiation_equations(t,X,flag,N,I,k)
% cell_differentiation_equations.m
% Contains the equations for example7_6

% Equations
dX(1)=I-k(1)*X(1);
for i=2:N
    dX(i)=k(i-1)*X(i-1)-k(i)*X(i);
end
% Convert to column vector
dX=dX';
```

Case (a) Results and discussion

The results of this case are plotted on Fig. 7.13. The top half of the plot shows the time profiles for stages one to nine. The constant input of cells into stage one (I = 5,000 cells/day) causes the first nine stages of differentiation to reach a steady state in less than ten days, with the number of cells in each stage given by:

$$X_i^* = \frac{I}{k_i} = \frac{5000 \text{ cells/day}}{2.2\,/\,day} = 2273 \text{ cells}$$

This result is obtained mathematically by setting the derivatives of the first nine differential equations to zero (steady state) and solving for X_i^* (the steady-state level of the cells). However, the differential equation for the final stage does not have a steady state because of the no death assumption ($k_{10} = 0$). Setting $\frac{dX_{10}}{dt} = 0$ yields $I = 0$, which we know is incorrect. For this reason, we call this case a *pseudo* steady state. The number of cells in the final stage is 29,547 in 10 days and continues to increase, as shown in the bottom half of Fig. 7.13. This final stage of the process is the intended goal of the differentiation; therefore it is reasonable to expect that cells will continue to accumulate in this stage.

```
Case (a)
The pseudo steady state number of cells in stages 1 to 9 = 2273
The number of cells in stage 10, at 10 days = 29547
```

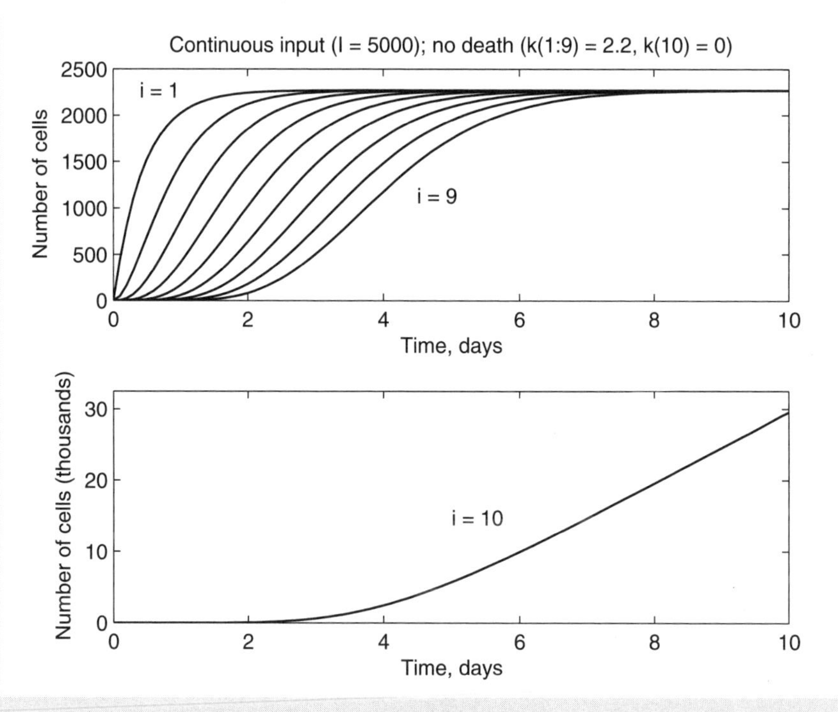

Figure 7.13 Time profiles of cells with continuous input and no death.

Case (b) Results and discussion

The results of this case are plotted on Fig. 7.14. There is no input of new cells and no death occurs in the last stage of differentiation. Therefore, the cells differentiate completely from one stage to the next, without any renewal from new cells entering, and finally accumulate in the last compartment of the process, as is clearly shown by Fig. 7.14. Since $I = 0$, the steady states for stages one to nine are all zero, i.e.,

$$X_i^* = \frac{I}{k_i} = \frac{0 \text{ cells/day}}{2.2\,/\,day} = 0 \text{ cells}$$

The final number of cells in stage 10 is ~5,000, as expected, remembering that the initial number of cells was 5,000, and there is no death of cells anywhere in this pathway.

```
Case (b)
The steady state number of cells in stages 1 to 9 =     0
The final number of cells in stage 10 at 10 days = 4997
```

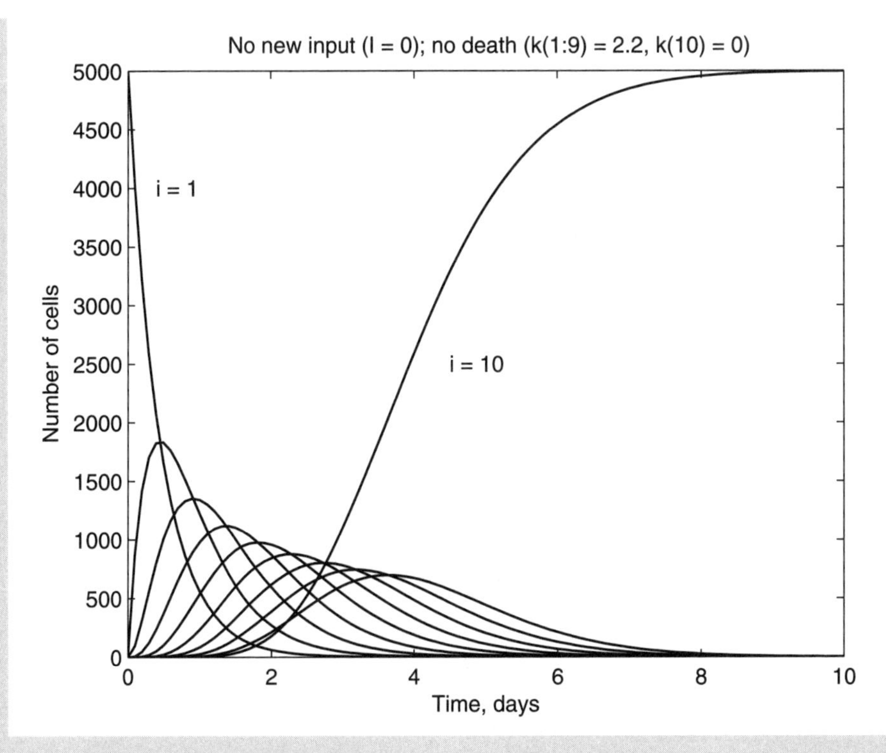

Figure 7.14 Time profiles of cells with no new input and no death.

Case (c) Results and discussion

In this case, there is a continuous rate of new stem cells, I = 5,000 cells/day, that commit to the differentiation process. There is also the occurrence of death at the completion of stage 10. The results of this case are plotted on Fig. 7.15. Under these circumstances, all 10 stages reach their steady states at:

$$X_i^* = \frac{I}{k_i} = \frac{5000 \text{ cells/day}}{2.2 / day} = 2273 \text{ cells}$$

The cells continue to differentiate from one stage to the next, with death occurring after the last stage. It is theoretically possible that this cell differentiation process may continue for the duration of the lifetime of the individual.

```
Case (c)
The steady state number of cells in all stages = 2273
```

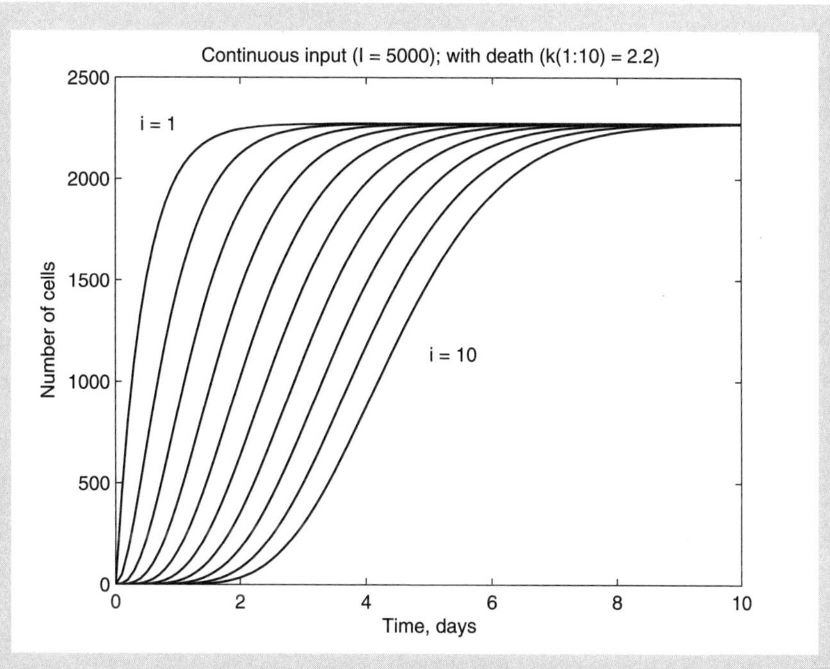

Figure 7.15 Time profiles of cells with continuous input and with death.

Example 7.7 Tissue engineering: models of epidermal cell migration.

Introduction

One aspect of tissue engineering is the proper design and manufacture of porous matrices (membranes) that imitate the properties of the epidermis and may be used as prosthetic scaffolding to promote dermal regeneration, thus enhancing the healing process of wounded or burned skin. During the healing process, cell migration is necessary for cells to repopulate a healing wound, imbedding themselves in an implanted scaffold for successful tissue regeneration. Cellular migration is known to depend on the interaction of specific cell surface receptors with cell-internalizable ligands that are present on the extracellular matrix. The formation of ligand-receptor bonds between skin epidermal cells (keratinocytes) and ligand presenting micro-carriers may initiate and promote the process of endocytosis—the ingestion of matrix molecules by the cells—thus significantly enhancing the levels of cell motility.

The dynamics of cell-ligand interactions and endocytically-coupled cell motility have been modeled from a kinetic-mechanistic point of view (Tjia and Moghe, 2002c) using diffusion-reaction descriptions and equations similar to those in the traditional Michaelis-Menten kinetics.

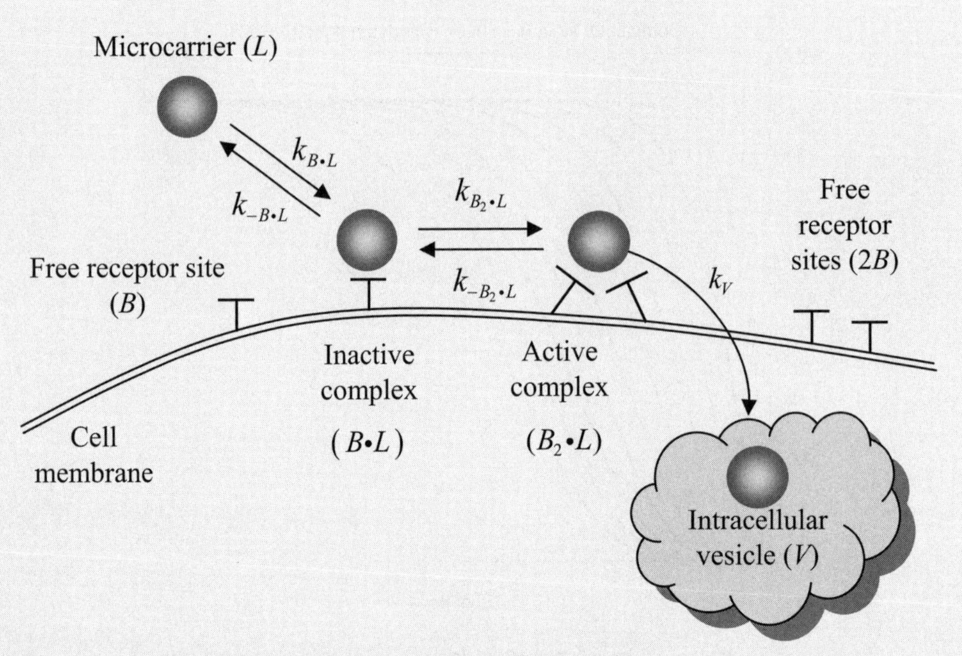

Figure 7.16 Cell-ligand interactions.

Formulation of the mechanism

Fig. 7.16 shows the mechanism of cell-ligand interactions schematically. The individual steps of this process are described below:

1. A ligand-adsorbed microcarrier, (L), interacts with a free receptor site, (B), on the surface of a cell to form an inactive complex, ($B{\bullet}L$):

$$L + B \underset{k_{-B\bullet L}}{\overset{k_{B\bullet L}}{\rightleftarrows}} B{\bullet}L \tag{7.86}$$

2. The inactive complex, in turn, binds reversibly with a second receptor to form the active complex ($B_2{\bullet}L$):

$$B{\bullet}L + B \underset{k_{-B_2\bullet L}}{\overset{k_{B_2\bullet L}}{\rightleftarrows}} B_2{\bullet}L \tag{7.87}$$

3. The active complex is ingested by the cell to produce an intracellular vesicle, (V). Once ingested, the microcarrier dissociates itself from the membrane receptors, thus freeing the receptors to recycle back to the cell surface. For the purposes of this model, the rates of ingestion and binding site recycling are

lumped into one parameter, k_V:

$$B_2{\bullet}L \xrightarrow{k_V} V + 2B \tag{7.88}$$

Formulation of the mathematical model

Cell migration will affect the degree of exposure of microcarriers to the cell, as migration would make new microcarriers available for internalization. The rate of cell migration has been derived, based on an analogy to molecular diffusion in a semi-infinite plane, to be:

$$\left.\frac{d[L]}{dt}\right|_{\text{Migration}} = \frac{\mu L_0}{A_{\text{cell}}} \tag{7.89}$$

where L = the effective ligand density encountered by the cell
μ = the random motility coefficient
A_{cell} = the spread area of the cell
L_0 = the overall density of microcarriers

The composite model, which includes both cell migration and ligand-receptor adhesion, is given below:

1. The density of local extracellular microcarriers encountered by the cell, $[L]$, changes according to the following rate equation:

$$\frac{d[L]}{dt} = -k_{B{\bullet}L}[L][B] + k_{-B{\bullet}L}[B{\bullet}L] + \frac{\mu L_0}{A_{\text{cell}}} \tag{7.90}$$

The first term in this equation corresponds to the forward rate in reaction (7.86), the second term corresponds to the reverse rate, and the third term reflects the rate of the cell migration given by Eq. (7.89).

2. The balance of the density of inactive microcarrier-receptor complex, $[B{\bullet}L]$, gives the rate equation (7.91):

$$\frac{d[B{\bullet}L]}{dt} = k_{B{\bullet}L}[L][B] - k_{-B{\bullet}L}[B{\bullet}L] - k_{B_2{\bullet}L}[B{\bullet}L][B] + k_{-B_2{\bullet}L}[B_2{\bullet}L] \tag{7.91}$$

3. The rate of change of the density of activated microcarrier-receptor complex, $[B_2{\bullet}L]$, is:

$$\frac{d[B_2{\bullet}L]}{dt} = k_{B_2{\bullet}L}[B{\bullet}L][B] - k_{-B_2{\bullet}L}[B_2{\bullet}L] - k_V[B_2{\bullet}L] \tag{7.92}$$

4. The density of the ingested microcarrier, $[V]$, changes at the following rate:

$$\frac{d[V]}{dt} = k_V[B_2{\bullet}L] \tag{7.93}$$

5. The total number of binding sites on the cell, $[B_T]$, is assumed to be constant:

$$[B_T] = [B] + [B{\bullet}L] + [B_2{\bullet}L] \tag{7.94}$$

The net effects of cell migration may be measured in terms of the rate at which cells effectively clear an area covered with ingestible microcarriers. For a given initial surface particle density, the rate of area clearance by a cell is equal to the sum of the rates of bounding and ingesting, divided by the initial particle density:

$$\frac{d[clearance]}{dt} = \frac{1}{L_o}\left(\frac{d[B \bullet L]}{dt} + \frac{d[B_2 \bullet L]}{dt} + \frac{d[V]}{dt}\right) \tag{7.95}$$

It should be noted that that the density terms in this model are in units of particle per surface area of cell. The $[clearance]$ term is in $(\text{min})^{-1}$.

In order to simplify the model, we make the following assumptions:

(a) The rate constant of decomposition of the inactivated microcarrier-receptor complex , $k_{-B{\bullet}L}$, is very small in comparison to the rate of initial binding, $k_{B{\bullet}L}$, and can be neglected. This effectively makes reaction (7.86) irreversible.

(b) The fully activated complex, $[B_2{\bullet}L]$, once formed, is highly reactive and is quickly ingested. Thus, this complex would be present only in low densities and may be assumed to be at pseudo steady state. This assumption causes the rate in Eq. (7.92) to be equal to zero, enabling us to solve for $[B{\bullet}L]$, as follows:

$$[B_2L] = \frac{[B{\bullet}L][B]}{K_m} \tag{7.96}$$

where K_m is a dissociation constant of the Michaelis-Menten type, defined as:

$$K_m = \frac{k_{-B_2{\bullet}L} + k_V}{k_{B_2{\bullet}L}} \tag{7.97}$$

The above two assumptions are used in Eqs. (7.90) through (7.95) to eliminate $[B{\bullet}L]$ and $[B]$, which simplifies the model to the following set of equations:

$$\frac{d[L]}{dt} = -k_{B\bullet L}[L]([B_T]-[B{\bullet}L]) + \frac{\mu L_0}{A_{\text{cell}}}$$

$$\frac{d[B{\bullet}L]}{dt} = k_{B\bullet L}[L]([B_T]-[B{\bullet}L]) - k_V\left(\frac{([B_T]-[B{\bullet}L])[B{\bullet}L]}{K_m + 2[B{\bullet}L]}\right)$$

$$\frac{d[V]}{dt} = k_V\left(\frac{([B_T]-[B{\bullet}L])[B{\bullet}L]}{K_m + 2[B{\bullet}L]}\right) \tag{7.98}$$

$$\frac{d[clearance]}{dt} = \frac{1}{L_o}\left(\frac{d[B\bullet L]}{dt} + \frac{d[V]}{dt}\right)$$

Equations (7.98) define the mathematical model of the dynamics of cellular migration enhanced by the presence of ligand-associated microcarriers. This is a set of ordinary differential equations that may be integrated to yield the temporal behavior of this process. For this problem, perform the following tasks:

(a) Evaluate and plot the time profiles, and discuss the results of the integration for the period of 300 minutes, using the following initial conditions and constants, based on the experimental work of Tjia and Moghe (2002c):

Initial conditions:

$$L_0 = 1.0 \text{ particle}/\mu\text{m}^2 \qquad [B{\bullet}L]|_0 = 0$$

$$[clearance]|_0 = 0 \qquad [V]|_0 = 0$$

Constants:

$$B_T = 3.74 \text{ particles}/\mu\text{m}^2 \qquad \mu = 10\ \mu\text{m}^2/\text{min} \qquad A_{\text{cell}} = 3400\ \mu\text{m}^2$$

$$K_m = 0.73 \text{ particles}/\mu\text{m}^2 \qquad k_V = 1.3\times10^{-3}\ \text{min}^{-1}$$

$$k_{B\bullet L} = 2.0\times10^{-3}\ \mu\text{m}^2/(\text{particle.min})$$

(b) Define the term "Sampling rate" as:

$$(\text{Sampling rate}) = \frac{d[B{\bullet}L]}{dt} - \frac{d[V]}{dt}$$

and show its effect on the internalization rate, d[V]/dt, and the clearance rate, d[clearance]/dt.

Solution

The program, `example7_7.m`, and the function, `cell_migration_equations.m`, that solve this problem are listed below:

Program

```
% example7_7.m - Solution of the epidermal cell migration
% model using MATLAB function ode45.m to integrate the
% differential equations that are contained in the file:
% cell_migration_equations.m

clc; clear all;
% Set the time span
tspan=[0:1:300];
% Set the constants
BT=3.74; mu=10; A_cell=3400;
Km=0.73; kV=1.3e-3; kBL=2.0e-3;
% Set the initial conditions
yzero=[1, 0, 0, 0];
L0=yzero(1);
% Integrate the equations
[t,y]=ode45('cell_migration_equations',tspan,yzero,[],...
    BT,mu,A_cell,Km,kV,kBL,L0);

% Plot concentration profiles
figure(1); plot(t,y(:,1),'-',t,y(:,2),':',t,y(:,3),'-.',...
    t,y(:,4),'--')
title('Time profiles of epidermal cell migration')
xlabel('Time, min'); ylabel('Densities, number/\mum^2');
legend('L','B{\bf\cdot}L','V','clearance',2)
n=length(y);

% Evaluate the derivatives
for i=1:n
dy(:,i)=feval('cell_migration_equations',t(i),y(i,:),flag,...
    BT,mu,A_cell,Km,kV,kBL,L0);
end
dy=dy';
rate_BL=dy(:,2);
rate_V=dy(:,3);
```

```
clearance_rate=dy(:,4);
sampling_rate=rate_BL-rate_V;

% Show the effect of microcarrier sampling rate on internalization
% and clearance rates
figure(2);
plot(sampling_rate*1e3,clearance_rate*1e3)
title('The effect of microcarrier sampling rate on clearance
rate')
ylabel('Clearance rate, d[clearance]/dt  x  10^3')
xlabel('Sampling rate, (d[B{\bf\cdot}L]/dt - d[V]/dt)  x  10^3')
```

Function that contains the equations (`cell_migration_equations.m`)

```
function dy=cell_migration_equations(t,y,flag,BT,mu,A_cell,...
    Km,kV,kBL,L0)
% cell_migration_equations.m
% Contains the equations for example7_7

% Equations
dy=[-kBL*y(1)*(BT-y(2))+mu*L0/A_cell
    kBL*y(1)*(BT-y(2))-kV*((BT-y(2))*y(2))/(Km+2*y(2))
    kV*((BT-y(2))*y(2))/(Km+2*y(2))
    (kBL*y(1)*(BT-y(2))-kV*((BT-y(2))*y(2))/(Km+2*y(2))...
     +kV*((BT-y(2))*y(2))/(Km+2*y(2)))/L0];
```

Discussion of results

In Fig. 7.17a, the dynamics of ligand interactions with skin epidermal cells are plotted as a function of time. The free ligand concentration decreased slowly over time, indicating the steady depletion of instantaneous ligand concentration due to cell internalization of the ligand. Concurrently, the concentrations of membrane-bound ligand complex $[B{\bullet}L]$ as well as internalized ligand $[V]$ increase over time. The instantaneous clearance of ligands increases over time as well, indicating the cells are not yet saturated with ligand-microcarriers.

In Fig. 7.17b, the cell clearance rate is graphed versus the net rate of ligand sampling by the cells (defined as the on-rate of ligand binding minus the off-rate of ligand internalization). It is assumed that internalized ligands can no longer activate the intracellular signaling necessary for increased cell migration. A monotonic increase in cell clearance rate of the ligands was observed with increased ligand sampling rate. This suggests that the migration may be a strong function of the dynamics of ligand sampling processes.

Results

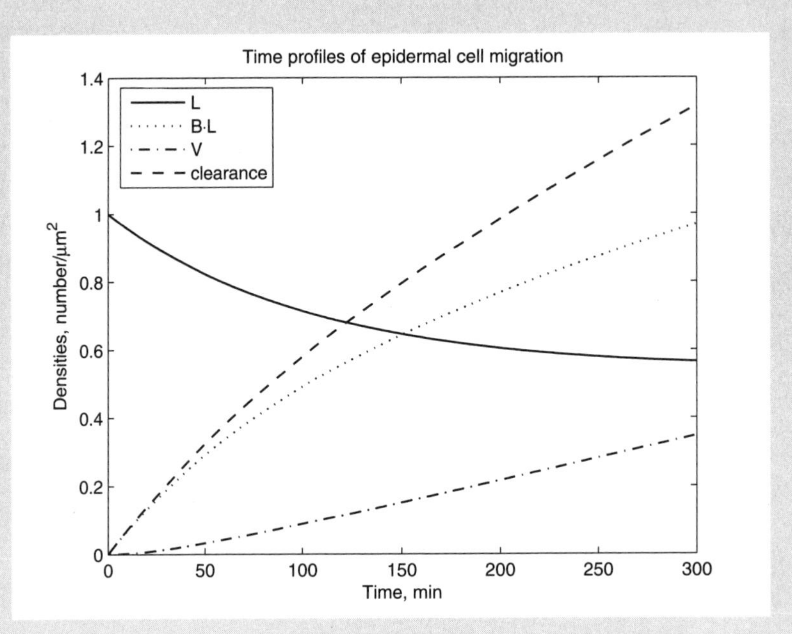

Figure 7.17a Time profiles of epidermial cell migration.

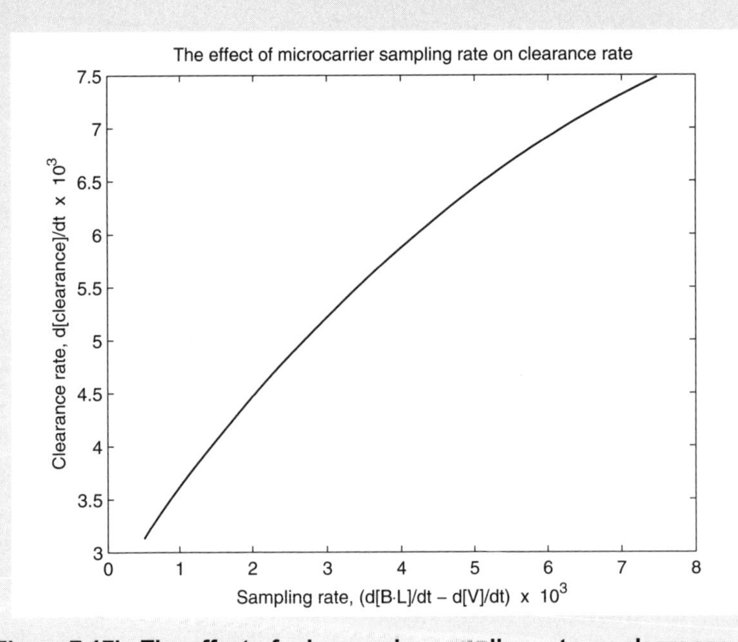

Figure 7.17b The effect of microcarrier sampling rate on clearance rate.

7.9 Lessons Learned in this Chapter

After studying this chapter, the student should have learned the following:

- The dynamics of physiological systems may be modeled using ordinary differential equations.
- Ordinary differential equations may be classified as:
    - First-, second-, third-order, etc.
    - Linear or nonlinear
    - Homogeneous or nonhomogeneous
    - Initial value or boundary value
- Second-order and higher ordinary differential equations may be converted to sets of first-order differential equations for numerical integration by the methods discussed in this chapter.
- The solution of linear ordinary differential equations depends on the eigenvalues and eigenvectors of the equations.
- Nonlinear differential equations (as well as linear ones) may be integrated numerically using methods that are based on finite differences.
- Integrating differential equations is like climbing a mountain: You move in the direction of the slope (or the weighted average of the slope at different points), taking many small steps (carefully), until you reach the destination.
- The stability of nonlinear differential equations depends on the eigenvalues of the Jacobian matrix of the equations.
- The stability of the numerical solution depends on the form of the equations, the method of solution, and the step size of integration.

7.10 Problems

7.1 Integrate the Hodgkin-Huxley model (see Example 7.5) for the period 0 to 50 ms using a constant current of 10 $\mu A/cm^2$. Examine and explain the results thoroughly.

7.2 The pool of fluid in the body of a patient undergoing dialysis has been modeled by Enderle et al. (2005) as a two-compartment system, as shown diagrammatically in Figure 7.18, where R is the rate of production of urea by the patient's body; V_1 is the volume of the intracellular fluid; V_2 is the volume of the extracellular fluid (blood and interstitial fluids); C_1 and C_2 are the concentrations of urea in the fluids of the two compartments, respectively; k_{12} and k_{21} are the mass transport parameters between the two compartments; and k_2 is the clearance rate constant for the dialysis unit.

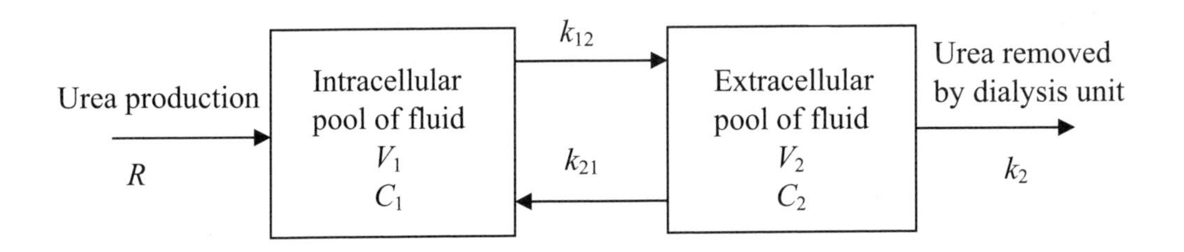

Figure 7.18 A two-compartment model of the fluid of a patient undergoing dialysis.

An unsteady-state mass balance of urea on each of the compartments yields the following two differential equations:

$$\begin{aligned} V_1 \frac{dC_1}{dt} &= R - k_{12}C_1 + k_{21}C_2 \\ V_2 \frac{dC_2}{dt} &= k_{12}C_1 - k_{21}C_2 - k_2C_2 \end{aligned} \tag{1}$$

For Patient X, the following parameters apply:

$$R = 100 \text{ mg/h} \qquad k_{12} = 33 \text{ liters/h} \qquad k_{21} = 33 \text{ liters/h}$$
$$V_1 = 10 \text{ liters} \qquad V_2 = 25 \text{ liters}$$

The dialysis unit clearance rate constant is $k_2 = 8$ liters/h.

When Patient X arrives at the dialysis unit, his blood urea nitrogen (BUN) is 150 mg/liter. Integrate the differential equations (1) to obtain answers to the following:

(a) How many hours of dialysis will the patient require in order to reduce the level of BUN to 75 mg/liter?
(b) After the completion of the treatment, how long will it take for the BUN of the patient to rise back to the 150 mg/liter level?
(c) Experiment with setting the values of k_{12} and k_{21} to be unequal to each other (say $k_{21} = 0.7\, k_{12}$, i.e., slower transfer from the extracellular pool to the intracellular one) and interpret the results.

Show clearly how you obtain your answers, and illustrate this by showing the concentrations vs. time profiles of C_1 and C_2 in all parts of the problem.

7.3 A computer simulation of the physiological human knee jerk reflex has been developed by Huang (1994). A strong tap on the patellar ligament of the leg elicits a knee jerk reflex, which follows closely the oscillations of the pendulum. The jerk of the patellar tendon stretches the muscle that sends a barrage of neural impulses to the spinal cord. The reflex signal passes back to the quadriceps muscle via the alpha motor neuron to produce a sudden contraction and forces the leg to move forward with a jerk. As the muscle relaxes, the leg system acts as a damped compound pendulum, swinging back and forth for a few oscillations. Eventually the leg returns to the normal position.

In his analysis, Huang assumed that the extensor and flexor muscles are identical and opposite in action. The numbers of primary and secondary nerve endings are considered equal, and the nervous signals are instantaneous when compared to the system's response. Small deflection angles are considered with constant damping coefficient within the range. Based on the equation of the pendulum, and for small oscillations, Huang developed a second order differential equation:

$$J\frac{d^2\theta}{dt^2}+c\frac{d\theta}{dt}+(\frac{mgL\theta}{2}-T)\theta=0 \quad (1)$$

that describes the angular position, θ, of the leg during the knee jerk reflex, where m is the mass of the leg, g is the gravitational acceleration constant, L is the length of the leg, J is the moment of inertia of the leg, and T is the gain produced by the isometric torque of the muscle. The natural frequency, ω_n, of the system is calculated by:

$$\omega_n=\sqrt{\frac{mgL}{2J}-\frac{T}{J}} \quad (2)$$

and the damping factor, α, is given by:

$$\alpha=\frac{c}{2\sqrt{J(\frac{mgL}{2}-T)}} \quad (3)$$

The values of ω_n and α are obtained experimentally. Solve the above equations with the following values of m, g, L, J, ω_n and α, and plot the time profile of the angular position of the leg during the jerk reflex.

$m=4$ kg	$g=9.81$ m/s^2	$L=0.34$ m
$J=0.154$ kg.m^2	$\omega_n=6.28$ rad/s	$\alpha=0.228$

Use the following initial conditions $\theta(0) = 0$ rad and $d\theta(0)/dt = 2\pi$ rad/s for the solution of the differential equation (1). HINTS: Use the MATLAB `solve` command for the algebraic equations and the `dsolve` command for the differential equation.

7.4 The pool of fluid in the body of a patient undergoing dialysis was modeled in Problem 7.2 (above) as a two-compartment system. Change this analysis to a one-compartment model by treating the total fluid of the patient as one unit of volume V_T (see Fig. 7.19).

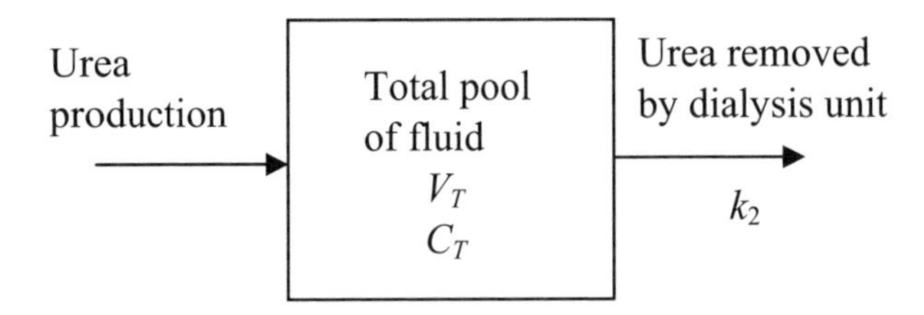

Figure 7.19 A one-compartment model of the fluid of a patient undergoing dialysis.

Derive the unsteady-state mass balance of urea for this one-compartment model. For patient X, the following parameters apply for the one-compartment analysis:

$$R = 100 \text{ mg/h} \quad V_T = 35 \text{ liters} \quad k_2 = 8 \text{ liters/h}$$

When patient X arrives at the dialysis unit his blood urea nitrogen (BUN) is 150 mg/liter. Integrate the differential equation to obtain answers to the following:

(a) How many hours of dialysis will the patient require in order to reduce the level of BUN to 75 mg/liter?
(b) After the completion of the treatment, how long will it take for the BUN of the patient to rise back to the 150 mg/liter level?

Compare these results with those of the two-compartment model (Problem 7.2).

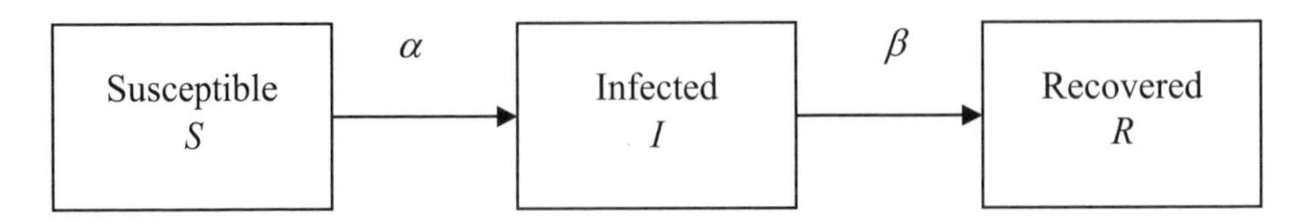

Figure 7.20 A simple model of an epidemic

7.5 A simple model of an epidemic is shown in Fig. 7.20, where S is the number of persons susceptible to the disease, I is the number infected with it, R is the number that have already been affected but have recovered (or died), α is the rate constant for infection, and β is the rate constant for recovery. Those who have recovered develop immunity to the infection. A dynamic model of the interactions between these three groups is given by Edelstein-Keshet (1988), as follows:

$$\begin{aligned} \frac{dS}{dt} &= -\alpha SI \\ \frac{dI}{dt} &= \alpha SI - \beta I \\ \frac{dR}{dt} &= \beta I \end{aligned} \tag{1}$$

One person, highly contagious with a new influenza virus, enters a small community that has a population of 1,000 individuals that are susceptible to the infection. The virus epidemic spreads quickly and eventually infects all susceptible individuals. The rate constants for this epidemic are

$$\begin{aligned} \alpha &= 0.005 \quad (\text{person})^{-1}(\text{week})^{-1} \\ \beta &= 1 \quad\quad\;\; (\text{week})^{-1} \end{aligned}$$

Integrate the differential equations (1) and determine the following:

(a) How many weeks does it take for this epidemic to reach its peak?
(b) What is the maximum number of persons sick at the peak of the epidemic?
(c) In how many weeks will the epidemic subside (when less than 0.5 % of the susceptible population is still infected)?

7.6 Modify the epidemic model in Problem 7.5 to allow loss of immunity that causes recovered individuals to become susceptible to the virus again (see Fig. 7.21). The loss of immunity rate constant has the following value (α and β remain the same as in Problem 7.5):

$$\gamma = 0.1 \qquad (\text{week})^{-1}$$

Integrate the modified set of differential equations for this epidemic and determine the following:

(a) How many weeks does it take for the epidemic to approach steady state?
(b) How many people will remain infected during steady state?
(c) Show phase plots and discuss the stability of the solutions with respect to the eigenvalues of the Jacobian matrix.

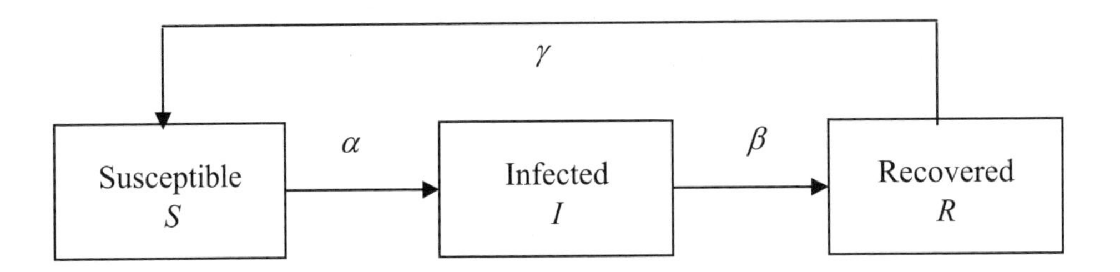

Figure 7.21 A modified model of an epidemic to account for loss of immunity.

7.7 The well-known van der Pol oscillator is the second-order nonlinear differential equation shown below:

$$\frac{d^2u}{dt^2} - k\left(1-u^2\right)\frac{du}{dt} + au = 0 \tag{1}$$

The solution of this equation exhibits stable oscillatory behavior. Van der Pol realized the parallel between the oscillations generated by this equation and certain biological rhythms, such as the heartbeat, and proposed this as a model of an oscillatory cardiac pacemaker. Integrate the van der Pol equation with the following value of k and initial conditions:

$$k = 1.0 \text{ s}^{-1} \qquad u(0) = 2 \text{ dimensionless} \qquad \left.\frac{du}{dt}\right|_0 = 0 \text{ s}^{-1}$$

and determine the value of a that would give a heart rate of 1.25 beats/second (75 beats/minute, which is a typical heart rate in a resting adult).

7.8 It is well known that most living cells—bacteria cells, stem cells, yeasts, etc.—replicate themselves by cell division. The growth of an organism is accompanied by an orderly increase in its mass and all of its chemical constituents, followed by division of the cell into two identical daughter cells or a mother and daughter cell, as in the case of yeasts. In Example 7.6, we simulated the process of stem cell differentiation without cell division. That was a rather simplistic model of cell differentiation, because most stages of differentiation have cell replication activity. The act of replication marks the completion of one stage of differentiation and the beginning of the next.

The human body produces and consumes approximately 200 billion red blood cells daily. The process of turning a bone marrow stem cell to a red blood cell is called erythropoiesis. The differentiation from the early precursor stage (pronormoblast) to a fully mature enucleated erythrocyte takes approximately one week (Palsson and Bhatia, 2004). This concept is depicted diagrammatically in Fig. 7.22.

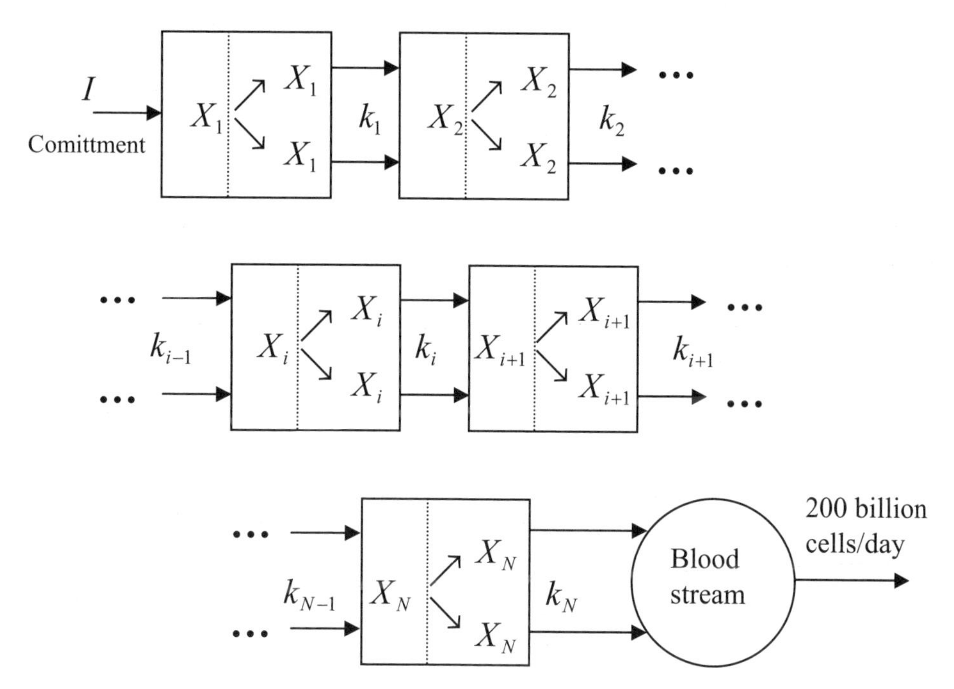

Figure 7.22 Stem cell differentiation with replication.

With the assumptions that a single cell that leaves compartment (i-1) splits into two cells that enter compartment i, we derive the mass balances for the N compartments representing differentiation, as follows:

$$\begin{aligned}
\frac{dX_1}{dt} &= I - k_1 X_1 \\
\frac{dX_2}{dt} &= 2k_1 X_1 - k_2 X_2 \\
&\vdots \\
\frac{dX_i}{dt} &= 2k_{i-1} X_{i-1} - k_i X_i \\
&\vdots \\
\frac{dX_N}{dt} &= 2k_{N-1} X_{N-1} - k_N X_N
\end{aligned} \tag{1}$$

In addition, we make the assumption that all the red blood cells that are formed by this process enter the blood stream, where they serve their purpose and die, at the rate of 200 billion cells per day. The number of red blood cells in a healthy individual remains relatively constant, i.e., there is steady state. Then the balance on the blood stream results in the following equation:

$$\frac{dX_{\text{Blood}}}{dt} = 2k_N X_N - 200 \times 10^9 = 0 \qquad (2)$$

Using the above differential equations, simulate numerically the erythropoiesis process and answer the following questions:

(a) Analyze the steady state condition of these equations. Express the steady state level of cells in each compartment, $X^*(i)$, in terms of I and k_i.

(b) What is the total number of stem cells per day, I, that need to commit to the erythropoiesis process in order to produce the required 200 billion red blood cells per day? Assume that the stem cells undergo a total of 10 stages of differentiation (N = 10) and that the initial conditions and transition rate constants are:

$$X_i(0) = 0 \text{ cells, for } i = 1, \ldots, N$$
$$k_i = 2.2 \text{ day}^{-1}, \text{ for } i = 1, \ldots, N$$

Explain carefully how you calculate the value of I.

(c) Show and discuss thoroughly the time profiles in the N stages of the differentiation/replication process and compare these results with those of Example 7.6 Case (c).

7.11 References

Burden, R. L., Faires, J. D., and Reynolds, A. C. 1981. *Numerical Analysis*. Boston, MA: Prindle, Weber & Schmidt.

Constantinides, A., and Mostoufi, N. 1999. *Numerical Methods for Chemical Engineers with MATLAB Applications*. Upper Saddle River, NJ; Prentice Hall PTR.

Edelstein-Keshet, L. 1988. *Mathematical Models in Biology*. New York: McGraw-Hill Book Company.

Enderle, J. D., Blanchard, S. M., and Bronzino, J. D. 2005. *Introduction to Biomedical Engineering*. 2nd ed. San Diego, CA: Academic Press.

Fournier, R. L. 1999. *Basic Transport Phenomena in Biomedical Engineering*. Philadelphia, PA: Taylor & Francis.

Hairer, E., Lubich, C., and Roche, M. 1980. *The Numerical Solution of Differential-Algebraic Systems by Runge-Kutta Methods*. Berlin: Springer-Verlag.

Hairer, E., and Wanner, G. 1991a. *Solving Ordinary Differential Equations I*. Berlin: Springer.

Hairer, E., and Wanner, G. 1991b. *Solving Ordinary Differential Equations II*. Berlin: Springer.

Hodgkin, A. L., and Huxley, A. F. 1952. A Quantitative Description of Membrane Current and its Application to Conduction and Excitation in Nerve. *J. Physiol,*. **117**:500-544.

Huang, B. K. 1994. *Computer Simulation Analysis of Biological and Agricultural Systems*. Boca Raton, FL: CRC Press.

Keener, J., and Sneyd, J. (1998). *Mathematical Physiology*. New York: Springer-Verlag.

Kubícek, M. and Hlavàcek, V. 1975. *Numerical Solution of Nonlinear Boundary Value Problems with Applications*. New York: Prentice Hall.

Lauffenburger, D. A., and A.F. Horwitz (1996). Cell Migration: A Physically Integrated Process. *Cell*. **84**:359–369.

Palsson, B. Ø., and Bhatia, S. N. 2004. *Tissue Engineering*. Upper Saddle River, NJ: Pearson Prentice Hall.

Tjia, J. S., and Moghe, P. V. 2002a. Regulation of Cell Motility on Polymer Substrates via Dynamic, Cell-Internalizable, Ligand Microinterfaces. *Tissue Eng.*, **8**:247-259.

Tjia, J. S. and Moghe, P. V. 2002b. Cell-Internalizable Ligand Microinterfaces on Biomaterials: Design of Regulatory Determinants Of Cell Migration in *Biomimetic Materials and Design: Interactive Biointerfacial Strategies for Drug Delivery and Tissue Engineering*. A. Dillow, T. Lowman (Eds.), Marcel-Dekker, 335-373.

Tjia, J. S., and Moghe, P. V. 2002c. Cell Migration on Cell-Internalizable Ligand Microdepots: A Phenomenological Model. *Annals of Biomedical Engineering*, **30**:851-866.

Tortora, G. J., and Grabowski, S. R. 2001. *Introduction to the Human Body*, 5th ed. Hoboken, NJ: John Wiley & Sons, Inc.

Chapter 8
Dynamic Systems: Partial Differential Equations

8.1 Introduction

Transport processes are essential to the function of biological systems. Fluids constitute a large portion of body weight and provide the conduit for transfer of nutrients and energy to and from tissues throughout the body. In order to successfully analyze the physiological and cellular processes in the body, the biomedical engineer needs to understand the mechanism of transport processes, and to have the ability to solve the mathematical models that describe these mechanisms. In addition, the design and operation of many biomedical devices for diagnostics and therapeutics depend on the flow of fluids and transport of nutrients.

The laws of conservation of mass, momentum, and energy form the basis of the field of transport phenomena. These laws, applied to the flow of fluids, result in the *equations of change,* which describe the change of velocity, temperature, and concentration with respect to time and position in the system. The dynamics of such systems, which have more than one independent variable, are modeled by *partial differential equations* (PDEs). The objective of this chapter is to present the methods for the numerical solution of partial differential equations.

The most commonly encountered partial differential equations in biomedical engineering are of first and second order. Our discussion in this chapter focuses on these two categories. In the following section, we give some examples of biological systems whose models result in partial differential equations. In Section 8.3, we classify these equations and their boundary conditions, and in the remainder of the chapter we develop the numerical methods, using finite difference analysis, for the solution of first-order and second-order partial differential equations. We apply these techniques in the examples throughout this chapter to obtain the solution of several models of physiological systems.

The material in this chapter will enable the student to accomplish the following:

- Model the dynamics of physiological systems using partial differential equations
- Express the partial differential equations into finite difference approximations
- Obtain numerical solutions of the partial differential equations, plot the numerical results, and interpret the dynamic behavior of the biosystems under a variety of conditions
- Appreciate the accuracy and stability of the models and the numerical solutions obtained from these models

8.2 Examples of PDEs in Biomedical Engineering

8.2.1 Diffusion across biological membranes

Cells are composed mostly of organic compounds and water, with more than 60% of the weight of the human body coming from water (Enderle et al., 2005). The cell membrane allows movement of chemicals in and out of the cell by several mechanisms of diffusion (see Section 7.1.4). For example, consider the diffusion of a nutrient (glucose or oxygen) across the membrane of a cell, and examine a segment of the membrane, as shown in Fig. 8.1. For simplicity, we assume that the membrane is flat and that the concentration of nutrient is higher in the fluid outside the cell, therefore diffusion occurs from left to right in Fig. 8.1. The thickness of the membrane (in the x direction) is finite of value L, while the length (in the y direction) and height (in the z direction) of the membrane are infinitely large, with respect to the thickness. We choose a differential element within the membrane, the control volume, of thickness Δx and surface area A perpendicular to the x direction, and we make a mass balance on this control volume. The general form of the unsteady-state mass balance is:

$$\begin{pmatrix}\text{Rate of} \\ \text{mass in}\end{pmatrix} = \begin{pmatrix}\text{Rate of} \\ \text{mass out}\end{pmatrix} + \begin{pmatrix}\text{Rate of mass} \\ \text{accumulation}\end{pmatrix} \tag{8.1}$$

The flux of material into the membrane is given by *Fick's first law of diffusion*:

$$J = -D\frac{dC}{dx} \tag{8.2}$$

where J is flux in moles/cm^2/s, C is the concentration of nutrient in moles/cm^3, and D is the diffusivity of nutrient in the membrane in cm^2/s. Fick's law simply states that diffusion of mass proceeds in the direction of the negative gradient of concentration with D as the proportionality constant. We apply Eq. (8.1) to the control volume over an interval of time, Δt, to obtain:

$$J_x A = J_{x+\Delta x} A + \frac{C_{x,t+\Delta t} - C_{x,t}}{\Delta t} A\Delta x \tag{8.3}$$

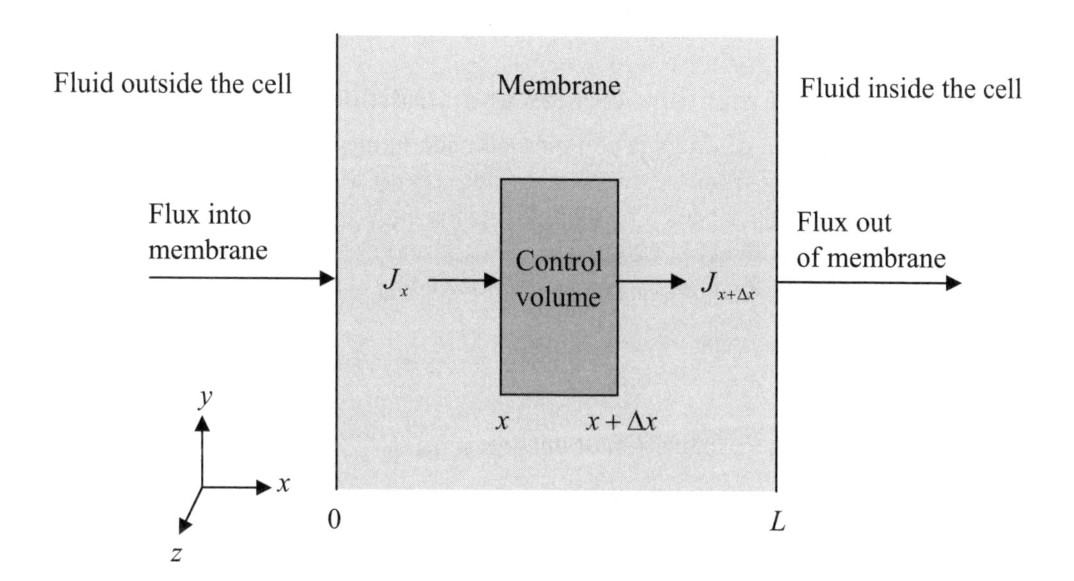

Figure 8.1 Diffusion of nutrient across a cell membrane.

Combine (8.2) and (8.3), rearrange:

$$\frac{C_{x,\,t+\Delta t}-C_{x,\,t}}{\Delta t}=\frac{\left[D\left.\frac{dC}{dx}\right|_{x+\Delta x}-D\left.\frac{dC}{dx}\right|_{x}\right]}{\Delta x} \tag{8.4}$$

and take the limit of Eq. (8.4) as Δx and Δt go to zero to obtain the partial differential equation:

$$\frac{\partial C}{\partial t}=D\frac{\partial^2 C}{\partial x^2} \tag{8.5}$$

The derivative on the left side of Eq. (8.5) gives the time change of the concentration while the derivative on the right side gives the spatial change of concentration in the membrane. This is *Fick's second law of diffusion*. It is a one-dimensional unsteady-state partial differential equation, whose numerical solution will be developed in Section 8.5.2 and demonstrated as Example 8.3. The three-dimensional analogue of Ficks' second law of diffusion, where diffusion occurs in all three directions, x, y, and z, is:

$$\frac{\partial C}{\partial t}=D\left(\frac{\partial^2 C}{\partial x^2}+\frac{\partial^2 C}{\partial y^2}+\frac{\partial^2 C}{\partial z^2}\right) \tag{8.6}$$

8.2.2 Diffusion of macromolecules and controlled release of drugs

The topical application of drugs is common practice in medicine. The mechanism of movement of the drug from the medium of delivery to the surface of the skin and across the skin has been modeled by using diffusion equations, such as Fick's second law of diffusion (Kubota et al., 2002). The concentration of the drug, C_m, through the medium of delivery of thickness, L_m, may be modeled by:

$$\frac{\partial C_m}{\partial t}=D_m\frac{\partial^2 C_m}{\partial x^2} \qquad -L_m \le x \le 0, \quad t>0 \tag{8.7}$$

and the concentration, C_s, in the skin of thickness, L_s, by:

$$\frac{\partial C_s}{\partial t}=D_s\frac{\partial^2 C_s}{\partial x^2} \qquad 0 \le x \le L_s, \quad t>0 \tag{8.8}$$

where D_m is the diffusivity in the medium of delivery, and D_s is the diffusivity in the skin. These two equations are coupled at the boundary between the delivery device (medium) and the skin. A very parallel problem to this is encountered in the area of delivery of drugs subcutaneously using polymeric materials that release precisely controlled quantities of macromolecules (Randomsky et al., 1990). This type of problem is presented in Problem 8.5.

8.2.3 Cell migration on vascular prosthetic materials

The successful design of vascular prosthetic materials is an important aspect of tissue engineering and tissue repair. These prosthetic materials, implanted during the reconstruction of damaged or occluded blood vessels, can be compromised by the incidence of bacterial infection. Because leukocytes serve as the key acute inflammatory mediators, the control of leukocyte motility on prosthetic material surfaces may be a critical factor in promoting implant infection resistance.

Rosenson-Schloss et al. (2002) have described cell motility on prosthetic materials under flow exposure using a diffusion-convection model shown below:

$$\frac{\partial C}{\partial t} = \mu_D \frac{\partial^2 C}{\partial z^2} - v_{eff} \frac{\partial C}{\partial z} \tag{8.9}$$

with initial and boundary conditions as follows:

$$\begin{array}{llll} t = 0 & \text{all} & z > 0 & C = 0 \\ t > 0 & & z = 0 & C = C_0 \\ & & z = z_r & C = 0 \end{array} \tag{8.10}$$

where C is the concentration of cells on the surface of the prosthetic material. Eq. (8.9) is similar to Fick's second law of diffusion based on μ_D, the random migration coefficient, but it includes, in addition, the transfer-due-to-convection term, and v_{eff} is the directional migratory velocity of the cells. This problem will be treated further in Example 8.2.

8.2.4 Fluid flow in physiological and extracorporeal vessels

Biological fluids are quite complex, exhibiting solid-like and liquid-like behavior and deforming in a time-dependent fashion. Examples of biological fluids include blood; synovial fluid in joints; lymph, which is produced by the filtration of blood plasma through tissues; and the vitrious fluid of the eye (Truskey et al., 2004). A quantitative characterization of the flow of blood is important in understanding physiological and pathological processes and in the design and operation of devices that treat blood.

In the simplest case, the blood flow in an artery may be modeled as the laminar flow of an incompressible Newtonian fluid in a rigid, circular blood vessel exposed to an oscillatory pressure field. If flow is only in the axial direction, with velocity v_z, and is fully developed, then the *Navier-Stokes* equation in cylindrical coordinates may be used:

$$\rho\frac{\partial v_z}{\partial t} = -\frac{\partial p}{\partial z} + \frac{\mu}{r}\frac{\partial}{\partial r}\left(r\frac{\partial v_z}{\partial r}\right) \tag{8.11}$$

The pressure gradient oscillates in time with frequency, ω, to simulate the pumping action of the heart:

$$-\frac{\partial p}{\partial z} = \frac{\Delta p}{L}\cos(\omega t) \tag{8.12}$$

This model is presented for solution in Problem 8.6.

8.3 Classification of Partial Differential Equations

Partial differential equations are classified according to their *order*, *linearity*, and *boundary conditions*.

The order of a partial differential equation is determined by the highest-order partial derivative present in that equation. Examples of first-, second-, and third-order partial differential equations are:

First order:
$$\frac{\partial u}{\partial x} - \alpha\frac{\partial u}{\partial y} = 0 \tag{8.13}$$

Second order:
$$\frac{\partial^2 u}{\partial x^2} + u\frac{\partial u}{\partial y} = 0 \tag{8.14}$$

Third order:
$$\left(\frac{\partial^3 u}{\partial x^3}\right)^2 + \frac{\partial^2 u}{\partial x \partial y} + \frac{\partial u}{\partial y} = 0 \tag{8.15}$$

Partial differential equations are categorized into *linear*, *quasilinear*, and *non-linear* equations. Consider, for example, the following second-order equation:

$$a(\cdot)\frac{\partial^2 u}{\partial y^2} + 2b(\cdot)\frac{\partial^2 u}{\partial x \partial y} + c(\cdot)\frac{\partial^2 u}{\partial x^2} + d(\cdot) = 0 \tag{8.16}$$

If the coefficients are constants or functions of the independent variables only, that is, if $(\cdot) \equiv (x, y)$, then Eq. (8.16) is linear. If the coefficients are functions of the dependent variable and/or any of its derivatives of lower order than that of the differential equation, that is, if $(\cdot) \equiv (x, y, u, \partial u/\partial x, \partial u/\partial y)$, then the equation is quasilinear. Finally, if the coefficients are functions of derivatives of the same order as that of the equation, that is, if $(\cdot) \equiv (x, y, u, \partial^2 u/\partial x^2, \partial^2 u/\partial y^2, \partial^2 u/\partial x \partial y)$, then the equation is nonlinear. In accordance with these definitions, Eq. (8.13) is linear, (8.14) is quasilinear, and (8.15) is nonlinear.

Linear second-order partial differential equations in two independent variables are further classified into three canonical forms: *elliptic*, *parabolic*, and *hyperbolic*. The general form of this class of equations is:

$$a\frac{\partial^2 u}{\partial x^2} + 2b\frac{\partial^2 u}{\partial x \partial y} + c\frac{\partial^2 u}{\partial y^2} + d\frac{\partial u}{\partial x} + e\frac{\partial u}{\partial y} + fu + g = 0 \tag{8.17}$$

where the coefficients are either constants or functions of the independent variables only. The three canonical forms are determined by the following criteria:

$$b^2 - ac < 0, \qquad \text{elliptic} \tag{8.18}$$

$$b^2 - ac = 0, \qquad \text{parabolic} \tag{8.19}$$

$$b^2 - ac > 0, \qquad \text{hyperbolic.} \tag{8.20}$$

If $g = 0$, then Eq. (8.17) is a *homogeneous* differential equation.

The classic examples of second-order partial differential equations that conform to the three canonical forms are:

Laplace's equation (elliptic):

$$\frac{\partial^2 u}{\partial x^2} + \frac{\partial^2 u}{\partial y^2} = 0 \tag{8.21}$$

Diffusion or heat conduction equation (parabolic):

$$\frac{\partial u}{\partial t} = \alpha \frac{\partial^2 u}{\partial x^2} \tag{8.22}$$

Wave equation (hyperbolic):

$$\frac{\partial^2 u}{\partial t^2} = a^2 \frac{\partial^2 u}{\partial x^2} \tag{8.23}$$

The methods of solution of partial differential equations depend on their canonical form, as will be demonstrated in the rest of this chapter. Because the coefficients of these equations can be functions of the independent variables, it is possible that an equation may shift from one canonical form to another over the range of integration of (x, y).

8.4 Initial and Boundary Conditions

In order to obtain unique numerical solutions to partial differential equations, the initial and boundary conditions associated with the equations must be specified. Boundary conditions for partial differential equations are divided into three categories. These are demonstrated below, using as example the one-dimensional unsteady-state diffusion equation formulated in Sec. 8.2.1:

$$\frac{\partial C}{\partial t} = D\frac{\partial^2 C}{\partial x^2} \tag{8.24}$$

Eq. (8.24) essentially describes the change in concentration within a solid or fluid slab of material (e.g., a membrane), where transfer takes place in the x-direction (see Fig. 8.2). Following are the three categories of boundary conditions:

Dirichlet conditions (first kind): The values of the dependent variable are given at fixed values of the independent variable(s). Examples of initial conditions of the Dirichlet type for the diffusion equation are:

$$C\big|_{t=0} = f(x) \qquad \text{at } t = 0 \text{ and } 0 \le x \le 1$$

or:

$$C\big|_{t=0} = C_0 \qquad \text{at } t = 0 \text{ and } 0 \le x \le 1$$

These are alternative initial conditions that specify that the initial concentration inside the membrane is a function of position, $f(x)$, or a constant, C_0 (Fig. 8.2a). Boundary conditions of the first kind are expressed as:

$$C\big|_{x=0} = f(t) \qquad \text{at } x = 0 \text{ and } t > 0$$

and:

$$C\big|_{x=1} = C_1 \qquad \text{at } x = 1 \text{ and } t > 0$$

These boundary conditions specify the value of the dependent variable at the left boundary as a function of time $f(t)$ (this may be the condition on the left side of the membrane that varies according to a preprogrammed concentration profile), and at the right boundary as a constant C_1 (e.g., the concentration in a large amount of fluid that is maintained at constant value) (Fig. 8.2a).

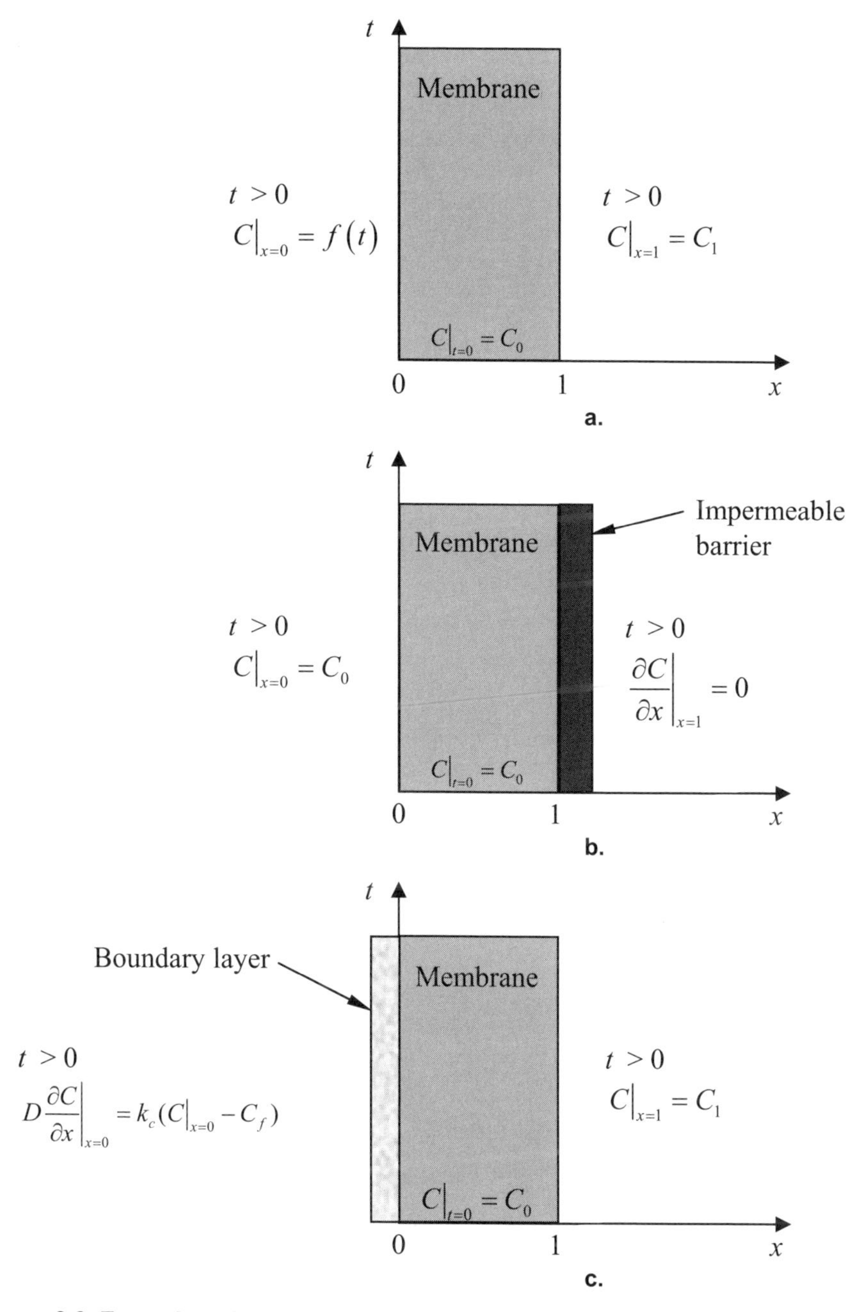

Figure 8.2 **Examples of initial and boundary conditions for the diffusion problem: a. Dirichlet conditions; b. Cauchy conditions (Dirichlet and Neumann): c. Robbins condition (on left boundary) and Dirichlet (on the right boundary). Time increases along the vertical axis.**

Neumann conditions (second kind): The derivative of the dependent variable is given as a constant or as a function of the independent variable. For example:

$$\left.\frac{\partial C}{\partial x}\right|_{x=1} = 0 \qquad \text{at } x = 1 \text{ and } t > 0$$

This condition specifies that the concentration gradient at the right boundary is zero. In the diffusion across a membrane problem, this can be theoretically accomplished by attaching an impermeable material at the right boundary (Fig. 8.2b).

Cauchy conditions: A problem that combines both Dirichlet and Neumann conditions is said to have Cauchy conditions (Fig. 8.2b).

Robbins conditions (third kind): The derivative of the dependent variable is given as a function of the dependent variable itself. For the diffusion problem, the rate of transfer at the membrane-fluid interface may be controlled by convection across a boundary layer; thus, it may be related to the difference between the concentration at the interface and that in the fluid, that is:

$$D\left.\frac{\partial C}{\partial x}\right|_{x=0} = k_c(C|_{x=0} - C_f) \qquad \text{at } x = 0 \text{ and } t > 0$$

where k_c is the convection mass transfer coefficient of the fluid boundary layer (Fig. 8.2c).

On the basis of their initial and boundary conditions, partial differential equations may be further classified into *initial-value* or *boundary-value* problems. In the first case, at least one of the independent variables has an *open region.* In the unsteady-state diffusion problem, the time variable has the range $0 \le t \le \infty$, where no condition has been specified at $t = \infty$; therefore, this is an initial-value problem. When the region is *closed* for all independent variables, and conditions are specified at all boundaries, then the problem is of the boundary-value type. An example of this is the three-dimensional steady-state heat conduction problem described by the equation:

$$\frac{\partial^2 T}{\partial x^2} + \frac{\partial^2 T}{\partial y^2} + \frac{\partial^2 T}{\partial z^2} = 0 \tag{8.25}$$

with the boundary conditions given at all six boundaries:

$$\left.\begin{matrix} T(0,y,z) \\ T(1,y,z) \end{matrix}\right\} = \text{specified}, \quad \left.\begin{matrix} T(x,0,z) \\ T(x,1,z) \end{matrix}\right\} = \text{specified}, \quad \left.\begin{matrix} T(x,y,0) \\ T(x,y,1) \end{matrix}\right\} = \text{specified} \tag{8.26}$$

8.5 Solution of Partial Differential Equations

In Chapter 6, we developed the methods of finite differences and demonstrated that ordinary derivatives can be approximated with any degree of desired accuracy by replacing the differential operators with finite difference operators.

In this section, we apply similar procedures in expressing partial derivatives in terms of finite differences. Since partial differential equations involve more than one independent variable, we first establish two-dimensional and three-dimensional grids in two and three independent variables, respectively, as shown in Fig. 8.3.

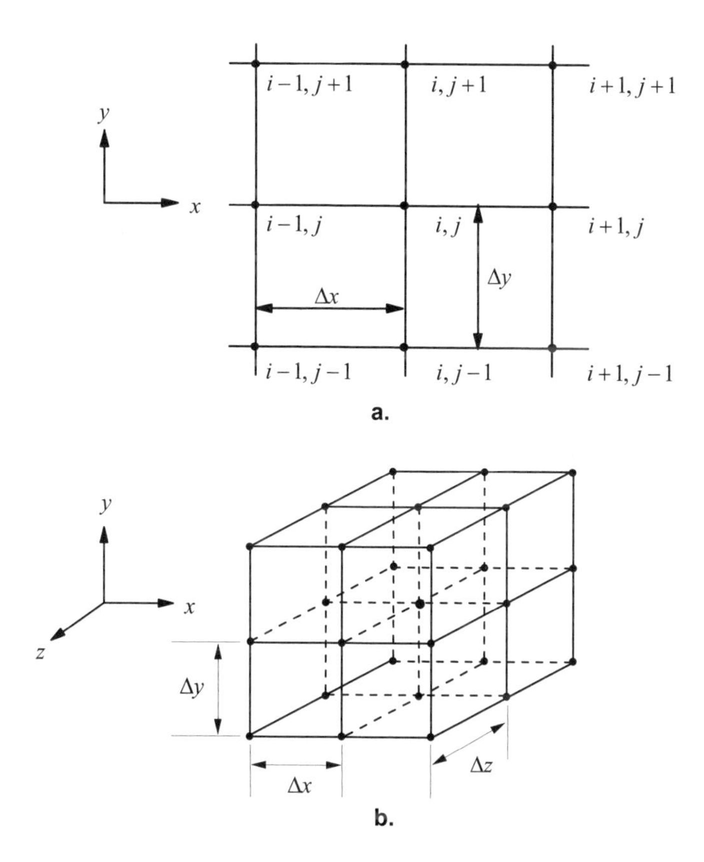

Figure 8.3 Finite difference grids: a. Two-dimensional grid; b. Three-dimensional grid.

The notation (i, j) is used to designate the pivot point for the two-dimensional space and (i, j, k) for the three-dimensional space, where $i, j,$ and k are the counters in the x, y, and z directions, respectively. For unsteady-state problems, in which *time* is one of the independent variables, the counter n is used to designate the time dimension, that is, (i, j, k, n). In order to keep the notation as simple as possible, we add subscripts only when needed. The distances between grid points are designated as Δx, Δy, and Δz. When time is one of the independent variables, the time step is shown by Δt.

We now express first, second, and mixed partial derivatives in terms of finite differences. We show the development of these approximations using central differences, and, in addition, we summarize in tabular form the formulas obtained from using forward and backward differences.

The partial derivative of u with respect to x implies that y and z are held constant; therefore:

$$\left.\frac{\partial u}{\partial x}\right|_{i,j,k} \equiv \left.\frac{du}{dx}\right|_{i,j,k} \tag{8.27}$$

Using Eq. (6.40), which is the approximation of the first-order derivative in terms of central differences, and converting it to the three-dimensional space, we obtain

$$\left.\frac{\partial u}{\partial x}\right|_{i,j,k} = \frac{1}{2\Delta x}\left(u_{i+1,j,k} - u_{i-1,j,k}\right) + O\left(\Delta x^2\right) \tag{8.28}$$

Similarly, the first-order partial derivatives in the y- and z-directions are given by

$$\left.\frac{\partial u}{\partial y}\right|_{i,j,k} = \frac{1}{2\Delta y}\left(u_{i,j+1,k} - u_{i,j-1,k}\right) + O\left(\Delta y^2\right) \tag{8.29}$$

$$\left.\frac{\partial u}{\partial z}\right|_{i,j,k} = \frac{1}{2\Delta z}\left(u_{i,j,k+1} - u_{i,j,k-1}\right) + O\left(\Delta z^2\right) \tag{8.30}$$

In an analogous manner, the second-order partial derivatives are expressed in terms of central differences by using Eq. (6.41):

$$\left.\frac{\partial^2 u}{\partial x^2}\right|_{i,j,k} = \frac{1}{\Delta x^2}\left(u_{i+1,j,k} - 2u_{i,j,k} + u_{i-1,j,k}\right) + O\left(\Delta x^2\right) \tag{8.31}$$

$$\left.\frac{\partial^2 u}{\partial y^2}\right|_{i,j,k} = \frac{1}{\Delta y^2}\left(u_{i,j+1,k} - 2u_{i,j,k} + u_{i,j-1,k}\right) + O\left(\Delta y^2\right) \tag{8.32}$$

$$\left.\frac{\partial^2 u}{\partial z^2}\right|_{i,j,k} = \frac{1}{\Delta z^2}\left(u_{i,j,k+1} - 2u_{i,j,k} + u_{i,j,k-1}\right) + O\left(\Delta z^2\right) \tag{8.33}$$

Finally, the mixed partial derivative is developed as follows:

$$\left.\frac{\partial^2 u}{\partial y \partial x}\right|_{i,j,k} = \frac{\partial}{\partial y}\left[\left.\frac{\partial u}{\partial x}\right|_{i,j,k}\right] \tag{8.34}$$

The operation of $\partial / \partial y$ applied on $\partial u / \partial x$ at (i, j, k) is equivalent to evaluating $\partial u / \partial x$ at points $(i, j + 1, k)$ and $(i, j - 1, k)$, so:

$$\begin{aligned} &\left.\frac{\partial^2 u}{\partial y \partial x}\right|_{i,j,k} \\ &= \frac{1}{2\Delta y}\left[\frac{1}{2\Delta x}\left(u_{i+1,j+1,k} - u_{i-1,j+1,k}\right) - \frac{1}{2\Delta x}\left(u_{i+1,j-1,k} - u_{i-1,j-1,k}\right)\right] + O\left(\Delta x^2 + \Delta y^2\right) \\ &= \frac{1}{4\Delta x \Delta y}\left(u_{i+1,j+1,k} - u_{i-1,j+1,k} - u_{i+1,j-1,k} + u_{i-1,j-1,k}\right) + O\left(\Delta x^2 + \Delta y^2\right) \end{aligned} \tag{8.35}$$

The above central difference approximations of partial derivatives are summarized in Table 8.1. The corresponding approximations obtained from using forward and backward differences are shown in Table 8.2 and Table 8.3, respectively. The nature of the problem, the geometry of the object being modeled, and the type of boundary conditions will dictate whether to use central, forward, or backward differences in replacing the partial derivatives. We know from the discussion in Chapter 6 that central finite differences are more accurate than either backward or forward, therefore central differences will be our first choice, however, there are situations where the use of central differences may result in numerical solutions that have instabilities (Section 8.5.2). Such cases may dictate the use of a forward or backward difference instead. Also, if the boundary conditions are of the Neumann or Robbins type, it may be preferable, due to the geometry of the system, to replace their partial derivatives with forward or backward differences (Section 8.5.1).

Equivalent sets of formulas, which are more accurate than the above, may be developed by using finite difference approximations that have higher accuracies, such as Eqs. (6.44) and (6.45) for central differences, Eqs. (6.31) and (6.32) for forward differences, and Eqs. (6.18) and (6.19) for backward differences. However, the more accurate formulas are not commonly used, because they involve a larger number of terms and require more extensive computation times.

The use of finite difference approximations is demonstrated in the following sections of this chapter in setting up the numerical solutions of elliptic, parabolic, and hyperbolic partial differential equations.

Table 8.1 Finite difference approximations of partial derivatives using central differences

Derivative	Central Difference	Error
$\left.\dfrac{\partial u}{\partial x}\right\|_{i,j,k}$	$\dfrac{1}{2\Delta x}\left(u_{i+1,j,k}-u_{i-1,j,k}\right)$	$O\left(\Delta x^2\right)$
$\left.\dfrac{\partial u}{\partial y}\right\|_{i,j,k}$	$\dfrac{1}{2\Delta y}\left(u_{i,j+1,k}-u_{i,j-1,k}\right)$	$O\left(\Delta y^2\right)$
$\left.\dfrac{\partial u}{\partial z}\right\|_{i,j,k}$	$\dfrac{1}{2\Delta z}\left(u_{i,j,k+1}-u_{i,j,k-1}\right)$	$O\left(\Delta z^2\right)$
$\left.\dfrac{\partial^2 u}{\partial x^2}\right\|_{i,j,k}$	$\dfrac{1}{\Delta x^2}\left(u_{i+1,j,k}-2u_{i,j,k}+u_{i-1,j,k}\right)$	$O\left(\Delta x^2\right)$
$\left.\dfrac{\partial^2 u}{\partial y^2}\right\|_{i,j,k}$	$\dfrac{1}{\Delta y^2}\left(u_{i,j+1,k}-2u_{i,j,k}+u_{i,j-1,k}\right)$	$O\left(\Delta y^2\right)$
$\left.\dfrac{\partial^2 u}{\partial z^2}\right\|_{i,j,k}$	$\dfrac{1}{\Delta z^2}\left(u_{i,j,k+1}-2u_{i,j,k}+u_{i,j,k-1}\right)$	$O\left(\Delta z^2\right)$
$\left.\dfrac{\partial^2 u}{\partial y \partial x}\right\|_{i,j,k}$	$\dfrac{1}{4\Delta x\Delta y}\left(u_{i+1,j+1,k}-u_{i-1,j+1,k}-u_{i+1,j-1,k}+u_{i-1,j-1,k}\right)$	$O\left(\Delta x^2+\Delta y^2\right)$

Table 8.2 Finite difference approximations of partial derivatives using forward differences

Derivative	Forward Difference	Error
$\left.\dfrac{\partial u}{\partial x}\right\|_{i,j,k}$	$\dfrac{1}{\Delta x}\left(u_{i+1,j,k}-u_{i,j,k}\right)$	$O(\Delta x)$
$\left.\dfrac{\partial u}{\partial y}\right\|_{i,j,k}$	$\dfrac{1}{\Delta y}\left(u_{i,j+1,k}-u_{i,j,k}\right)$	$O(\Delta y)$
$\left.\dfrac{\partial u}{\partial z}\right\|_{i,j,k}$	$\dfrac{1}{\Delta z}\left(u_{i,j,k+1}-u_{i,j,k}\right)$	$O(\Delta z)$
$\left.\dfrac{\partial^2 u}{\partial x^2}\right\|_{i,j,k}$	$\dfrac{1}{\Delta x^2}\left(u_{i+2,j,k}-2u_{i+1,j,k}+u_{i,j,k}\right)$	$O(\Delta x)$
$\left.\dfrac{\partial^2 u}{\partial y^2}\right\|_{i,j,k}$	$\dfrac{1}{\Delta y^2}\left(u_{i,j+2,k}-2u_{i,j+1,k}+u_{i,j,k}\right)$	$O(\Delta y)$
$\left.\dfrac{\partial^2 u}{\partial z^2}\right\|_{i,j,k}$	$\dfrac{1}{\Delta z^2}\left(u_{i,j,k+2}-2u_{i,j,k+1}+u_{i,j,k}\right)$	$O(\Delta z)$
$\left.\dfrac{\partial^2 u}{\partial y \partial x}\right\|_{i,j,k}$	$\dfrac{1}{\Delta x \Delta y}\left(u_{i+1,j+1,k}-u_{i,j+1,k}-u_{i+1,j,k}+u_{i,j,k}\right)$	$O(\Delta x+\Delta y)$

Table 8.3 Finite difference approximations of partial derivatives using backward differences

Derivative	Backward Difference	Error
$\left.\frac{\partial u}{\partial x}\right\|_{i,j,k}$	$\frac{1}{\Delta x}\left(u_{i,j,k}-u_{i-1,j,k}\right)$	$O(\Delta x)$
$\left.\frac{\partial u}{\partial y}\right\|_{i,j,k}$	$\frac{1}{\Delta y}\left(u_{i,j,k}-u_{i,j-1,k}\right)$	$O(\Delta y)$
$\left.\frac{\partial u}{\partial z}\right\|_{i,j,k}$	$\frac{1}{\Delta z}\left(u_{i,j,k}-u_{i,j,k-1}\right)$	$O(\Delta z)$
$\left.\frac{\partial^2 u}{\partial x^2}\right\|_{i,j,k}$	$\frac{1}{\Delta x^2}\left(u_{i,j,k}-2u_{i-1,j,k}+u_{i-2,j,k}\right)$	$O(\Delta x)$
$\left.\frac{\partial^2 u}{\partial y^2}\right\|_{i,j,k}$	$\frac{1}{\Delta y^2}\left(u_{i,j,k}-2u_{i,j-1,k}+u_{i,j-2,k}\right)$	$O(\Delta y)$
$\left.\frac{\partial^2 u}{\partial z^2}\right\|_{i,j,k}$	$\frac{1}{\Delta z^2}\left(u_{i,j,k}-2u_{i,j,k-1}+u_{i,j,k-2}\right)$	$O(\Delta z)$
$\left.\frac{\partial^2 u}{\partial y \partial x}\right\|_{i,j,k}$	$\frac{1}{\Delta x \Delta y}\left(u_{i,j,k}-u_{i,j-1,k}-u_{i-1,j,k}+u_{i-1,j-1,k}\right)$	$O(\Delta x+\Delta y)$

8.5.1 Elliptic partial differential equations

Elliptic partial differential equations are often encountered in steady-state diffusion and heat conduction operations. For example, for two-dimensional steady-state diffusion, Fick's second law (Eq. (8.6)) simplifies to:

$$\frac{\partial^2 C}{\partial x^2}+\frac{\partial^2 C}{\partial y^2}=0 \tag{8.36}$$

A partial differential equation of this form is known as the Laplace equation. We begin our discussion of numerical solutions of elliptic differential equations by first examining this two-dimensional problem in its general form:

$$\frac{\partial^2 u}{\partial x^2}+\frac{\partial^2 u}{\partial y^2}=0 \tag{8.37}$$

We replace each second-order partial derivative by its approximation in central differences, Eqs. (8.31) and (8.32), to obtain:

$$\frac{1}{\Delta x^2}\left(u_{i+1,j,k}-2u_{i,j,k}+u_{i-1,j,k}\right)+\frac{1}{\Delta y^2}\left(u_{i,j+1,k}-2u_{i,j,k}+u_{i,j-1,k}\right)=0 \tag{8.38}$$

which rearranges to:

$$-2\left(\frac{1}{\Delta x^2}+\frac{1}{\Delta y^2}\right)u_{i,j}+\left(\frac{1}{\Delta x^2}\right)u_{i+1,j}+\left(\frac{1}{\Delta x^2}\right)u_{i-1,j}+\left(\frac{1}{\Delta y^2}\right)u_{i,j+1}+\left(\frac{1}{\Delta y^2}\right)u_{i,j-1}=0 \tag{8.39}$$

This is a linear algebraic equation involving the value of the dependent variable at five adjacent grid points.

A rectangular-shaped object (Fig. 8.4) of length L and width W may be divided into n_x segments in the x-direction and n_y segments in the y-direction. Thus, this object will have $(n_x + 1) \times (n_y + 1)$ total grid points and $(n_x - 1) \times (n_y - 1)$ internal grid points. Eq. (8.39), written for each of the internal points, constitutes a set of $(n_x - 1) \times (n_y - 1)$ simultaneous linear algebraic equations in $(n_x + 1) \times (n_y + 1) - 4$ unknowns (the four corner points do not appear in these equations). The boundary conditions provide the additional information for the solution of the problem. If the boundary conditions are of Dirichlet type, the values of the dependent variable are known at all the external grid points. On the other hand, if the boundary conditions at any of the external surfaces are of the Neumann or Robbins type, which specify partial derivatives at the boundaries, these conditions must also be replaced by finite difference approximations.

We demonstrate this by specifying the Neumann and Robbins conditions in a unified form as:

$$\frac{\partial u}{\partial x} = \beta + \gamma u \tag{8.40}$$

Eq. (8.40) is a Neumann condition when $\gamma = 0,$ and a Robbins condition when $\gamma \neq 0.$ The value of β could be any number, including zero. To implement these conditions, we replace the partial derivative in Eq. (8.40) with a finite difference approximation. If central difference is used to approximate the derivative, Eq. (8.40) becomes:

$$\frac{1}{2\Delta x}\left(u_{i+1,j} - u_{i-1,j}\right) = \beta + \gamma u_{i,j} \tag{8.41}$$

This approximation is valid only at the boundary where either a Neumann or a Robbins condition exists. However, the use of central differences on the boundaries introduces fictitious points that are located outside the object, thus making the problem a little more complex to solve. On the other hand, the use of forward and backward finite differences at the boundaries introduces no fictitious points. The forward/backward difference formula used at the boundary should have the same order of accuracy as the finite difference approximations used to replace the derivatives in the differential equation. So in this case, the forward and backward difference equations of $O(h^2)$ for the first derivative (Eqs. (6.31) and (6.18), respectively) should be used. We demonstrate, in the next few paragraphs, the use of these equations at the four boundaries of a rectangular-shaped object.

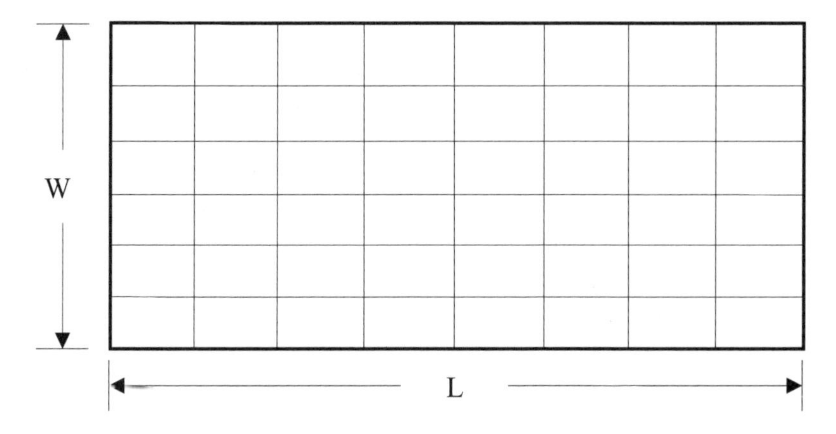

Figure 8.4 A grid for a rectangular-shaped object witn n_x = 8 and n_y = 6. This has 63 total and 35 internal grid points.

At the lower x boundary, where $x = 0$ and $i = 1$, we replace the derivative in the Neumann/Robbins condition (Eq. (8.40)) with the forward finite difference (Eq. (6.31)), and solve for $u_{i,j}$:

$$u_{i,j} = \frac{4u_{i+1,j} - u_{i+2,j} - 2\Delta x\beta}{(3 + 2\Delta x\gamma)} \quad \text{where} \quad i = 1 \tag{8.42}$$

Eq. (8.42) is valid only at the lower x boundary when the Neumann/Robbins condition exists.

At the upper x boundary, where $x = L$ and $i = n_x + 1$, we replace the derivative in the Neumann/Robbins condition with the backward finite difference (Eq. (6.18)), and solve for $u_{i,j}$:

$$u_{i,j} = \frac{4u_{i-1,j} - u_{i-2,j} + 2\Delta x\beta}{(3 - 2\Delta x\gamma)} \quad \text{where} \quad i = n_x + 1 \tag{8.43}$$

Eq. (8.43) is valid only at the upper x boundary where the Neumann/Robbins condition exists.

At the lower y boundary, where $y = 0$ and $j = 1$, we replace the derivative in the Neumann/Robbins condition with the forward finite difference (Eq. (6.31)), and solve for $u_{i,j}$:

$$u_{i,j} = \frac{4u_{i,j+1} - u_{i,j+2} - 2\Delta y\beta}{(3 + 2\Delta y\gamma)} \quad \text{where} \quad j = 1 \tag{8.44}$$

Eq. (8.44) is valid only at the lower y boundary where the Neumann/Robbins condition exists.

At the upper y boundary, where $y = W$ and $j = n_y + 1$, we replace the derivative in the Neumann/Robbins condition with the backward finite difference (Eq. (6.18)), and solve for $u_{i,j}$:

$$u_{i,j} = \frac{4u_{i,j-1} - u_{i,j-2} + 2\Delta y\beta}{(3 - 2\Delta y\gamma)} \quad \text{where} \quad j = n_y + 1 \tag{8.45}$$

Eq. (8.45) is valid only at the upper y boundary where the Neumann/Robbins condition exists.

Eq. (8.39) and the appropriate boundary conditions constitute a set of linear algebraic equations, so the Gauss methods for the solution of such equations may be used. Eq. (8.39) is a "borderline" predominantly diagonal system; therefore, the Gauss-Seidel method (see Section 4.5.2) may be used for the solution of this problem. Rearranging Eq. (8.39) to solve for $u_{i,j}$:

$$u_{i,j} = \frac{\frac{1}{\Delta x^2}\left(u_{i+1,j} + u_{i-1,j}\right) + \frac{1}{\Delta y^2}\left(u_{i,j+1} + u_{i,j-1}\right)}{2\left(\frac{1}{\Delta x^2} + \frac{1}{\Delta y^2}\right)} \tag{8.46}$$

which can be used in the iterative Gauss-Seidel substitution method. An initial estimate of all $u_{i,j}$ is needed, but this can be easily obtained from averaging the Dirichlet boundary conditions.

When an equidistant grid can be used, that is, when $\Delta x = \Delta y$, Eq. (8.46) simplifies to:

$$u_{i,j} = \frac{u_{i+1,j} + u_{i-1,j} + u_{i,j+1} + u_{i,j-1}}{4} \tag{8.47}$$

which simply shows that the value of the dependent variable at the pivotal point (i, j) in the Laplace equation is the arithmetic average of the values at the grid points to the right and left of and above and below the pivot point. This is demonstrated by the computational representation of Fig. 8.5, which is sometimes referred to as a "five-point star."

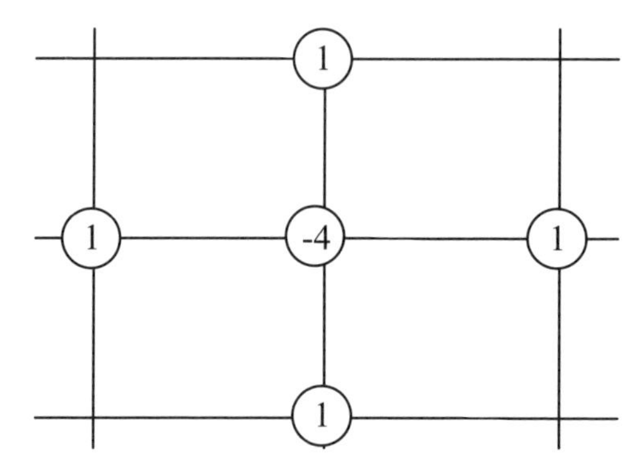

Figure 8.5 Computational representation for the Laplace equation using equidistant grid. The number in each circle is the coefficient of that point in the difference equation (8.47).

The three-dimensional elliptic partial differential equation:

$$\frac{\partial^2 u}{\partial x^2} + \frac{\partial^2 u}{\partial y^2} + \frac{\partial^2 u}{\partial z^2} = 0 \tag{8.48}$$

can be similarly converted to linear algebraic equations using finite difference approximations in three-dimensional space. Applying Eqs. (8.31), (8.32), and (8.33) to replace the three partial derivatives of Eq. (8.48), we obtain:

$$\frac{1}{\Delta x^2}\left(u_{i+1,j,k} - 2u_{i,j,k} + u_{i-1,j,k}\right) + \frac{1}{\Delta y^2}\left(u_{i,j+1,k} - 2u_{i,j,k} + u_{i,j-1,k}\right) + \frac{1}{\Delta z^2}\left(u_{i,j,k+1} - 2u_{i,j,k} + u_{i,j,k-1}\right) = 0 \tag{8.49}$$

For the equidistant grid ($\Delta x = \Delta y = \Delta z$), the above equation reduces to:

$$u_{i,j,k} = \frac{u_{i+1,j,k} + u_{i-1,j,k} + u_{i,j+1,k} + u_{i,j-1,k} + u_{i,j,k+1} + u_{i,j,k-1}}{6} \tag{8.50}$$

In parallel with the two-dimensional case, the value of the dependent variable at the pivot point (i, j, k) is the arithmetic average of the values at the six grid points adjacent to the pivot point. The computational representation for the three-dimensional elliptic equation is shown in Fig. 8.6.

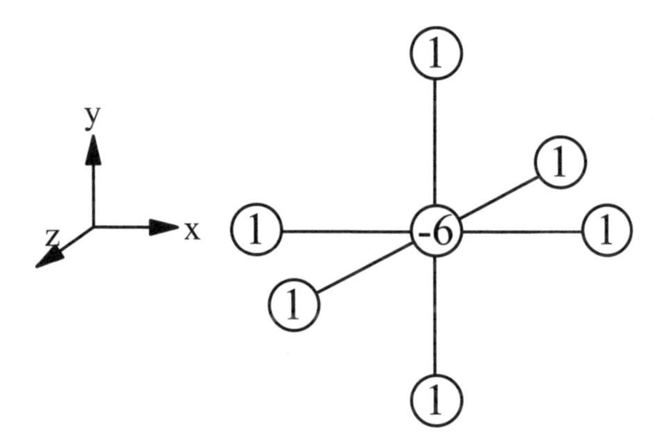

Figure 8.6 Computational representation for the three-dimensional elliptic partial differential equation using equidistant grid. The number in each circle is the coefficient of that point in the difference equation (8.50).

The *nonhomogeneous* form of the Laplace equation is the *Poisson* equation:

$$\frac{\partial^2 u}{\partial x^2}+\frac{\partial^2 u}{\partial y^2}=f(x,y) \tag{8.51}$$

which also belongs to the class of elliptic partial differential equations. A form of the Poisson equation:

$$\frac{\partial^2 \phi}{\partial x^2}+\frac{\partial^2 \phi}{\partial y^2}=-\frac{p}{T} \tag{8.52}$$

is used to describe the displacement, ϕ, of a highly stretched membrane of tension T that is subjected to a uniform pressure p. The finite difference formulation of the Poisson equation is:

$$-2\left(\frac{1}{\Delta x^2}+\frac{1}{\Delta y^2}\right)u_{i,j}+\left(\frac{1}{\Delta x^2}\right)u_{i+1,j}+\left(\frac{1}{\Delta x^2}\right)u_{i-1,j}+\left(\frac{1}{\Delta y^2}\right)u_{i,j+1}+\left(\frac{1}{\Delta y^2}\right)u_{i,j-1}=f_{i,j} \tag{8.53}$$

or, solving for $u_{i,j}$:

$$u_{i,j}=\frac{\frac{1}{\Delta x^2}\left(u_{i+1,j}+u_{i-1,j}\right)+\frac{1}{\Delta y^2}\left(u_{i,j+1}+u_{i,j-1}\right)}{2\left(\frac{1}{\Delta x^2}+\frac{1}{\Delta y^2}\right)}-\frac{f_{i,j}}{2\left(\frac{1}{\Delta x^2}+\frac{1}{\Delta y^2}\right)} \tag{8.54}$$

The numerical solution of the Laplace and Poisson elliptic partial differential equations is demonstrated in Example 8.1.

Example 8.1 Solution of the Laplace and Poisson equations. Dynamics of a membrane subject to tension and pressure.

Statement of the problem

The cochlea, which is part of the inner ear, is a small fluid-filled chamber that contains the biological structures that convert mechanical acoustic signals into neural signals. The cochlea has two compartments filled with fluid separated by the basilar membrane. The upper compartment corresponds to the *vestibular* canal, and the lower compartment to the *timpanic* canal. External pressure applied to the ear may cause displacement of the basilar membrane. The partial differential equation that describes displacement, ϕ, of membranes that are subjected to a tension, T, and uniform pressure, p, is the Poisson equation:

$$\frac{\partial^2 \phi}{\partial x^2} + \frac{\partial^2 \phi}{\partial y^2} = -\frac{p}{T} \tag{8.52}$$

Consider a 1-cm-square membrane that is firmly fastened to a support frame on all four sides, so that there is no displacement along the four boundaries, that is,

$$\phi(0, y) = 0, \qquad \phi(1, y) = 0, \qquad \phi(x, 0) = 0, \qquad \phi(x, 1) = 0$$

Solve for the displacement, ϕ, of this membrane using the following values of the parameters:

$$\frac{p}{T} = 0.5 \quad \text{cm}^{-1}$$

Plot the results as a three-dimensional surface.

Solution

The program listed below evaluates the finite difference solution of the two-dimensional Laplace or Poisson equation (Eq. (8.46) or (8.54)), using the iterative Gauss-Seidel substitution method with boundary conditions that can be of the Dirichlet, Neumann, or Robbins type. Eqs. (8.42) to (8.45) are used to implement the Neumann or Robbins conditions.

Program

```
% example8_1.m - This program solves the two-dimensional
% Laplace or Poisson equation with Dirichlet, Neumann,
% or Robbins boundary conditions. The program uses finite
% differences and the Gauss-Seidel method. It is applied to
% the calculation of the displacement of a stretched membrane.

clc; clear all;
bcdialog = [' Lower x boundary condition:'
            ' Upper x boundary condition:'
            ' Lower y boundary condition:'
            ' Upper y boundary condition:'];
bc = zeros(4,3);  % zero the boundary conditions
disp('       Solution of the Laplace or Poisson equation')
disp('(Two-dimensional elliptic partial differential equation)')
```

```
disp(' '); disp(' ')
disp('                     Upper y boundary            '         )
disp('         _______________________________________ '         )
disp('        |                                      |'          )
disp('        |                                      |'          )
disp('        |                                      |'          )
disp('        |                                      |'          )
disp('Lower x |                                      | Upper x' )
disp('boundary|                                      | boundary')
disp('        |                                      |'          )
disp('        |                                      |'          )
disp('        y                                      |'          )
disp('        |___x________ length (L) ______________|'          )
disp('       0                                         '         )
disp('                     Lower y boundary            '         )
disp(' ');
L = input(' Length of the object (x-direction) (cm) = ');
W = input(' Width of the object  (y-direction) (cm) = ');
nx = input(' Number of divisions in x-direction = ');
ny = input(' Number of divisions in y-direction = ');
disp(' ');
f = input('Right-hand side of the Poisson equation (f) = ');
disp(' '); guess = input('Initial starting guess = ');
disp(' '); disp(' Boundary conditions:')
for k = 1:4
   disp(' ')
   disp(bcdialog(k,:))
   disp(' 1 - Dirichlet')
   disp(' 2 - Neumann')
   disp(' 3 - Robbins')
   bc(k,1) = input(' Enter your choice : ');
   if bc(k,1) == 1
      bc(k,2) = input(' Value of Dirichlet boundary = ');
   end
   if bc(k,1) == 2
      bc(k,2) = input(' Value of Neummann boundary = ');
   end
   if bc(k,1) == 3
      disp(' Value of Robbins boundary:');
      disp(' u'' = (beta) + (gamma)*u')
      bc(k,2) = input(' Constant    (beta)  = ');
      bc(k,3) = input(' Coefficient (gamma) = ');
   end
end
dx = L/nx;  dy = W/ny;     % Calculate the increments
x = [0:nx]*dx; y = [0:ny]*dy;
```

```
u = guess*ones(nx+1,ny+1); %Set initial guess for all values of u
U=u;
count = 0; iter=0;          % Zero the counters
total_points = (nx+1)*(ny+1);
while count < (total_points)
count = 0; iter = iter + 1;
% Set the boundary conditions
% Lower x boundary condition
i = 1;
switch bc(1,1)
case 1
    u(i,:) = bc(1,2);
case {2, 3}
    u(i,:) = (4*u(i+1,:)-u(i+2,:)-2*dx*bc(1,2))/(3+2*dx*bc(1,3));
end
% Upper x boundary condition
i = nx+1;
switch bc(2,1)
case 1
    u(i,:) = bc(2,2);
case {2, 3}
    u(i,:) = (4*u(i-1,:)-u(i-2,:)+2*dx*bc(2,2))/(3-2*dx*bc(2,3));
end
% Lower y boundary condition
j = 1;
switch bc(3,1)
case 1
    u(:,j) = bc(3,2);
case {2, 3}
    u(:,j) = (4*u(:,j+1)-u(:,j+2)-2*dy*bc(3,2))/(3+2*dy*bc(3,3));
end
% Upper y boundary condition
j = ny+1;
switch bc(4,1)
case 1
    u(:,j) = bc(4,2);
case {2, 3}
    u(:,j) = (4*u(:,j-1)-u(:,j-2)+2*dy*bc(4,2))/(3-2*dy*bc(4,3));
end
% Evaluate all internal points using Gauss-Seidel method
for i = 2:nx
    for j = 2:ny
        u(i,j)=((u(i+1,j)+u(i-1,j))/dx^2+(u(i,j+1)...
            +u(i,j-1))/dy^2 - f)/(2*((1/dx^2)+(1/dy^2)));
    end
end
% Examine convergence
for i = 1:nx+1
```

```
        for j = 1:ny+1
            if u(i,j) ~= 0
                if abs((U(i,j)-u(i,j))/u(i,j)) <= 1e-12
                    count = count+1;
                end
            else
                if abs((U(i,j)-u(i,j))) <= 1e-12
                    count = count+1;
                end
            end
        end
    end
    U=u;
    end

    fprintf('\n Iterations = %g  ',iter)
    fprintf('\n Points converged = %g/%g \n\n', count, total_points)
    disp(' The matrix of results is:')
    u'
    %Plot the final results
    clf; figure(1); surf(x,y,u')
    xlabel('Length (cm)'); ylabel('Width (cm)');
    zlabel('Displacement (cm)')
    shading interp
    title('Displacement of the Stretched Membrane')
```

Input

```
>> example8_1

         Solution of the Laplace or Poisson equation
 (Two-dimensional elliptic partial differential equation)


                         Upper y boundary
              _______________________________________
              |                                      |
              |                                      |
              |                                      |
              |                                      |
 Lower x      |                                      | Upper x
 boundary     |                                      | boundary
              |                                      |
              |                                      |
              y                                      |
              |___x________ length (L) ______________|
             0
```

```
                    Lower y boundary

 Length of the object (x-direction) (cm) = 1
 Width of the object  (y-direction) (cm) = 1
 Number of divisions in x-direction = 10
 Number of divisions in y-direction = 10

Right-hand side of the Poisson equation (f) = -0.5

Initial starting guess = 0.02

 Boundary conditions:

 Lower x boundary condition:
 1 - Dirichlet
 2 - Neumann
 3 - Robbins
 Enter your choice : 1
 Value of Dirichlet boundary = 0

 Upper x boundary condition:
 1 - Dirichlet
 2 - Neumann
 3 - Robbins
 Enter your choice : 1
 Value of Dirichlet boundary = 0

 Lower y boundary condition:
 1 - Dirichlet
 2 - Neumann
 3 - Robbins
 Enter your choice : 1
 Value of Dirichlet boundary = 0

 Upper y boundary condition:
 1 - Dirichlet
 2 - Neumann
 3 - Robbins
 Enter your choice : 1
 Value of Dirichlet boundary = 0
```

Results

```
Iterations = 240
 Points converged = 121/121
 The matrix of results is:
```

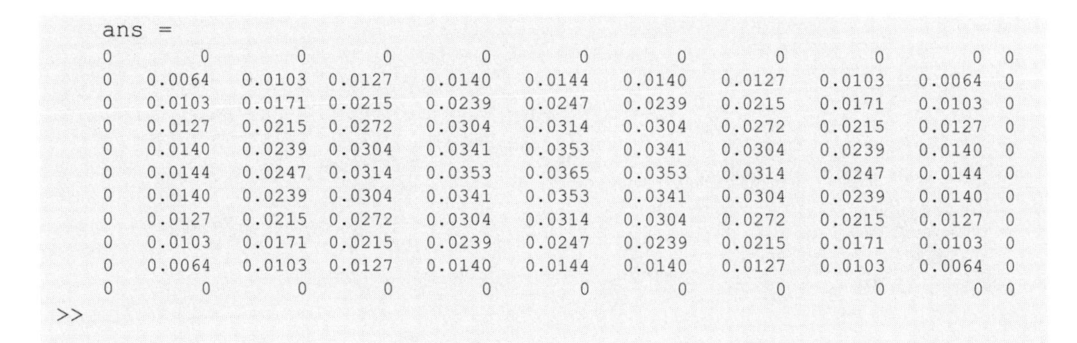

```
ans =
0        0        0        0        0        0        0        0        0        0  0
0   0.0064   0.0103   0.0127   0.0140   0.0144   0.0140   0.0127   0.0103   0.0064  0
0   0.0103   0.0171   0.0215   0.0239   0.0247   0.0239   0.0215   0.0171   0.0103  0
0   0.0127   0.0215   0.0272   0.0304   0.0314   0.0304   0.0272   0.0215   0.0127  0
0   0.0140   0.0239   0.0304   0.0341   0.0353   0.0341   0.0304   0.0239   0.0140  0
0   0.0144   0.0247   0.0314   0.0353   0.0365   0.0353   0.0314   0.0247   0.0144  0
0   0.0140   0.0239   0.0304   0.0341   0.0353   0.0341   0.0304   0.0239   0.0140  0
0   0.0127   0.0215   0.0272   0.0304   0.0314   0.0304   0.0272   0.0215   0.0127  0
0   0.0103   0.0171   0.0215   0.0239   0.0247   0.0239   0.0215   0.0171   0.0103  0
0   0.0064   0.0103   0.0127   0.0140   0.0144   0.0140   0.0127   0.0103   0.0064  0
0        0        0        0        0        0        0        0        0        0  0
>>
```

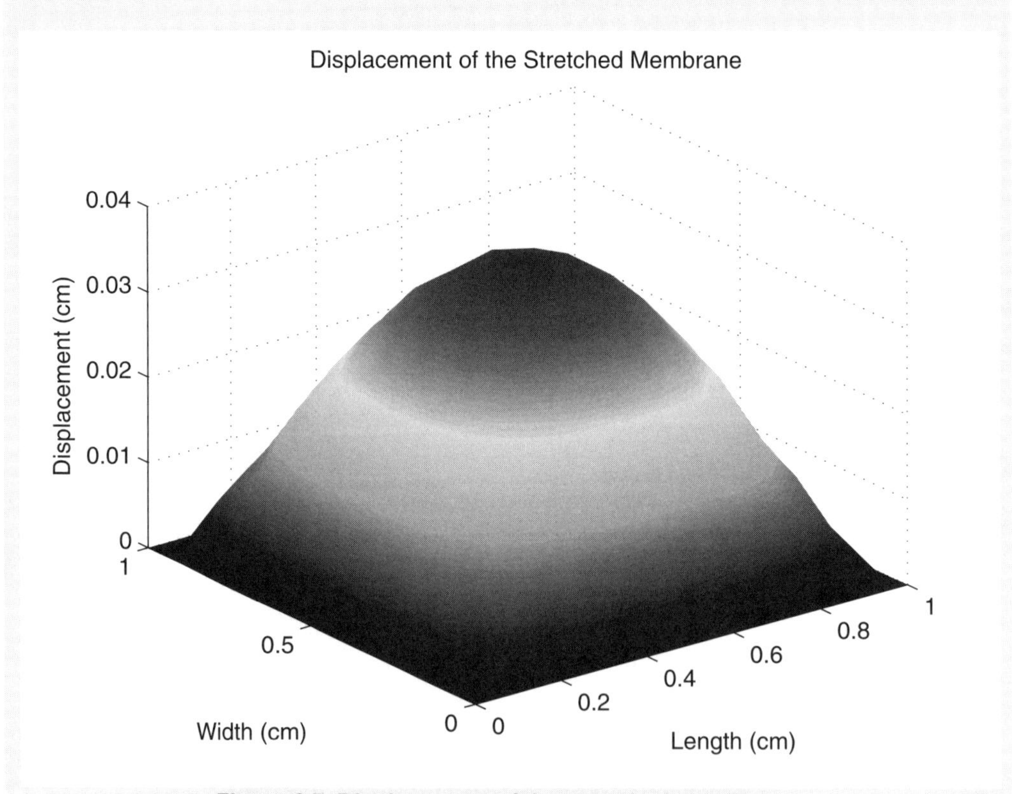

Figure 8.7 Displacement of the stretched membrane.

Discussion of results

The above results indicate that the displacement of the membrane is maximum at the geometric center, as would be expected. **Note**: The above program solves the Laplace equation when the value of f is set to zero.

8.5.2 Parabolic partial differential equations

Classic examples of parabolic differential equations are the one-dimensional unsteady state diffusion equation (Fick's second law of diffusion):

$$\frac{\partial C}{\partial t} = D\frac{\partial^2 C}{\partial x^2} \tag{8.24}$$

and the one-dimensional unsteady-state heat conduction equation:

$$\frac{\partial T}{\partial t} = \alpha\frac{\partial^2 T}{\partial x^2} \tag{8.55}$$

These equations may have Dirichlet, Neumann, or Cauchy boundary conditions.

Let us consider this class of equations in the general one-dimensional form:

$$\frac{\partial u}{\partial t} = \alpha\frac{\partial^2 u}{\partial x^2} \tag{8.22}$$

In this section, we develop several methods of solving Eq. (8.22) using finite differences.

Explicit methods: We express the derivatives in terms of central differences around the point (i, n), using the counter i for the x-direction and n for the t-direction:

$$\left.\frac{\partial^2 u}{\partial x^2}\right|_{i,n} = \frac{1}{\Delta x^2}\left(u_{i+1,n} - 2u_{i,n} + u_{i-1,n}\right) + O\left(\Delta x^2\right) \tag{8.56}$$

$$\left.\frac{\partial u}{\partial t}\right|_{i,n} = \frac{1}{2\Delta t}\left(u_{i,n+1} - u_{i,n-1}\right) + O\left(\Delta t^2\right) \tag{8.57}$$

Combining Eqs. (8.22), (8.56), (8.57), and rearranging:

$$u_{i,n+1} = u_{i,n-1} + \frac{2\alpha\Delta t}{\Delta x^2}\left(u_{i+1,n} - 2u_{i,n} + u_{i-1,n}\right) + O\left(\Delta x^2 + \Delta t^2\right) \tag{8.58}$$

This is an *explicit* algebraic formula, which calculates the value of the dependent variable at the next time step ($u_{j,n+1}$) from values at the current and earlier time steps. Once the initial and boundary conditions of the problem are specified, solution of an explicit formula is usually straightforward. However, this particular explicit formula is *unstable*, because it contains negative terms on the right side. A rigorous discussion of stability analysis is given in Appendix E. As a rule of thumb, when all the known values are arranged on the right side of the finite difference formulation, if there are any negative coefficients, the solution is unstable. This is stated more precisely by the positivity rule: "For

$$u_{i,n+1} = Au_{i+1,n} + Bu_{i,n} + Cu_{i-1,n} \tag{8.59}$$

if A, B, and C are positive, and $A + B + C \leq 1$, then the numerical scheme is stable."

In order to eliminate the instability problem, we replace the first-order derivative in Eq. (8.22) with the forward difference:

$$\left.\frac{\partial u}{\partial t}\right|_{i,n} = \frac{1}{\Delta t}\left(u_{i,n+1} - u_{i,n}\right) + O(\Delta t) \tag{8.60}$$

Combining Eqs. (8.22), (8.56), and (8.60) we obtain the explicit formula:

$$u_{i,n+1} = \left(\frac{\alpha \Delta t}{\Delta x^2}\right)u_{i+1,n} + \left(1 - 2\frac{\alpha \Delta t}{\Delta x^2}\right)u_{i,n} + \left(\frac{\alpha \Delta t}{\Delta x^2}\right)u_{i-1,n} + O\left(\Delta x^2 + \Delta t\right) \tag{8.61}$$

For a stable solution, the positivity rule requires that:

$$\left(1 - 2\frac{\alpha \Delta t}{\Delta x^2}\right) \geq 0 \tag{8.62}$$

Rearranging Eq. (8.62), we get:

$$\frac{\alpha \Delta t}{\Delta x^2} \leq \frac{1}{2} \tag{8.63}$$

This inequality determines the relationship between the two integration steps, Δx in the x-direction and Δt in the t-direction. As Δx gets smaller, Δt becomes much smaller, thus requiring longer computation times.

If we choose to work with the equality part of Eq. (8.62) or (8.63), that is,

$$\frac{\alpha \Delta t}{\Delta x^2} = \frac{1}{2} \tag{8.64}$$

then Eq. (8.61) simplifies to

$$u_{i,n+1} = \frac{1}{2}\left(u_{i+1,n} + u_{i-1,n}\right) + O\left(\Delta x^2 + \Delta t\right) \tag{8.65}$$

This explicit formula calculates the value of the dependent variable at position i of the next time step $(n + 1)$ from values to the right and left of i at the present time step n. The computational representation for this equation is shown in Fig. 8.8.

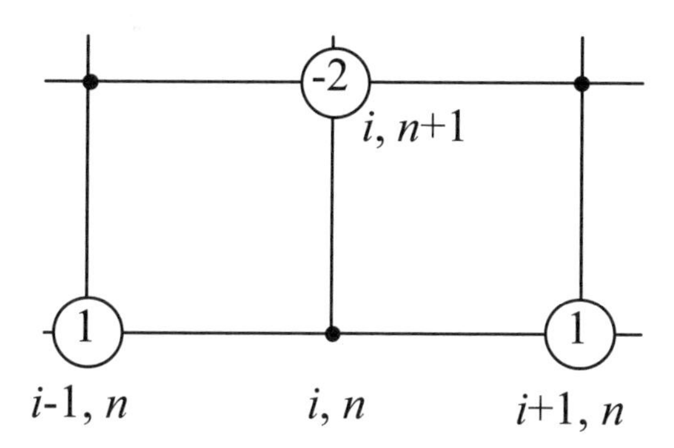

Figure 8.8 Computational representation of Eq. (8.65).

It should be emphasized that using the forward difference for the first-order derivative introduces the error of order $O(\Delta t)$; therefore, Eq. (8.61) is of order $O(\Delta t)$ in the time direction and $O(\Delta x^2)$ in the x-direction. However, the advantage of gaining stability outweighs the loss of accuracy in this case.

The finite difference solution of the nonhomogeneous parabolic equation:

$$\frac{\partial u}{\partial t} = \alpha \frac{\partial^2 u}{\partial x^2} + f(x,t) \tag{8.66}$$

is given by the following explicit formula:

$$u_{i,n+1} = \left(\frac{\alpha \Delta t}{\Delta x^2}\right) u_{i+1,n} + \left(1 - 2\frac{\alpha \Delta t}{\Delta x^2}\right) u_{i,n} + \left(\frac{\alpha \Delta t}{\Delta x^2}\right) u_{i-1,n} + (\Delta t) f_{i,n} + O\left(\Delta x^2 + \Delta t\right) \tag{8.67}$$

In diffusion, we encounter equations of the type in Eq. (8.66) when there is a source or sink of the solute present in the physical problem.

The same treatment for the two-dimensional parabolic formula:

$$\frac{\partial u}{\partial t} = \alpha \left(\frac{\partial^2 u}{\partial x^2} + \frac{\partial^2 u}{\partial y^2}\right) + f(x,y,t) \tag{8.68}$$

results in:

$$\begin{aligned} u_{i,j,n+1} &= \left(\frac{\alpha\Delta t}{\Delta x^2}\right)\left(u_{i+1,j,n} + u_{i-1,j,n}\right) + \left(\frac{\alpha\Delta t}{\Delta y^2}\right)\left(u_{i,j+1,n} + u_{i,j-1,n}\right) \\ &+ \left(1 - 2\frac{\alpha\Delta t}{\Delta x^2} - 2\frac{\alpha\Delta t}{\Delta y^2}\right)u_{i,j,n} + (\Delta t) f_{i,j,n} + O\left(\Delta x^2 + \Delta y^2 + \Delta t\right) \end{aligned} \quad (8.69)$$

The following stability condition is obtained from the positivity rule:

$$1 - 2\alpha\Delta t\left(\frac{1}{\Delta x^2} + \frac{1}{\Delta y^2}\right) \geq 0 \quad (8.70)$$

This can be rearranged to:

$$\frac{1}{\Delta x^2} + \frac{1}{\Delta y^2} \leq \frac{1}{2\alpha\Delta t} \quad (8.71)$$

Some algebraic manipulation is involved (see Constantinides and Mostoufi, 1999) to get the stability condition expressed as:

$$\frac{\alpha\Delta t}{\Delta x^2 + \Delta y^2} \leq \frac{1}{8} \quad (8.72)$$

The formula for the three-dimensional parabolic equation can be derived by adding to Eq. (8.69) the terms that come from $\partial^2 u / \partial z^2$. The right-hand side of the stability condition in this case is $1/18$.

Parabolic partial differential equations can have initial and boundary conditions of the Dirichlet, Neumann, Cauchy, or Robbins type. These were discussed in Section 8.4. Examples of these conditions for the diffusion problem are demonstrated in Fig. 8.3. The boundary conditions must be discretized using the same finite-difference grid as used for the differential equation. For Dirichlet conditions, this simply involves setting the values of the dependent variable along the appropriate boundary equal to the given boundary condition. For Neumann and Robbins conditions, the gradient at the boundaries must be replaced by finite difference approximations, as was done in Section 8.5.1, resulting in additional algebraic equations that must be incorporated into the overall scheme of solution of the resulting set of algebraic equations.

Example 8.2 Migration of human leukocytes on prosthetic materials.

Statement of the problem

The flow and migration of human leukocytes on prosthetic material surfaces may be modeled by the diffusion-convection equation (see Section 8.2.3):

$$\frac{\partial C}{\partial t} = \mu_D \frac{\partial^2 C}{\partial z^2} - v_{eff} \frac{\partial C}{\partial z}$$

Solve this equation numerically by expressing the partial derivatives into finite difference approximations. Apply the following conditions:

initial condition:	$t = 0$	all	$z > 0$	$C = 0$
boundary conditions:	$t > 0$		$z = 0$	$C = 1 \times 10^6$
			$z = 0.2$	$C = 0$

and constants:

random migration coefficient: $\mu_D = 1 \times 10^{-4}\ \text{cm}^2/s$

directional migratory velocity of cells: $v_{eff} = 1 \times 10^{-5}\ \text{cm}/s$

Solution

Express the time derivative into forward finite difference:

$$\left.\frac{\partial C}{\partial t}\right|_{i,n} = \frac{1}{\Delta t}\left(C_{i,n+1} - C_{i,n}\right)$$

Express the first-order and second-order spatial derivatives in terms of central finite differences:

$$\left.\frac{\partial C}{\partial z}\right|_{i,n} = \frac{1}{2\Delta z}\left(C_{i+1,n} - C_{i-1,n}\right)$$

$$\left.\frac{\partial^2 C}{\partial z^2}\right|_{i,n} = \frac{1}{\Delta z^2}\left(C_{i+1,n} - 2C_{i,n} + C_{i-1,n}\right)$$

Combine the above into Eq. (8.9), and rearrange to solve for $C_{i,n+1}$:

$$C_{i,n+1} = \left[\left(\frac{\mu_D \Delta t}{\Delta z^2}\right) - \left(\frac{v_{eff}\Delta t}{2\Delta z}\right)\right] C_{i+1,n} + \left[1 - \left(\frac{2\mu_D \Delta t}{\Delta z^2}\right)\right] C_{i,n} + \left[\left(\frac{\mu_D \Delta t}{\Delta z^2}\right) - \left(\frac{v_{eff}\Delta t}{2\Delta z}\right)\right] C_{i-1,n}$$

Note that when $v_{eff} = 0$, the above solution is identical to Eq. (8.61). According to the positivity rule, the square-bracketed terms in the above solution must be positive. For this to happen, the following inequality must be true:

$$\left(\frac{v_{eff}\Delta t}{2\Delta z}\right) < \left(\frac{\mu_D \Delta t}{\Delta z^2}\right) \leq \frac{1}{2}$$

This example has Dirichlet boundary conditions, which are straightforward to apply. The MATLAB program listed below implements the solution of this problem and verifies that the above inequality is observed.

Program

```
% example8_2.m - This program calculates and plots the
% concentration profiles of migrating human leukocytes
% on prosthetic materials by solving the parabolic
% partial differential equation using finite differences

clc; clear all;

disp(' Solution of parabolic partial differential equation.')
disp(' ');
h = input(' Total distance of migration (cm) = ');
tmax = input(' Maximum integration time (s) = ');
nz = input(' Number of divisions in z-direction = ');
nt = input(' Number of divisions in t-direction = ');
mu_D = input(' Random migration coefficient, mu_D, (cm^2/s) = ');
v_eff = input(' Directional migratory velocity, v_eff, (cm/s) =
');
disp(' '); disp(' Boundary conditions:'); disp(' ');
b_initial = input(' Density of leukocytes at initial time
(cells/cm^2) = ');
b_left = input(' Density of leukocytes at left boundary
(cells/cm^2) = ');
b_right = input(' Density of leukocytes at right boundary
(cells/cm^2) = ');
% Calculate the increments
dz = h/nz;
z = ([0:nz]*dz)';
% Apply stability criteria
if tmax/nt < dz^2/(2*mu_D)
```

```
        dt = tmax/nt;
    else
        disp(' ')
        disp(' Resetting dt to confirm with stability criteria')
        dt = dz^2/(2*mu_D)
    end
    if (v_eff*dt)/(2*dz) >= (mu_D*dt)/(dz^2)
        error(' Stability criterion violated')
    end

    nt = ceil(tmax/dt);
    t = [0:nt]*dt;
    % Create the solution matrix
    C = zeros(nz+1,nt+1);
    % Enter the initial conditions into the solution matrix
    C(:,1) = b_initial;
    % Enter the boundary conditions into the solution matrix
    C(1,2:nt+1) = b_left;
    % Enter the boundary conditions into the solution matrix
    C(nz+1,2:nt+1) = b_right;
    % Iteration on t
    for n = 2:nt
    % Calculate the concentration profile
        for i=2:nz
            A=(mu_D*dt)/(dz^2)-(v_eff*dt)/(2*dz);
            B=1-2*(mu_D*dt)/(dz^2);
            C(i,n+1)=A*C(i+1,n)+B*C(i,n)+A*C(i-1,n);
        end
    end
    % Plot k concentration profiles
    k=5;
    new_C(:,1)=C(:,1)/b_left;
    count=fix((length(t)-1)/k);
    for i=1:k
        new_C(:,i+1)=C(:,1+i*count)/b_left;
    end
    clf; figure(1); plot(z, new_C)
    xlabel('Distance, cm'); ylabel('Normalized cell density, C/C_0')
    title('Normalized Cell Density Profiles')
    % Create textarrow
    annotation1 = annotation(figure(1),'textarrow',...
        [0.2967 0.4695],[0.2643 0.4713],'String',{'Time'});
```

Input

```
>> example8_2
Solution of parabolic partial differential equation.

 Total distance of migration (cm) = 0.2
 Maximum integration time (s) = 40
```

```
  Number of divisions in z-direction = 100
  Number of divisions in t-direction = 100
  Random migration coefficient, mu_D, (cm^2/s) = 1e-4
  Directional migratory velocity, v_eff, (cm/s) = 1e-5

  Boundary conditions:

  Density of leukocytes at initial time (cells/cm^2) = 0
  Density of leukocytes at left boundary (cells/cm^2) = 1e6
  Density of leukocytes at right boundary (cells/cm^2) = 0

  Resetting dt to confirm with stability criteria
dt =
    0.0200
>>
```

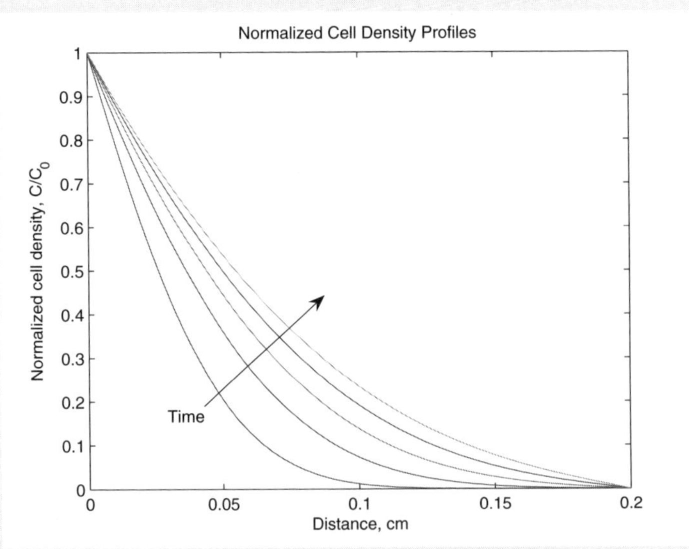

Figure 8.9 Normalized cell density profiles.

Discussion of results

In Fig. 8.9, we plot the normalized density profile, C/C_0, along the distance of migration. Originally (at t =0) the density of cells on the surface of the prosthetic material is zero. A high density of cells (1×10^6 cell/cm^2) is seeded at the left boundary (at z =0). As time progresses, the cells migrate to the right, as demonstrated by the solution of the differential equation. The right boundary has a Dirichlet condition of zero. This is equivalent to maintaining the concentration at zero; that is, having an infinite sink that will take away the cells at the right boundary, therefore preventing accumulation. The student is encouraged to solve this problem with v_{eff} = 0, and compare the results. Homework problem 8.4 solves this model with a Neumann boundary condition.

Implicit methods: Let us now consider some implicit methods for the solution of parabolic equations. We use the grid of Fig. 8.10, in which the half point in the t-direction (i, n+½) is shown. This arrangement enables us to use central differences, thus increasing the accuracy of the method. Instead of expressing $\partial u / \partial t$ in terms of forward difference around (i, n), as was done in the explicit form, we express this partial derivative in terms of central difference around the half point:

$$\left.\frac{\partial u}{\partial t}\right|_{i,n+½} = \frac{1}{\Delta t}\left(u_{i,n+1} - u_{i,n}\right) \tag{8.73}$$

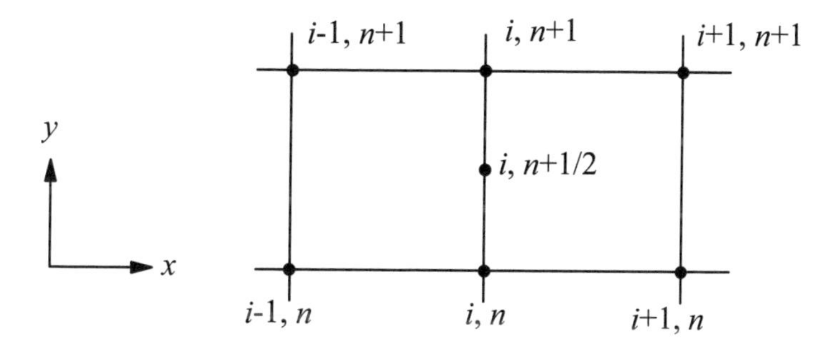

Figure 8.10 Finite difference grid for derivation of implicit formulas.

In addition, the second-order partial derivative is expressed at the half point as a weighted average of the central differences at points (i, n + 1) and (i, n):

$$\begin{aligned}\left.\frac{\partial^2 u}{\partial x^2}\right|_{i,n+½} &= \theta\left.\frac{\partial^2 u}{\partial x^2}\right|_{i,n+1} + (1-\theta)\left.\frac{\partial^2 u}{\partial x^2}\right|_{i,n} \\ &= \theta\left[\frac{1}{\Delta x^2}\left(u_{i+1,n+1} - 2u_{i,n+1} + u_{i-1,n+1}\right)\right] + (1-\theta)\left[\frac{1}{\Delta x^2}\left(u_{i+1,n} - 2u_{i,n} + u_{i-1,n}\right)\right]\end{aligned} \tag{8.74}$$

where θ is in the range $0 \le \theta \le 1$. By substituting Eqs. (8.73) and (8.74) into Eq. (8.22)

$$\frac{\partial u}{\partial t} = \alpha\frac{\partial^2 u}{\partial x^2} \tag{8.22}$$

we obtain the *variable-weighted implicit* approximation of the parabolic partial differential equation:

$$\alpha\theta\left[\frac{1}{\Delta x^2}\left(u_{i+1,n+1}-2u_{i,n+1}+u_{i-1,n+1}\right)\right]-\frac{1}{\Delta t}u_{i,n+1}$$
$$=-\alpha(1-\theta)\left[\frac{1}{\Delta x^2}\left(u_{i+1,n}-2u_{i,n}+u_{i-1,n}\right)\right]-\frac{1}{\Delta t}u_{i,n} \tag{8.75}$$

This formula is implicit because the left-hand side involves more than one value at the $(n + 1)$ position of the difference grid (that is, more than one unknown at any step in the time domain).

When $\theta = 0$, Eq. (8.75) becomes identical to the classic explicit formula Eq. (8.67). When $\theta = 1$, Eq. (8.75) becomes:

$$-\left(\frac{\alpha\Delta t}{\Delta x^2}\right)u_{i-1,n+1}+\left(1+2\frac{\alpha\Delta t}{\Delta x^2}\right)u_{i,n+1}-\left(\frac{\alpha\Delta t}{\Delta x^2}\right)u_{i+1,n+1}=u_{i,n} \tag{8.76}$$

This is called the *backward implicit* approximation, which can also be obtained by approximating the first-order partial derivative using the backward difference at $(i, n + 1)$ and the second-order partial derivative by the central difference at $(i, n + 1)$.

Finally, when $\theta = ½$, Eq. (8.75) yields the widely-used *Crank-Nicolson implicit formula*:

$$-\left(\frac{\alpha\Delta t}{\Delta x^2}\right)u_{i-1,n+1}+2\left(1+\frac{\alpha\Delta t}{\Delta x^2}\right)u_{i,n+1}-\left(\frac{\alpha\Delta t}{\Delta x^2}\right)u_{i+1,n+1}$$
$$=\left(\frac{\alpha\Delta t}{\Delta x^2}\right)u_{i-1,n}+2\left(1-\frac{\alpha\Delta t}{\Delta x^2}\right)u_{i,n}+\left(\frac{\alpha\Delta t}{\Delta x^2}\right)u_{i+1,n} \tag{8.77}$$

For an implicit solution to the following nonhomogeneous parabolic equation:

$$\frac{\partial u}{\partial t}=\alpha\frac{\partial^2 u}{\partial x^2}+f(x,t) \tag{8.66}$$

by the above method, we also need to calculate the value of f at the midpoint $(i, n + ½)$ which we take as the average of the value of f at grid points $(i, n + 1)$ and (i, n):

$$f_{i,n+½}=\frac{1}{2}\left(f_{i,n+1}+f_{i,n}\right) \tag{8.78}$$

Putting Eqs. (8.73), (8.74) (considering $\theta = ½$), and (8.78) into Eq. (8.66) results in

$$-\left(\frac{\alpha\Delta t}{\Delta x^2}\right)u_{i-1,n+1}+2\left(1+\frac{\alpha\Delta t}{\Delta x^2}\right)u_{i,n+1}-\left(\frac{\alpha\Delta t}{\Delta x^2}\right)u_{i+1,n+1}-(\Delta t)f_{i,n+1}$$
$$=\left(\frac{\alpha\Delta t}{\Delta x^2}\right)u_{i-1,n}+2\left(1-\frac{\alpha\Delta t}{\Delta x^2}\right)u_{i,n}+\left(\frac{\alpha\Delta t}{\Delta x^2}\right)u_{i+1,n}+(\Delta t)f_{i,n} \tag{8.79}$$

Eq. (8.79) is the Crank-Nicolson implicit formula for the solution of the nonhomogeneous parabolic partial differential equation (8.66).

When written for the entire difference grid, implicit formulas generate sets of simultaneous linear algebraic equations whose matrix of coefficients is usually a tridiagonal matrix. This type of problem may be solved using a Gauss elimination procedure (Chapter 4), or more efficiently using the Thomas algorithm (Lapidus and Pinder, 1982), which is a variation of Gauss elimination.

Implicit formulas of the type described above are unconditionally stable. It can be generalized that most explicit finite difference approximations are conditionally stable, whereas most implicit approximations are unconditionally stable. The explicit methods, however, are computationally easier to solve than the implicit techniques.

8.5.3 Hyperbolic partial differential equations

Physical problems connected with vibration processes yield models consisting of second-order partial differential equations of the hyperbolic type. For example, the one-dimensional wave equation:

$$\rho\frac{\partial^2 u}{\partial t^2}=T_0\frac{\partial^2 u}{\partial x^2}+f(x,t) \tag{8.80}$$

describes the transverse motion of a vibrating string that is subjected to tension T_0 and external force $f(x, t)$. In the case of constant density, ρ, the equation is written in the form:

$$\frac{\partial^2 u}{\partial t^2}=a^2\frac{\partial^2 u}{\partial x^2}+F(x,t) \tag{8.81}$$

where:

$$a^2=\frac{T_0}{\rho} \quad \text{and} \quad F(x,t)=\frac{1}{\rho}f(x,t)$$

If no external force acts on the string, Eq. (8.81) becomes a homogeneous equation:

$$\frac{\partial^2 u}{\partial t^2}=a^2\frac{\partial^2 u}{\partial x^2} \tag{8.82}$$

To find the numerical solution of Eq. (8.82), we expand each second-order derivative in terms of central finite differences to obtain:

$$\frac{u_{i,n+1} - 2u_{i,n} + u_{i,n-1}}{\Delta t^2} = a^2 \left(\frac{u_{i+1,n} - 2u_{i,n} + u_{i-1,n}}{\Delta x^2} \right) + O\left(\Delta x^2 + \Delta t^2\right) \tag{8.83}$$

Rearranging to solve for $u_{i,n+1}$,

$$u_{i,n+1} = 2\left(1 - \frac{a^2 \Delta t^2}{\Delta x^2}\right) u_{i,n} + \frac{a^2 \Delta t^2}{\Delta x^2}\left(u_{i+1,n} + u_{i-1,n}\right) - u_{i,n-1} + O\left(\Delta x^2 + \Delta t^2\right) \tag{8.84}$$

This is an *explicit* numerical solution of the hyperbolic equation (8.82).

The positivity rule, Eq. (8.59), applied to Eq. (8.84) shows that this solution is stable if the following inequality limit is obeyed:

$$\frac{a^2 \Delta t^2}{\Delta x^2} \le 1 \tag{8.85}$$

Similarly, the homogeneous form of the two-dimensional hyperbolic equation:

$$\frac{\partial^2 u}{\partial t^2} = a^2 \left(\frac{\partial^2 u}{\partial x^2} + \frac{\partial^2 u}{\partial y^2} \right) \tag{8.86}$$

is expanded using central finite difference approximation to yield:

$$\begin{aligned} &\frac{u_{i,j,n+1} - 2u_{i,j,n} + u_{i,j,n-1}}{\Delta t^2} \\ &\quad = a^2 \left(\frac{u_{i+1,j,n} - 2u_{i,j,n} + u_{i-1,j,n}}{\Delta x^2} \right) + a^2 \left(\frac{u_{i,j+1,n} - 2u_{i,j,n} + u_{i,j-1,n}}{\Delta y^2} \right) \\ &\quad + O\left(\Delta x^2 + \Delta y^2 + \Delta t^2\right) \end{aligned} \tag{8.87}$$

Rearranging this equation to the explicit form, using an equidistant grid in x- and y-directions, results in:

$$\begin{aligned} u_{i,j,n+1} &= 2\left[1 - 2\left(\frac{a^2 \Delta t^2}{\Delta x^2}\right)\right] u_{i,j,n} - u_{i,j,n-1} \\ &+ \frac{a^2 \Delta t^2}{\Delta x^2}\left(u_{i+1,j,n} + u_{i-1,j,n} + u_{i,j+1,n} + u_{i,j-1,n}\right) \end{aligned} \tag{8.88}$$

This solution is stable when:

$$\frac{a^2 \Delta t^2}{\Delta x^2} \le \frac{1}{2} \tag{8.89}$$

Implicit methods for solution of hyperbolic partial differential equations can be developed using the *variable-weight* approach, where the space partial derivatives are weighted at $(n + 1)$, n, and $(n - 1)$. The implicit formulation of Eq. (8.82) is:

$$\begin{aligned} \frac{u_{i,n+1} - 2u_{i,n} + u_{i,n-1}}{\Delta t^2} &= \frac{a^2}{\Delta x^2}\Big[\theta\left(u_{i+1,n+1} - 2u_{i,n+1} + u_{i-1,n+1}\right) \\ &+ (1-2\theta)\left(u_{i+1,n} - 2u_{i,n} + u_{i-1,n}\right) + \theta\left(u_{i+1,n-1} - 2u_{i,n-1} + u_{i-1,n-1}\right)\Big] \end{aligned} \tag{8.90}$$

where $0 \le \theta \le 1$. When $\theta = 0$, Eq. (8.90) reverts back to the explicit method, Eq. (8.83). When $\theta = ½$, Eq. (8.90) is a Crank-Nicolson-type approximation. Implicit methods yield tridiagonal sets of linear algebraic equations whose solutions can be obtained using Gauss elimination methods (Chapter 4).

8.6 Polar Coordinate Systems

All the methods we have developed so far in this chapter were based on Cartesian coordinate systems. Quite often, however, the objects whose properties are being modeled by the partial differential equations may have circular, cylindrical, or spherical shapes. The finite difference approximations may be modified to handle such geometries.

Cylindrical-shaped objects are more conveniently expressed in polar coordinates. The transformation from Cartesian coordinate to polar coordinate systems is performed using the following relationships, which are based on Fig. 8.11:

$$\begin{aligned} x &= r\cos\theta \qquad & y &= r\sin\theta \\ r &= \sqrt{x^2 + y^2} \qquad & \theta &= \tan^{-1}\frac{y}{x} \end{aligned} \tag{8.91}$$

The Laplacian operator in polar coordinates becomes:

$$\frac{\partial^2 u}{\partial x^2} + \frac{\partial^2 u}{\partial y^2} = \frac{\partial^2 u}{\partial r^2} + \frac{1}{r}\frac{\partial u}{\partial r} + \frac{1}{r^2}\frac{\partial^2 u}{\partial \theta^2} \tag{8.92}$$

Fick's second law of diffusion in polar coordinates is:

$$\frac{\partial C}{\partial t} = D\left(\frac{\partial^2 C}{\partial r^2} + \frac{1}{r}\frac{\partial C}{\partial r} + \frac{1}{r^2}\frac{\partial^2 C}{\partial \theta^2} + \frac{\partial^2 C}{\partial z^2}\right) \tag{8.93}$$

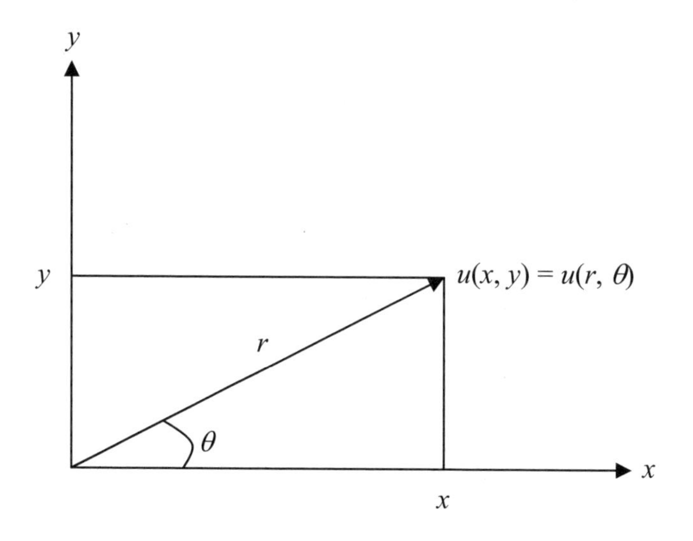

Figure 8.11 Transformation to polar coordinates.

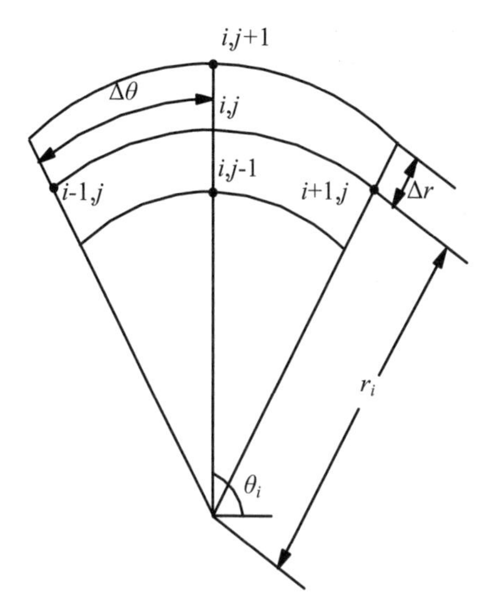

Figure 8.12 Finite difference grid for polar coordinates.

Using the finite difference grid for polar coordinates shown in Fig. 8.12, the partial derivatives are approximated by:

$$\left.\frac{\partial^2 u}{\partial r^2}\right|_{i,j} = \frac{1}{\Delta r^2}\left(u_{i,j+1} - 2u_{i,j} + u_{i,j-1}\right) \tag{8.94}$$

$$\left.\frac{\partial^2 u}{\partial \theta^2}\right|_{i,j} = \frac{1}{\Delta \theta^2}\left(u_{i+1,j} - 2u_{i,j} + u_{i-1,j}\right) \tag{8.95}$$

$$\left.\frac{\partial u}{\partial r}\right|_{i,j} = \frac{1}{2\Delta r}\left(u_{i,j+1} - u_{i,j-1}\right) \tag{8.96}$$

where i and j are counters in θ- and r-directions, respectively. Partial derivatives in z- and t-dimensions (not shown in Fig. 8.12) would be similarly expressed through the use of additional subscripts.

8.7 Stability Analysis

In Chapter 7, we discussed the stability considerations in the application of numerical integration methods for ordinary differential equations. Equally important are the stability problems encountered in the numerical solution of partial differential equations. In Appendix E of this book, we discuss the stability of finite difference approximations using the well-known von Neumann procedure.

8.8 PDE Toolbox in MATLAB

MATLAB has a powerful toolbox for solution of linear and nonlinear partial differential equations that is called the *Partial Differential Equation* (or *PDE*) *Toolbox*. However, this toolbox must be purchased separately, and may not be available to all users. The PDE Toolbox uses the finite element method for solution of partial differential equations in two space dimensions. The basic equation of this toolbox is the equation:

$$-\nabla \cdot (c\nabla u) + au = f \tag{8.97}$$

where c, a, and f are complex-valued functions in the solution domain and may also be functions of u. The PDE Toolbox can also solve the following equations:

$$d\frac{\partial u}{\partial t} - \nabla \cdot (c\nabla u) + au = f \tag{8.98}$$

and:

$$d\frac{\partial^2 u}{\partial t^2} - \nabla \cdot (c\nabla u) + au = f \tag{8.99}$$

where d, c, a, and f are constants or complex-valued functions in the solution domain and also can be functions of time. The symbol ∇ is the *vector differential operator*, or *gradient* (not to be confused with ∇, the backward difference operator).

In the PDE Toolbox, Eqs. (8.97) to (8.99) are named elliptic, parabolic, and hyperbolic, respectively, regardless of the values of the coefficients and boundary conditions.

In order to solve a partial differential equation using the PDE Toolbox, one may simply use the *graphical user interface* by employing the `pdetool` command. In this separate environment, the user is able to define the two-dimensional geometry, introduce the boundary conditions, solve the partial differential equation, and visualize the results. Use of the PDE toolbox is demonstrated in Example 8.3 for the solution of Fick's second law of diffusion in one-dimensional form (Eq. (8.24)).

Example 8.3 Solution of Fick's second law of diffusion using the PDE Toolbox.

Statement of the problem

Simple diffusion of molecules across a flat membrane may be modeled by the one-dimensional form of Fick's second law:

$$\frac{\partial C_A}{\partial t} = D_{AB}\frac{\partial^2 C_A}{\partial x^2} \tag{8.24}$$

where C_A is the concentration of component A in the membrane, D_{AB} is the diffusion coefficient of component A in the membrane, and x is the direction of diffusion. This is a one-dimensional unsteady state partial differential equation of the parabolic type. Solve this equation using the PDE Toolbox for the following initial and boundary conditions:

initial condition:	$t = 0$	all	$x > 0$	$C = 0$
boundary conditions:	$t > 0$		$x = 0$	$C = 1$
			$x = 0.2$	$C = 0$

and constants:

Diffusion coefficient: $D_{AB} = 1\times 10^{-4}$ cm^2/s

Solution

From the MATLAB command window give the command:

```
>>pdetool
```

A blank PDE Toolbox window appears, as shown in Fig. 8.13.

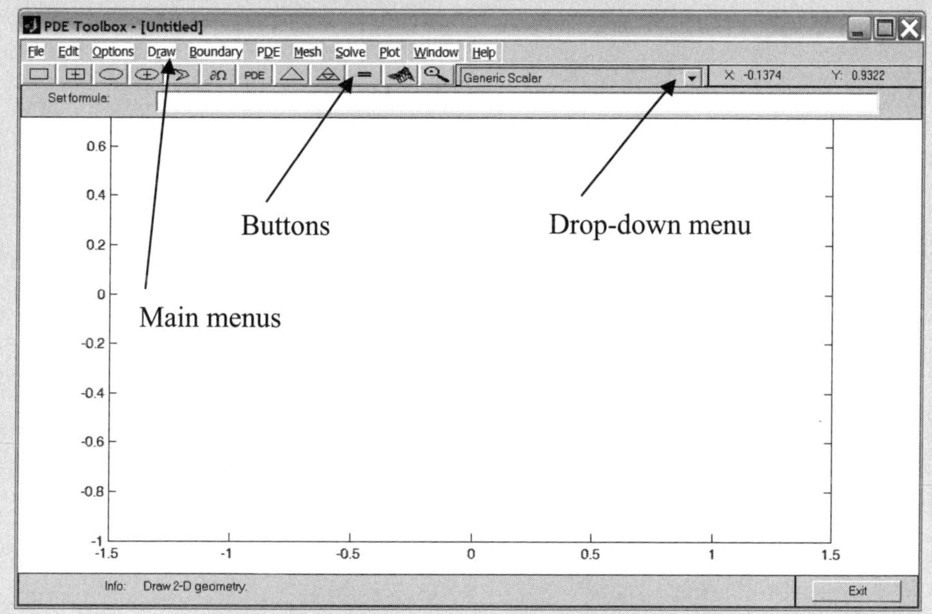

Figure 8.13 PDE Toolbox window and menus.

Please notice the main menus, the series of buttons, and the drop-down menu in the above window. From the drop-down menu, choose Diffusion. From the series of buttons, choose the one labeled PDE, set the type of PDE to Parabolic, and enter the coefficients as shown in Fig. 8.14.

PDE Specification

Equation: c'-div(D*grad(c))=Q, c=concentration

Type of PDE: Elliptic, Parabolic, Hyperbolic, Eigenmodes

Coefficient	Value	Description
D	1e-4	Diffusion coefficient
Q	0.0	Volume source

OK Cancel

Figure 8.14 PDE specification dialog.

Next, choose the rectangle button (extreme left) and draw a small rectangle on the screen, as shown in Fig. 8.15.

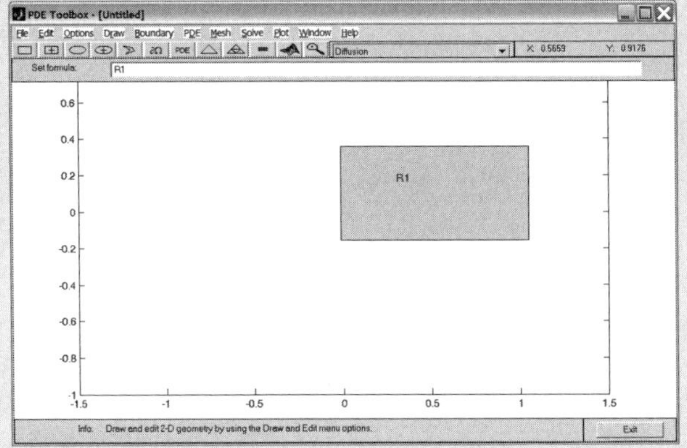

Figure 8.15 Drawing the geometry of the problem.

Double click on the rectangle and set the coordinates of the object in the Object Dialog box (Fig. 8.16).

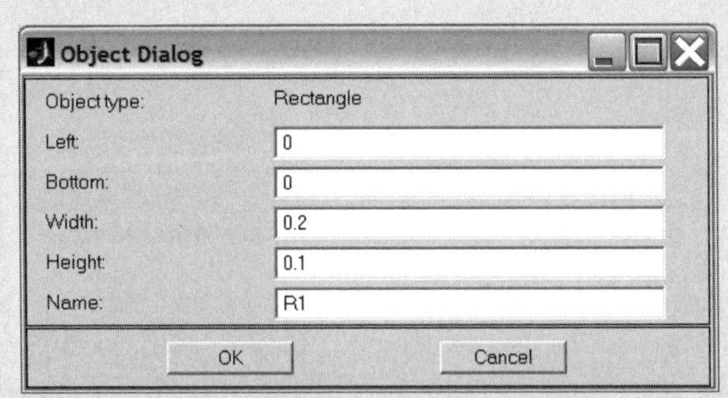

Figure 8.16 Object Dialog box: setting the coordinates and size of object.

From the Options menu, choose Axes Limits… and set the X-axis and Y-axis ranges as shown in Fig. 8.17.

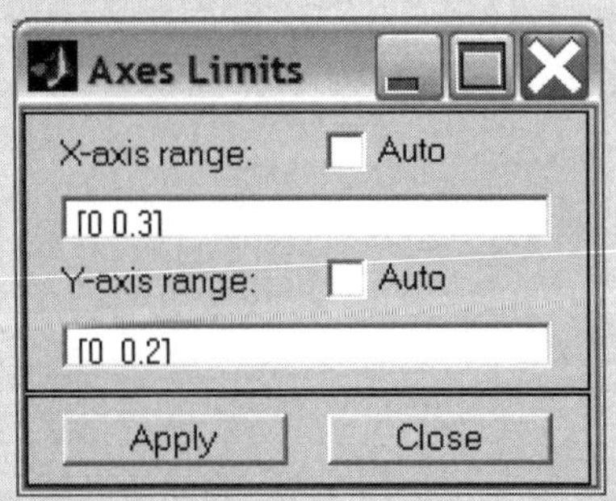

Figure 8.17 Axes Limits dialog box: specifying the axes ranges.

Click on the $\partial\Omega$ button. This highlights in red the boundaries of the rectangle, and enables you to specify the boundary conditions. Double-click on each side of the rectangle and specify the boundary conditions as shown in Figs. 8.18, 8.19, 8.20, and 8.21.

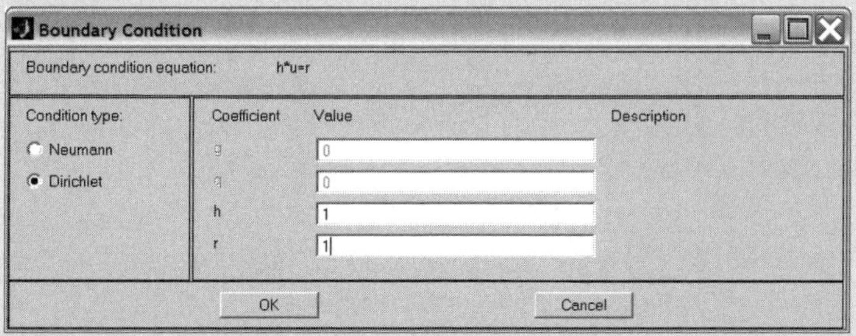

Boundary Condition

Boundary condition equation: h*u=r

Condition type: Neumann / Dirichlet (selected)

Coefficient	Value	Description
g	0	
q	0	
h	1	
r	1	

OK Cancel

Figure 8.18 Boundary Condition dialog: setting the left boundary.

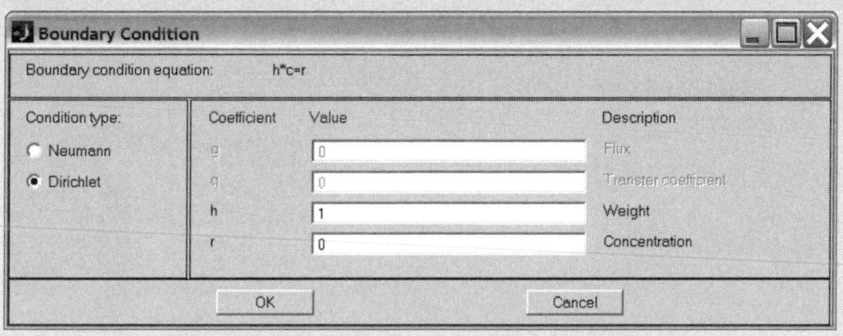

Boundary Condition

Boundary condition equation: h*c=r

Condition type: Neumann / Dirichlet (selected)

Coefficient	Value	Description
g	0	Flux
q	0	Transfer coefficient
h	1	Weight
r	0	Concentration

OK Cancel

Figure 8.19 Setting the right boundary.

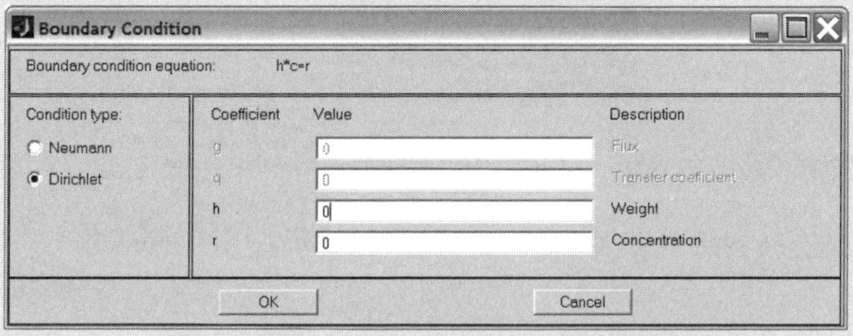

Boundary Condition

Boundary condition equation: h*c=r

Condition type: Neumann / Dirichlet (selected)

Coefficient	Value	Description
g	0	Flux
q	0	Transfer coefficient
h	0	Weight
r	0	Concentration

OK Cancel

Figure 8.20 Setting the lower boundary.

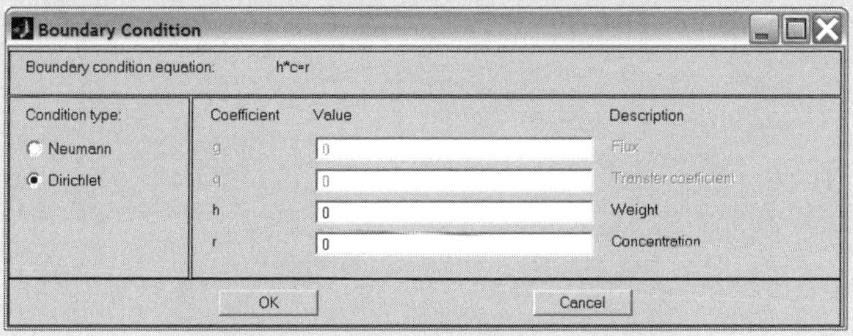

Boundary Condition

Boundary condition equation: h*c=r

Condition type: Neumann / Dirichlet (selected)

Coefficient	Value	Description
g	0	Flux
q	0	Transfer coefficient
h	0	Weight
r	0	Concentration

OK Cancel

Figure 8.21 Setting the upper boundary.

From the Solve menu, choose Parameters and set the time of integration and the initial condition as shown in Fig. 8.22.

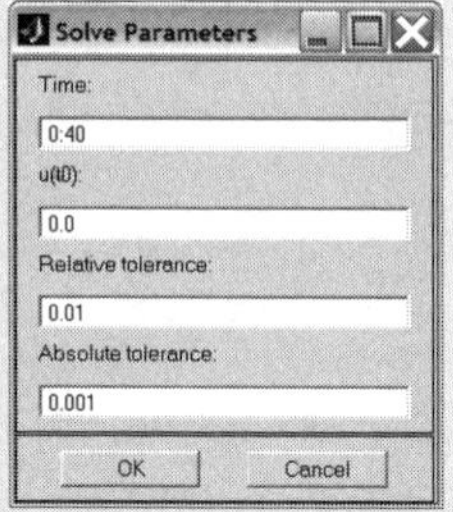

Figure 8.22 Solve Parameters dialog box.

Finally, press the = button to solve the differential equation, and observe the results in Fig. 8.23, which shows the diffusion from the left side to the right side of the object:

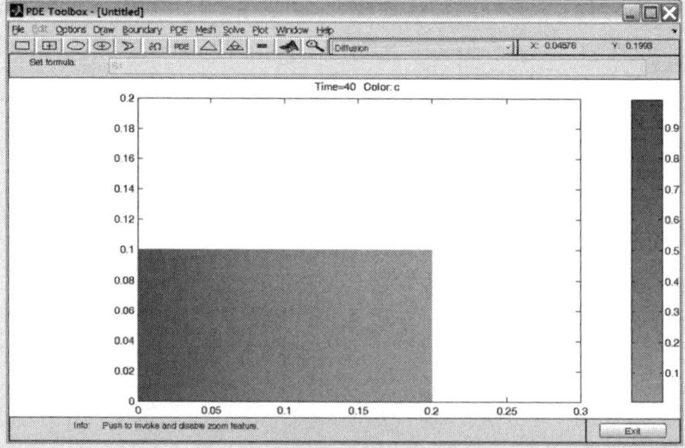

Figure 8.23 Solution of the differential equation: diffusion results.

To observe the diffusion as it takes place, press the Plot Selection button (second button from the right), choose Animation, and press Plot as shown in Fig. 8.24).

Plot Selection

Plot type:	Property:	User entry:	Plot style:
☑ Color	concentration		interpolated shad.
☐ Contour			
☐ Arrows	concentration gradient		proportional
☐ Deformed mesh	concentration gradient		
☐ Height (3-D plot)	concentration		continuous
☑ Animation	Options...		

☐ Plot in x-y grid Contour plot levels: 20 ☑ Plot solution automatically
☐ Show mesh Colormap: cool Time for plot: 0

Plot Done Cancel

Figure 8.24 Plot Selection dialog box.

To see a 3-D plot, press the Plot Selection button, choose Height (3-D plot), and press Plot (as shown in Fig. 8.25).

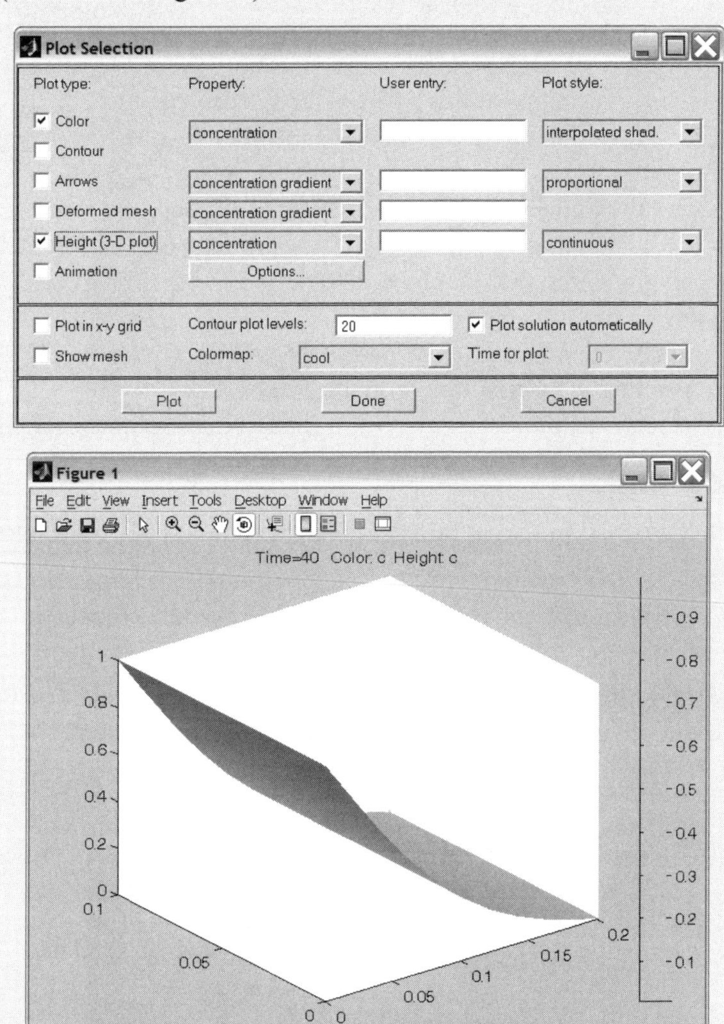

Figure 8.25 Plot Selection choice and resulting 3-D plot.

Choosing both Height (3-D plot) and Animation would generate a three-dimensional movie of the diffusion sequence. The student is encouraged to experiment with other plotting options, longer integration times, and different boundary conditions.

The results of the numerical solution may be exported to the MATLAB workspace by selecting Solve/Export Solution in the PDE Toolbox main menu.

8.9 Lessons Learned in this Chapter

After studying this chapter, the student should have learned the following:

- The dynamics of biological systems that have more than one independent variable may be modeled using partial differential equations.
- Partial differential equations may be classified as elliptic, parabolic, or hyperbolic.
- To obtain the numerical solution of a artial differential equation, the partial derivatives may be replaced by finite difference approximations, thus converting the problem of solving a PDE to that of solving a set of algebraic equations.
- The boundary conditions of the partial differential equation play a major role in determining the form of the solution.

8.10 Problems

8.1 Modify the program in Example 8.1 to solve the three-dimensional problem:

$$\frac{\partial^2 u}{\partial x^2}+\frac{\partial^2 u}{\partial y^2}+\frac{\partial^2 u}{\partial z^2}=f$$

Apply this program to calculate the distribution of the dependent variable within a solid body that is subject to the following boundary conditions:

$$u(0,y,z)=1 \qquad u(1,y,z)=1$$
$$u(x,0,z)=1 \qquad u(x,1,z)=1$$
$$u(x,y,0)=1 \qquad u(x,y,1)=1$$

and the value of *f*: $f=5$

8.2 Solve Laplace's equation with the following boundary conditions and discuss the results:

$$u(0,y)=100 \qquad \left.\frac{\partial u}{\partial x}\right|_{10,y}=10$$

$$\left.\frac{\partial u}{\partial y}\right|_{x,0}=0 \qquad \left.\frac{\partial u}{\partial y}\right|_{x,1}=0$$

8.3 Develop the finite difference approximation of Fick's second law of diffusion in polar coordinates. Write a MATLAB program that can be used to solve the following problem: A wet cylinder of agar gel at 278 K with a uniform concentration of urea of 100 mol/m^3 has a diameter of 0.03 m and is 1.0 m long with flat parallel ends. The diffusivity is 5×10^{-10} m^2/s. Calculate the concentration at the midpoint of the cylinder after 100 h for the following cases if the cylinder is suddenly immersed in turbulent pure water. Since the cylinder is long, you may ignore axial diffusion.

8.4 The flow and migration of human polymorphonuclear leukocytes on prosthetic material surfaces was solved in Example 8.2 by the diffusion-convection equation:

$$\frac{\partial C}{\partial t} = \mu_D \frac{\partial^2 C}{\partial z^2} - v_{eff} \frac{\partial C}{\partial z}$$

Assume that the value of the directional migratory velocity of cells is insignificant; that is, $v_{\text{eff}} = 0.0$ cm/s, and that the random migration coefficient is $\mu_D = 1\times10^{-4}$ cm^2/s. Solve this problem with a Neumann boundary condition on the right boundary; that is:

Initial condition:	$t = 0$	all	$z > 0$	$C = 0$
Boundary conditions:	$t > 0$		$z = 0$	$C = 1\times10^6$
			$z = 0.2$	$\frac{\partial C}{\partial z} = 0$

Integrate the equation for a period of 100 seconds and compare the results with those of Example 8.2.

8.5 The mechanism of movement of the molecules during the percutaneous application of a finite dose of a drug to the surface of the skin and across the skin has been modeled by using diffusion equations, such as Fick's second law of diffusion (Kubota et al., 2002, Simon and Loney, 2003). This process is shown diagrammatically on Fig. 8.26. The drug first diffuses through the medium of delivery of thickness, L_m. The concentration of the drug, C_m, through the medium of delivery may be modeled by:

$$\frac{\partial C_m}{\partial t} = D_m \frac{\partial^2 C_m}{\partial x^2} \qquad -L_m \le x \le 0, \quad t > 0$$

Subsequently, the drug enters and diffuses through the skin membrane of thickness, L_s. The concentration, C_s, in the skin is modeled by

$$\frac{\partial C_s}{\partial t} = D_s \frac{\partial^2 C_s}{\partial x^2} \qquad 0 \le x \le L_s, \quad t > 0$$

where D_m is the effective drug diffusivity in the medium of delivery, and D_s is the diffusivity in the skin. These two equations are coupled at the boundary between the medium of delivery and the skin. The drug is finally absorbed in the receptor cell.

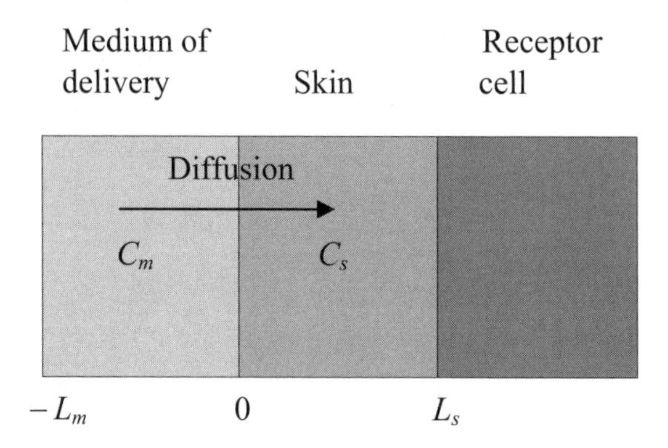

Figure 8.26 Percutaneous diffusion of drug.

Solve the above equations simultaneously, with the following initial and boundary conditions:

Initial conditions:

$$t = 0 \qquad C_m(x,0) = C_{m0} \qquad -L_m \le x < 0$$

$$C_s(x,0) = 0 \qquad 0 \le x \le L_s$$

Boundary conditions:

$$x = -L_m \qquad \frac{\partial C_m(-L_m,t)}{\partial x} = 0 \qquad 0 \le t \le T$$

$$x = 0 \qquad -D_m \frac{\partial C_m(0,t)}{\partial x} = -D_s \frac{\partial C_s(0,t)}{\partial x} \qquad 0 \le t \le T$$

$$x = 0 \qquad K_m C_m(0,t) = C_s(0,t) \qquad 0 \le t \le T$$

$$x = L_s \qquad -D_s \frac{\partial C_s(L_s,t)}{\partial x} = K_{cl} C_s(L_s,t) \qquad 0 < t \le T$$

where K_m is the medium-skin partition coefficient, and K_{cl} is the clearance per unit area of the drug per unit concentration excess at skin-receptor boundary. The constants to be used in this problem are:

$L_m = 0.004$ cm $\quad L_s = 0.025$ cm $\quad D_m = 9.72 \times 10^{-6}$ cm²/h

$D_s = 11.25 \times 10^{-6}$ cm²/h $\quad K_m = 0.5 \quad K_{cl} = 25$ cm/h

$C_{m0} = 0.2$ g/cm³ $\quad T = 6$ h

8.6 The blood flow in an artery may be modeled as the laminar flow of an incompressible Newtonian fluid in a rigid, circular blood vessel, of radius R, exposed to an oscillatory pressure field. Flow is only in the axial direction (that is, $u_r = u_\theta = 0$), and is fully developed; therefore, u_z, is a function of radial position only. The Navier-Stokes equation in cylindrical coordinates may be used to model the velocity (Truskey et al., 2004):

$$\frac{\partial u_z}{\partial t} = \nu \frac{1}{r}\frac{\partial}{\partial r}\left(r\frac{\partial u_z}{\partial r}\right) - \frac{1}{\rho}\frac{\partial p}{\partial z} \quad (1)$$

where ν is the kinematic viscosity and ρ is the density of the fluid. The position along the radius of the blood vessel is measured by r. The pressure gradient oscillates in time with frequency, ω (rad/s), to simulate the pumping action of the heart:

$$-\frac{\partial p}{\partial z} = \frac{\Delta p}{L}\cos(\omega t) \quad (2)$$

The initial and boundary conditions for this problem are as follows:

Initial condition: $\quad t = 0 \quad u_z = u_{z0}$

Boundary conditions: $\quad t > 0 \quad r = 0 \quad \frac{\partial u_z}{\partial r} = 0 \quad u_z$ is finite

$\quad r = R \quad u_z = 0$

The following constants may be used for the solution of this problem. These are characteristic of the human left main artery, human blood, and heart pumping action:

$R = 0.425$ cm $\quad \nu = 0.09$ cm²/s $\quad \rho = 1$ g/cm³

$\omega = 3$ cycles per second $= 6\pi$ rad/s

$$\frac{\Delta p}{L} = 1 \text{ dyne/cm}^3 = 1\ (\text{g.cm/s}^2)/\text{cm}^3$$

The above constants correspond to a Womersley Number:

$$\alpha = R\sqrt{\frac{\omega}{\nu}} = 6.15$$

which is consistent with that reported in the literature for the human left main artery (see Truskey et al., 2004).

8.7 The use of lasers for the irradiation of tissue for therapeutic purposes is common practice in surgery. When tissue is irradiated with laser light, the photons penetrate into the target and the energy is distributed within the tissue. A portion of this energy is absorbed by the tissue and converted into thermal energy, making the laser act as a distributed heat source. As a result of this process, the temperature in the biological tissue rises, but this cannot continue indefinitely. A threshold temperature can be reached beyond which intense vaporization of the water content of the tissue occurs. This results in the creation of vacuoles, vacuolization, followed by pyrolysis at higher temperatures. The combined process of intense vaporization, vacuolization, and pyrolysis of tissue is the laser ablation process.

Enderle et al. (2005) give the model equation for the temperature, T, within the tissue during the preablation heating stage for a one-dimensional semi-infinite and purely absorbing medium, as follows:

$$\frac{\partial(\rho c T)}{\partial t} = k\frac{\partial^2 T}{\partial x^2} + \mu_a I_0 e^{-\mu_a x}$$

where ρ is the density, c is the heat capacity, μ_a is the absorption coefficient, and k is the conductivity of the tissue. I_0 is the intensity of the laser beam. Solve this problem and plot the temperature profiles as a function of depth and time. Determine the time at which the tissue temperature exceeds 100°C (ablation temperature). The initial and boundary conditions for this problem are as follows:

Initial condition:	$t = 0$	all x	$T = 37^\circ\text{C}$
Boundary conditions:	$t > 0$	$x = 0$	$-k\frac{\partial T}{\partial x} = \frac{q}{A}$
		$x = 3.0$ mm	$\frac{\partial T}{\partial x} = 0$

Use the following values of the constants for the solution of this problem:

$\rho = 1$ g/cm^3 $\quad c = 3.14$ J/(g.K) $\quad \mu_a = 0.25$ cm^{-1}

$k = 0.0054$ W/(cm.K) $\quad \frac{q}{A} = 5$ J/s/cm^2 $\quad I_0 = 100$ W/cm^2

8.11 References

Constantinides, A., and Mostoufi, N. 1999. *Numerical Methods for Chemical Engineers with MATLAB Applications*. Upper Saddle River, NJ: Prentice Hall PTR.

Enderle, J. D., Blanchard, S. M., and Bronzino, J. D. 2005. *Introduction to Biomedical Engineering.* 2nd ed. San Diego, CA: Academic Press.

Kubota, K., Dey, F., Matar, S. A., and Twizell, E. H. 2002. A repeated dose model of percutaneous drug absorption. *Appl. Math Modelling*, **26**, 529-544.

Lapidus, L., and Pinder, G. F. 1982. *Numerical Solution of Partial Differential Equations in Science and Engineering*, New York: Wiley.

Randomsky, M. L., Whaley, K. J., Cone, R. A., and Saltzman, W. M. 1990. Macromolecules released from polymers: diffusion into unstirred fluids. *Biomaterials*, **11**, 619-624.

Rosenson-Schloss, R. S., Chang, C. C., Constantinides, A., Moghe, P. V. 2002. Alteration of Leukocyte Migration on Prosthetic Material, ePTFE, Under Flow: Role of CD43, an Adhesion Regulator Molecule. *Journal of Biomed. Mater. Res.*, **60**, 8-19.

Simon, L. and Loney, N. W. 2003. An analytical solution for percutaneous drug absorption. *Proceedings of IEEE 29th Annual Bioengineering Conference*, 22-23 March, 2003, pp. 303-304.

Truskey, G. A., Yuan, F., and Katz, D. F. 2004. *Transport Phenomena in Biological Systems*. Upper Saddle River, NJ: Pearson Prentice Hall.

Chapter 9
Measurements, Models and Statistics

9.1 The Role of Numerical Methods

Numerical methods are more than "number crunching;" they are techniques for gleaning insights from data. Numerical methods guide how we organize and interpret data. Biomedical engineers frequently use numerical methods for mining or searching for patterns or clues within biomolecular sequences, and for determining whether these patterns are trustworthy and can be generalized.

Statistics is the branch of mathematics used to characterize measurement error and biological variability, and for (1) describing the process underlying a set of measurements and (2) drawing conclusions about the world from the measurements (or the observations). Consequently, statistical models are integral to the experimental design, measurement methods, and interpretations of solutions to problems in biomedical engineering.

The methods presented in this chapter are central to an understanding of both the searching (i.e., detection) and interpretation (i.e., evaluation) tasks alluded to above. Any data collected will have variation, due either to the underlying biological process and/or the data collection equipment (the "noise"). The tasks laid out for the biomedical engineer are:

- To describe and characterize the noise
- To interpret the data in the presence of noise
- To determine general trends in the presence of noise
- To develop general principles of finding trends that are periodicities

In this chapter, tools for summarizing the data and tools for fitting models to data are introduced.

When describing models or statistics, each of the processes or patterns being measured is referred to as a *variable*. An *independent variable* is a variable that is manipulated or can be controlled. A *dependent variable* is a variable that is measured. For example, in a recording of an electrocardiogram, the independent variable is time and the dependent variable is the voltage.

The material in this chapter will enable the student to accomplish the following:

- Compute descriptive statistics of measured data
- Compute simple inferential statistics of samples of data
- Use linear least squares to fit polynomial models to a set of data.
- Fit polynomial models that are based on using exact data
- Fit models that are based on periodicities

9.2 Measurements, Errors and Uncertainty

Measurements are how data is gathered; they are made using instruments that are imperfect, and thus contain error due to the measuring device. The "truth" of a measurement is that value that a measurement would have were it free of the error introduced by the measuring instrument.

However, there is no true value in biological systems. For the biomedical engineer, "truth" of a measurement takes on a somewhat different meaning: the inherent variability in a biological process should not be confounded with measurement error and requires a description that summarizes this variation. The relationship between a measurement and the underlying process has three components:

measurement = summary measure + biological variability + measurement error

The detection problem is to find or determine this summary measure of the biological process in the presence of the measurement error. There are two important questions to keep in mind as measurements are made and the data is used to answer a question about the underlying biology:

- Do the measurements have any systematic or random error that affects the data?
- Do the data make sense?

The answers to these questions are crucial to drawing conclusions from data. First though, it is important to be able to summarize the data and, if there is error, to characterize the variation in the data that is presumably due to measurement error. *Measurement* is how we gather data. *Measurement error* can be classified as either systematic or random, as elaborated further below.

Systematic error causes a measurement to be skewed in a certain direction, i.e., to be consistently larger or consistently smaller. For example, weighing yourself repeatedly on a bathroom scale that has an initial reading of 20 lbs. will result in the reading of your weight to be greater than your "true" weight. The addition of 20 lbs. is a systematic error. This form of error can be reduced once identified (calibration). A systematic measurement error may also be called a measurement bias. Other sources of bias include one-sided errors arising from various sources, such as flaws in study design or data collection methods.

Random error is statistical fluctuation in the measured data due to limitations on the precision of the measurement device. It is bidirectional (as opposed to the unidirectional nature of a bias). If you weighed yourself on 10 different bathroom scales (or weighed yourself 10 times on one bathroom scale) over the period of a few minutes, you would record a range of values even though you know your weight is essentially not changing. This sort of error cannot be eliminated, but can be managed by taking many measurements to get a closer estimate of the mean.

Precision and accuracy are two generally accepted terms used to characterize measurements made with error,. Students should carefully note their distinction.

Precision is the degree of agreement among a series of individual measurements or *observations*. If you weighed yourself 10 times on a bathroom scale, the readings could vary by as much as 5 pounds. However, on a scale at the doctor's office, the same measurement may only vary by half a pound; we say that the latter measurement is more *precise*. Precision is limited by random error. Precision can be expressed as the standard deviation of a set of measurements.

Accuracy is the degree of conformity of a measured value to the "truth" (as defined above). Measurement A is more accurate than Measurement B if the value of A is closer to the "truth" than B is. When you zeroed your bathroom scale, you improved the accuracy of the measurement. Accuracy is limited by systematic error. Note that you can have a high-precision, low-accuracy measurement, such as a doctor's scale that hasn't been calibrated. Conversely, you can have a low-precision, high-accuracy scale, such as a cheap bathroom scale that happens to yield an average weight close to your true weight. Fig. 9.1 gives a representation of the permutations of accuracy and precision.

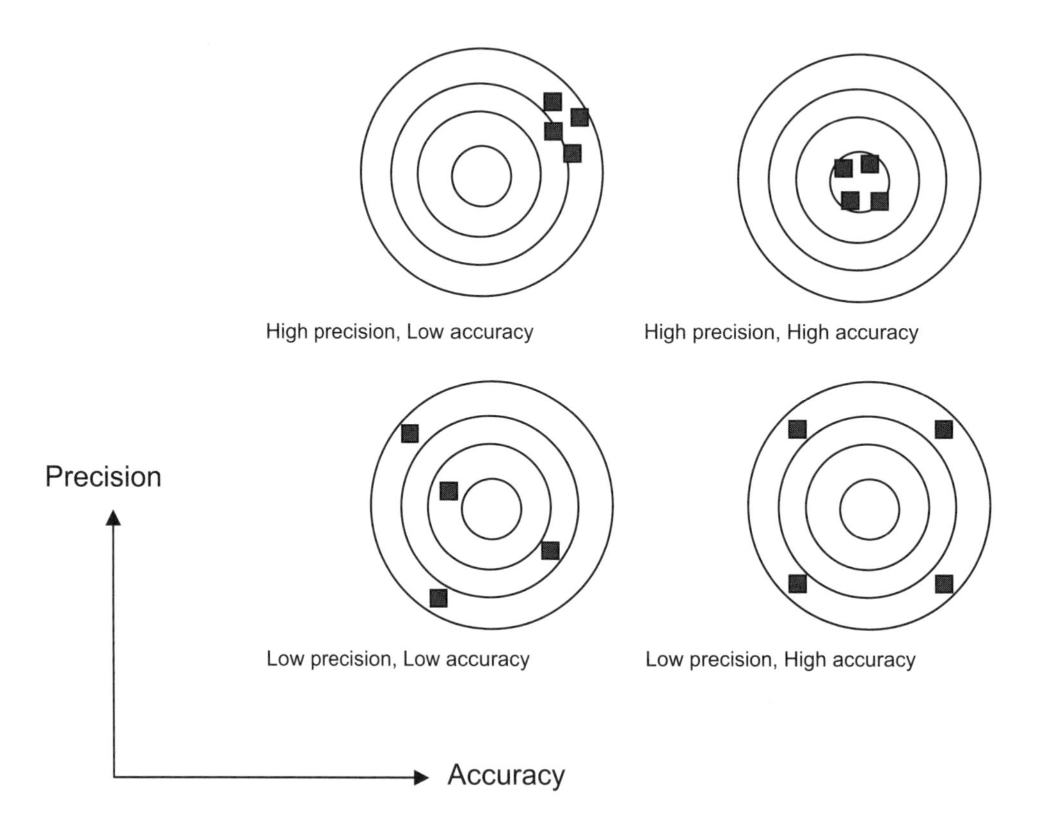

Figure 9.1 Accuracy and precision of measurements.

The *reproducibility* of a measurement is the measure of the ability to yield the same or consistent results in different observations. This means that when different people with different measuring equipment (or with the same equipment at different times) perform the same measurement, they get values that are within the range of the measurements of previous trials. The *resolution* is the minimum difference between two measurements that can be distinguished by a measuring device.

9.3 Descriptive Statistics

A *statistic* is an algebraic expression combining observations into a single number. Statistics serve to describe the data and estimate parameters of the underlying patterns or processes, and they are used to generalize or draw inference, as shown in the next section.

There are two ways to summarize a set or distribution of measurements: a histogram or a numerical approach. A *histogram* is a graphical summary of the distribution in which the independent variable is the range of possible values of the measurement, and the dependent variable is the frequency of occurrence of each possible measurement.

The second approach for summarizing the distribution of a set of measurements is called a *numerical approach.* In a numerical approach, the distribution is characterized by *descriptive statistics*, a small set of numbers that characterize where the distribution is located, how spread the distribution is, and how symmetric and how peaked the distribution is, for example. Descriptive statistics characterize properties of the distribution as follows:

Measures of central tendency describe the location of the distribution, i.e., where the distribution is located along the range of possible values of the random variable. The most common statistic used to describe central tendency is the *average value*. For a set of measured data, the average is computed using the formula:

$$\overline{x} = \frac{\sum_{n=1}^{N} x_i}{N} \tag{9.1}$$

Two other common measures of central tendency are the *median* and the *mode*. The median is the middle observation, and the mode is the most frequent value of the set of observations.

Measures of variation about the mean describe the degree to which the data differ. The most common statistic that characterizes this variation is called the *variance*, the degree to which the data differ with respect to the central tendency. For a set of measured data, the variance is the average of the deviation about the average and is computed using the following formula:

$$s^2 = \frac{\sum_{n=1}^{N} \left(x_i - \overline{x}\right)^2}{N-1} \tag{9.2}$$

The *standard deviation*, s_x, is the square root of the variance. The *coefficient of variation,* cv, is a measure of the spread of the data, but normalized to the average:

$$\text{cv} = \frac{s_x}{\bar{x}} \cdot 100\% \tag{9.3}$$

A large spread around a small average has a higher coefficient of variation than a large spread around a large average.

Skewness is the measure of the lack of symmetry in the variation about the mean. It is computed as the third central moment:

$$m_3 = \frac{\sum_{n=1}^{N} (x_i - \bar{x})^3}{(N-1)} \tag{9.4}$$

divided by the standard deviation cubed:

$$skewness = \frac{m_3}{s^3} \tag{9.5}$$

Kurtosis is a measure of the peakedness of the distribution. It is the ratio of the fourth central moment to the fourth power of the standard deviation:

$$m_4 = \frac{\sum_{n=1}^{N} (x_i - \bar{x})^4}{(N-1)} \tag{9.6}$$

$$kurtosis = \frac{m_4}{s^4} \tag{9.7}$$

Higher order statistics are computed in a similar fashion. All are central moments, i.e., moments about the mean:

$$m_k = \frac{\sum_{n=1}^{N} (x_i - \bar{x})^k}{(N-1)} \tag{9.8}$$

normalized by the same power of the standard deviation s^k. For example, a descriptive statistic of the fifth-order characteristic of the distribution is given by:

$$\frac{1}{s^5} \frac{\sum_{n=1}^{N} (x_i - \bar{x})^5}{(N-1)} \tag{9.9}$$

Only a small set of these statistics are frequently used in practice. The MATLAB commands that correspond to these statistics are shown in Table 9.1.

Table 9.1 Selected MATLAB commands for statistics.

Statistic	Equation	MATLAB command(s)
Mean	9.1	`mean(x)`
Median		`median(x)`
Variance	9.2	`var(x)`
Standard Deviation		`std(x)`
Skewness	9.5	`moment(x,3)/std(x).^3`
Kurtosis	9.7	`moment(x,4)/std(x).^4`

Example 9.1 Computing statistics of MRI and CT image intensitites.

The purpose of the Visible Human Project (VHP), which is sponsored by the National Library of Medicine (NLM), is to have complete, detailed 3D representations of normal male and female bodies. The long-term goal of the project is to be able to relate detailed knowledge of the anatomy, such as that found in Computed Tomography (CT) and Magnetic Resonance images (MRI), to symbolic and textual knowledge of the anatomy, such as that found in books and other resources. The VHP datasets include both Computed Tomography and Magnetic Resonance images for each of the male and female bodies. The CT images highlight differences between the hard tissues (e.g., bones and teeth) and the soft tissues; the MR images highlight soft tissue differences.

Write a MATLAB program that displays CT and MR images from the NIH Visible Human Project database and summarizes the distribution of image intensities in each image (http://www.nlm.nih.gov/research/visible/visible_human.html). Use the descriptive statistics of the CT image to identify an anatomical region such as the sinuses.

In this example, the two images referred to as head_mri_small and head_fresh_ct can be found and downloaded from the VHP Web site.

In MATLAB, images are displayed as figures using several special commands. First, the command `imread` is used to read the image file and store it in the workspace as an array. The command `image` is used to display this array as a figure. The command `colormap` is used to relate how the image intensities stored in the array are displayed in the figure window.

The CT and MR images in the VHP database are JPEG images that are (256 × 256) arrays with intensities in the range 0 to 255. A colormap to display images with 256 shades of gray is created with the following MATLAB commands:

```
v=0:(1/255):1;
map=[v; v; v]';
colormap(map);
```

An MR image from the VHP male database is read, stored in the workspace and displayed (Fig. 9.2) with the commands:

```
I=imread('head_mri_small','JPEG');
image(I)
```

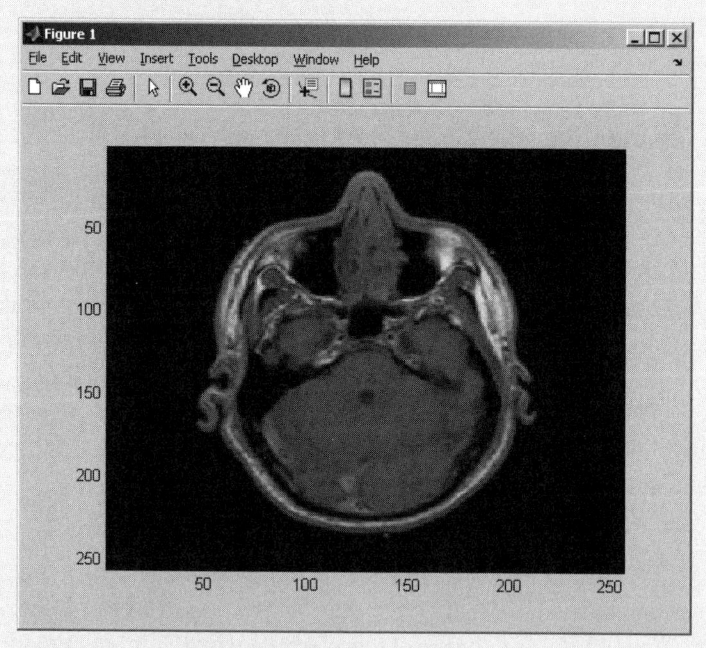

Figure 9.2 An MRI image displayed via MATLAB commands.

There are MATLAB functions to compute some descriptive statistics as shown in Table 9.1. However, the image is stored as a 2D matrix, and the commands operate on 2D matrices differently than on vectors. Therefore, the data in the image matrix has to be recast as a vector in order to compute the descriptive statistics:

```
[rI,cI]=size(I);
II=reshape(I,rI*cI,1);
meanI=mean(II);
varI=var(II);
stdI=std(II);
fprintf('\nAverage intensity\t%f\nVariance\t%f\nStandard
Deviation\t%f\n',meanI,varI,stdI);
```

The VHP images are 256 rows by 256 columns, and the image data is recast as a (65536×1) vector for computing the mean, variance and standard deviation above. The three statistics are printed in a formatted fashion using `fprintf`.

Recall that a histogram is a graphical summary of the distribution of image intensities. The histogram, is easily computed and displayed by the commands:

```
h=histc(II,0:255);
figure(2)
plot(0:255,h,'.',[meanI;meanI;meanI],[100; 500; 1000],'s')
```

In the plot command, the location of the average value is shown as three squares on top of each other (Fig. 9.3).

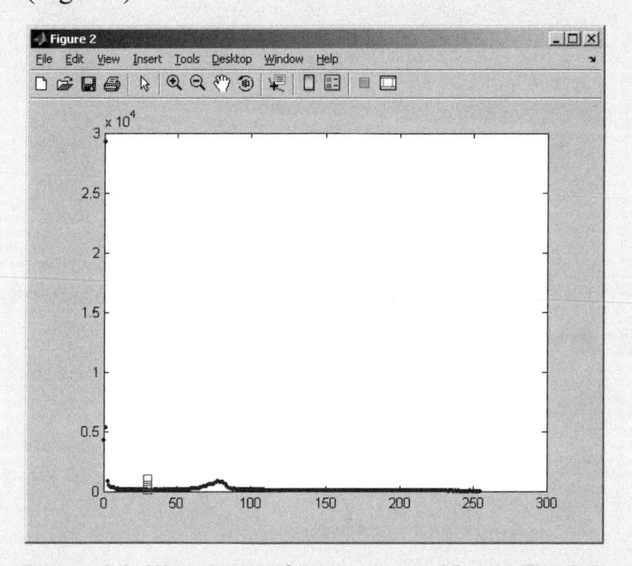

Figure 9.3 Histogram of image intensities in Fig. 9.2.

A CT image can be displayed and processed in a similar fashion, using the same colormap (Figs. 9.4, 9.5):

```
J=imread('head_fresh_ct','JPEG');
figure(3)
colormap(map)
image(J)
[rJ,cJ]=size(J);
JJ=reshape(J,rJ*cJ,1);
meanJ=mean(JJ);
varJ=var(JJ);
stdJ=std(JJ);
h=histc(JJ,0:255);
figure(4)
plot(0:255,h,'.',[meanJ;meanJ;meanJ],[100; 500; 1000],'s')
```

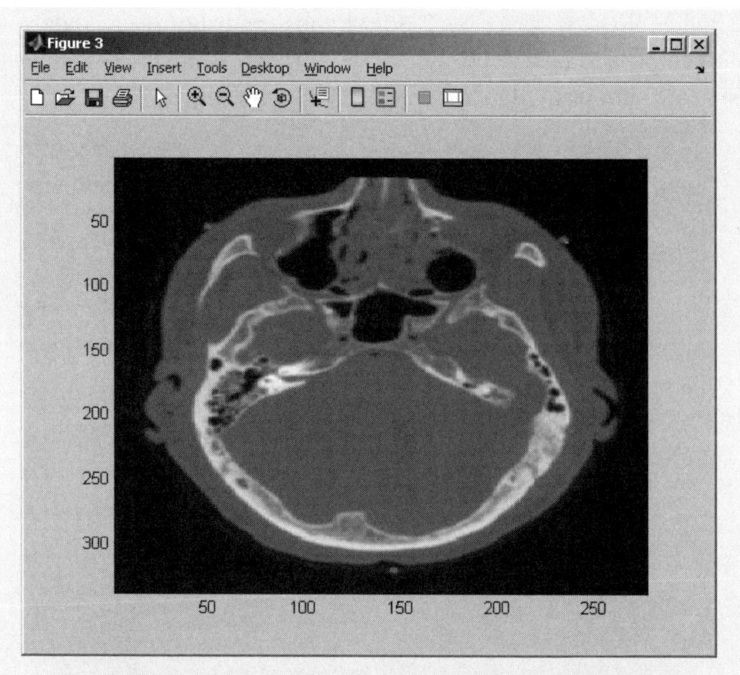

Figure 9.4 A CT image displayed via MATLAB commands.

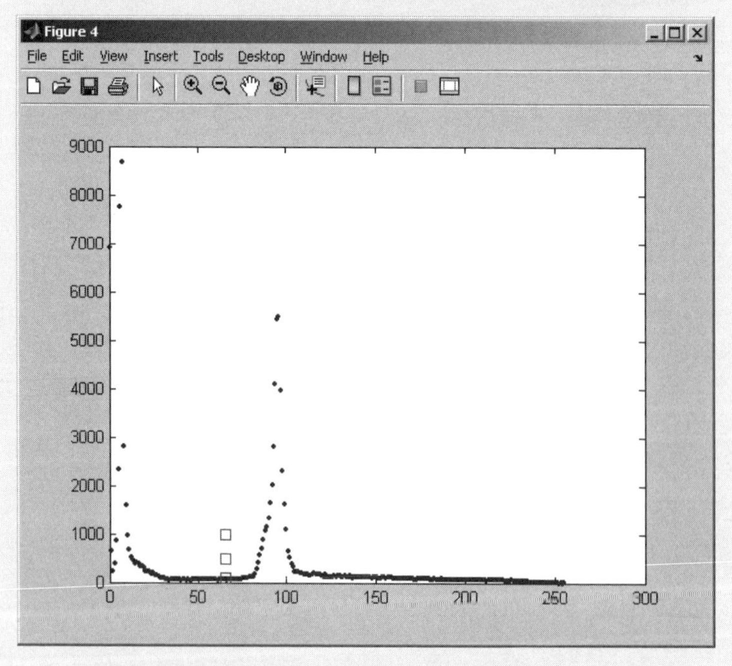

Figure 9.5 Histogram of image intensities in Fig. 9.4.

Sometimes descriptive statistics are used to locate or identify image regions of importance or interest. For example, in the CT image, very dark areas, i.e., intensities at or near zero, represent air space. The mean image intensity in the CT image can be used as a threshold on image intensity to try to segment or separate out air spaces that may correspond to the sinus cavities:

```
figure(5)
colormap(map)
TJ=J;
TJ(J<meanJ)=0;
TJ(TJ>0)=255;
image(TJ)
```

In this sequence of commands, logical indexing is used to create a new matrix, `TJ`, which represents a thresholded version of the CT image `J`.

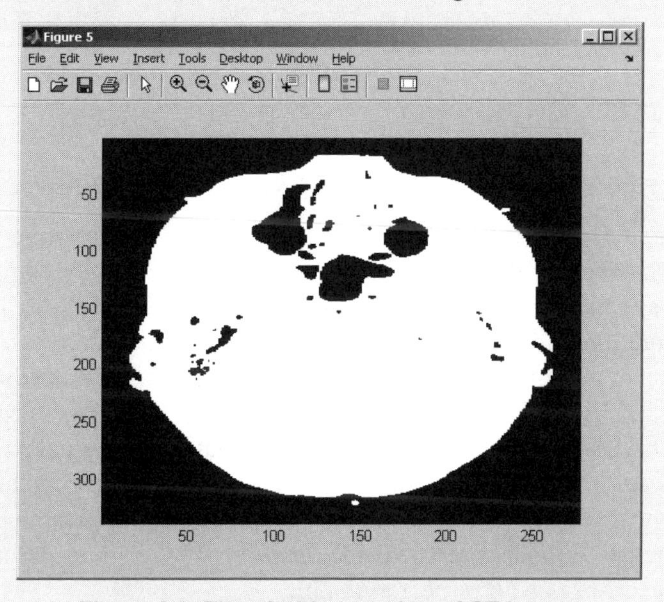

Figure 9.6 Thresholded version of CT image.

The three large black regions in Fig. 9.6 correspond to three sinus cavities: the maxillary sinuses (towards the top of the image) and the sphenoid sinus, which is below the two maxillary sinuses. A comparison of Figs. 9.4 and 9.6 shows that thresholding using the average image intensity to identify the sinus cavities is not perfect. Problems like these are in the broad area of image processing and medical image analysis.

More details about the Visible Human Project can be found at the NLM Web site (http://www.nlm.nih.gov/).

9.4 Inferential Statistics

Statistics can also be used to draw conclusions or make generalizations about a population via data collected from a sample drawn from the larger population. In practice, it is neither feasible nor possible to collect data from every member of a population (e.g., all cells, animals, or humans). Instead, data is collected from a sample, and the results are used to make generalizations about the population, as shown diagrammatically in Fig. 9.7.

Figure 9.7 Statistical inference from sample to population.

where measurements are collected from the sample, descriptive statistics are computed to characterize the sample, and *inferential statistics* are computed in order to make generalizations about the population.

Some additional terms are used to describe inferential statistics. A *population* is the collection of all objects or subjects, such as people, animals, cells or molecules. A *sample* is a subset of the population that is used to collect measurements. A *population parameter* is a characteristic of the population and is unknown at the time measurements are made. A *sample statistic* is an estimate of the population parameter computed from the data collected on the sample. Since the sample statistic can change if different samples are used, the *sampling distribution* describes how the sample statistic changes with respect to sample.

Inferential statistics are used to determine whether a sample statistic is a "good" estimate of the population parameter. Each and every estimate has two properties: amount and assuredness. The "amount" of the relationship tells how well the sample statistic predicts the population parameter. The "assuredness" is a quantity that indicates how well one can say that the sample statistic applies to all possible sets of similar measurements. If the measurements are collected and attention is given to a well-defined standard procedure, then this quantity represents how sure we are that the relationship seen in the sample of measurements holds in all possible samples of measurements. It would then be safe to conclude that the observed variations would be due to biological variability or measurement error, but not due to a relationship that that has not yet been uncovered.

There are two procedures for inferring relationships between variables: *parameter estimation* and *hypothesis testing*. In parameter estimation, a quantity that characterizes the relationship is computed from the sample (amount) and a *confidence interval* (assuredness) is constructed that tells the possible range of values of this parameter. A large confidence interval suggests that it is more likely that the range includes the population parameter. A small confidence interval suggests that the true population parameter can be determined very well.

A general formula for computing a confidence interval for a population parameter from sample statistics is:

$$\text{statistic} \pm z \cdot s_{statistic} \tag{9.10}$$

where z is a quantity based on sample data and the desired degree of assuredness. The values of z can be found in tables in texts or online. For example, the confidence interval for the population mean is given by:

$$M - z \cdot s_M \leq \mu \leq M + z \cdot s_M \tag{9.11}$$

where M is the sample average. Under certain circumstances, the quantity z is replaced by t. If the standard deviation has to be computed as well as the average, then t is used, as in:

$$M - t \cdot s_M \leq \mu \leq M + t \cdot s_M \tag{9.12}$$

Example 9.2 Estimating the mean value of a population from a sample.

A normal or Gaussian distribution is characterized by two population parameters: the mean, μ, and the standard deviation, σ. Given a sample of normally distributed random variables, compute a confidence interval for the mean, assuming that the standard deviation is known. The confidence interval should be constructed so that we are 95% sure that the mean is in the interval.

The so-called standard normal distribution is a normal distribution where $\mu = 0$ and $\sigma = 1$. One hundred (100) samples from a population of standard normal distribution are generated in MATLAB with the command:

```
d=random('normal',0,1,[100 1])
```

For a 95% confidence interval, with σ known to be 1, the value of z is 1.96. The confidence interval is computed as:

```
M=mean(d);
sM=std(d);
low=M-1.96;
high=M+1.96;
fprintf('The confidence interval is (%f, %f)\n',low,high)
```

Repeating this sequence of commands for different samples will yield different results. For one particular sample,

```
The confidence interval is (-2.121312, 1.798688)
```

For larger and larger samples, the confidence interval will approach (-1.96, 1.96).

If, on the other hand, σ is not known but is estimated from the sample, the MATLAB code to compute the confidence interval is:

```
M=mean(d);
s=std(d);
sM=s/sqrt(100);
low=M-1.9842*sM;
high=M+1.9842*sM;
fprintf('The confidence interval is (%f, %f)\n',low,high)
```

where the value of $t = 1.9842$ is taken from a table.

```
The confidence interval is (-0.235331, 0.151654)
```

Hypothesis testing is a comparison of two statements of truth about the population. It might appear that an independent variable has some effect on measurements from a sample; however, it is important to state with some assuredness that the effect is really due to the independent variable and not just chance. Hypothesis testing is used to reject one statement (termed the *null hypothesis*) in favor of a second statement about the population (termed the *alternate hypothesis*).

A null hypothesis is a statement about a population parameter that is often the opposite of what is believed. The null hypothesis is posited to see if the data can contradict it. The alternate hypothesis is a similar statement about the population parameter, but it is given an assumption that the independent variable does have some effect on the dependent variable. There are four parts to hypothesis testing:

1. Specify the null hypothesis and alternate hypothesis.
2. Choose the desired level of assuredness, which is called the *significance level*. The assuredness is an indication of how confident we are in assuming that the alternate hypothesis holds instead of the null hypothesis.
3. Calculate the sample (or test) statistic that is the estimate of the population parameter.

4. Determine the probability value (or *p value*) that the test statistic could appear if the null hypothesis was true. The p value is compared to the significance level and if it is less than the significance level, then the null hypothesis is rejected in favor of the alternate hypothesis.

This is a very brief introduction to statistical inference and hypothesis testing. For purposes of illustration, two tests that are commonly found in biomedical engineering are presented here: The first is called *Student's t-test* and the second is called *correlation*.

The general formulae for constructing confidence intervals given above are used, but a bit differently, in hypothesis testing. The population parameter is determined once the null hypothesis is chosen. After the significance level is chosen and the experiment is conducted, the test statistic M and standard error s_M (if necessary) are computed from the sample. Depending on whether the sample variance is known, the p value comes from either the table of z or t values. If the inequalities in the expression are not true, then the null hypothesis is rejected.

Student's t-test (or simply the *t-test*) is a hypothesis test that is used to describe a relation between the means of two populations. That is, the relationship between the two populations is that either the means are the same (the null hypothesis) or that the means differ (the alternate hypothesis). The test statistic is the difference between the two sample averages. The null hypothesis is often that the two means are the same then their difference is zero. The alternate hypothesis could take the form that the means are not the same, or that one is greater than the other. For example, the hypothesis may be:

$$\begin{aligned} H_0 &: \mu_1 = \mu_2 \\ H_1 &: \mu_1 \neq \mu_2 \end{aligned} \tag{9.13}$$

for which the test statistic is $M_1 - M_2$. The p value is computed by rearranging the formula for the confidence interval:

$$t = \frac{(M_1 - M_2) - (\mu_1 - \mu_2)}{s_{M_1 - M_2}} \tag{9.14}$$

if the null hypothesis is that the means are the same:

$$t = \frac{(M_1 - M_2)}{s_{M_1 - M_2}} \tag{9.15}$$

where $s_{M_1 - M_2}$ is computed by:

$$s_{M_1-M_2} = \sqrt{\frac{s_1^2}{N_1} + \frac{s_2^2}{N_2}} \tag{9.16}$$

Correlation is a hypothesis test about a linear relationship between an independent and a dependent variable. The correlation coefficient, r, varies from -1 (negative linear relationship) to 0 (no linear relationship) to 1 (positive linear relationship). The magnitude is often reported as r^2, rather than r, and has the interpretation of the percentage of the variance in the dependent variable explained by the relationship with the independent variable.

If X and Y are the independent and dependent variables, respectively, then the correlation coefficient r is given by:

$$r = \frac{\sum_{i=1}^{N} X_i Y_i - \sum_{i=1}^{N} X_i \cdot \sum_{i=1}^{N} Y_i}{\sigma_X \sigma_Y} \tag{9.17}$$

if the sample variances are known. If the variances are not known:

$$r = \frac{\sum_{i=1}^{N} X_i Y_i - \frac{\sum_{i=1}^{N} X_i \cdot \sum_{i=1}^{N} Y_i}{N}}{\sqrt{\left(\sum_{i=1}^{N} X_i^2 - \frac{\left(\sum_{i=1}^{N} X_i\right)^2}{N}\right)\left(\sum_{i=1}^{N} X_i^2 - \frac{\left(\sum_{i=1}^{N} X_i\right)^2}{N}\right)}} \tag{9.18}$$

Notice that the random variables X and Y are interchangeable in the above expression. This implies that correlation, as measured by the correlation coefficient, is a symmetric relationship. This means that a strong correlation does not necessarily imply causality, i.e., changes in one variable being responsible for changes in the other.

Although this is the most common formula for correlation, it does not reflect nonlinear relationships between the variables. There are other types of correlation that handle other types of variables, nonlinear relationships or multiple correlations, or relationships between two or more independent variables and a dependent variable.

Example 9.3 Hypothesis testing in DNA microarray analysis.

DNA microarrays are used to study the relative expression of genes under different experimental conditions. One of the commonly studied systems is the yeast during fermentation and/or respiration.

DNA microarray datasets can be found from many authors, or can be found on the NIH gene expression Web site (http://www.ncbi.nlm.nih.gov/genome/guide/human/resources.shtml). A sample dataset is included in the MATLAB bioinformatics toolbox. Choose two genes from the dataset and test the hypothesis that their expression profiles are correlated

With the bioinformatics toolbox, load the MATLAB variable `yeastdata.mat`. If the toolbox is not available, the full dataset can be found at the NIH Web site with accession number GSE28:
http://www.ncbi.nlm.nih.gov/entrez/query.fcgi?db=gds&cmd=search&term=GSE28

```
load yeastdata.mat
```

The microarray expression levels are the array `yeastdata` in the workspace, and the cell array `genes` contains information on each of the 6,400 genes in this dataset. Each row of the array contains the relative expression levels over the seven time points of the experiment; the actual time points are stored in the workspace variable times. Selecting two genes, 20 and 21:

```
figure(1)
plot(times,yeastvalues(20:21,:))
```

This plot of the relative expression profiles for genes 20 and 21 (Fig. 9.7) suggests that they are related. The MATLAB function `corr` can be used to test this hypothesis against a null hypothesis that the expression profiles are not linearly related. The two profiles are arguments to the function `corr`:

```
corr(yeastvalues(20,:)', yeastvalues(21,:)')
```

The value of the correlation coefficient r is 0.8693. Under a null hypothesis that the correlation coefficient is zero, compute the t statistic using the formula:

$$t = \frac{r}{\sqrt{(1-r^2)/(N-2)}}$$

using:

```
N=length(times);
t=r/sqrt((1-r^2)/(N-2))
```

which, for genes 20 and 21, is 3.9331. Now, using a significance level of 0.05, and since there are seven time points, the number of degrees of freedom is five. From a table for *t* values to find the critical value for the desired level of significance and the number of free parameters, the *t* value is 2.5706. Since 3.9331 is greater than 2.5706, the null hypothesis is rejected in favor of the alternate, that the two expression profiles are related.

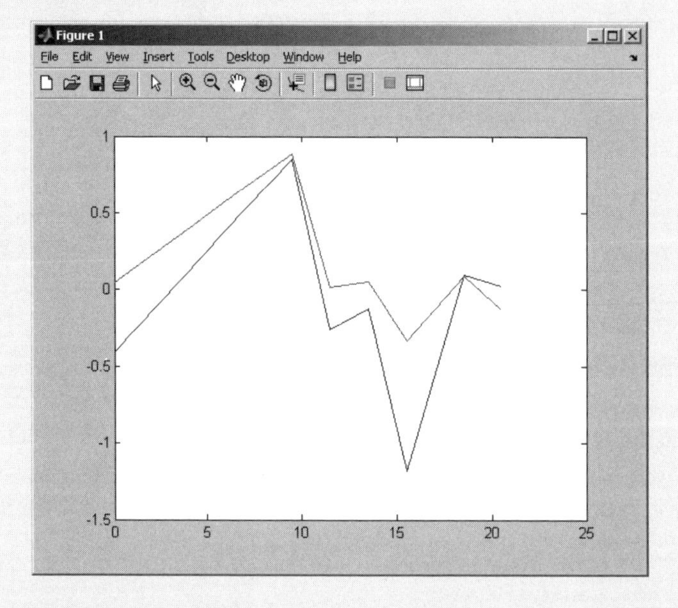

Figure 9.7 Plot of relative expression profiles for genes 20 and 21.

Now choose genes 20 and 30. The plot of the profiles is shown in Fig. 9.8, and:

```
r=corr(yeastvalues(20,:)', yeastvalues(30,:)')
t=r/sqrt((1-r^2)/(N-2))
```

The correlation coefficient is -0.0150. The negative sign means that any linear relationship is inversely, not directly related to the time sequence. The t statistic is -0.0335, which is less than 2.5706. Thus the null hypothesis, that there is no correlation, cannot be rejected.

Failure to reject a null hypothesis as in this case does not mean that the null hypothesis is necessarily acceptable. Rejecting the null hypothesis means that the probability or likelihood that the correlation in the sample is due to chance is less than 0.05.

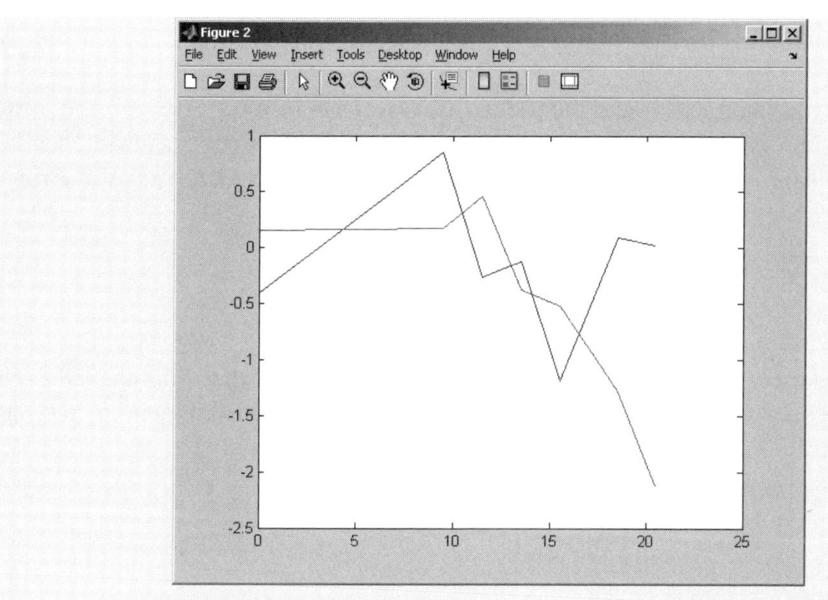

Figure 9.8 Plot of relative expression profiles for genes 20 and 30.

9.5 Least Squares Modeling

Least squares is a numerical method for fitting an analytical model to data that is noisy. The procedure is used to compute the parameters of the model from the measurements data.

Assume that measurements of a dependent variable are made as a function of an independent variable (such as voltage as a function of time). The model is fitted to the measured data under the following assumption:

$$Y = f(X) + \varepsilon \tag{9.19}$$

for an independent variable X, dependent variable Y and noise ε.

Assume that a set of measurements (x_i, y_i) is collected and that a linear relationship is assumed between the independent variable X and dependent variable Y. Thus, the model:

$$Y = aX + b \tag{9.20}$$

could be fitted exactly, if no noise or variability were present.

Least squares is a technique to calculate a set of values of the parameters of the model that minimizes the total error between the model and the measured data. In the presence of noise in the measurement, there is a difference:

$$d_i = y_i - (ax_i + b_i) \tag{9.21}$$

between the measurement y_i and the prediction given by the model ($ax_i + b_i$). In order to find the total error in the model fit, it is best not to use a signed difference but rather the square of the difference, which, when summed, better reflects the error in the fit. That is:

$$E_T = \sum_{i=1}^{n} \left(y_i - \left(ax_i + b\right)\right)^2 \tag{9.22}$$

and the numerical problem is to find the parameters (a, b) that minimize E_T. The minimum occurs where the derivatives, with respect to each of the parameters, are equal to zero:

$$\begin{aligned} \frac{\partial E_T}{\partial a} &= -2\sum_{i=1}^{n}(y_i - (ax_i + b))x_i = 0 \\ \frac{\partial E_T}{\partial a} &= -2\sum_{i=1}^{n}(x_i y_i - ax_i^2 - bx_i) = 0 \end{aligned} \tag{9.23}$$

and:

$$\frac{\partial E_T}{\partial b} = -2\sum_{i=1}^{n}(y_i - (ax_i + b)) = 0 \tag{9.24}$$

The pair of equations must be solved simultaneously, since both parameters are in each equation. The pair can be rewritten as:

$$\begin{aligned} \frac{\partial E_T}{\partial a} &= -2\sum_{i=1}^{n}(x_i y_i - ax_i^2 - bx_i) &= 0 \\ \frac{\partial E_T}{\partial b} &= -2\sum_{i=1}^{n}(y_i - ax_i - b) &= 0 \end{aligned} \tag{9.25}$$

and, after dropping the coefficients of -2 and rearranging:

$$\begin{aligned} a\sum_{i=1}^{n} x_i^2 + b\sum_{i=1}^{n} x_i &= \sum_{i=1}^{n} x_i y_i \\ a\sum_{i=1}^{n} x_i + b\sum_{i=1}^{n} 1 &= \sum_{i=1}^{n} y_i \end{aligned} \tag{9.26}$$

which can be rewritten as:

$$\begin{aligned} b\sum_{i=1}^{n} 1 + a\sum_{i=1}^{n} x_i &= \sum_{i=1}^{n} y_i \\ b\sum_{i=1}^{n} x_i + a\sum_{i=1}^{n} x_i^2 &= \sum_{i=1}^{n} x_i y_i \end{aligned} \tag{9.27}$$

or, in matrix-vector form:

$$\begin{bmatrix} n & \sum_{i=1}^{n} x_i \\ \sum_{i=1}^{n} x_i & \sum_{i=1}^{n} x_i^2 \end{bmatrix} \begin{bmatrix} b \\ a \end{bmatrix} = \begin{bmatrix} \sum_{i=1}^{n} y_i \\ \sum_{i=1}^{n} x_i y_i \end{bmatrix} \tag{9.28}$$

These equations are referred to as the *normal equations*. The normal equations are a standard form for setting up a least squares problem. They constitute a set of linear algebraic equations, which may be solved by the methods described in Chapter 4 of this book.

Example 9.4 Least square fit of a first-order polynomial (straight line).

Use least squares regression to fit a straight line to a set of measurements (x_i, y_i):

(1,4), (3,5), (5,6), (7,5), (10,8), (12,7), (13,6), (16,9), (18,12), (20,11)

and solve for the coefficients of the linear equation using MATLAB.

The measurements can be stored in two arrays, one each for the values of the independent variable and the dependent variable:

```
x =[1 3 5 7 10 12 13 16 18 20];
y =[4 5 6 5 8 7 6 9 12 11];
```

The normal equations can be easily computed and represented as a MATLAB matrix. For the given problem, the matrix is:

```
A =[length(x),sum(x); sum(x), sum(x.^2)]
A =
         10         105
        105        1477
```

and the right-hand side of Eq. (9.28) is given by:

```
c =[sum(y);sum(x.*y)]
c =
    73
   906
```

Now the coefficient vector can be computed:

```
z=A^-1*c
z =
     3.3888
     0.3725
```

That is, the intercept is 3.3888 and the slope is 0.3725.

The data and the model fit to the data can both be plotted easily. The following MATLAB command:

```
plot(x,y,'o',x,z(1)+z(2)*x)
```

will plot two curves. For the given problem, the measured data are shown as circles, and the line is the model fitted to the data. The first two arguments of plot are the independent and dependent variable for the measured data. The third argument 'o' is the symbol used to represent the data points. The last two arguments are the independent and dependent variables used to plot the model curve. Notice that the second dependent variable is not the measured data, but rather the model fitted to the data. The equation `a(1)+a(2)*x` is a vector where each element is multiplied by the estimated slope and added to the estimated intercept. The graph in Fig. 9.9 shows the MATLAB output.

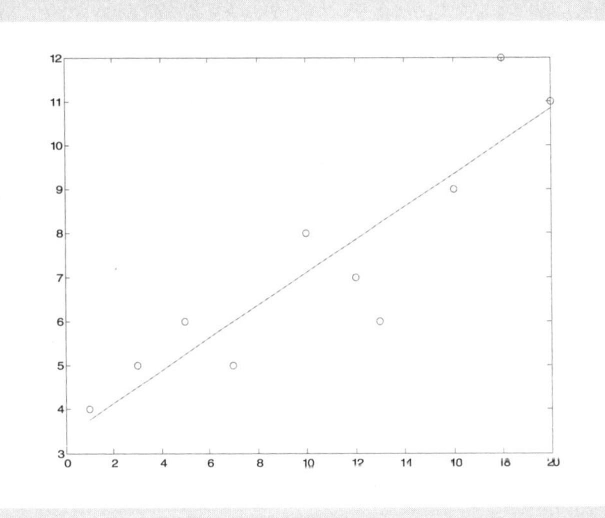

Figure 9.9 Plot of measured data and model fitted to data.

Example 9.5 Least squares fit of a cubic polynomial.

Set up the least squares problem to fit a cubic polynomial to a set of data (x_i, y_i).

The cubic polynomial can be written in the form:

$$y_i = a_0 + a_1 x_i + a_2 x_i^2 + a_3 x_i^3$$

for each of the data. Using a procedure similar to that above, write an expression for the squared error:

$$(y_i - (a_0 + a_1 x_i + a_2 x_i^2 + a_3 x_i^3))^2$$

and, after differentiating with respect to each of the parameters a_i, the normal equations are:

$$\begin{bmatrix} n & \sum x_i & \sum x_i^2 & \sum x_i^3 \\ \sum x_i & \sum x_i^2 & \sum x_i^3 & \sum x_i^4 \\ \sum x_i^2 & \sum x_i^3 & \sum x_i^4 & \sum x_i^5 \\ \sum x_i^3 & \sum x_i^4 & \sum x_i^5 & \sum x_i^6 \end{bmatrix} \begin{bmatrix} a_0 \\ a_1 \\ a_2 \\ a_3 \end{bmatrix} = \begin{bmatrix} \sum y_i \\ \sum x_i y_i \\ \sum x_i^2 y \\ \sum x_i^3 y \end{bmatrix}.$$

The parameters a_i are solved by:

$$\begin{bmatrix} a_0 \\ a_1 \\ a_2 \\ a_3 \end{bmatrix} = \begin{bmatrix} n & \sum x_i & \sum x_i^2 & \sum x_i^3 \\ \sum x_i & \sum x_i^2 & \sum x_i^3 & \sum x_i^4 \\ \sum x_i^2 & \sum x_i^3 & \sum x_i^4 & \sum x_i^5 \\ \sum x_i^3 & \sum x_i^4 & \sum x_i^5 & \sum x_i^6 \end{bmatrix}^{-1} \begin{bmatrix} \sum y_i \\ \sum x_i y_i \\ \sum x_i^2 y \\ \sum x_i^3 y \end{bmatrix}$$

The same procedure can be used to fit a nonlinear model by first applying a linear transformation to the equation(s).

Example 9.6 Least squares fit of a nonlinear model.

Set up the least squares problem to fit the model $y = ae^{bx}$ to a set of data (x_i, y_i).

First, take the natural logarithm of both sides of the equation to recast the problem:

$$\ln y = \ln a + bx$$

which is a linear least squares problem with $a_0 = \ln a$ and $a_1 = b$. The normal equations are:

$$\begin{bmatrix} n & \sum_{i=1}^{n} x_i \\ \sum_{i=1}^{n} x_i & \sum_{i=1}^{n} x_i^2 \end{bmatrix} \begin{bmatrix} \ln a \\ b \end{bmatrix} = \begin{bmatrix} \sum_{i=1}^{n} \ln y_i \\ \sum_{i=1}^{n} x_i \ln y_i \end{bmatrix}$$

which can be solved first for the coefficients ln a and b. The coefficient a of the nonlinear form can then easily be determined.

In a similar fashion, a multivariate model can be fitted, as shown in Example 9.7.

Example 9.7 Least squares fit of a multivariate model.

Fit a multivariate (multivariable) model to a set of measurements ($x_{1,i}$, $x_{2,i}$, y_i). There are two independent variables x_1 and x_2 and a single dependent variable.

The model has the form:

$$y = a_0 + a_1 x_1 + a_2 x_2 + \varepsilon$$

where ε represents the noise in the measurements. The normal equations have the form:

$$\begin{bmatrix} n & \sum x_{1i} & \sum x_{2i} \\ \sum x_{1i} & \sum x_{1i}^2 & \sum x_{1i} x_{2i} \\ \sum x_{2i} & \sum x_{1i} x_{2i} & \sum x_{2i}^2 \end{bmatrix} \begin{bmatrix} a_0 \\ a_1 \\ a_2 \end{bmatrix} = \begin{bmatrix} \sum y_i \\ \sum x_{1i} y_i \\ \sum x_{2i} y_i \end{bmatrix}$$

9.6 Curve Fitting

In the least squares technique, described in the previous section, the model can take many forms. The examples shown in Section 9.5:

$$\begin{aligned} y &= a_0 + a_1 x \\ y &= a_0 + a_1 x + a_2 x^2 + a_3 x^3 \\ y &= a e^{bx} \\ y &= a_0 + a_1 x_1 + a_2 x_2 \end{aligned} \tag{9.29}$$

are only a small subset of the possible forms that an analytical model can take.

Models are also used to predict the output for a value of the input or independent variable that is not one collected. There are two kinds of prediction problems: *interpolation* and *extrapolation*.

Suppose that we have a set of data (x_1, y_1) to (x_n, y_n). Interpolation is the problem of predicting the output y given a value of x where $x_1 < x < x_n$ and is not one of the x_i. Extrapolation is the problem of predicting the output y given a value of x where $x < x_1$ or $x > x_n$.

If the data is sampled data (that is, assumed to have been measured in the presence of noise), then the least squares procedure from Section 9.4 should be used to compute the parameters or coefficients of the model. On the other hand, if the data is known to be exact (that is, the data comes from a table of exact values) then the least squares procedure is not necessary. In fact, one really wants to have a model that goes through the data **exactly** if the data does not contain any noise. The *Gregory-Newton* and the *Lagrange* interpolating polynomials (see Chapter 6) are models that go through the data exactly and can be used for interpolation and extrapolation problems.

9.6.1 Lagrange interpolating polynomials

The basic principle is that there is only one polynomial of order n that goes through n+1 data points. For example, to estimate the first-order polynomial:

$$y = a_0 + a_1 x \tag{9.30}$$

two points are needed to estimate the values of the two coefficients, a_0 and a_1.

Given the two data points (x_0, y_0) and (x_1, y_1), the first-order polynomial has to go through these points exactly, i.e., the goal is to find the interpolating polynomial $P(x)$ such that $P(x_0) = y_0$ and $P(x_1)\,) = y_1$. The polynomial:

$$P_1(x) = \frac{(x - x_1)}{(x_0 - x_1)} y_0 + \frac{(x - x_0)}{(x_1 - x_0)} y_1 \tag{9.31}$$

has exactly this property. When $x = x_0$, the second term is 0, and when $x = x_1$, the first term is zero, so $P_1(x_0) = y_0$ or $P_1(x_1) = y_1$, respectively. The interpolating polynomial $P_2(x)$, which passes through the three points(x_0, y_0), (x_1, y_1), and (x_2, y_2), has the form:

$$P_2(x) = \frac{(x-x_1)(x-x_2)}{(x_0-x_1)(x_0-x_2)}y_0 + \frac{(x-x_0)(x-x_2)}{(x_1-x_0)(x_1-x_2)}y_1 + \frac{(x-x_0)(x-x_1)}{(x_2-x_0)(x_2-x_1)}y_2 \qquad (9.32)$$

The polynomial $P_2(x)$ is a linear combination of three terms, each of which is a second-order polynomial that has zeroes at two of the three known points and has magnitude 1 at the other point. These three polynomials can be written as:

$$\begin{aligned} L_{2,0}(x) &= \frac{(x-x_1)(x-x_2)}{(x_0-x_1)(x_0-x_2)} \\ L_{2,1}(x) &= \frac{(x-x_0)(x-x_2)}{(x_1-x_0)(x_1-x_2)} \\ L_{2,2}(x) &= \frac{(x-x_0)(x-x_1)}{(x_2-x_0)(x_2-x_1)} \end{aligned} \qquad (9.33)$$

and the interpolating polynomial, $P_2(x)$, is:

$$P_2(x) = \sum_{k=0}^{2} y_k L_{2,k}(x) \qquad (9.34)$$

Polynomials of this form are called *Lagrange interpolating polynomials*. The general form is:

$$P_n(x) = \sum_{k=0}^{n} y_k L_{n,k}(x) \qquad (9.35)$$

for an order n interpolating polynomial where:

$$L_{n,k}(x) = \frac{(x-x_0)(x-x_1)\cdots(x-x_{k-1})(x-x_{k+1})\cdots(x_k-x_n)}{(x_k-x_0)(x_k-x_1)\cdots(x_k-x_{k-1})(x_k-x_{k+1})\cdots(x_k-x_n)} \qquad (9.36)$$

9.6.2 Newton divided difference interpolating polynomials

Lagrange interpolation is not the only method for determining the representation of an interpolating polynomial. The methods for determining an interpolating polynomial from a table of data are called *divided difference methods*. These methods can also be used to approximate derivatives and integrals of functions, as well as to approximate solutions to differential equations.

Divided differences will be denoted using special notation as follows. Let $f[x_i] = f(x_i)$ and let

$$f[x_i, x_{i+1}] = \frac{f(x_{i+1}) - f(x_i)}{(x_{i+1} - x_i)} \tag{9.37}$$

denote the first-order finite difference on the two points x_i and x_{i+1}. For two divided differences $f[x_i, x_{i+1}, x_{i+2}, \ldots, x_{i+k-1}]$ and $f[x_{i+1}, x_{i+2}, \ldots, x_{i+k}]$, the kth divided difference on the points $x_i, x_{i+1}, x_{i+2}, \ldots, x_{i+k}$ is:

$$f[x_i, x_{i+1}, x_{i+2}, \ldots, x_{i+k}] = \frac{f[x_i, x_{i+1}, x_{i+2}, \ldots, x_{i+k-1}] - f[x_{i+1}, x_{i+2}, \ldots, x_{i+k}]}{x_{i+k} - x_i} \tag{9.38}$$

The Lagrange interpolating polynomial $P_n(x)$ can also be written in the form:

$$P_n(x) = a_0 + a_1(x - x_0) + a_2(x - x_0)(x - x_1) + \cdots + a_n(x - x_0)\ldots(x - x_{n-1}) \tag{9.39}$$

Since $P_n(x_0) = f(x_0)$, the coefficient a_0 is given by $a_0 = f(x_0) = f[x_0]$. Similarly:

$$P_n(x_1) = f(x_1) = a_0 + a_1(x_1 - x_0) \tag{9.40}$$

from which:

$$a_1 = \frac{f(x_1) - f(x_0)}{(x_1 - x_0)} = f[x_0, x_1] \tag{9.41}$$

and the interpolating polynomial can be rewritten as:

$$P_n(x) = f[x_0] + f[x_0, x_1](x - x_0) + a_2(x - x_0)(x - x_1) + \cdots + a_n(x - x_0)\ldots(x - x_{n-1}) \tag{9.42}$$

The remaining coefficients can be determined in a similar fashion. As might be expected:

$$a_k = f[x_0, x_1, \ldots, x_k] \tag{9.43}$$

Then, the interpolating polynomial $P_n(x)$ is:

$$P_n(x) = f[x_0] + \sum_{k=1}^{n} f[x_0, \ldots, x_k](x - x_0)\ldots(x - x_{k-1}) \tag{9.44}$$

which is known as the *Newton divided difference interpolating polynomial.* The formulae for computing the divided differences are given in Chapter 6.

9.6.3 Splines

The Lagrange and Newton divided difference interpolating polynomials are based on the principle that $n+1$ points of data are needed to approximate a function of order n. In fact, some of the $n+1$ data points can be replaced by other conditions or constraints, so that fewer than $n+1$ data points are needed. *Splines* are polynomial approximations where some data points are replaced by constraints on data and the interpolating polynomials.

Sometimes it is preferable to fit a lower order polynomial to the $n+1$ data. The higher-order polynomial captures all of the variation, and in situations such as medical imaging and graphics, where smooth surfaces are desirable, small variations detract from the presentation. Splines are low-order polynomials that are used to interpolate subsets of the data. The lower-order polynomial is smoother and a set of low order polynomials, each using a subset of the data, would yield a more visually appealing presentation. The constraints implicitly used to fit a Lagrange or Newton divided difference polynomial are that the polynomial has to pass through the $n+1$ data. Suppose now that only two data points are used. The principle would suggest that only a linear function could be interpolated between the pair of points. However, quadratic or even cubic functions can be fitted to a pair of data points by using constraints on the smoothness of the curve instead of using additional data points.

Suppose that a quadratic function (or quadratic spline) is fit to each of the n intervals between pairs of the $n+1$ data. To fit a quadratic polynomial to the interval between each pair of points, the three unknowns, ($a_{i,0}$, $a_{i,1}$, and $a_{i,2}$) in the equation:

$$f_i(x) = a_{i,0} + a_{i,1}x + a_{i,2}x^2 \tag{9.45}$$

have to be determined. There are $3n$ unknowns, thus $3n$ equations or conditions are required to determine these unknowns. There are some general rules for splines that are used, in addition to the data, to determine the coefficients of each polynomial:

1. **The function values of adjacent polynomials at the interior data points must be equal.** This rule applies to a pair of polynomials that pass through the n-1 interior data points, for a total of $2n$-2 conditions. This rule can be expressed as:

$$\begin{aligned} f(x_{i-1}) &= a_{i-1,0} + a_{i-1,1}x_{i-1} + a_{i-1,2}x_{i-1}^2 \\ f(x_{i-1}) &= a_{i,0} + a_{i,1}x_{i-1} + a_{i,2}x_{i-1}^2 \end{aligned} \tag{9.46}$$

2. **The first and last polynomials must pass through the end points.** This rule applies only to the first and last polynomial and provides only two conditions. This rule is expressed as:

$$f_1(x_0) = a_{1,0} + a_{1,1}x_0 + a_{1,2}x_0^2$$
$$f_n(x_n) = a_{n,0} + a_{n,1}x_n + a_{n,2}x_n^2 \tag{9.47}$$

3. **The first derivative at the interior data points must be continuous**. The n-1 rules are of the form:

$$a_{i-1,1} + a_{i-1,2}x_{i-1} = a_{i,1} + a_{i,2}x_{i-1} \tag{9.48}$$

4. **The second derivative at one of the end points is zero**. The choice is arbitrary, but if the second derivative of the first polynomial is taken to be zero, the condition is expressed as:

$$a_{1,2} = 0 \tag{9.49}$$

The total number of constraints or conditions is then:

Rule	Conditions
1	$2n$-2
2	2
3	n-1
4	1
Total	**3n**

Therefore, there are enough conditions to be able to solve for the coefficients of each quadratic polynomial or spline. Given the data (x_0, y_0) to (x_n, y_n), the equations above can be used to solve for the coefficients of each spline from the linear system:

$$\begin{bmatrix} 1 & x_0 & 0 & 0 & 0 & \cdots & 0 & 0 & 0 \\ 1 & x_1 & 0 & 0 & 0 & \cdots & 0 & 0 & 0 \\ 0 & 0 & 1 & x_1 & x_1^2 & \cdots & 0 & 0 & 0 \\ 1 & x_1 & -1 & -x_1 & 0 & \cdots & 0 & 0 & 0 \\ \vdots & \vdots & \vdots & \vdots & \vdots & \cdots & \vdots & \vdots & \vdots \\ \vdots & \vdots & \vdots & \vdots & \vdots & \cdots & \vdots & \vdots & \vdots \\ \vdots & \vdots & \vdots & \vdots & \vdots & \cdots & \vdots & \vdots & \vdots \\ \vdots & \vdots & \vdots & \vdots & \vdots & \cdots & \vdots & \vdots & \vdots \\ 0 & 0 & 0 & 0 & 0 & \cdots & 1 & x_n & x_n^2 \end{bmatrix} \begin{bmatrix} a_{1,0} \\ a_{1,1} \\ a_{2,0} \\ a_{2,1} \\ a_{2,2} \\ \vdots \\ a_{n,0} \\ a_{n,1} \\ a_{n,2} \end{bmatrix} = \begin{bmatrix} f(x_0) \\ f(x_1) \\ f(x_1) \\ 0 \\ \vdots \\ \vdots \\ \vdots \\ \vdots \\ f(x_n) \end{bmatrix} \tag{9.50}$$

Notice that the coefficient $a_{1,2}$ does not appear in the linear system, since Rule 4 specifies that $a_{1,2} = 0$. The first and last equations in the system are from Rule 2. The second and third equations are from Rule 1, and the fourth equation is from Rule 4.

The equations from Rules 2, 3 and 4 are repeated for each interior point, x_1 to x_{n-1}. This linear system can be solved for the coefficients $a_{i,j}$ using one of the methods from Chapter 4.

The coefficients for cubic splines can be determined in a similar fashion, except that there is an additional rule that specifies that the second derivatives at the interior points must be equal. The rules for cubic splines are:

1. The function values of adjacent polynomials at the interior data points must be equal.
2. The first and last polynomials must pass through the end points.
3. The first derivative at the interior data points must be continuous.
4. The second derivative at the interior data points must be continuous.
5. The second derivative at the end points is zero.

The total number of constraints or conditions is then:

Rule	Conditions
1	$2n$-2
2	2
3	n-1
4	n-1
5	2
Total	**4*n***

which is exactly what is needed to fit a cubic polynomial, or spline, to the interval between each pair of points.

MATLAB has built-in functions that compute the coefficients of, and generate, splines.

Example 9.8 Resampling and baseline correction of MALDI-TOF mass spectra data.

Mass spectrometry is an important technique for analyzing and characterizing large molecules. The Matrix Assisted Laser Desorption/Ionization Time of Flight (MALDI-TOF) mass spectrometry technique, when used in conjunction with standard biochemical assays, can be used to determine the mass and primary sequence information of large biomolecules (Kim et al., 2004).

These time of flight mass spectrometers operate by the principle that when a spatially and temporally well-defined group of ions is subject to a constant electric field and allowed to drift in a region of constant electric field, the ions will traverse the region in a time proportional to their mass-to-charge (m/z) ratio.

One problem with the interpretation of MALDI-TOF spectra is that signals from low mass compounds can be confounded with signal from fragments of large molecular weight components. In order to accurately interpret the spectra, the contribution due to fragmentation must be removed computationally.

Write a MATLAB program that removes the baseline variation due to the fragmentation component in the MALDI-TOF spectra.

The contribution of the mass spectra contributed by the fragments is a slowly varying component and not a sharp peak in the spectrum. Furthermore, the (*m*/*z*) sampling is entirely dependent on the molecule being studied and the matrix being used and may not be uniform. Therefore, two key post-processing tasks are: (1) sample the spectra uniformly and (2) remove the contributions to the spectra from fragmentation.

Both of these tasks can be done easily in MATLAB by fitting a spline to the raw data. Resampling the spectra is interpolating uniformly spaced samples in (*m*/*z*). Removing the signal due to fragmentation is subtracting the baseline component. The MATLAB program to display and process the data is as follows:

The MALDI-TOF data set `clear.xls`, shown graphically in Fig. 9.10, is available from the book website. It has 16,727 samples, with a peak at 4817.2, which is due to the substrate used in the experiment. The program so far is:

```
data=xlsread('clear');
figure(1);
plot(data(:,1),data(:,2))
```

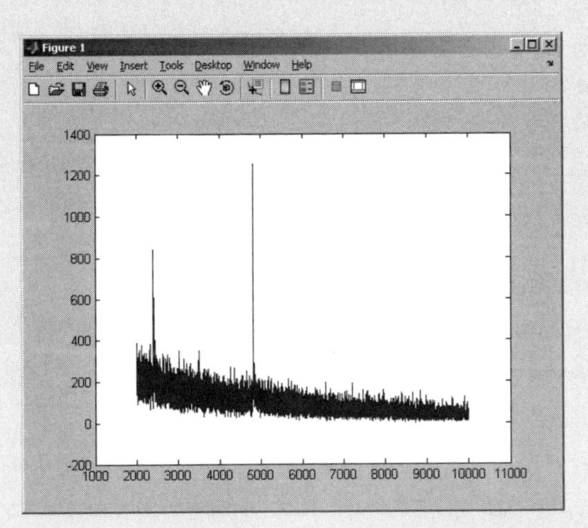

Figure 9.10 Graphic representation of the MALDI-TOF data set `clear.xls`.

For purposes of this example, the interpolation will be shown for the first 200 samples in the spectra, over the range of (*m*/*z*) from 2,000 to 2,060 (Fig. 9.11).

```
% work with the first 200 samples
mz=data(1:200,1);
y=data(1:200,2);
figure(2);
plot(mz,y);
```

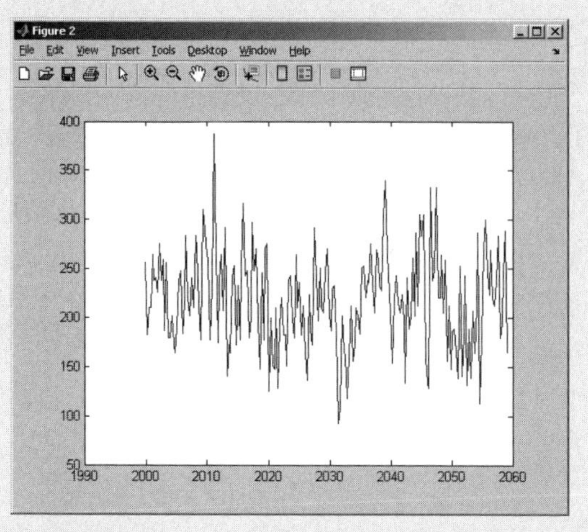

Figure 9.11 Interpolation for first 200 samples from `clear.xls`.

Now subsample the data so that the spline is created from 100 of these 200 points. This means that the spline polynomial will go through these 100 points exactly, and the magnitude at the other 100 points is estimated.

```
% decimate the MALDI data to estimate the baseline
mz_dec=mz(1:2:200);
y_dec=y(1:2:200);
% compute the spline estimate from the decimated data
% interpolate on the orginal m/z range
% pp is the piecewise polynomial
% force the endpoint conditions to be zero slope
% as described in the textbook
y_dec=[y(1); y(1:2:200); y(200)]
pp=spline(mz_dec,y_dec);
```

The MATLAB function `spline` returns an object that is a representation of the spline—a piecewise polynomial, `pp`. The variable `mz_dec` has the 100 decimated (m/z) values, and `y_dec` has the frequency counts for these (m/z) values.

Now interpolate the spectra at the remaining 100 points. The original set of 200 m/z frequencies is used, but the spline used 100 of these in the construction, so only 100 are really interpolated (Fig. 9.12).

```
% evaluate the polynomial on the original m/z range
yy=ppval(pp,mz);
% plot both:
```

```
% original data in blue dots
% spline estimate of baseline in red dashes
figure(3)
plot(mz,y,'b.',mz,yy,'r--')
```

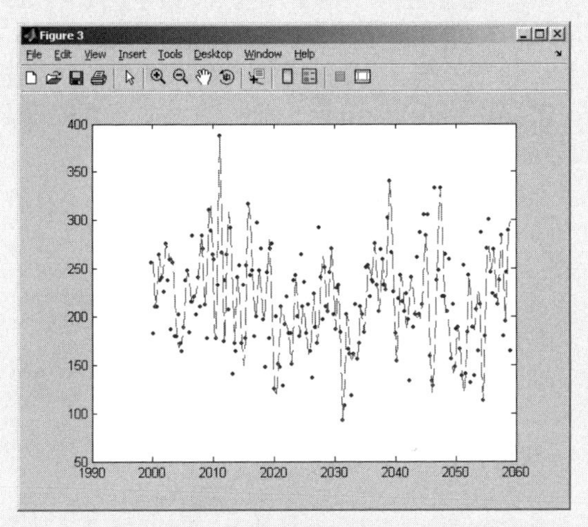

Figure 9.12 Interpolated baseline for `clear.xls`.

The original spectrum is represented by the dots and the interpolated baseline is the dashed line. Now the baseline can be removed by subtracting the interpolated spline from the original data.

```
% do the baseline removal:
% subtract the spline estimate from the raw data
figure(4)
plot(mz,y-yy)
```

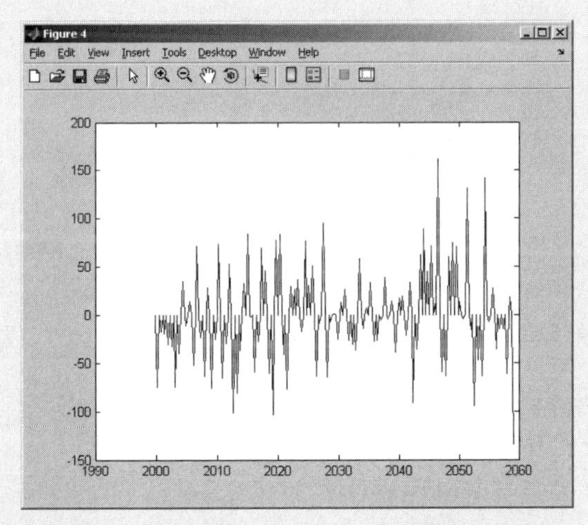

Figure 9.13. Baseline removed from the sample.

Notice that in Fig. 9.13 the baseline is zero. The remaining frequency counts are representative of the molecular components in the sample being tested.

The whole process can be applied to the entire spectra, with one small change. The baseline can be estimated with less than 50% of the data. In the example below, the baseline is estimated with less than 10% of the data, linearly spaced over the extent of the spectra using the MATLAB function `linspace`. The indices are set to be integers using the function `fix`.

```
mz=data(:,1);
y=data(:,2);
indicies=floor(linspace(1,length(mz),1000));
mz_dec=mz(indicies);
y_dec=[y(1); y(indicies); y(length(mz))];
pp=spline(mz_dec,y_dec);
yy=ppval(pp,mz);
figure(5)
plot(mz,y,'b.',mz,yy,'r--')
figure(6)
plot(mz,y-yy)
```

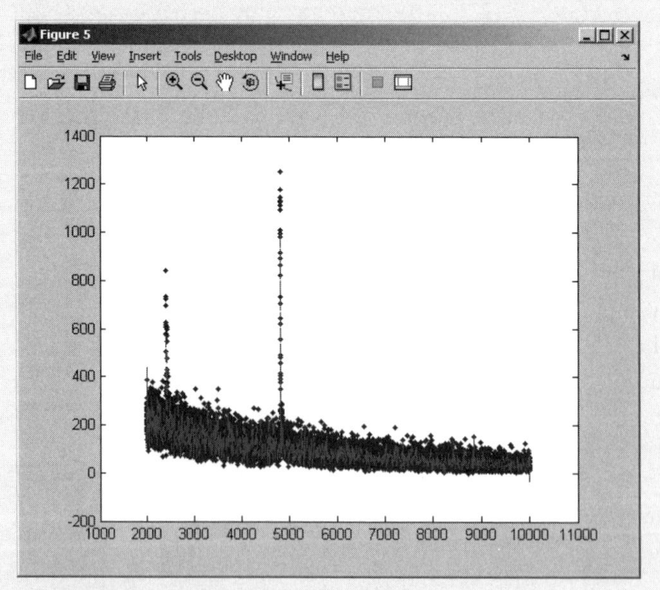

Figure 9.14 Baseline variation estimated from less than 10% of the data.

c

The baseline variation, shown in Fig. 9.14 above as the dashed line, is removed by subtraction. Notice in Fig. 9.15 that the peak at 4817.2 remains, but the baseline component due to fragmentation is removed.

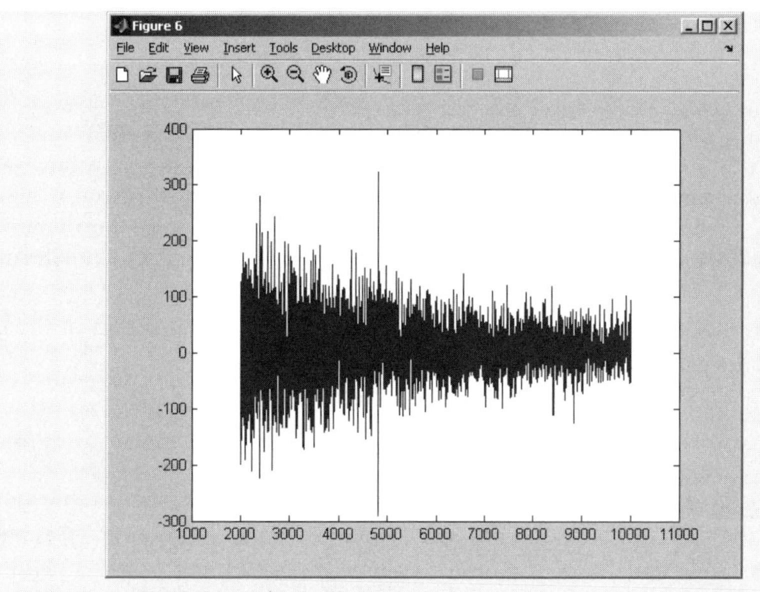

Figure 9.15 Baseline removed from MALDI-TOF data shown in Fig. 9.14.

9.7 Fourier Transforms

Time frequency analysis is an important component of the toolset of any biomedical engineer. The reader is referred to Bracewell (1999) and James (2002) that explore the subject in depth, so the material is not covered here. However, a rudimentary introduction is given, since MATLAB includes the basic tools. Some examples are given in Chapter 10.

The Fourier transform is a special type of curve fitting where a model of periodicity is fitted to the data.

The Fourier transform is an example of changing the representation of a signal or a set of signals from one independent variable to another. Most often, the data is represented as a function of time, and the Fourier transform is used to re-represent the data in terms of the periodicities in the data. These periodicities are typically represented as sets of sinusoids.

If the Fourier transform is being used to change the representation of a signal, then the two independent variables are *time* and *frequency*. Recall that there is a difference between continuous and discrete variables, so there must be a difference between continuous and discrete *time* and/or *frequency*. It is these differences that give rise to the different forms of the Fourier transforms and series.

It is probably easiest to organize the relationships in terms of the properties of the variables, as follows:

		FREQUENCY	
		Continuous	Discrete
TIME	Continuous	Fourier Transform	Fourier Series
	Discrete	Discrete Fourier Transform	Finite Fourier Transform

The transform that is introduced in systems analysis courses and used in theoretical analysis is the forward Fourier transform:

$$F(\omega) = \int f(t) e^{-j\omega t} dt \tag{9.51}$$

which represents a function of time in terms of its periodicities, represented as sinusoids. Inside the integral is the product of two continuous functions: $f(t)$ and $e^{-j\omega t}$, which is the exponential form of pairs of sinusoids, $\sin(\omega)$ and $\cos(\omega)$. The product of any pair of functions is a measure of the similarity of the two functions; the integral of this product is the sum of this similarity over the range of the independent variable. This integral of similarity is the coefficient $F(\omega)$. That is, the set of coefficients $F(\omega)$ are the measures of the degree to which the frequency ω is present in the function $f(t)$.

The inverse Fourier transform:

$$f(t) = \frac{1}{2\pi} \int F(\omega) e^{j\omega t} d\omega \tag{9.52}$$

has an analogous meaning. Again, there is a product of two functions $F(\omega)e^{j\omega t}$ inside an integral. This integral, which sums the similarity, is the measure of the contribution of the frequency ω to the function of time, $f(t)$.

These two transforms, termed the Fourier Transform pair, are used to analyze signals in the domain most appropriate for the task at hand. However, both the independent and the dependent variable are continuous, and only discrete analogs of the transforms can be used in a computational environment, such as MATLAB.

In the discrete analogs, both of the variables are sampled, and the continuous integral must be replaced by a summation. A function of a continuous variable, t, is replaced by a function that is sampled at a set of time points $n\tau$, where n is the index or sample number and τ is the time interval between samples. By convention, the first sample collected will be at index $n = 0$ and the last at index $n = N - 1$.

The Nyquist theorem answers the question of how finely the signal must be sampled in order to represent the signal accurately. By the Nyquist criterion, if the maximum frequency is ω_N, then the signal must be sampled with a rate at least τ:

$$\omega_N \leq \frac{2\pi}{\tau} \tag{9.53}$$

The Nyquist theorem is commonly interpreted to mean that if a signal is sampled at a rate τ, then the maximum frequency that can be measured is ω_N.

The discrete Fourier transform represents functions that are sampled data in terms of continuous frequency. Since the data is discrete, the integral over time must be replaced by a summation over sample indices, but the frequency variable is still continuous:

$$F(\omega) = \sum f(n\tau) e^{-j\omega n\tau} \tag{9.54}$$

The inverse transform, from continuous frequency to discrete time, is still an integral:

$$f(n\tau) = \frac{\tau}{2\pi} \int F(\omega) e^{j\omega n\tau} d\omega \tag{9.55}$$

but the limits of integration are bounded to a 2π interval. An appreciation of the Nyquist theorem makes these limits easier to understand. Recall that for a sampling rate of τ, the maximum frequency that can be measured is ω_N. Thus, with sampled data, the maximum frequency that could be measured is ω_N; then, implicitly, there is no frequency content below zero or above ω_N, so the limits of integration can be changed. Also, limits of integration between zero and ω_N are equivalent to limits of $(-\omega_N/2)$ to $(\omega_N/2)$. With the relation $\omega = 2\pi$, and knowing that the sampling frequency is $f_s = 1/\tau$, the limits of the inverse transform are $(-\pi/\tau)$ to $(-\pi/\tau)$.

Discretizing frequency, on the other hand, is a different matter. There are two forms of discrete frequency: a fundamental frequency and its harmonics, or a set of frequencies up to and including the Nyquist limit. Each form of discrete frequency representation leads to a transform.

The Fourier series is a representation of a function $f(t)$ in terms of a fundamental frequency and its harmonics. The forward transform, from time to frequency, is:

$$F_n = \frac{1}{T} \int_0^T f(t) e^{-j\omega_0 nt} dt \tag{9.56}$$

and the inverse transform, from frequency to time, is:

$$f(t) = \sum_{n=-\infty}^{\infty} F_n e^{j\omega_0 nt} \tag{9.57}$$

where $\omega_0 = 2\pi f_0$ is the *fundamental frequency*. The index n in each transform is the multiple, or harmonic, of the fundamental frequency. This representation of $f(t)$ assumes that there are no frequency components between $n\omega_0$ and $(1+n)\omega_0$. A consequence of this representation is that the signal represented by a Fourier series is actually a periodic version of the original signal where the periodicities are outside the limits of integration 0 to T.

The finite Fourier transform is the transform mapping functions of discrete time to discrete frequency. The forward transform is:

$$F_u = \sum_{n=0}^{N-1} f(n\tau) e^{-j\omega_0 nu} \tag{9.58}$$

and the inverse transform is:

$$f(n\tau) = \sum_{u=0}^{N-1} F_n e^{j\omega_0 nu} \tag{9.59}$$

Unlike the Fourier series, however, the discrete set of frequencies are cycles per sequence length N and are not harmonics of the fundamental frequency $\omega_0 = 2\pi/N$. Both the time series and frequency representations are periodic beyond the limits of sampling and summation ($0 \le n,\ u < N$).

The acronym FFT is commonly used to mean fast Fourier transform, which is actually an efficient algorithm for computing the finite Fourier transform. The MATLAB commands `fft` and `ifft` compute the fast Fourier transform and the inverse, respectively. The MATLAB documentation for both commands incorrectly refers to the functions as implementing the discrete Fourier transform, when the documentation should refer to the finite Fourier transform.

Example 9.9 Separating EEG frequency components.

The electroencephalogram (EEG) is a recording *at the scalp* of the electrical activity of certain neuronal groups. These small changes are typically in response to some external event, and are therefore referred to as event-related potentials (ERPs).

EEG signals and their analysis are tools for understanding the dynamic processes in the brain that are the bases of physical and mental behavior. EEGs (and specifically, ERPs) are one method used to localize the source of the brain activity associated with specific tasks or behaviors.

Certain behaviors are associated with well-defined frequency components of EEG signals. The frequency components of EEG data are:

1. *delta* waves are from 0 to 2 Hz and are the dominant rhythm in infants up to one year and in stages 3 and 4 of sleep.
2. *theta* waves are from 2 to 7 Hz and are classified as "slow" activity. This is abnormal in awake adults but is perfectly normal in children up to 13 years and during sleep.
3. *alpha* waves are from 7 to 13 Hz. Alpha waves are brought out by closing the eyes and by relaxation, and abolished by opening the eyes or alerting by any mechanism (thinking, calculating). This is the major rhythm seen in normal relaxed adults—it is present during most of life, especially beyond the thirteenth year when it dominates the resting state.
4. *beta* waves are from 14 Hz, typically up to 64Hz; this is generally regarded as a normal rhythm. This is the dominant rhythm in patients who are alert or anxious or who have their eyes open.

Write a MATLAB program that will separate the frequency components (delta, theta, alpha, and beta) of the EEG signal and determine the power in each band.

Fig. 9.16, below, shows 10 seconds of an EEG signal recorded on the scalp right above the midbrain. The sampling frequency was 250 Hz, so there are 2,500 samples in the range: -20 microvolts to 20 microvolts. This data is from Garrett et al. (2003).

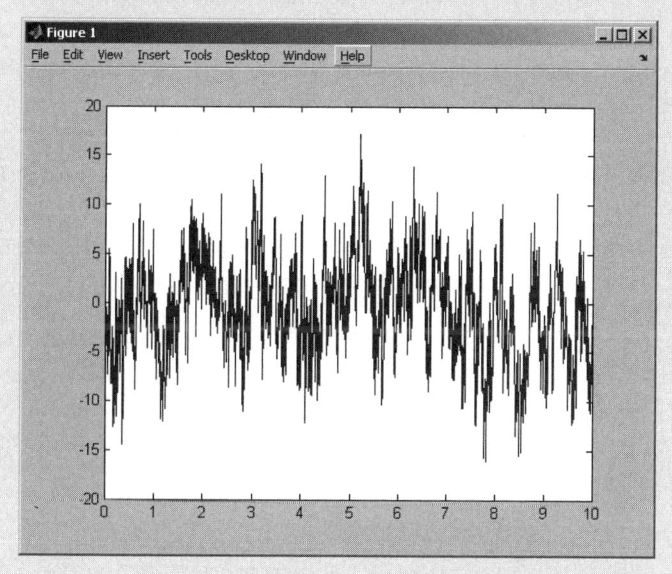

Figure 9.16 10 seconds of an EEG signal sampled at 250 Hz.

Suppose that the EEG data is stored in the variable `eegchannel`. The MATLAB commands to compute the finite Fourier transform and to display the results are:

```
eegfreq=abs(fft(eegchannel));
plot(eegfreq);
```

The result of the `fft` command is a vector of complex numbers that are the coefficients of the real and imaginary frequency content. The MATLAB function `abs` will return a vector with the magnitudes of the complex numbers. The frequency domain representation is shown in Fig. 9.17.

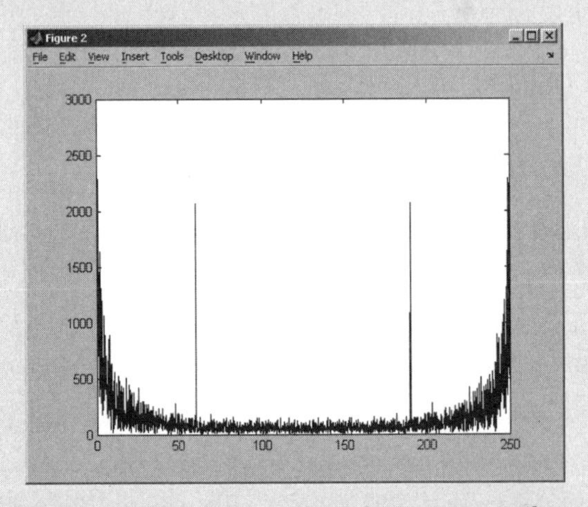

Figure 9.17 Frequency domain representation.

Recall from the presentation above that with the fast Fourier transform, there are periodicities in both time and frequency, meaning that there should be a symmetry or periodicity in the frequency spectra. Also, by the Nyquist criteria, the maximum frequency that can be sampled is half of the sampling rate, which is, in this case, 125 Hz. Since the maximum frequency is 125 Hz, and since the frequency domain is periodic, half of the frequency domain samples must overlap; in the spectra shown above, samples that appear to be at frequencies higher than 125 Hz are the same as those from -125 Hz to 0 Hz.

MATLAB provides a command to redisplay frequency data so that the DC component (0 Hz) is in the center of the graph, which is another way of saying that the positive frequency above the Nyquist frequency (125 Hz) is displayed as negative frequency. This command, `fftshift`, is used as follows:

```
eegfreqshift=fftshift(eegfreq);
plot(eegfreqshift);
```

The plot of the frequency spectra, in Fig. 9.18, is centered at 0 Hz with limits of -125

Hz to 125 Hz. The two large symmetric peaks correspond to ± 60 Hz, which is noise in the recordings due to AC power. The DC component and the 60 Hz components should be removed before any further analysis.

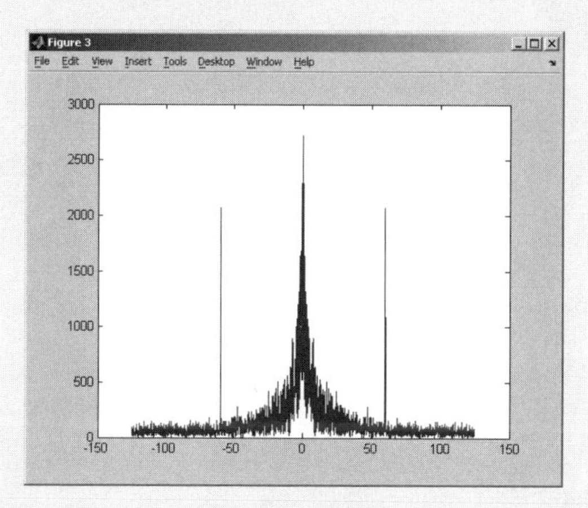

Figure 9.18 Frequency spectra centered at 0 Hz.

Although not elegant, a simple way to remove these components is to set the corresponding entries in the frequency spectra to zero.

```
% remove DC component
eegfreqshift(1240:1260)= 0;
% remove 60Hz
eegfreqshift(645:655)=0;
eegfreqshift(1845:1855)=0;
```

The frequency spectra now looks Fig. 9.19:

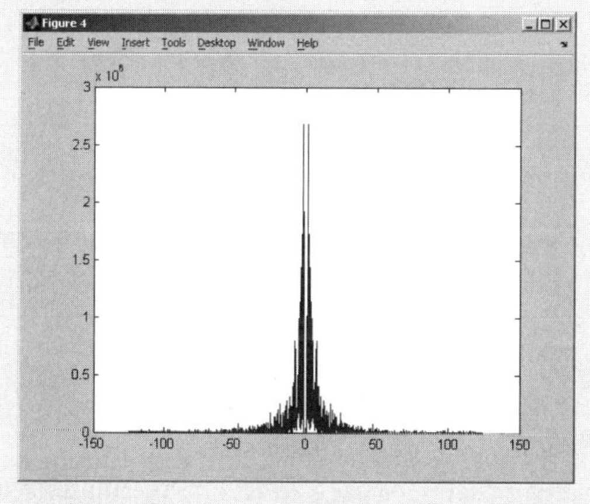

Figure 9.19 Frequency spectra centered at 0 Hz, with 60 Hz and DC component removed.

The percentage power in each frequency band is calculated by computing the power in each band and dividing by the total power in the signal.

```
power=eegfreqshift.^2;
delta=1:20;
theta=20:70;
alpha=70:130;
beta=140:640;
center=10*Nyquist;
power_total=sum(power(650:1850));
power_delta=(sum(power(center-
delta))+sum(power(center+delta)))/power_total
power_theta=(sum(power(center-
theta))+sum(power(center+theta)))/power_total
power_alpha=(sum(power(center-
alpha))+sum(power(center+alpha)))/power_total
power_beta=(sum(power(center-
beta))+sum(power(center+beta)))/power_total
```

The power in each band is:

```
power_delta =
    0.2453
power_theta =
    0.3883
power_alpha =
    0.1690
power_beta =
    0.2125
```

This suggests that the dominant band is the theta band (2 to 7 Hz), which is normal in adults that are not awake.

9.8 Lessons Learned in the Chapter

After studying this chapter, the student should have learned the following:

- All measurements have variation. Some of the variation is due to measurement error, and some can be due to biological variability.
- Descriptive statistics are used to characterize or describe a sample of measured data.
- Inferential statistics are used to draw conclusions or make generalizations about a population given only data from a sample.
- Least squares is a procedure used to fit a mathematical model to a set of data that is assumed to have variation.

- Interpolating polynomials are used to fit mathematical models to data that is assumed to be exact.
- The four Fourier transforms are used to fit models of periodicities to data. These periodicities are represented mathematically as sinusoids.
- There are built-in functions in MATLAB to compute some descriptive statistics, inferential statistics, interpolating splines, and Fourier transforms. Least squares can be easily implemented using matrix-vector arithmetic.

9.9 Problems

9.1 The Visible Human Project Web site (http://www.nlm.nih.gov/research/visible/visible_human.html) has CT images from both fresh and frozen cadavers. Choose any pair of corresponding images (such as the head or thorax) and compute the descriptive statistics for each image. Are there any differences between the distributions? Do these differences reflect the differences between fresh and frozen tissue?

9.2 Choose either image of the feet from the VHP CT datasets. Write a MATLAB script to read, display and threshold the image so that only the phalanges (toe bones) are white and the soft tissues (muscles and tendons) are black.

9.3 PhysioNet is an archive of physiologic signals maintained by the National Center for Research Resources of the NIH (NIH-NCRR). The database, which can be found at http://physionet.incor.usp.br/, includes samples of electrocardiograms that can be downloaded in text format from http://www.physionet.org/cgi-bin/rdsamp. Download one sample from the MIT-BIH Normal Sinus Rhythm database and write a MATLAB script to remove the 60Hz noise in the signal. Hint: see example 9.9.

9.4 Use the script written for Problem 9.3 for a sample from the MIT-BIH Arrhythmia Database, which is also part of the PhysioNet archive. Is the frequency content in the ECG with arrhythmias different than the ECG with normal sinus rhythm?

9.5 The MIMIC database (Moody and Mark, 1996), which is also part of PhysioNet, is a database of multiparameter (bedside) telemetry signals. First, select and download record number 055. Write a MATLAB script to determine whether any pair of signals is correlated using the correlation coefficient of Section 9.4. Is ambulatory blood pressure (ABP) significantly correlated with positive airway pressure (PAP)? Is respiration (RESP) significantly correlated with PAP?

9.10 References

Bracewell, R. N. 1999. The Fourier Transform & Its Applications. 3rd ed. New York: McGraw-Hill Book Company.

Garrett, D., Peterson, D. A., Anderson, C. W., and Thaut, M. H. 2003. Comparison of Linear and Nonlinear Methods for EEG Signal Classification. *IEEE Transactions on Neural Systems and Rehabilitative Engineering*, **11**(2):141-144.

Goldberger, A. L., Amaral, L. A. N., Glass L., Hausdorff, J. M., Ivanov, P. Ch., Mark, R. G., Mietus, J. E., Moody, G. B., Peng, C. K., and Stanley, H. E. 2000. PhysioBank, PhysioToolkit, and Physionet: Components of a New Research Resource for Complex Physiologic Signals. *CIRCULATION*, **101**(23):e215-e220.

James J. F. 2002. A Student's Guide to Fourier Transforms With Applications in Physics and Engineering. Cambridge: Cambridge University Press.

Kim, S., Ulz, M. E., Nguyen, T., Li, C., Tycko, B., and Ju, J. 2004. Digital high-throughput multiplex genotyping in human tumors. *Genomics*, **83**:924-931.

Moody, G. B., and Mark, R. G. 1996. A Database to Support Development and Evaluation of Intelligent Intensive Care Monitoring. *Computers in Cardiology*, **23**:657-660.

Web sites

Visible Human Project
http://www.nlm.nih.gov/research/visible/visible_human.html

NLM
http://www.nlm.nih.gov/

NIH Gene Expression Web site
http://www.ncbi.nlm.nih.gov/genome/guide/human/resources.shtml

GSE28
http://www.ncbi.nlm.nih.gov/entrez/query.fcgi?db=gds&cmd=search&term=GSE28

Chapter 10
Modeling Biosystems: Applications

10.1 Numerical Modeling of Bioengineering Systems

This chapter offers a number of examples designed to illustrate the range of applications of numerical methods and computing environments such as MATLAB and Simulink to biomedical engineering problems. Every attempt has been made to first state the clinical or physiological problem, introduce the modeling method and show how MATLAB and/or Simulink are used to create the computer solution using the numerical methods from this text.

A number of these examples are drawn from resources that are commonly available. Section 10.2 describes *PhysioNet*, a resource that includes a very large archive of bioelectric signals and software tools that can be used for research projects. The examples are a brief introduction as to how this resource can be used. The electroencephalogram (EEG) data used in Section 10.3 is available, as is the Simulink simulation PHYSBE used in Section 10.7. In all other cases, the problems were taken from the biomedical engineering literature.

There is no particular order in which to read or review these examples. As a guide, Table 10.1 is a summary of the examples presented in this chapter, in which the problems have been categorized by the underlying phenomena and the numerical method used. Simulink is indicated in the cases where it is used.

Table 10.1 A summary of the application examples in this chapter.

Section	Examples	Application	Phenomena	Models/Methods
10.2	10.1, 10.2	PhysioNet and PhysioBank	Bioelectric	Computing Model fitting ODEs*
10.3	10.3	Brain activation	Bioelectric	Model fitting Statistics
10.4	10.4	Diabetes and insulin regulation	Biochemical	Systems of ODEs Simulink
10.5	10.5	Renal clearance	Biochemical	ODEs
10.6	10.6	Gait and motion estimation	Biomechanical	Numerical linear algebra
10.7	10.7 to 10.12	PHYSBE	Biochemical	Systems of ODEs Simulink

* ODEs stands for Ordinary Differential Equations.

The material in this chapter will enable the student to accomplish the following:

- Learn how MATLAB can be used to solve biomedical engineering problems
- Learn how Simulink can be used to simulate models of physiological systems
- Write MATLAB programs to solve bioengineering problems
- Develop Simulink models of physiological systems
- Gain an understanding of the role that numerical methods play as the interface between mathematical model and computer solution

10.2 PhysioNet, PhysioBank, and PhysioToolkit

PhysioNet is an Internet resource for biomedical research and development sponsored by the National Center for Research Resources of the National Institute of Health (NIH). PhysioBank, which was described in Chapter 9, is an archive of contributed and standardized physiologic signals and annotations. In Chapter 9, several samples from PhysioBank were used as examples to show how models, specifically frequency domain models, can be used to answer questions about the physiology.

Another component of PhysioNet is PhysioToolkit, which is a repository of software for analyzing PhysioBank data, visualizing data (including PhysioBank data), signal processing, software development, and simulation. Several of these software tools can be used with MATLAB; some will be shown here as examples.

10.2.1 ECG simulation

In Chapter 2, the MATLAB script to generate a very crude simulation of a single QRS complex was given. In PhysioToolkit, there is a MATLAB script for generating both normal complexes and arrhythmia. The script, called ECGwaveGen, was contributed by Floyd Harriott and based on the article (Ruha 1997). In this script, the user can control the heart rate, signal duration, sampling frequency, QRS amplitude and duration, and T-wave amplitude of the synthesized signal.

Example 10.1 Using the MATLAB script ECGwaveGen to synthesize ECG data.

Download the `ECGwaveGen.m` and the `QRSpulse.m` MATLAB scripts from http://www.physionet.org/physiotools/matlab/ECGwaveGen/, and use these programs to synthesize five seconds each of normal and sinus tachycardial rhythms.

Normal rhythm, sometimes called sinus rhythm, is synthesized using `ECGwaveGen` by default. The help for `ECGwaveGen` is:

```
[QRSwave]=ECGwaveGen(bpm,dur,fs,amp) generates an artificial ECG/EKG
waveform
    Heart rate (bpm) sets the qrs event frequency (RR interval).
    Duration of the entire waveform (dur) is in units of seconds.
    Sample frequency (fs) sets the sample frequency in Hertz.
    Amplitude (amp) of the QRS event is measured in micro Volts. The
    waveform consists of a QRS complex and a T-wave. No attempt to
    represent a P-wave has been made.

    There are two additional parameters that can be changed from
    within the function. They are the parameters that set the QRS
    width (default 0.1 secs) and the t-wave amplitude (default 500
    uV).
```

The MATLAB commands:

```
Norm=ECGwaveGen(60,5,64,1000);
time=[(1:length(Norm))/64];
plot(time,Norm);
```

create a vector of time values (not sample numbers), the synthesized QRS complexes, and plot the ECG against the time, as shown in Fig. 10.1.

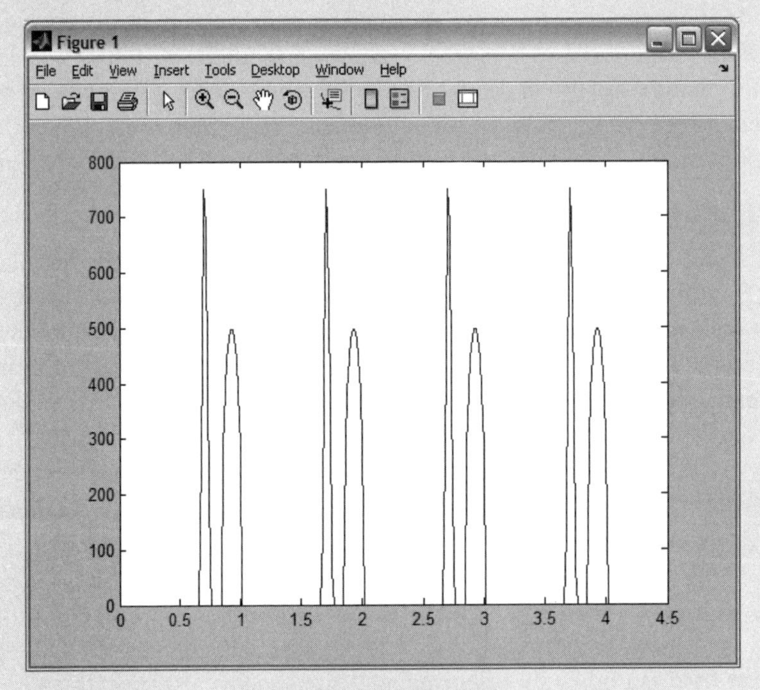

Figure 10.1 ECG profile.

where the QRS complex peak is approximately .75 mV and the T wave peak is .5 mV.

Sinus tachycardia is characterized by a heart rate greater than 100 beats per minute. The MATLAB code to simulate sinus tachycardia at 110 bpm is reproduced below, and the results shown in Fig. 10.2.

```
figure(2);
Norm=ECGwaveGen(110,5,64,1000);
time=[(1:length(Norm))/64];
plot(time,Norm);
title('ECGwaveGen synthesis of Sinus Tachycardia');
xlabel('Time in seconds');
ylabel('Amplitude in microvolts');
```

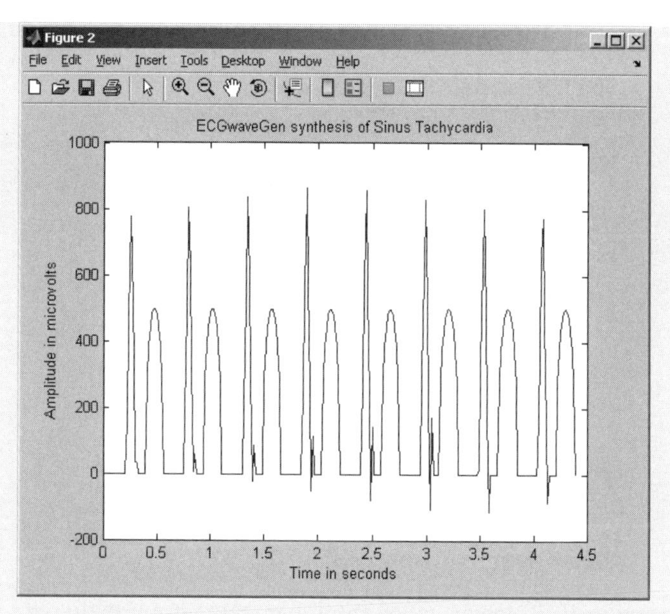

Figure 10.2 Simulation of sinus tachycardia.

Unfortunately, the `ECGwaveGen` script available at PhysioNet will not generate sinus tachycardia greater than 110 bpm. There is a second script at PhysioNet called `ecgsyn`. This script, based on the work of McSharry et al. (2003), allows the user to have more control over the morphology and frequency content. `ecgsyn` is a simulation of three coupled differential equations; see Chapter 7 for details on solving systems of simultaneous differential equations. The `ecgsyn.m` MATLAB script may be downloaded from http://www.physionet.org/physiotools/ecgsyn/.

With no arguments, the ecgsyn script will generate synthetic ecg with the following parameters:

```
ECG sampled at 256 Hz
Approximate number of heart beats: 256
Measurement noise amplitude: 0
Heart rate mean: 60 bpm
Heart rate std: 1 bpm
LF/HF ratio: 0.5
Internal sampling frequency: 512
```

To generate sinus rhythm of 10 beats sampled at 64 Hz with no noise:

```
s=ecgsyn(64,10,0,60);
```

Since the ECG is sampled at 64 Hz, the first 256 samples would cover 4 complexes (see Fig. 10.3).

```
plot(s(1:256))
```

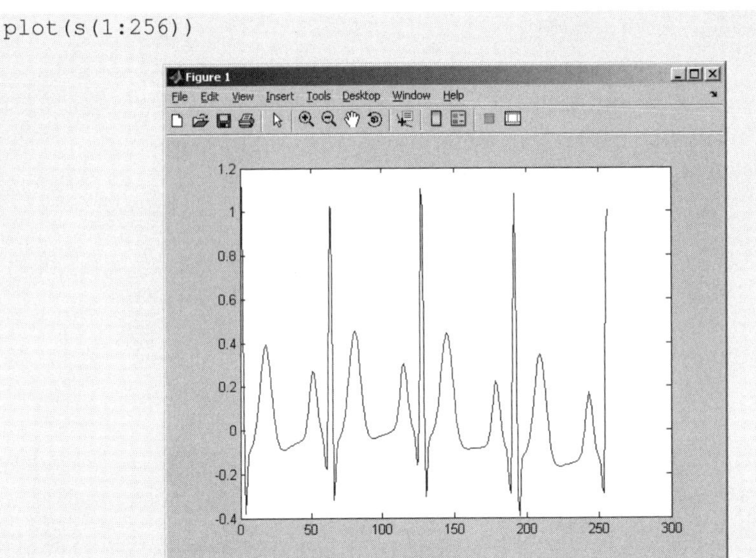

Figure 10.3 Normal sinus rhythm generated using ecgsyn.

The fourth argument is changed to vary the heart rate. To simulate tachycardia using `ecgsyn`:

```
s=ecgsyn(64,16,0,110);
```

and display the first 256 samples as before (Fig. 10.4).

```
plot(s(1:256))
```

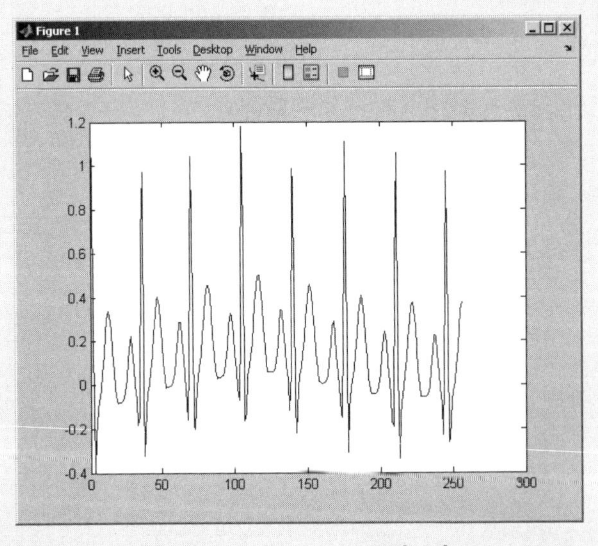

Figure 10.4 Tachycardia generated using ecgsyn.

There are roughly twice the number of PQRST complexes in Fig. 10.4 as there are in the same interval of time in Fig. 10.3.

Synthesized ECG is not meant to be realistic, as it is normally used to test biomedical instruments, using an ECG-like signal with well-defined characteristics. Nevertheless, the difference between the two simulators is obvious by visually comparing their output: the `ecgsyn` output is much more realistic, as it includes the P waves and an S-T depression. The tradeoff is obvious: the more realistic simulation required solving a system of three differential equations (Chapter 7) rather than a simulation based on curve fitting techniques (Chapter 9). A more realistic signal will still have more realistic frequency content.

10.2.2 Reading PhysioBank data

To solve Problems 9.3 to 9.5, one had to access the PhysioBank archives using the Chart-O-Matic Web browser (which uses the PhysioToolkit function `rdsamp`). The instructions were to download a text file and write a MATLAB script that would read the text file. There are also MATLAB scripts available to read the raw PhysioBank data files directly.

Example 10.2 Read and visualize PhysioBank signals and annotations.

The ECG data available through PhysioBank is stored in a unique format called the "212" format. Download the same sample used in Problem 9.3 and display it directly in MATLAB using the `rddata.m` script.

Throughout this text, we have stressed the value of using arguments to functions to preserve their flexibility. In the `rddata.m` script, the parameters of pathname, header, attribute, and data file names are not arguments, but are assignment statements that must be modified. The output, when using the MIT-BIH sample data file (in three parts 100.hea, 100.atr and 100.dat) is shown in Fig. 10.5.

Although difficult to dissect at first glance, the graph shows that the ECG recording is just over 80 seconds long. A small section of the data can be viewed by extracting the data (and attributes and annotations) from the MATLAB workspace variables. The MATLAB variable `M` contains the sampled data that is two signals, one in each column of the array.

The sampling frequency is stored in a separate MATLAB variable, `sfreq`. For the given sample, the sampling rate was 360. If the subject's heart rate was 60 bpm, at the given sample rate, 4 QRS complexes would be captured in 1440 samples.

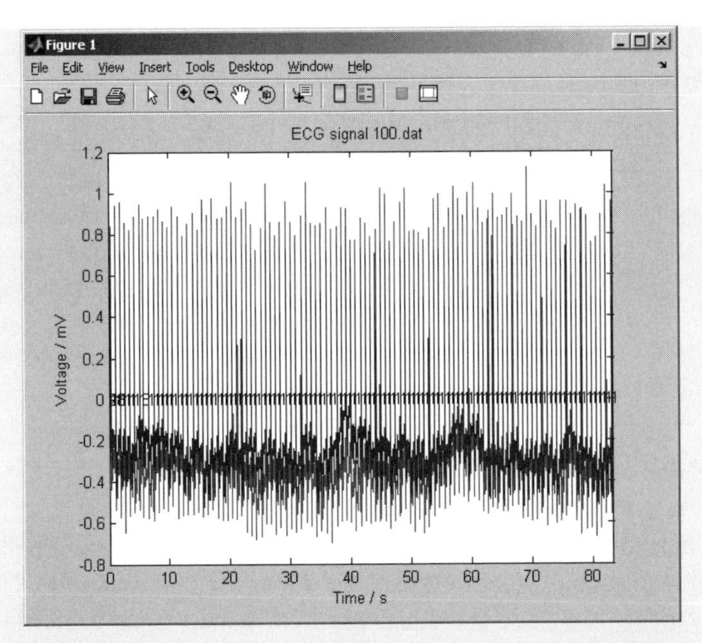

Figure 10.5 ECG sample data file from the MIT-BIH database.

Displaying the first 1440 samples of the first channel, as shown in Fig. 10.6,

```
plot(M(1:1440,1))
```

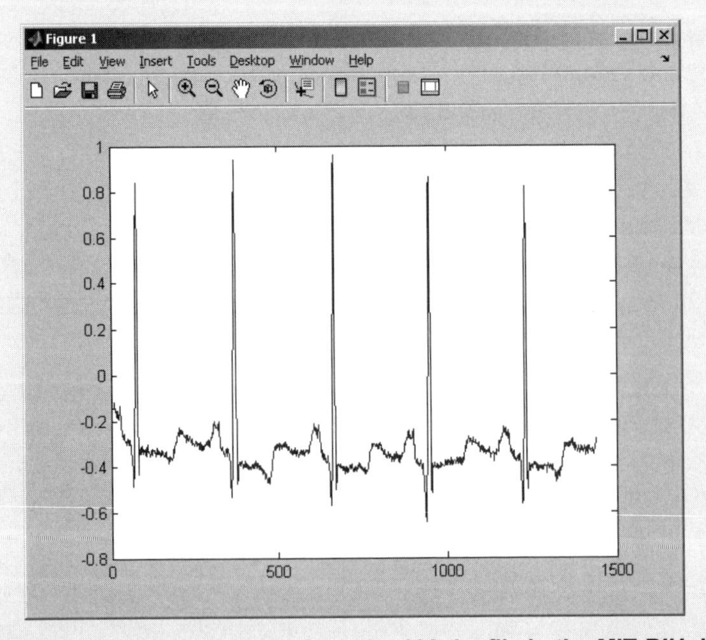

Figure 10.6 The first 1440 samples from the 100.dat file in the MIT-BIH database.

shows that the real heart rate is greater than 60 bpm. Now that the data is in a MATLAB variable, the true heart rate can be easily computed by finding the R waves (the peaks) and computing the R-R interval.

There are two QRS complexes in the first 512 samples: one in the first 256 samples, and the second somewhere between sample 256 and sample 512 (see the graph above). The R-R interval is computed by finding the maxima and their indices.

```
[p1,i1]=max(M(1:256,1));
[p2,i2]=max(M(256:512,1));
```

The amplitude and location of the peaks can be verified against the graph above. The R-R interval is:

```
RR=(i2+256)-i1;
```

Heart rate is the inverse of the R-R interval. Since the data is sampled at 360 samples per second, the heart rate can be estimated by

```
360*60/RR
ans =
   73.4694
```

This simple example illustrates how to read the raw PhysioBank data. Many of the processing and analysis techniques shown through the book can now be applied to determine properties of this and other physiologic signals.

10.3 Signal Processing: EEG Data

EEG signals and their analysis are tools for (1) understanding the dynamic processes in the brain that are the bases of physical and mental behavior and (2) localizing the source of the brain activity associated with specific tasks or behaviors. The purpose of this example is to determine whether or not there is differential brain activity between the right and left hemispheres during certain cognitive tasks.

The standard convention for recording EEG data is called the 10-20 system. The name 10-20 refers to the percentage of arc length from nasion to inion through the vertex, as shown in Figure 10.7.

The signals recorded from A1 and A2 are reference signals that record a signal called the electro-oculogram (EOG): muscle artifact in the EEG signals that is due to eye blinking.

The data for this project, from Keirn (1988), is publicly available from Professor Charles Anderson in the Department of Computer Science at Colorado State University. In addition to the EOG, signals were recorded at O1, O2, P3, P4, and C3, C4.

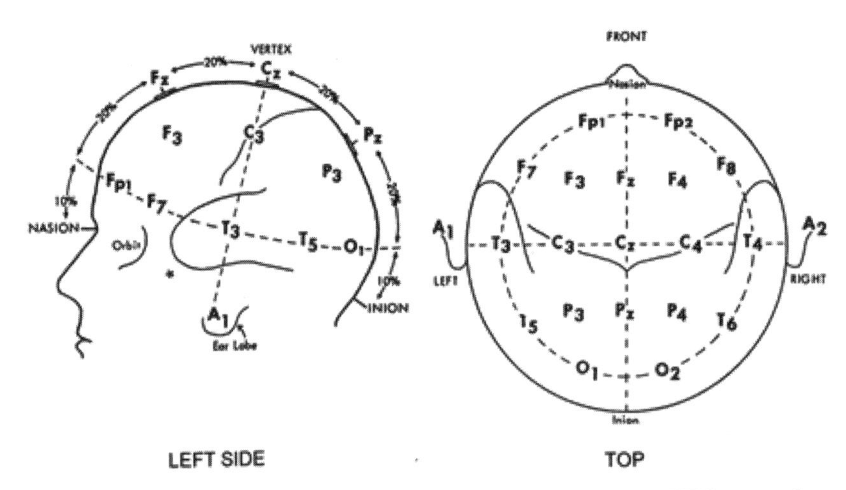

Figure 10.7 The International 10-20 system for lead placement in EEG recordings, from (Jasper 1958).

The subjects in this study underwent the following five tasks:

1. In the **resting** or **baseline** task, the subjects opened and closed their eyes and were asked to relax as much as possible. In a resting phase, alpha waves are produced and any left-right hemispheric asymmetries can be determined so that they can be subtracted from the measurements during cognitive tasks.
2. In the **arithmetic** task, subjects were asked to solve a complex multiplication problem without speaking or other muscle movements.
3. In the **geometric** task, subjects were given 30 seconds to study a drawing of a 3D block figure. After the figure was removed, the subjects were instructed to visualize the object rotating around an axis.
4. In the **letter composition** task, the subjects were instructed to mentally compose a letter to a friend or relative without speaking. In the given study, the tasks were repeated several times; during the repetitions of the letter composition task, the subjects were instructed to pick up where they left off in the previous trial.
5. In the **visual counting** task, subjects were asked to visualize numbers being written on a blackboard. The numbers were written sequentially, and each one was erased before the next was written. As with the other tasks, the subjects were asked not to speak, but simply to visualize the numbers. In the repetitions of the visual counting task, the subjects were told to pick up counting where they left off in the previous trial.

These five cognitive tests were administered to seven subjects as summarized on Table 10.2.

Table 10.2 Demographic information on subjects tested in Keirn (1988).

Subject	Age	Handedness	Gender	Sessions
1	48	Left	Male	2
2	39	Right	Male	1
3	< 30	Right	Male	2
4	< 30	Right	Male	2
5	< 30	Right	Female	3
6	< 30	Right	Male	2
7	< 30	Right	Male	1

To the greatest extent possible, the tests were performed without vocalization or physical movement. Each task was repeated five times, in each session. From the chart above, there were a total of 13 sessions; thus, there are

13 (sessions) x 5 (tasks) x 5 (trials per task) = 325 trials

in the complete dataset.

The complete dataset is available as a MATLAB variable that is organized as a cell array. Recall from Chapter 2 that a cell array is a data structure in MATLAB used to organize data that is related but may be of different types (such as floating-point numbers, integers, and character string). The data set includes a set of trials of EEG measurement and some annotation. Each trial has four elements in the cell:

- Subject number
- Type of the cognitive task
- Trial number
- The digitized EEG data

Each digitized EEG sample is 10 seconds long and was sampled at 250 Hz. Data was recorded from six channels: C3, C4, O1, O2, P3 and P4. The electro-oculogram (EOG) was measured at A1 and A2. Each trial has

250 samples/second x 10 seconds x 6 channels = 15,000 data

The sampled data has been bandpass filtered for 0.1 to 100 Hz. The frequency components of EEG data are:

- *delta* waves, from 0 to 2 Hz

- *theta* waves , from 2 to 7 Hz
- *alpha* waves, from 7 to 13 Hz
- *low beta* waves, from 14 to 20 Hz
- *high beta* waves, above 20 Hz, typically up to 64 Hz

The problem is to analyze the data and determine whether there is inter-hemispheric differential activity that shows up in any or all of these cognitive tasks. The potential sites of differential activity for each task:

1. For the baseline/resting task there should be no differential activity.
2. For the verbal tasks, it is possible that there is differential activity in the left hemisphere in the occipital area (O1 or O2).
3. For the mathematical tasks, it is possible that there is differential activity in the left hemisphere, in the parietal area (P3 or P4).

How does one model brain activation? For the analysis, one needs to find a signal representation that is as small as possible, but it should contain the information necessary to differentiate different mental states. The MATLAB program in Example 10.3 was written to analyze this large set of data and determine whether there are differences in activation, using a mathematical (and numerical) model of brain activation as frequency content.

If the frequency content of a particular signal is primarily in the alpha band (7 to 13 Hz), then the subject must be resting and not performing any tasks. If the subject is performing a task then the majority of the power shifts to the beta frequency band. Of course, artifact due to 60 Hz noise has to be removed (as in Example 9.9), and muscle artifact due to eye blinking has to be removed. The 60Hz noise is removed from high beta and the muscle artifact will show up in the delta and theta frequency bands.

Example 10.3 Differential brain activity in the left and right hemispheres.

Given the dataset described above, analyze the frequency content of the EEG signals to determine if there are differences in brain activity, between hemispheres, that can be attributed to cognitive task, frequency band, or any of the demographic characteristics.

```
% constants
TRIALS = 5;
EXPERIMENTS = 5;
SESSIONS = 13;
SR = 250; %sample rate
```

```
% subject/trial map
SUBJ{1} = 1:50;
SUBJ{2} = 51:75;
SUBJ{3} = 76:125;
SUBJ{4} = 126:175;
SUBJ{5} = 176:250;
SUBJ{6} = 251:300;
SUBJ{7} = 301:325;

load eegdata.mat
for i = 1:length(data) data{i}{4} = double(data{i}{4}); end;
datamag = data; % create copy of data

f = SR*(0:2500/2)/2500; % frequency data

for m = 1:length(data)
    kernel=fft(datamag{m}{4}(7,:));
    kernel=kernel/max(kernel);
    datamag{1};
    for (n = 1:6)
        z = abs(fft(data{m}{4}(n,:)) .*(1 - kernel)); % remove eog
        z = z(1:length(f));
        z = bpf(59.5, 60.5, SR, z, 'f', 1); % apply bandstop
filter at 60Hz (see bpf.m)
        datamag{m}{5}(n,:) = abs(z); % power series for trial
    end
end

clear avgdata;
% average over all experiments
for m = 1 : EXPERIMENTS
    trials = [];
   for n = 1:SESSIONS
        trials = [trials (m - 1) * TRIALS + (n - 1) * EXPERIMENTS
* TRIALS + (1:TRIALS)];
   end
   avgdata{m} = averagedata(datamag, trials);

   %by subject
   for i = 1 : length(SUBJ)
       subjdata{i,m} = averagedata(datamag, intersect(trials,
SUBJ{i}));
   end

   %by age
   for i = 1 : 3
       if (i == 1)
           z = [SUBJ{3} SUBJ{4} SUBJ{5} SUBJ{6} SUBJ{7}]; %age < 30
       elseif (i == 2)
           z = SUBJ{2}; % age 30 - 40
```

```
        else
            z = SUBJ{1}; % age 40 - 50
        end
        agedata{i,m} = processavg(averagedata(datamag,
intersect(trials, z)), SR);
    end

    %by handedness
    for i = 1:2
        if (i == 1)
            z = SUBJ{1}; % LH
        else
            z = setdiff(1:EXPERIMENTS*SESSIONS*TRIALS, SUBJ{1}); %
RH, all not in LH
        end
        handdata{i,m} = processavg(averagedata(datamag,
intersect(trials, z)), SR);
    end

    % by gender
    for i = 1:2
        if (i == 1)
            z = SUBJ{5}; % female
        else
            z = setdiff(1:EXPERIMENTS*SESSIONS*TRIALS, SUBJ{5}); %
male (setdiff ==> everyone else)
        end
        genderdata{i,m} = processavg(averagedata(datamag,
intersect(trials, z)), SR);
    end
end
trialdiff = processavg(avgdata, SR);
disp('Done');
end
```

There are several functions, such as bandpass filtering, that are in separate MATLAB scripts. These scripts are not included here, but are available on the companion Web site for this text.

Notice that most of the references here are with respect to the cell array called data. The reader is encouraged to study Chapter 2 and the MATLAB help files for more information on how to use cell arrays.

Most of the signal processing steps involved in the filtering are shown in the program in Example 9.9 and will not be repeated here. The EOG is removed using a technique called a matched filter: the frequency spectra of the EOG signal is computed by the FFT of the A1-A2 difference signal. This frequency content is then subtracted from the frequency content of the other six signals in the recordings for a given trial.

The amount of data is too large to present or analyze here, but an example of the type of analysis is shown in Table 10.3, where positive differences are to the left hemisphere and negative differences to the right. The task numbers are references to the list above.

This data suggests that there is a large activation in the left occipital lobe during the geometry task (task 3) and a significant but less left occipital activation during the other mathematics tasks. Although these differences were expected in the parietal region, there were only six electrodes used in the study and it is possible that the signals measured at O1 and O2 had components from the parietal region.

10.4 Diabetes and Insulin Regulation

Most, but not all, diabetics are required to use insulin to manage their glucose (blood sugar) levels. The insulin, administered either as a tablet or injection, acts as a feedback control system to stabilize the blood sugar or keep it within an acceptable range.

Glucose is typically monitored with a glucose tolerance test (GTT). In a GTT, blood samples are taken from a fasting subject at regular intervals of time, following a single intravenous injection of glucose. The blood samples are then analyzed for glucose and insulin content.

To determine what is "normal," or to be expected, requires comparison of the measured data to a model of normality. Since glucose regulation is a dynamic process, this model will take the form of a set of differential equations that express how the concentration of plasma glucose changes over time. If the observed concentrations are comparable to those predicted by the model, then the results can be interpreted by the clinician as normal.

These models can vary in level of detail from the very simple (called minimal) to the complex. The level of detail changes with the number of compartments or systems: Either glucose alone or glucose and insulin concentrations in the blood can be modeled; and the abdomen, kidneys and pancreas can be modeled alone or as a combined system.

Van Riel (2004) describes a minimal model with three compartments:

- Plasma Glucose, $G(t)$, in units of mg/dL,
- Plasma Insulin, $I(t)$, in units of μU/mL, and
- Interstitial Insulin, $X(t)$ in units min^{-1}

The latter is a parameter of a single compartment that accounts for insulin in the abdomen, kidneys and pancreas. The variable $X(t)$ is not a physiological quantity, but

Table 10.3 **Summary results of hemispheric activation differences in 5 tests on 7 subjects from Keirn (1988). These quantities are expressed in percent differences between the left and right hemispheres.**

	Central (C3, C4)					Parietal (P3, P4)					Occipital (O1, O2)				
Freq Band	δ	θ	α	βl	βh	δ	θ	α	βl	βh	δ	θ	α	βl	βh
Task	**COMPOSITE DIFFERENCES**														
1	-1.66	-0.07	1.64	-2.30	-6.12	-0.64	-0.05	1.24	-1.71	-2.70	-0.40	1.56	4.81	1.72	2.84
2	0.72	0.75	-3.86	-3.60	-0.35	0.97	0.39	-5.59	-7.35	1.67	0.70	1.71	2.09	0.76	4.60
3	-1.62	-1.33	-4.12	-3.44	-1.78	-0.62	-0.99	-1.44	-2.19	0.48	0.75	2.68	6.93	4.89	5.73
4	-0.83	1.13	-2.49	-2.64	-0.38	-0.60	0.54	-1.42	-4.87	0.32	1.89	2.09	3.33	1.27	3.74
5	0.06	0.16	-1.84	-2.66	-2.36	-0.14	0.65	-1.32	-3.72	-0.62	0.46	1.95	4.24	2.17	0.93

is used to model insulin activity. The pair of differential equations that govern this system are

$$\begin{aligned} \frac{dG(t)}{dt} &= k_1(G_b - G(t)) - X(t)G(t) \\ \frac{dX(t)}{dt} &= k_2(I(t) - I_b) - k_3X(t) \end{aligned} \tag{10.1}$$

where $G(t_0) = G_0$, the initial concentration of plasma glucose, and $X(t_0) = 0$, that is, there is no interstitial insulin. The term G_b represents the basal level of glucose in the blood; if the glucose concentration is less than the basal level, glucose enters the blood and if the concentration rises above this level then glucose leaves the plasma compartment. The basal levels are typically measured before the administration of the glucose in a GTT.

The second equation expresses that insulin enters, or leaves, the interstitial tissues at rates k_2 or k_3, respectively. Insulin enters the interstitial tissues if the plasma insulin concentration rises above the basal level I_b, and insulin leaves the interstitial tissues if the plasma level falls below the basal level.

This system of equations can be easily solved in either MATLAB or Simulink.

Example 10.4 Simulink model of glucose regulation.

Use Simulink to solve the system of equations (10.1) for glucose as a function of time. Use the constants k_1, k_2 and k_3 and insulin profile $I(t)$ from Van Riel (2004).

To use Eqs. (10.1), four unknowns must be determined before using MATLAB or Simulink to solve for $G(t)$. Van Riel (2004) rewrites the second of Eqs. (10.1) as

$$\frac{dX(t)}{dt} = k_3(S_I(I(t) - I_b) - X(t))$$

where $S_I = k_2/k_3$ is called the insulin sensitivity.

Van Riel solves this problem using the Runge-Kutta solver `ode45` in a MATLAB script. While there is flexibility in using the script to vary the parameters or the insulin profile, a set of MATLAB scripts and a Simulink implementation make for easier documentation of the different model systems. The MATLAB script for normal glucose kinetics is

```
% Set the workspace variables for the
% Simulink model VanRiel.mdl
% Basal Glucose and Insulin
```

```
Gb=92;
Ib=11;
G0=279;
% Model constants
SI=5e-4;
k3=0.025;
k1=2.6e-2;
```

where the constants and insulin profile are taken from (Van Riel 2004). Notice that the constants G_b and I_b are MATLAB variables and the initial value of the integrator labeled dG/dt is the MATLAB variable `G0`. The gain boxes serve as multipliers by the constants k_1, S_I and k_3, respectively, rather than using a multiplier box and a constant for each. The Simulink model is shown in Fig. 10.8.

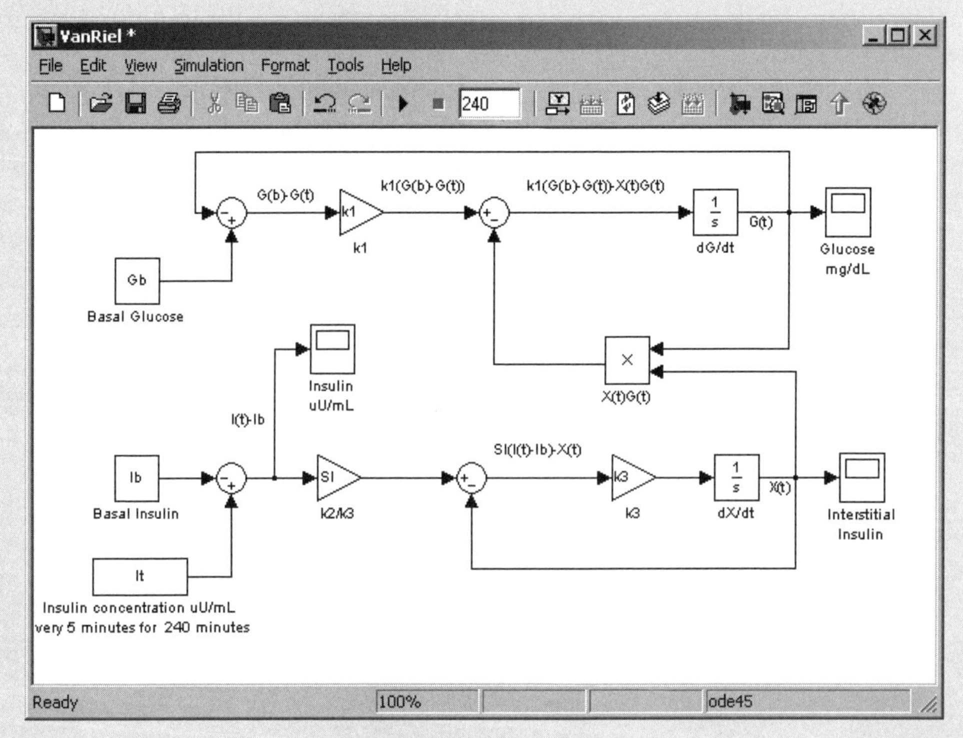

Figure 10.8 Simulink model of glucose regulation.

The key to making this simulation work is the "From Workspace" block, which is in source of the insulin profile $I(t)$. The source for a "From Workspace" block is a 2D array, with the time points in the first column and the values in the second. For this problem, the array is

```
%
% Time course of insulin I(t)
%
t_insu=[0 5 10 15 20 25 30 40 60 80 100 120 140 160 180 240];
u=Ib+[0 100 100 100 100 0 0 0 0 0 0 0 0 0 0 0 ];
It=[t_insu' u'];
```

Notice that the first seven samples are every five minutes and they are spaced further apart until 240 minutes. Since the time step of the simulator must be a constant, it is necessary to indicate to Simulink that it must interpolate the missing values. Also, in the event that the input is shorter than the simulation, the "From Workspace" block provides the ability to specify what to do for the missing data. These source sampling parameters are specified in the block parameters, shown in Fig. 10.9.

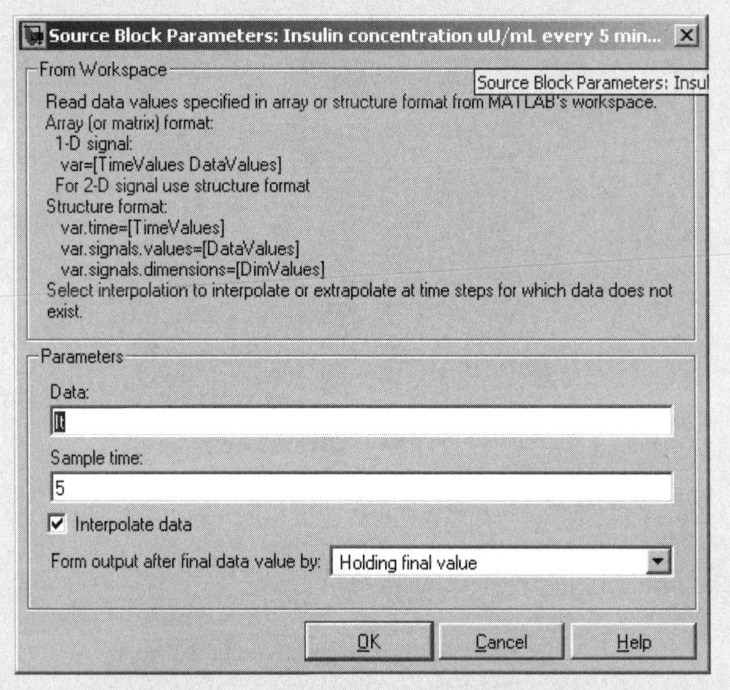

Figure 10.9 Parameters for the source block labelled insulin concentration in Fig. 10.8.

where the data is the MATLAB variable `It`, the sample time is 5 (meaning 5 minutes), the missing data is interpolated, and the final value is held as the output after the final data element.

Lastly, change the simulation parameters to start at 0.0 and end at 240.0 (minutes). The simulation output is shown in the three scopes (Figs. 10.10, 10.11, 10.12).

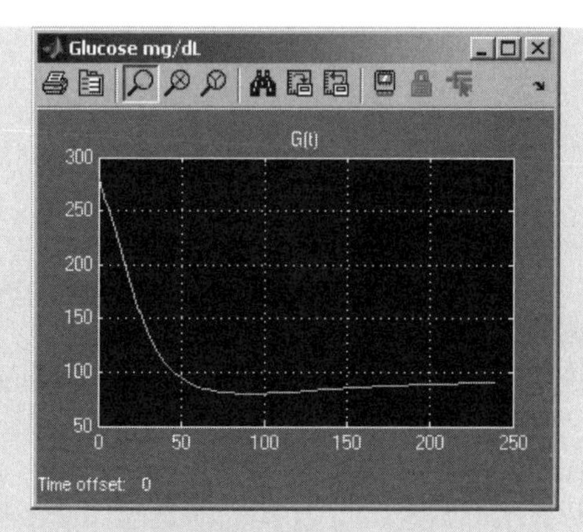

Figure 10.10 Glucose.

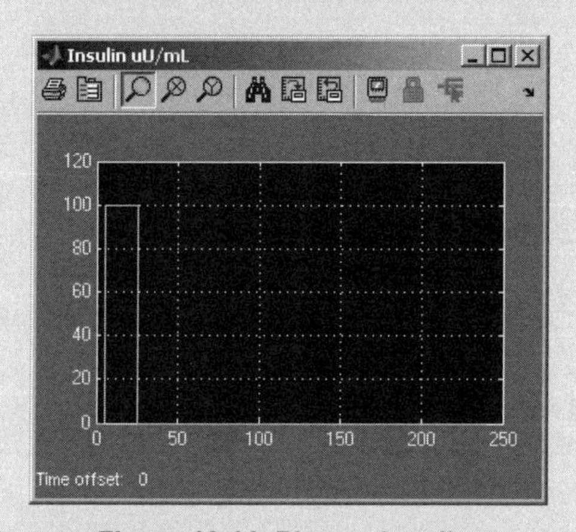

Figure 10.11 Plasma insulin.

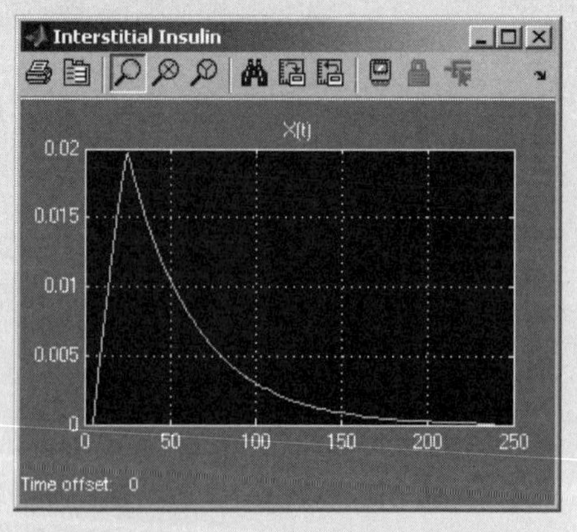

Figure 10.12 Interstitial insulin.

In a GTT, the samples measured are compared to a glucose profile that is similar to the simulation results above. Notice that the glucose concentration dips below the basal level at approximately one hour after the glucose intake and subsequent insulin release. For this simulation of normal conditions, the minimum glucose concentration is 80 mg/dL.

In normal healthy adults, cognitive function is impaired when plasma glucose falls below 65 mg/dL. If the plasma glucose dips below this level and the patient has altered mood and/or impaired cognitive function, a diagnosis of reactive hypoglycemia is often made. If the glucose level is normal, but the patient exhibits these signs, then the condition is referred to as *postprandial syndrome*.

The same model can also be used to simulate glucose regulation in a diabetic. The MATLAB script for the model parameters is

```
% Set the workspace variables for the
% Simulink model VanRiel.mdl
% Basal Glucose and Insulin
Gb=92;
Ib=11;
G0=365;
% Model constants
SI=0.7e-4;
k3=0.01;
k1=1.7e-2;
```

and if the simulation is run for 240 minutes with the same insulin profile, the glucose concentration is shown in Fig. 10.13.

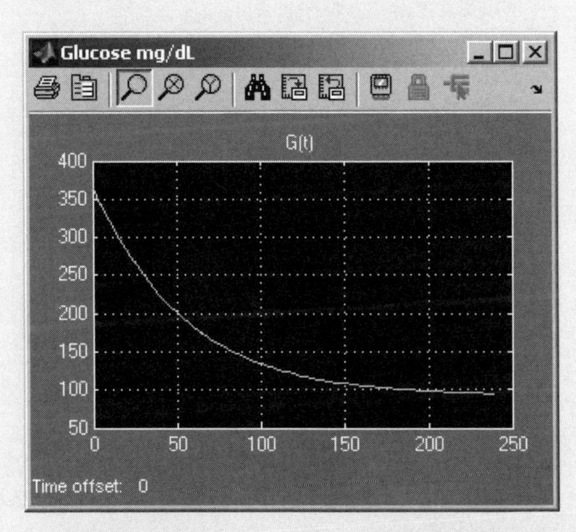

Figure 10.13 Glucose concentration.

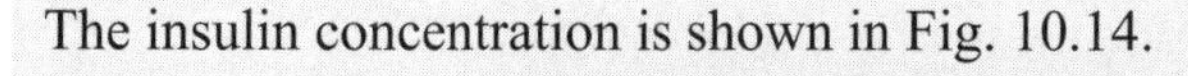

The insulin concentration is shown in Fig. 10.14.

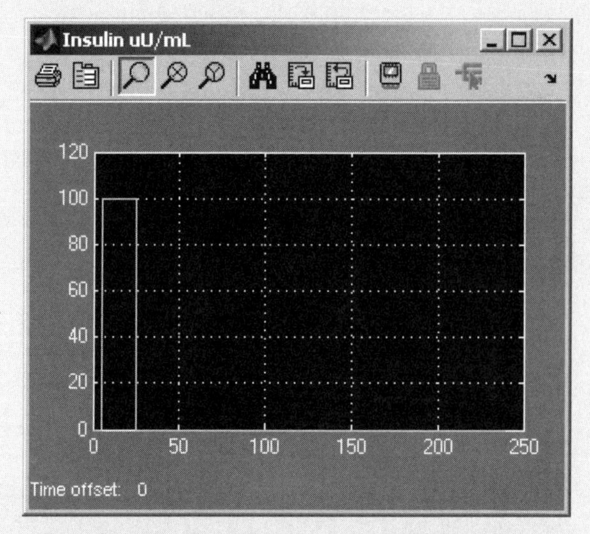

Figure 10.14 Insulin concentration.

The interstitial insulin concentration is shown in Fig. 10.15.

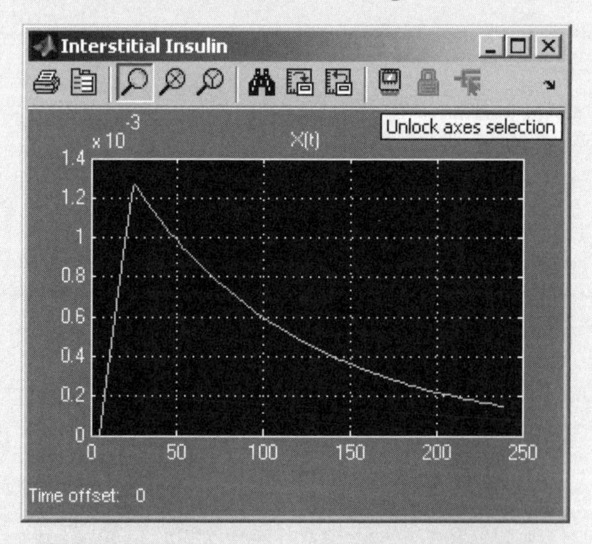

Figure 10.15 Interstitial insulin concentration.

The NIH-NIDDK criterion for diabetes is a glucose concentration of 200 mg/dL or more, 2 hours after the glucose administration. This simulation shows a profile that would be diagnosed as pre-diabetes, or impaired glucose tolerance.

10.5 Renal Clearance

Renal clearance (Cl_R) is a measure of kidney transport in units of volume of plasma per unit time. The volume of plasma measured is that volume for which a given substance (e.g., urea or drugs) is completely removed per minute. There are a number of specific forms of renal clearance: Glomerular Filtration Rate, Effective Renal Plasma Flow, or Tubular Extraction Rate.

There are a number of techniques for evaluating renal kinetics, some of which are based on measuring radionuclide concentration over time from urine and plasma; plasma alone; or using a gamma camera to measure isotope concentration in the kidneys as the isotope is extracted from the plasma. Renal function is assessed by comparing clinical measurements to simulation results based on a model of normal kinetics. The last two of the three cases are easiest to model with only two compartments, central and peripheral.

One such two-compartment model was described by Estelberger and Popper (2002). In that model, there are two functionally separated spaces, a well-perfused central volume and a less-perfused peripheral compartment. The marker kinetics, measured indirectly from the time course of radionuclide concentration in the two compartments, is the result of the isotope injection, the transport between the two compartments, and the renal elimination process. Fig. 10.16 is a diagram of the Estelberger/Popper model.

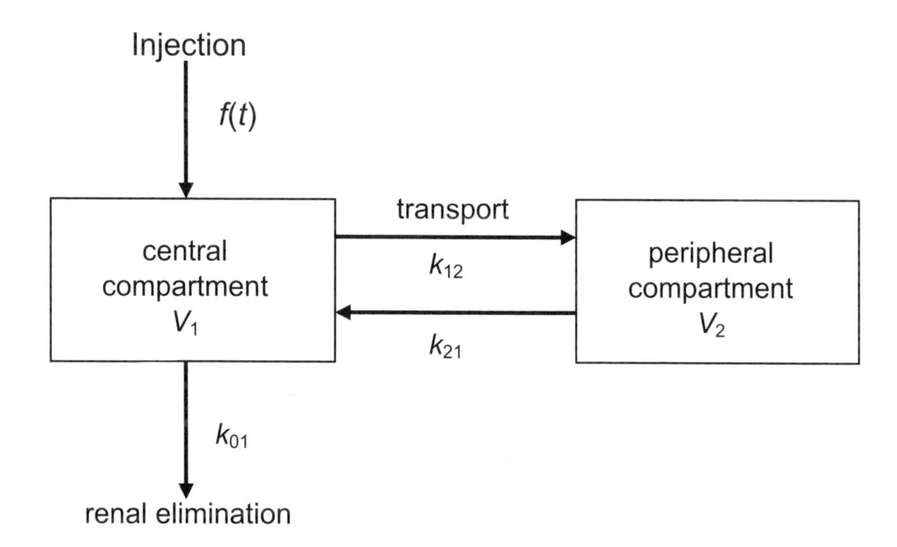

Figure 10.16 A two compartment model of renal clearance, from Estelberger and Popper (2002).

The transport model can be formulated by a set of two simultaneous differential equations describing the rates of change of radionuclide concentration in the two compartments:

$$\begin{aligned} \frac{dc}{dt} &= d(t) - (k_{01} + k_{21})c + k_{12}p \\ \frac{dp}{dt} &= k_{21}c - k_{12}p \end{aligned} \tag{10.2}$$

where

$$d(t) = \frac{D}{T}, \qquad 0 \le t < T \tag{10.3}$$

that is, the bolus isotope injection is delivered over T seconds. For purposes of this exercise, assume that the isotope is delivered uniformly during this period of time. Assume also that the central volume V_1 is unknown.

The brief paper by Estelberger and Popper shows how the group estimates the parameters of the model from empirical data. The optimization procedure is beyond the scope of this text and rather than detract from the example, the reader is referred to Estelberger and Popper (2002) for details.

In comparison to the simulation given in Section 10.3, this set of differential equations will be solved using MATLAB, rather than Simulink.

Example 10.5 Renal clearance.

Solve the system of two differential equations (10.2) and the initial value condition (10.3) for renal clearance using the constants $k_{01} = 0.0041$, $k_{12} = 0.0585$, $k_{21} = 0.0498$, and $V_1 = 7.3$, $c(0) = p(0) = 0$.

This problem is designed to illustrate the use of global variables in MATLAB. The mathematical problem is formulated as a conservation of mass with the isotope injection beginning at time zero; the length of the bolus injection varies with trial.

The parameters of the model and the parameters of the bolus injection are set as global variables, which can then be used by the script that generates the graph and the script that evaluates the system of differential equations, called `renal.m`.

First, the system of equations (10.2) is coded as a MATLAB function per the MATLAB convention, with arguments time t and the vector of function values, y. All other model parameters are global variables.

```
function yprime=renal(t,y)
% separate the components of the state:
% central compartment concentration and
% peripheral compartment concentration
c=y(1);
p=y(2);
%
global D tau;
global k01 k12 k21 V1;
% let time t be in minutes
% now compute f(t)
%
if t <= tau
    f=D/tau;
else f=0;
end
%
dcdt=(f-(k01+k21)*c+k12*p)/V1;
dpdt=k21*c-k12*p;
yprime=[dcdt, dpdt]';
```

There is a single script that solves the system of equations three times—once for each of the different injections of isotopes. A single figure with the three concentrations in the central compartment is plotted with different symbols. The succeeding plots are added to the same figure by using the `hold on` command.

```
% Example 10.5 Renal Clearance
global D tau;
global k01 k12 k21 V1;
% parameters
k01=0.0041;
k12=0.0585;
k21=0.0498;
V1=7.3;
% initial value conditions
c0=0;
p0=0;
%
D=2500;
tau=0.5;
% use ode45
[t,y]=ode45(@renal,[0:0.5:240],[c0 p0]);
plot(t,y(:,1),'b.');
xlabel('Time t (min)')
ylabel('Isotope concentration c(t)')
hold on
%
D=2500;
tau=10.0;
```

```
% use ode45
[t,y]=ode45(@renal,[0:0.5:240],[c0 p0]);
plot(t,y(:,1),'g:');
%
D=2500;
tau=240;
% use ode45
[t,y]=ode45(@renal,[0:0.5:240],[c0 p0]);
plot(t,y(:,1),'r--');
hold off
```

In each case, the solution is run for 240 minutes in time steps of 0.5 minutes. The set of solutions is shown in the graph in Fig. 10.17. Notice that the first two solutions have very similar trajectories; the traces of isotope concentration are the impulse response of the system to the bolus radionuclide input, whereas the solution to the ramp input (the dashed line) does not exhibit the feedback control at 240 minutes.

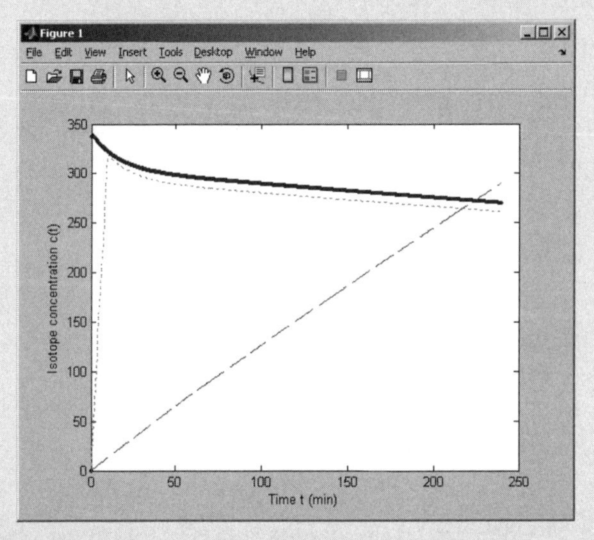

Figure 10.17 Solutions to renal clearance program. Solid line: bolus injection over 0.5 minutes; dotted line: injection over 10 minutes; dashed line: injection over 240 minutes.

10.6 Correspondence Problems and Motion Estimation

A key concept in estimating motion from image sequences is that the relative motion between a pair of images can be estimated from the correspondence of a number of markers or features on the object(s). The features must be uniquely identifiable, and it is often assumed that the object is rigid or articulated.

Gait, the analysis of human body motion, is estimated by recording the location of features on the hips and legs. These features can be, for example, light emitting diodes (LEDs), and the motion is recorded by having the subject walk in a dark room. Many rehabilitation facilities have gait laboratories where such recordings take place, and the motion recorded can be used for diagnostic purposes.

This example, courtesy of Professor Charles Krousgrill at Purdue University, illustrates how rotation matrices can be estimated from the correspondence of markers on the hips and legs (see Fig. 10.18).

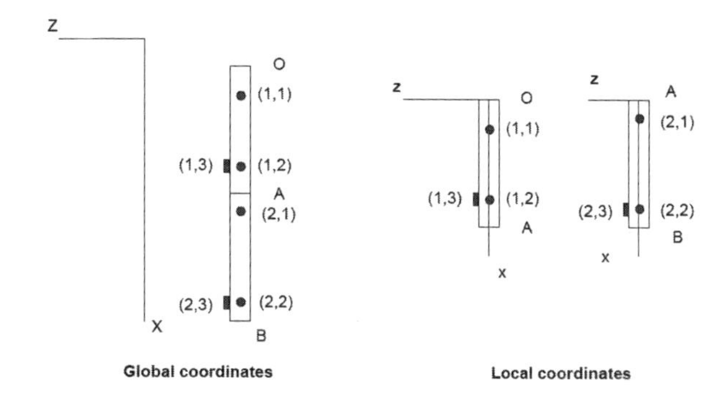

Figure 10.18 Coordinate systems for gait markers on the hip and legs, from Krousgrill (2005).

Example 10.6 Estimating motion from features on a rigid body.

Six markers have been attached to the hip, thigh, and lower leg, as shown in Fig. 10.18. The coordinates for these markers are given in Table 10.4.

Table 10.4 Local coordinates for gait markers on the hip and legs, from Krousgrill (2005).

Marker	x-coordinates	y-coordinates	z-coordinates
(1,1)	6	4	0
(1,2)	12	4	0
(1,3)	12	0	4
(2,1)	6	3	0
(2,2)	12	3	0
(2,3)	12	0	4

Position data have been collected for these markers and the hip from one frame of motion. These coordinates are given in terms of global coordinates in 3D in Table 10.5.

Table 10.5 Global coordinates of gait markers on the hip and legs after rigid body motion, from Krousgrill (2003).

Marker	x-coordinates	y-coordinates	z-coordinates
Hip (O)	2	1	0
(1,1)	3.18	7.13	3.60
(1,2)	7.16	10.47	6.60
(1,3)	8.89	5.88	9.41
(2,1)	9.11	15.45	10.44
(2,2)	6.99	20.66	12.52
(1,1)	9.36	19.88	16.86

Assuming a segment (thigh or lower leg) length of 18.13 inches, use the above data to determine the rotation matrices for the thigh and lower leg. Determine the 3D global coordinates for the position of the ankle B.

The position of the markers on the hip and legs are specified in *local* coordinates, 3D position with respect to one of the markers. For this exercise, the hip is the reference marker for the markers on the thigh, and the knee (position A) is the reference for the lower leg, all shown in Table 10.5. The local coordinates are given in Table 10.4.

The motion cannot be recorded in a local coordinate system, only in the *global* coordinate system of the gait laboratory. The cameras record images of the subject while walking, and the position of the markers in these images can be used to determine the motion *with respect to the position of the camera(s) in the laboratory, not with respect to the hip or leg*. The problem of determining the rotations of the thigh and lower leg then means switching between the local and global coordinates.

The general strategy to solve this problem is to (1) convert the recorded global coordinates to local coordinates for each segment, and (2) determine the rotation from the correspondence of the markers when referenced to the initial or resting position. The rotation of the thigh must be computed first, since it is required when solving for the rotation of the lower leg.

Bookkeeping is always a challenge when solving for motion. There are three sets of coordinates to keep track of: the local coordinates of each segment before motion (at rest), the global coordinates of each marker after motion, and the local coordinates of each segment, which are computed.

The first step is to initialize some MATLAB variables with the data from Tables 10.4 and 10.5.

```
% Example 10.6
% At rest (in local coordinates)
% Each column vector is the 3D coordinates
% of a marker (initial conditions)
ThighRestLocal=[ 6 4 0; 12 4 0; 12 0 4]';
LegRestLocal=[6 3 0; 12 3 0; 12 0 4 ]';
ThighFinalGlobal=[3.18 7.13 3.60; 7.16 10.47 6.6; 8.89 5.88
9.41]';
LegFinalGlobal=[9.11 15.45 10.44; 6.99 20.66 12.52; 9.36 19.88
16.86]';
% Column vectors of positions of O, A and B
HipPos=[2 1 0]';
ThighLen=[18.13 0 0]';
LegLen=[18.13 0 0]';
```

The final position, in global coordinates of the thigh after motion, is given by the equation

$$\mathbf{X}_{\text{thigh, motion}} = \mathbf{R}_{\text{thigh}} \cdot \mathbf{X}_{\text{thigh, initial}} + \mathbf{O} \tag{10.4}$$

where $\mathbf{X}_{\text{initial}}$ and $\mathbf{X}_{\text{motion}}$ are matrices whose column vectors are the positions of the markers, $\mathbf{R}_{\text{thigh}}$ is the rotation about the hip, and $\mathbf{O}$ is the 3D position of the hip after motion. $\mathbf{R}_{\text{thigh}}$ can be computed from

$$\mathbf{R}_{\text{thigh}} = (\mathbf{X}_{\text{thigh, motion}} - \mathbf{O}) \cdot (\mathbf{X}_{\text{thigh, initial}})^{-1} \tag{10.5}$$

The MATLAB implementation is

```
% Find rotation of thigh about hip
% First, determine coordinates of thigh markers
% with respect to the hip
ThighFinalLocal=ThighFinalGlobal-[HipPos HipPos HipPos];
% Rotation computed by correspondence matching
% and is in the local coordinate system centered at the Hip
RotThigh=ThighFinalLocal*inv(ThighRestLocal);
```

Computing the rotation of the lower leg is a little more challenging. The lower leg is rotated about the knee, which rotates about the hip, so there are two rotations to the motion:

$$\mathbf{X}_{\text{leg, motion}} = \mathbf{R}_{\text{thigh}} \cdot \mathbf{R}_{\text{leg}} \cdot \mathbf{X}_{\text{leg, initial}} + \mathbf{A} \tag{10.6}$$

where

$$\mathbf{A} = \mathbf{O} + \mathbf{R}_{\text{thigh}} \cdot [18.13 \;\; 0 \;\; 0]^{\text{T}} \tag{10.7}$$

After substitution,

$$\mathbf{R}_{\text{leg}} = \mathbf{R}_{\text{thigh}} \cdot (\mathbf{X}_{\text{leg, motion}} - \mathbf{A}) \cdot (\mathbf{X}_{\text{leg, initial}})^{-1} \tag{10.8}$$

For which the MATLAB implementation is

```
% Find final position of knee
KneePos=ThighLen;
FinalKneePos=(RotThigh*KneePos)+HipPos;
% Find rotation of leg about knee
% First determine local coordinates of leg markers
% with respect to the knee
% The rotation is in the local coordinate system of the Knee
LegFinalLocal=LegFinalGlobal-[FinalKneePos FinalKneePos
FinalKneePos];
CombinedRotLeg=LegFinalLocal*inv(LegRestLocal);
RotLeg=inv(RotThigh)*CombinedRotLeg;
```

Lastly, the position of the ankle has to be computed, since there is no marker at the ankle. The global position after motion is given by rotating the end of the lower leg (position B in Figure 10.3) and translating it with respect to the position of the knee:

$$\mathbf{B} = \mathbf{R}_{\text{thigh}} \cdot \mathbf{R}_{\text{leg}} \cdot [18.13 \;\; 0 \;\; 0]^{\text{T}} + \mathbf{A} \tag{10.9}$$

which is, in MATLAB,

```
% Coordinates of the ankle are
% Ankle=Knee+Rot*Ankle at Rest
% Ankle is in the local coordinates of the
% Knee, which is the center of rotation
AnklePos=LegLen;
FinalAnklePos=FinalKneePos+CombinedRotLeg*AnklePos;
```

The rotation matrices are

```
>> RotThigh
RotThigh =
     0.6633   -0.7000   -0.2675
     0.5567    0.6975   -0.4500
     0.5000    0.1500    0.8525
```

```
>> RotLeg
RotLeg =
    0.4233   -0.8933    0.1581
    0.9063    0.4183   -0.0749
   -0.0011    0.1747    0.9864
```

and the position of the ankle is

```
>> FinalAnklePos
FinalAnklePos =
    7.6203
   26.8353
   15.3501
```

Rotation matrices should be orthogonal, since the transformation does not change the magnitude or length of the thigh or lower leg. A matrix $\mathbf{R}$ is orthogonal if $\mathbf{R}^T\mathbf{R}$ is the identity matrix $\mathbf{I}$

```
>> RotThigh'*RotThigh
ans =
    0.9999   -0.0011   -0.0017
   -0.0011    0.9990    0.0013
   -0.0017    0.0013    1.0008

>> RotLeg'*RotLeg
ans =
    1.0007    0.0007   -0.0020
    0.0007    1.0035   -0.0003
   -0.0020   -0.0003    1.0036
```

which are both the identity matrix $\mathbf{I}$ to within numerical error. Check the condition number of each matrix, and see Chapter 4 for how error is propagated when computing an inverse.

10.7 PHYSBE Simulations

PHYSBE is a model of the circulatory system for simulating the flow of oxygen, nutrients, heat, or chemical tracers within the bloodstream (McCleod, 1966, 1968). Although the fundamental work on PHYSBE dates from the 1960s, PHYSBE has since been implemented in Simulink, making it accessible for educational and research uses. The Simulink implementation of PHYSBE can be downloaded from the MathWorks MATLAB Central Web site; details about installation, and an introduction to using the Simulink version of PHYSBE, are in Appendix B.4.3. This section includes several examples of how to use PHYSBE to model cardiovascular system pathologies and predict the result on blood flow.

Example 10.7 Normal PHYSBE operation.

Use the procedure in Appendix B.4.3 to install and run one simulation of PHYSBE.

Start the PHYSBE simulation by running the MATLAB script `pctrl2` to open the PHYSBE control panel (Fig. 10.19). Do not change any of the default values, but be sure to save the parameter file.

PHYSBE Control Center

File Edit View Insert Tools Desktop Window Help

Save | **PHYSBE Control Center** | Reset

	Ri	Ro	C	W	A	V0	h0	fe	fd
R.Heart	0.0128	0.0111	75	600	0	150	0	0	0
Lung	0.1429	0	7.519	1000	0	120	0	0	0
L.Heart	0.0588	0.0125	80	600	0	150	0	0	0
Aorta	0	0	1.25	0	0	100	0	0	0
Arms	5.15	10	4.25	7000	3670	280	0	0	0
Head	2.58	5	1.21	4500	1400	80	0	0	0
Trunk	0.67	1.42	34	53000	6000	2250	0	0	0
Legs	2.58	5	11.1	18500	7000	730	0	0	0
V.Cava	0	0	250	0	0	500	0	0	0
Lung2*	0	0	30.3	0	0	240	0	0	0

Figure 10.19 PHYSBE control panel.

Next, start the PHYSBE Simulink model (Fig. 10.20) and start the simulation.

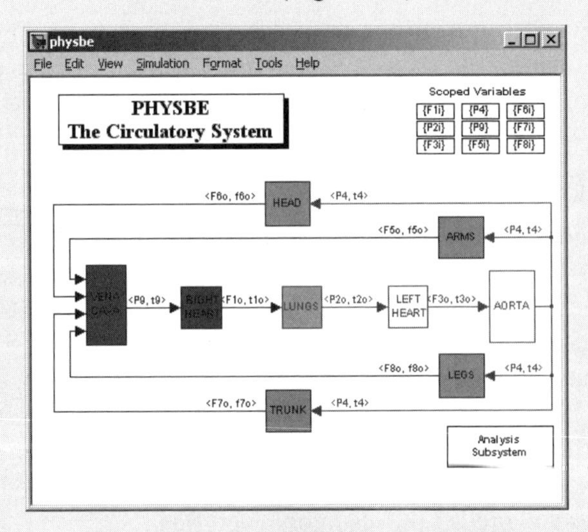

Figure 10.20 PHYSBE Simulink model.

The simulation results which are displayed on a Simulink scope are the heart pressure (Fig. 10.21) and heart volume (Fig. 10.22).

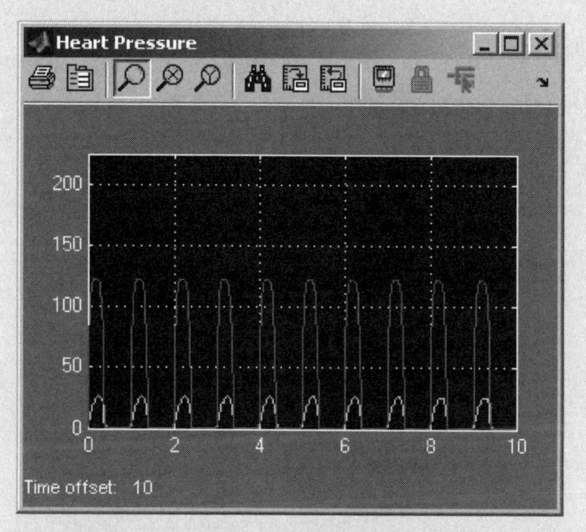

Figure 10.21 PHYSBE Simulink model: heart pressure.

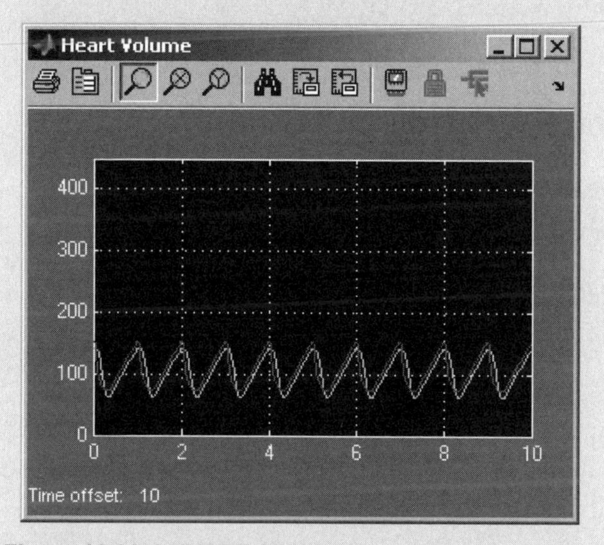

Figure 10.22 PHYSBE Simulink model: heart volume.

Both are easiest to differentiate by color. There is also a floating scope, which can be tied to other system parameters of interest.

These normal results will be compared to the pressure and volume relationships in each of the examples below.

10.7.1 Coarctation of the aorta

The affliction characterized by a short constriction of the aorta is known as a *coarctation*. Such narrowing creates a pressure difference whereby abnormally high blood pressure exists before the point of coarctation, and abnormally low blood pressure exists after the point of coarctation. Commonly, this narrowing is situated such that there is a pressure differential between the upper body and arms (high blood pressure) and the lower body and legs(low blood pressure) (Suk, 2001). Fig. 10.23 shows where a coarctation occurs.

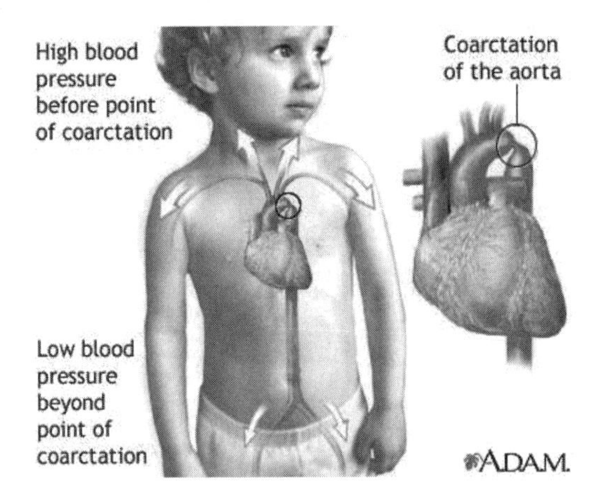

Figure 10.23 Coarctation of the aorta, from Suk (2001).

The untreated consequences of coarctation of the aorta are numerous and severe. Without repair, this condition results in high mortality from complications such as myocardial infarction, intracranial hemorrhage, infective endocarditis, coronary artery disease, and aortic rupture. There is a stunning mortality rate of 90% before the age of 50 when this condition is left untreated. Optimistically, the dire consequences of this ailment can be almost completely eliminated with modern medical techniques.

Coarctation is a sort-of "pinch" in the aorta located between significant arterial branchings (Sokolow and McIlroy, 1981). This pinch translates into a significantly reduced aortic radius. Because blood flow rate immediately before, within, and after the pinch must remain constant, velocity must increase (since $A = \pi r^2$ will decrease) and this translates into a pressure gradient that *reduces* pressure to the lower extremities and trunk. A reduction in pressure to the lower extremities and trunk is implemented in Simulink with the addition of a resistance to the flow entering these regions. The target change in pressure is generated by this resistance according to Ohm's relation $Q = \Delta P/R$ (Li, 2004).

Example 10.8 Simulink model of coarctation of the aorta.

Modify the Simulink implementation of PHYSBE to model a coarctation of the aorta and predict the effect on blood pressure and volume in the heart.

The only direct modifications to the PHYSBE model that need to be made are the addition of a resistance and a pulse delay to the descending aorta.

In the Simulink model, the major modifications were implemented in a subsystem format (for simplicity and ease of use). The pulse delay was implemented via a pulse delay block. The time-of-delay parameter input to this block was 0.08 seconds. Data for the pulse delay was in this range, and cited as anywhere between 0.1 and 0.08 seconds (Sokolow, 1981).

The resistance is implemented by adding a gain block with a value equal to $1/R$, following the pulse delay and in series with the vessels extending to legs and trunk. Because the value of this resistance is dependent upon individual physiology, conditional geometry, and severity, it can vary from case to case.

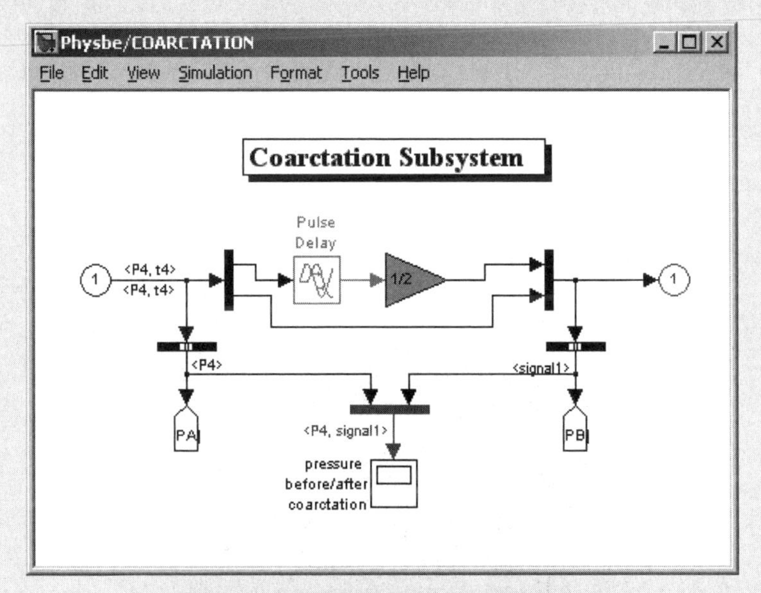

Figure 10.24 Coarctation subsystem.

The coarctation subsystem (Fig. 10.24) is between the ascending aorta and the legs (Fig. 10.25).

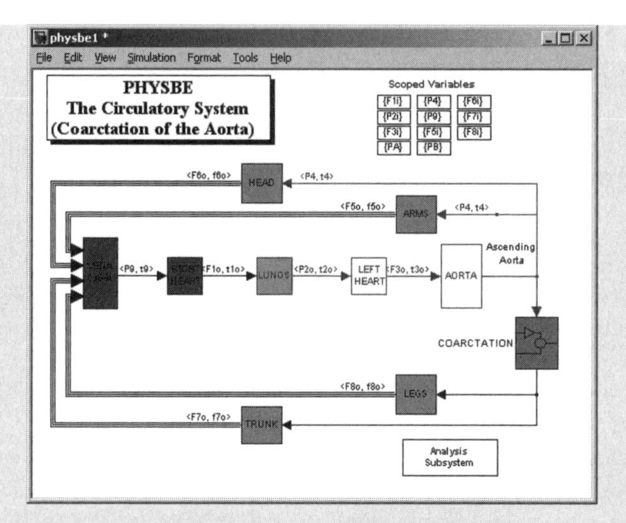

Figure 10.25 Location of coarctation subsystem in circulatory system.

Lastly, the pressures on either side of the coarctation are output to MATLAB objects for plotting and quantification purposes. These new MATLAB objects, called PA and PB (pressure above and pressure below), are visible in the coarctation subsystem in Fig. 10.24. These objects were used by the PHYSBE Control and Analysis subsection and shown as the 10th and 11th columns in the modified PHYSBE Control Center.

The simulation results are as shown in Figs. 10.26, 10.27, and 10.28. First the right and left ventricular pressures, as shown in Fig. 10.26.

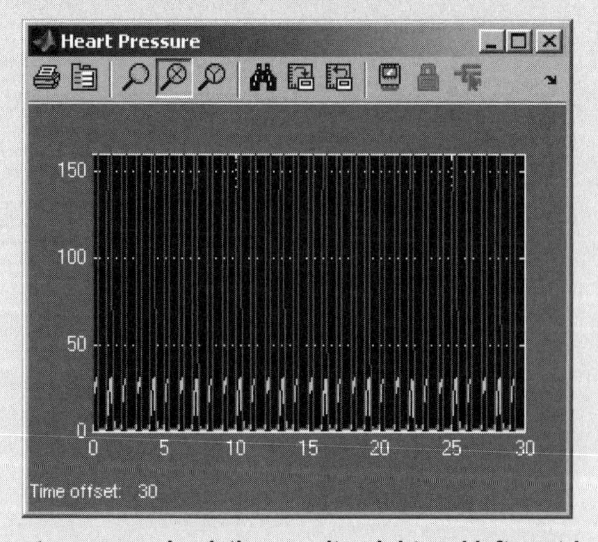

Figure 10.26 Heart pressure simulation results: right and left ventricular pressures.

The blood volumes in the right and left heart:

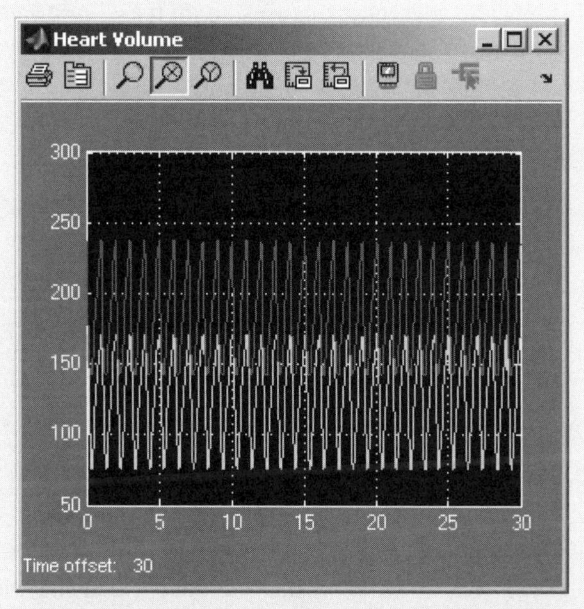

Figure 10.27 Blood volume simulation results, left and right heart.

The blood pressure before and after the coarctation:

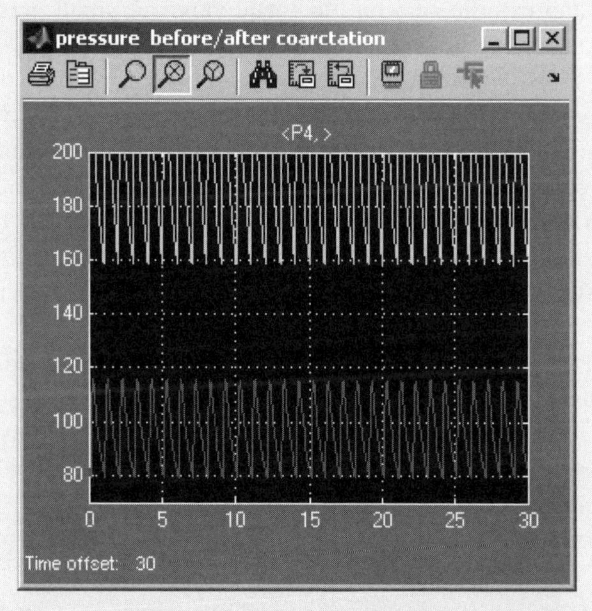

Figure 10.28 Blood pressure simulation results, before and after coarctation.

Notice that there is an increase in left ventricular pressure, while the right ventricular pressure is not affected. The coarctation causes an increase in left ventricular pressure as blood encounters greater resistance. By the time the blood has passed through the capillaries and pooled during venous return, the pressure drop caused by the coarctation is diminished. The same analysis could be traced in reverse over the pulmonary circuit.

These results suggest that a small change in resistance (that is, conductance) affected pressures throughout the body. However, the PHYSBE system is somewhat limited in that a pressure differential of 15 mmHg between the right and left arms suggests the narrowing of the artery (Braunwald, 1988). This incipient coarctation could not be modeled by PHYSBE, as the arms as well as the legs are lumped into one subsystem each.

10.7.2 Aortic stenosis

Aortic stenosis is a condition where deposits on the aortic valve cause it to become narrowed or blocked. As a result, the valve is unable to open properly, and blood flow out of the left ventricle into the aorta is reduced, causing the left ventricle to work harder to compensate and ensure that the body receives the necessary blood supply (Texas Heart Institute, 2004).

There are numerous causes of this condition, including congenital defects, rheumatic fever, and calcification on the aortic valve. A small percentage of people are born with two cusps on their aortic valve instead of three and, as a result of wear and tear over the years, the valve may become calcified or scarred, or its motility may be reduced. Rheumatic fever also damages the cusps of the aortic valve, causing the edges of the cusps to fuse together. As a person ages, the collagen in his/her body, including in the cusps of the aortic valve, is destroyed, and calcium deposits form. The calcium deposits on the valve reduce the cusps' motility and therefore increase the resistance of the blood flow. Symptoms resulting from a stenosis include fainting, shortness of breath, heart palpitations, angina, and coughing (Texas Heart Institute, 2004).

The increased resistance at the aortic valve results in an increase in left ventricular pressure and a decrease in aortic pressure. Severe stenosis results in decreased stroke volume, increased afterload, and increased end systolic volume. If the model of aortic stenosis is accurate, then the simulation results should mimic these pressure changes as shown in Figure 10.29.

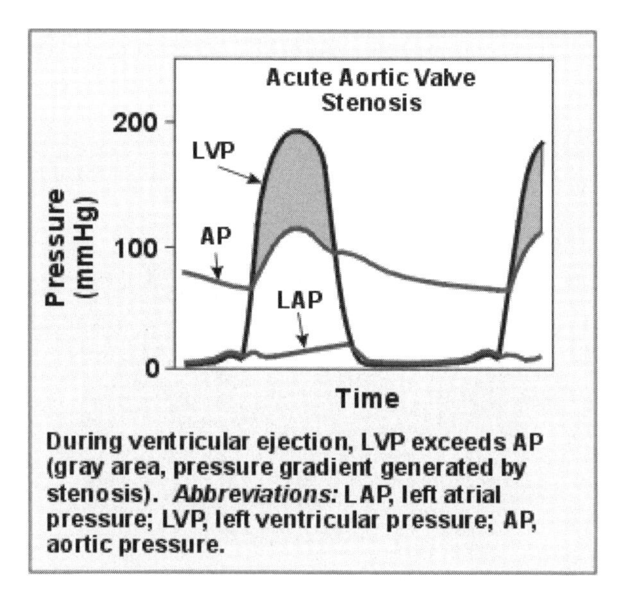

Figure 10.29 Pressure as a function of time during Aortic Valve Stenosis, from Klabunde (2004).

Example 10.9 Simulink model of aortic valve stenosis.

Modify the Simulink implementation of PHYSBE to model an aortic valve stenosis and predict the effect on blood pressure and volume in the heart.

The PHYSBE model does not allow for changing physical parameters such as the radius of the aortic valve, which would be the most accurate model. Therefore, in order to model an aortic valve stenosis using PHYSBE, the resistance (*Ro*) of the left heart was increased. When stenosis occurs and the radius of the valve decreases, the resistance to blood flow increases, since resistance is inversely proportional to radius as given by Poiseuille's Law (Germann, 2005). The physiological effect of a stenosis is a decrease in blood flow due to the increase in resistance (due to the decrease in radius) as modeled by the relationship $F = \Delta P/R$, where F is blood flow, ΔP is the change in pressure, and R is the resistance. This decrease in blood flow was used to show whether or not an aortic valve stenosis was modeled.

The resistance was changed from 0.0125 mmHg/mL/s to 0.135 mmHg/mL/s, which is a physiologically relevant value for an aortic stenosis. This value was obtained by converting the value for severe aortic stenosis, 180 dyne s, which was reported by Mascherbauer (2004).

When there is an aortic stenosis, maximum heart pressure within the left ventricle increases to over 160 mmHg (Fig. 10.30).

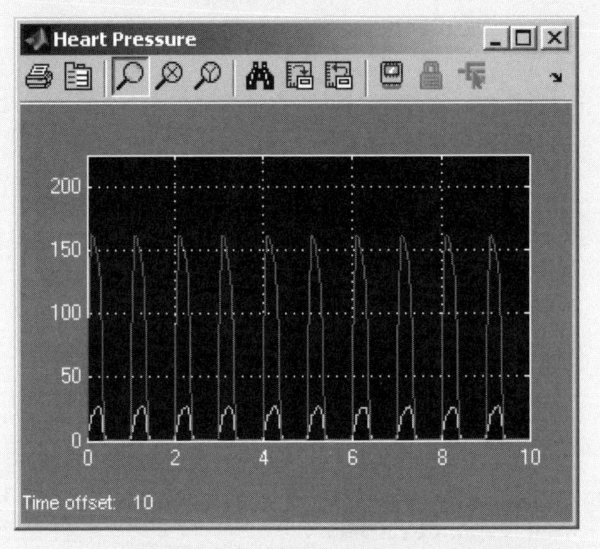

Figure 10.30 Aortic stenosis: heart pressure simulation results.

When an aortic valve stenosis occurs, the end systolic volume, or the volume of blood left in the heart after a single pump, increases to about 85 mL (Fig. 10.31).

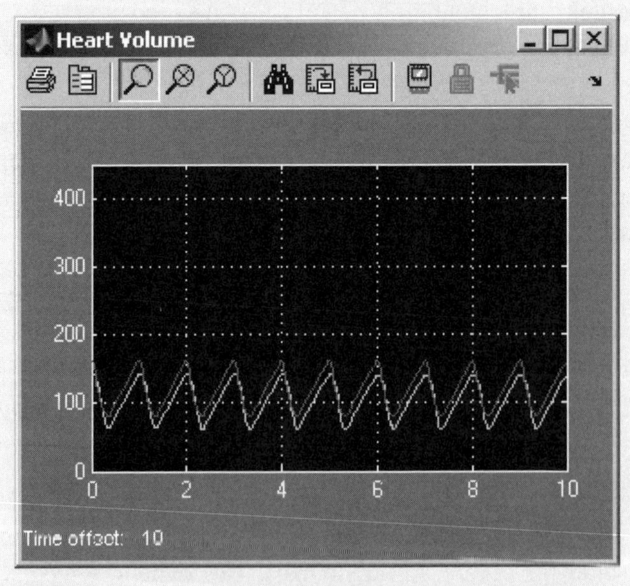

Figure 10.31 Aortic stenosis: heart volume simulation results.

In a normal heart, maximum flow out of the left ventricle is about 925 mL/min (Fig. 10.32).

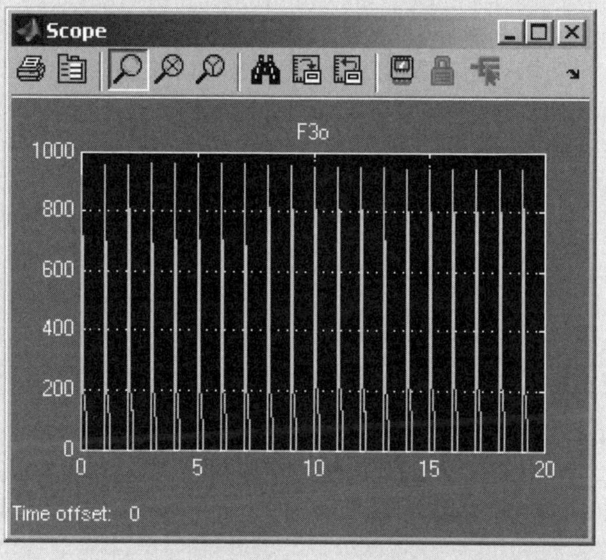

Figure 10.32 Normal heart: left ventrical outflow simulation results.

In the case of an aortic valve stenosis, the blood flow out of the heart is cut in half to 460 mL/min (Fig. 10.33).

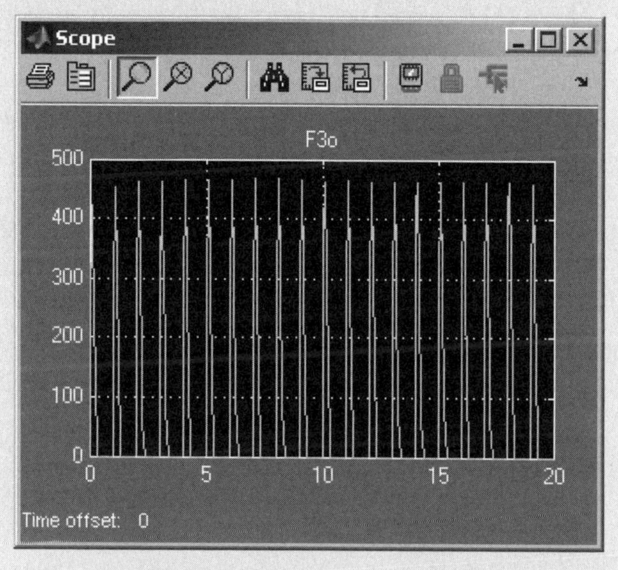

Figure 10.33 Aortic stenosis: heart blood outflow simulation results.

In this simulation, as in the pathophysiology, blockage of the aortic valve results in an increased resistance to blood flow, due to the inverse relationship of radius and resistance. Because the relationship of blood flow to resistance is $F = \Delta P/R$, an increase in resistance should result in a decrease in blood flow out of the heart. If not enough blood is being ejected from the heart, and if blood flow is not high enough, the necessary oxygen and other nutrients, which the tissues need to survive, will not be efficiently delivered via the vascular system. The decrease in flow also triggers a homeostatic response within the body to compensate for the loss of blood flow. The left ventricle has to pump harder and thus fatigues at a faster rate than a normal heart.

In addition, as a result of the increased resistance, blood cannot escape the heart, specifically the left ventricle, as quickly as under normal conditions. Thus the end systolic volume, or the amount of blood that remains in the heart after the heart has finished contracting, also increases.

10.7.3 Ventricular septal defect

Ventricular septal defect (VSD) is a heart malformation often arising at birth or in conjunction with other pathology, such as myocardial infarction. VSD is characterized by a hole in the septum between the two lower ventricles (Lue and Takao, 1986) and is the most common congenital heart defect. The significance of VSD varies with the size of the hole: the greater the hole, the higher the risk. The size of the hole can range from a pinpoint to almost the absence of the septum. In the case of a small hole, it can spontaneously close within the first three years of life. In the case of a large hole, the left ventricle shunts blood to the right ventricle via the septum hole. The extra blood causes the right ventricle to do more work and makes the lung receive too much blood, and this increases the pressure. If the problem is not resolved, it can result in a weakened or enlarged right ventricle caused by stress and overwork. Also, the lung can become crowded with the extra blood, causing clotting in the lungs that can lead to abnormal heartbeat or heart failure.

Example 10.10 Ventricular septal defect.

Modify the Simulink implementation of PHYSBE to model a ventricular septal defect and predict the effect on blood pressure and volume in the heart.

VSD can be crudely modeled in PHYSBE as blood flow from the left to right ventricles. The flow between the ventricles, stimulated by a pressure difference, can be modeled using Ohm's Law:

$$Q_s = \frac{P_{lv} - P_{rv}}{R} \quad \text{and} \quad R = \frac{8\eta L}{\pi r^4}$$

Blood has a viscosity, η, of 3 cP (Fournier, 1999), and the thickness of the ventricular wall is, in an adult, 4mm (Lue and Takao, 1986). Here Q represents the blood flow, R the resistance, and r the radius of the hole.

A ventricular septal defect is modeled in PHYSBE in the left heart. First, the right ventricular pressure is connected to the left heart to allow modeling through a global "goto" tag (Fig. 10.34).

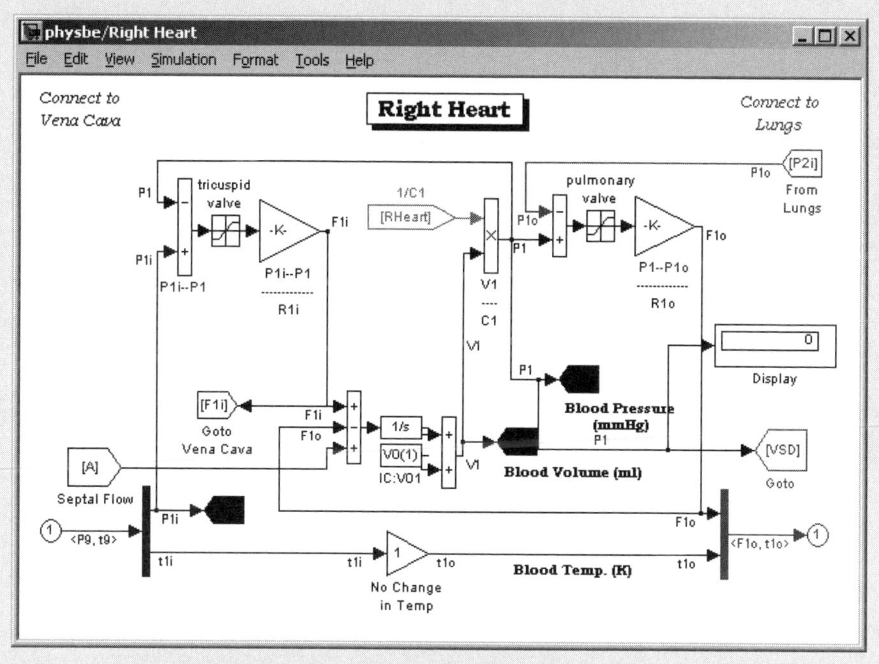

Figure 10.34 PHYSBE model of the right heart with ventricular septal defect.

This is subtracted from the left ventricular pressure to measure the pressure change (Fig. 10.35).

This difference is divided by resistance, as determined by the Bernoulli equation:

$$\frac{8(3\text{cP})(4\text{mm})}{\pi r^4} = \frac{0.2292}{r^4}\left(\frac{\text{mmHg}\cdot\text{s}}{\text{mL}}\right)$$

with r in millimeters. The reciprocal (since the formula calls for division) represents the constant C in Fig. 10.34. This difference is subtracted from the left heart flow rate, and added to the right heart flow rate, shown in Fig. 10.34 for the left heart, and appears as a source block in the model of the right heart, above.

This simulation of VSD predicts that the blood pressure in the left heart reaches a steady state that is less than the normal pressure, whereas the pressure in the right heart increased significantly. This pressure change is consistent with theory since the right has an increased flow from the left ventricle. (See Figs. 10.36 and 10.37.)

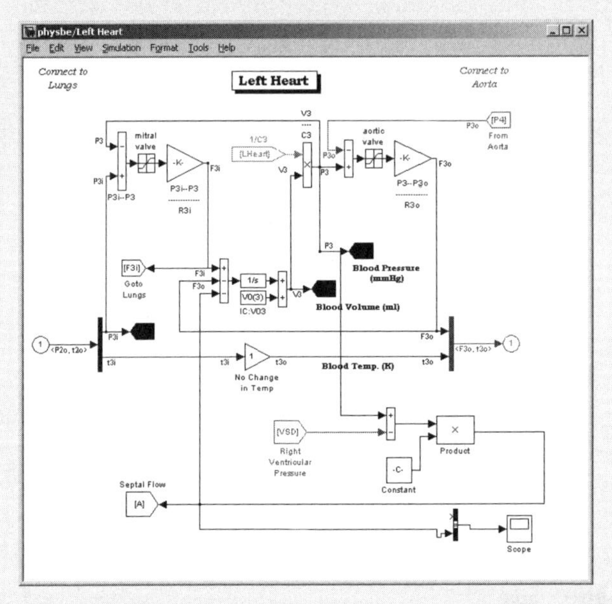

Figure 10.35 PHYSBE model of the left heart with ventricular septal defect.

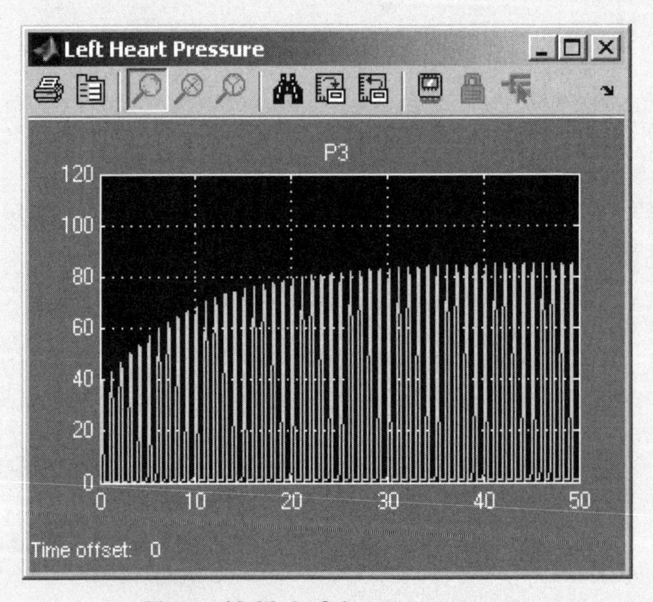

Figure 10.36 Left heart pressure.

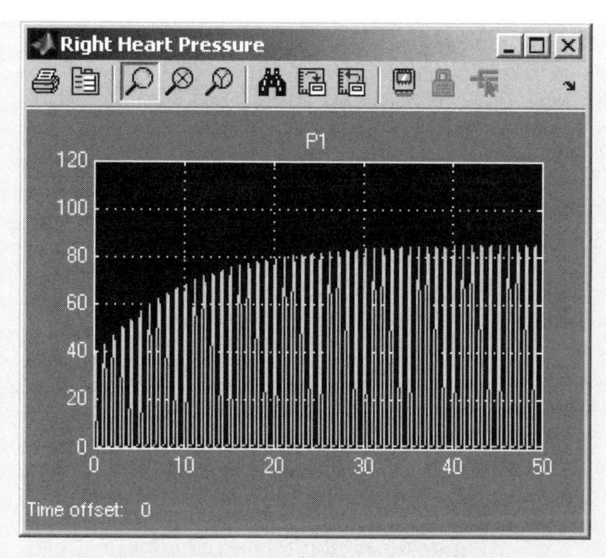

Figure 10.37 Right heart pressure.

In this simulation, a hole greater than one square centimeter was modeled. The theory would predict that with a hole this size, a considerable amount of blood would be in the right ventricle and the blood volume in the pulmonary arteries would increase. The increased blood volume in the pulmonary arteries causes the pressure in the pulmonary circuit to increase—this increased pressure in the lung is called *hypertension*. The plots in Figs. 10.38 and 10.39 show that the blood volume in both the right and left heart are greater than normal.

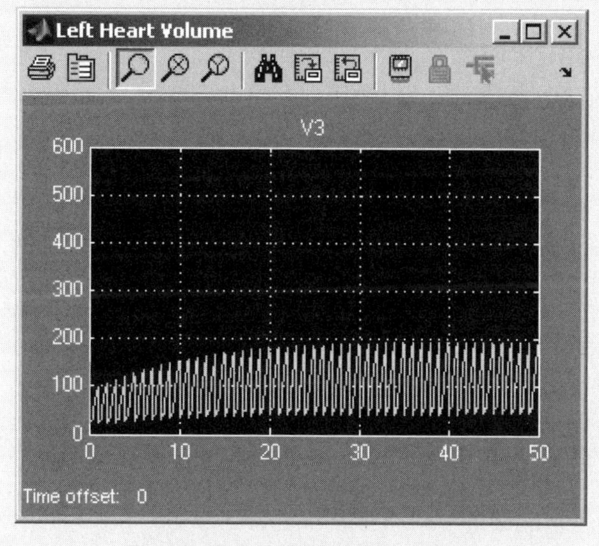

Figure 10.38 Left heart volume.

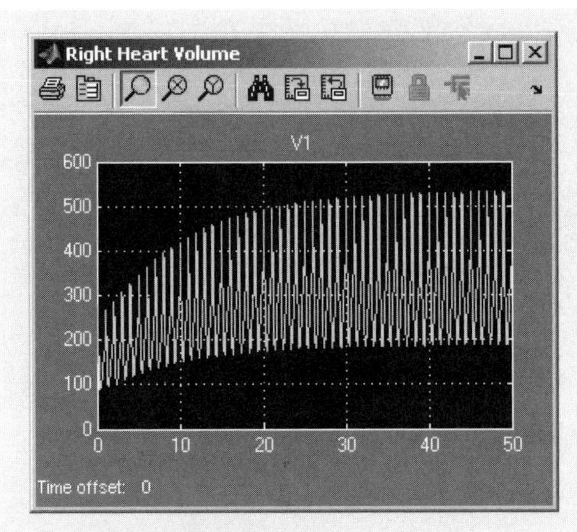

Figure 10.39 Right heart volume.

Ventricular septal defect is a problem that affects the pressure in the heart because of the blood flow between the left and right ventricles through the septum. If the hole is large, the blood from the left ventricle blood flows into the right ventricle because of the pressure difference. The additional blood volume also causes more blood in the lung, causing extra pressure that sometimes results in hypertension. If VSD is not treated, the arteries in the lungs can thicken up under the extra pressure and permanent damage can be done to the lung, leading to the weakening of the valve.

10.7.4 Left ventricular hypertrophy

The heart responds to prolonged inadequate myocardial contraction by enlarging to keep pace with the requirements for increased cardiac output (Hori et al., 1989). This increased workload causes an increase in the wall thickness, a condition commonly referred to as *left ventricular hypertrophy* (shown in Fig. 10.40).

This inward expansion of the ventricular wall reduces both the rate and amount of relaxation during diastole, leading to a decrease in the wall compliance. The decrease in compliance (increase in wall stiffness), coupled with the concomitant increase in pressure, leads to a reduction in the size of the left chamber. If the left chamber is too small, it cannot fill efficiently, which leads to blood backing up into the vessels of the lungs and less blood circulating to the vital organs. At times, hypertrophy is itself the primary disease; but more commonly it is the consequence of another disorder.

Frequently, chronic hypertension is found to be the underlying cause of left ventricular dysfunction in humans. When operating correctly, the aortic and pulmonary valves are responsible for creating the pressure differences that lead to proper blood flow and circulation. Abnormal functioning of the valves is the underlying factor that produces either (1) pressure overloading due to restricted opening; or (2) volume overloading due to inadequate closure (Legato, 1987).

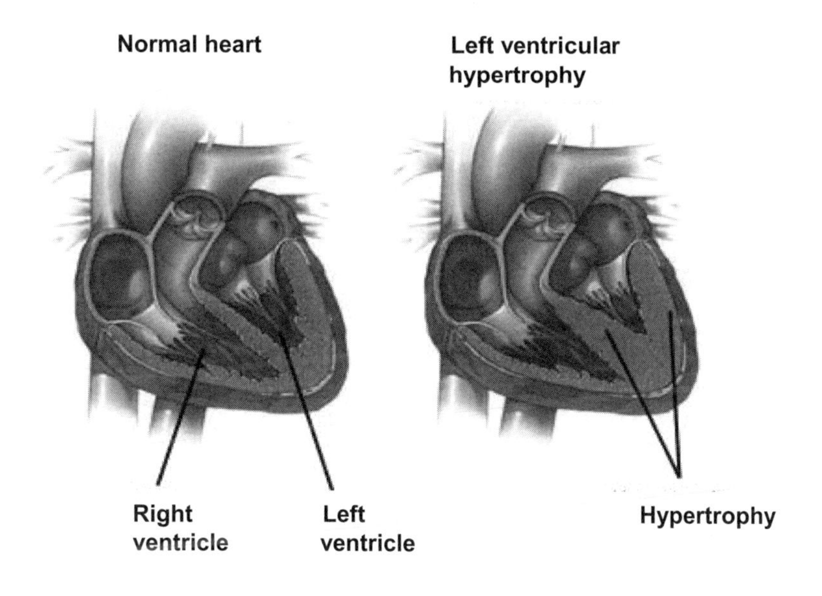

Figure 10.40 Overwork of the left ventricle causes an increase in wall thickness, a condition called left ventricular hypertrophy. Picture courtesy of the Mayo Educational Foundation.

A narrowing of the aortic or pulmonary valves induces a decrease in blood flow (by Ohm's Law). Assuming constant resistance in the blood transport vessels, the narrowing of each valve leads to a decrease in the corresponding pressure gradient. The arterial compliance ($C = \Delta V/\Delta P$) will decrease in response to an impaired pressure gradient caused by valve dysfunction, reflecting the change in blood volume due to the decreasing arterial pressure. This model then predicts that the blood volume in the transport vessels must increase, leading to a decrease in left ventricular blood volume.

PHYSBE can be used to model left ventricular hypertrophy indirectly by first modeling inadequate aortic and pulmonary valves. The weakened valves induce left ventricular pressure overloading and a reduction in blood flow through the valves. This condition gives rise to left ventricular hypertrophy.

Example 10.11 Left ventricular hypertrophy.

Modify the Simulink implementation of PHYSBE to model left ventricular hypertrophy.

Weakened aortic and pulmonary valves can be modeled by changing the upper limits of the aortic and pulmonary valves in the left heart (Fig. 10.41) and right heart (Fig. 10.42), respectively.

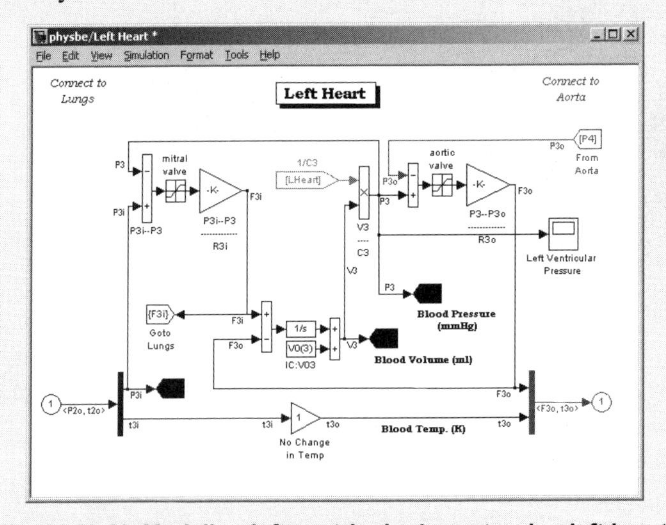

Figure 10.41 Modeling left ventricular hypertrophy: left heart.

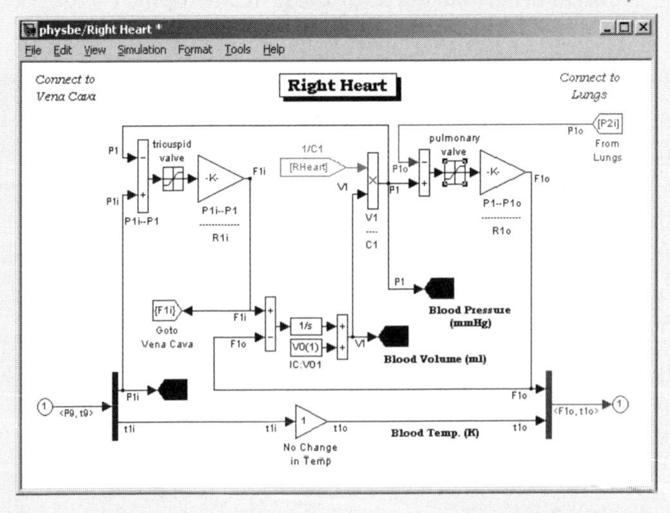

Figure 10.42 Modeling left ventricular hypertrophy: right heart.

In addition to the scope output of the left ventricular pressure in the left heart, a scope should be added to the aorta to show the effect on aortic blood volume (Fig. 10.43).

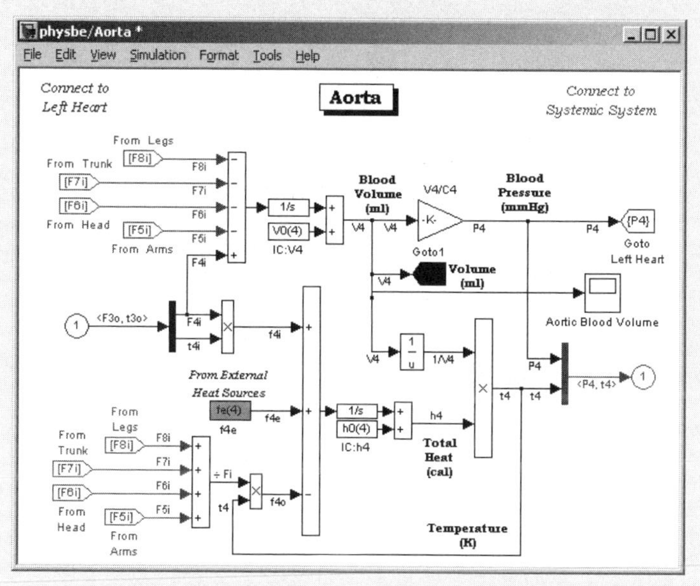

Figure 10.43 Modeling left ventricular hypertrophy: aorta.

Left ventricular hypertrophy is characterized by an increase in pressure in the left ventricle (Fig. 10.44).

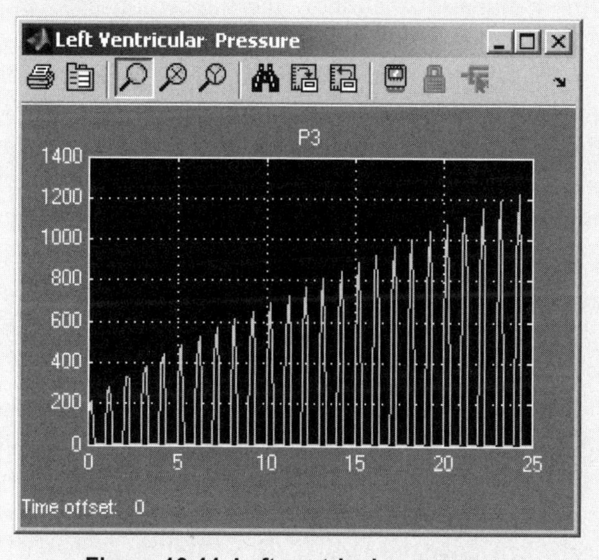

Figure 10.44 Left ventricular pressure.

The volume of blood in the aorta decreases commensurately with the increase in pressure in left ventricular hypertrophy (Fig. 10.45).

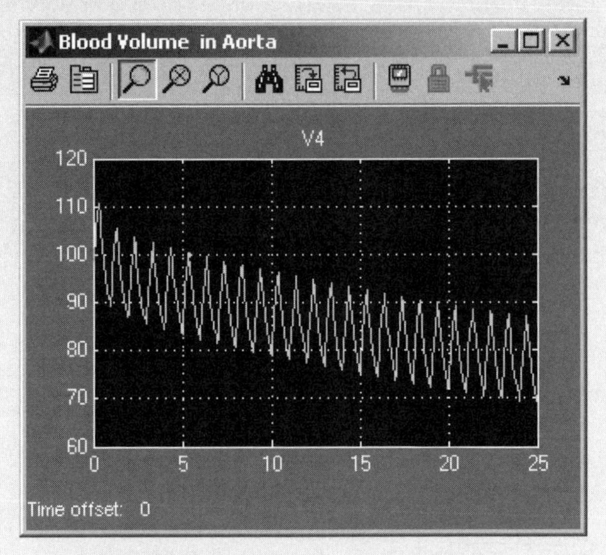

Figure 10.45 Blood volume in aorta.

The Pressure-Volume (PV) loop is a graphical tool for assessing the interplay of ventricular function and circulation.

Under normal conditions, the external work performed by the left ventricle is confined within the boundaries that show the interaction between the end-systolic pressure-volume relationship (ESPVR) and end-diastolic pressure-volume relationship (EDPVR). The place where the ESPVR is tangent to the PV-loop represents the point at which the aortic valve closes. The corresponding EDPVR tangent line is proportional to the reciprocal of ventricular compliance. In addition, the area under the loop represents the mechanical work performed by the ventricle, and the width of the loop corresponds to the difference between end diastolic volume and end systolic volume, defining the stroke volume (Li, 2004).

Example 10.12 Pressure-volume loops.

Modify the Simulink implementation of PHYSBE so that a PV-loop can be created in MATLAB. Use the PV-loop to illustrate the dynamic characteristics of left ventricular hypertrophy.

PV-loops are easily generated in MATLAB from PHYSBE data by plotting left ventricular pressure against left ventricular volume, measurements that were saved to the MATLAB variables LV_Volume and LV_Pressure, respectively. The PV-loops are generated with the MATLAB script:

```
plot(LV_Volume,LV_Pressure)
ylabel('Left Ventricular Pressure')
xlabel('Left Ventricular Volume')
axis([0 500 0 1200])
```

The Pressure Volume Loop diagram of the left ventricle is constructed based on the filling, contraction, ejection, and relaxation of the left ventricle. A PV-loop of normal cardiovascular conditions is shown in Fig. 10.46, and one representative of a patient with left ventricular hypertrophy is shown in Fig. 10.47.

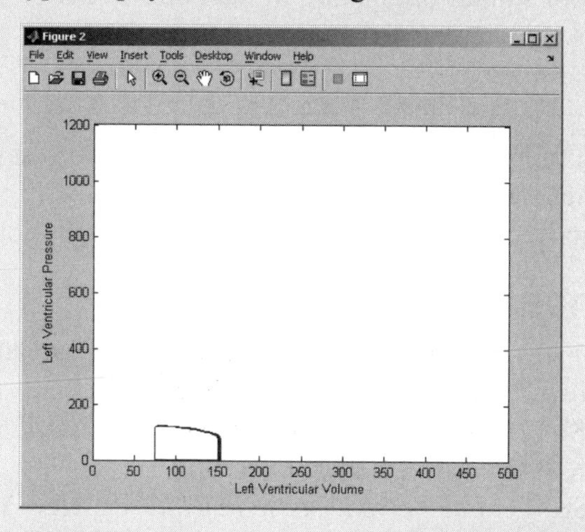

Figure 10.46 PV-loop of normal cardiovascular conditions.

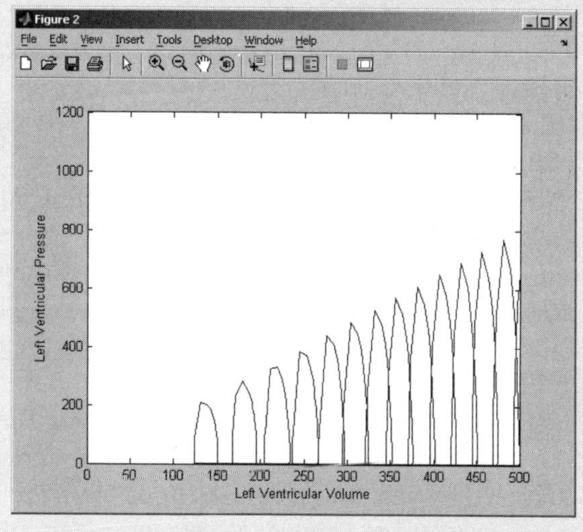

Figure 10.47 PV-loop of patient with left ventricular hypertrophy.

As expected in cases of left ventricular hypertrophy, there is a leftward and upward shift in the ESPVR, and a shift in the PV-loop itself. The total area enclosed by each loop was reduced, reflecting a decrease in stroke work and volume. If the stroke volume is decreased, the heart rate must increase in order to maintain a normal cardiac output. Eventually, if left untreated, this adaptive effect will fail and cardiac output will fall.

This PHYSBE simulation shows that left ventricular hypertrophy is a compensatory mechanism for maintaining normal cardiac output under abnormal increases in pressure and decreases in volume. A decrease in compliance and an increase in wall stiffness resulted from stimuli originating in the body's control center. Furthermore, reductions in stroke volume and work, computed from the PV-loops, are characteristic of the disease.

This set of examples is meant to illustrate to the reader the broad range of biomedical applications of numerical methods and show the reader how to use computing tools and numerical methods to solve problems in medicine and physiology.

10.8 References

Bhat, M. A. 2001. Fate of Hypertension after Repair of Coarctation of the Aorta in Adults. *British Journal of Surgery*, **88**:536-538.

Braunwald, E. 1988. *Heart Disease: a Textbook of Cardiovascular Medicine*. Philadelphia, PA: W. B. Saunders Company.

Dunn, J. M., ed. 1988. *Cardiac Valve Disease In Children*. New York: Elsevier.

Estelberger, W., and Popper, N. 2002. Comparison 15: Clearance Identification. *Simulation News Europe* **35/36**:65-66.

Fournier, R. L. 1999. *Basic Transport Phenomena in Biomedical Engineering*. Philadelphia, PA: Taylor & Francis.

Germann, W. J. 2005. *Principles of Human Physiology*, 2nd ed. San Francisco, CA: Pearson Education.

Goldberger, A. L., Amaral, L. A. N., Glass, L., Hausdorff, J. M., Ivanov, P. Ch., Mark, R. G., Mietus, J. E., Moody, G. B., Peng, C. K., and Stanley, H. E. 2000. PhysioBank, PhysioToolkit, and Physionet: Components of a New Research Resource for Complex Physiologic Signals. *Circulation* **101**(23):e215-e220. (Circulation Electronic Pages; http://circ.ahajournals.org/cgi/content/full/101/23/e215.)

Hori, M., Suga, J., and Yellin, E. L. 1989. *Cardiac Mechanics and Function in the Normal and Diseased Heart*. Tokyo, Japan: Springer-Verlag.

Jasper, H. H. 1958. The ten-twenty electrode system of the international federation. *Electroencephalography and Clinical Neurophysiology*, **10**:371-373.

Keirn, Z. 1988. *Alternative modes of communication between man and machine.* Master's thesis, Purdue University.

Klabunde, R. E. 2004. *Cardiovascular Physiology Concepts: Valvular Stenosis.* http://www.cvphysiology.com/Heart Disease/HD004.htm .

Krousgrill, C. M. 2005. Personal Communication.

Legato, M. J. 1987. *The Stressed Heart.* Boston, MA: Martinus Nijhoff Publishing.

Li, J. K-J. (2004). *Dynamics of the Vascular System.* Boston, MA: World Scientific Publishing.

Lue, H. C., and Takao A., eds. 1986. *Subpulmonic Ventricular Septal Defects.* Tokyo, Japan: Springer-Verlag.

Mascherbauer, J. 2004. Value and limitations of aortic valve resistance with particular consideration of low flow-low gradient aortic stenosis: an in vitro study. *Eur Heart J.* **25**(9):787-793.

McLeod, J. 1966. PHYSBE…a physiological simulation benchmark experiment. *SIMULATION*, **7**(6):324-329.

McLeod, J. 1968. PHYSBE…a year later, *SIMULATION*, **10**(1):37-45.

McSharry, P. E., Clifford, G. D., Tarassenko, L., and Smith, L. 2003. A dynamical model for generating synthetic electrocardiagram signals. *IEEE Transactions on Biomedical Engineering*, **50**(3):289-294.

Mohiaddin, R. H. 1993. Magnetic Resonance Volume Flow and Jet Velocity Mapping in Aortic Coarctation. *JACC*, **22**:1515-1521.

Ruha, A., and Nissila, S. 1997. A real-time microprocessor QRS detector system with a 1-ms timing accuracy for the measurement of ambulatory HRV. *IEEE Trans Biomed Eng* **44**(3):159-167.

Sokolow, M., and McIlroy, M. B. 1981. *Clinical Cardiology.* Los Altos, CA: LANGE Medical Publications.

Suk, J., ed. 2001. Coarctation of the Aorta. *Yahoo Health.* http://health.yahoo.com/health/ency/adam/000191/i18128.

Texas Heart Institute. 2004. *Leading With the Heart.* http://www.tmc.edu/thi/vaortic.html .

Van Riel, N. 2004. *Minimal Models for Glucose and Insulin Kinetics* - A MATLAB implementation; version February, 5, 2004. Technical Report, Technische Universiteit Eindhoven.

Web site links

http://www.physionet.org/physiotools/matlab/ECGwaveGen/

http://www.physionet.org/physiotools/ecgsyn/

http://www.emedicine.com/ped/topic2543.htm

http://circ.ahajournals.org/cgi/content/full/101/23/e215

http://www.cvphysiology.com/Heart Disease/HD004.htm

http://health.yahoo.com/conditions/
http://www.tmc.edu/thi/vaortic.html
http://physionet.incor.usp.br/
http://www.physionet.org/cgi-bin/rdsamp

Appendix A: Introduction to MATLAB

In this Appendix we introduce the MATLAB features that are essential to understanding the basics of the software developed throughout this text. Several good MATLAB self-study tutorial books are available; see Hanselman and Littlefield (2005), Hahn (2002), and Lyshevski (2003). Also, the manuals that accompany MATLAB are recommended as additional references. The MATLAB program has extensive built-in help and tutorials.

We strongly recommend that the readers go through this appendix and practice each command while sitting at a computer running MATLAB. This exercise may fail to convey its intended lessons if the user is not practicing these commands at the computer. The symbol “>>” is MATLAB’s prompt, which automatically appears in the command window. You do not need to type it. In this Appendix, and throughout this book, any line that starts with the symbol “>>” is a Command Window line.

A.1 The MATLAB Environment

The default working environment in MATLAB version 7.0 (R14) displays the four panels shown in Fig. A.1. These are:

Workspace: This browser displays the variables created and used during a MATLAB session. Variables are added to the workspace by defining them in the command window, using functions, running M-files (MATLAB language programs), and/or loading saved workspaces. To view the workspace and information about each variable, use the workspace browser, or use the commands `who` and `whos` in the Command Window. To view and/or edit any variable, double-click on the variable of interest in the workspace browser. This will open the Array Editor (see Fig. A.2) to show you the contents of that variable and enable you to edit it.

Current Directory (not open in Fig. A.1): This panel displays the contents of the current directory.

Command History: This panel displays all the commands that have been entered in the present and previous MATLAB sessions. Point the mouse on any of these commands and click the right button to see the options available: Copy, Evaluate Selection, Create M-file, and Delete. Any of the commands in the Command History window may be executed again by dragging the command from the Command History and dropping it on the Command Window.

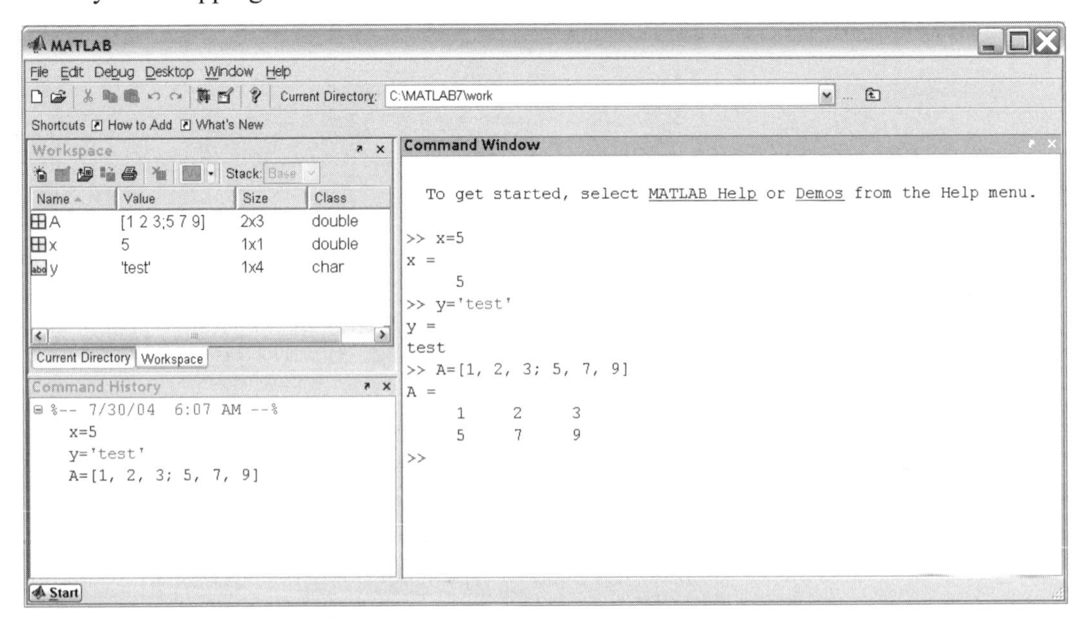

Figure A.1 The MATLAB default environment.

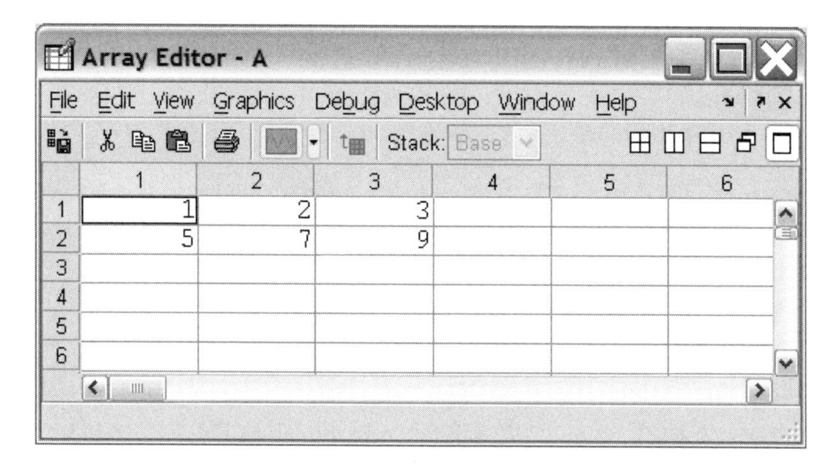

Figure A.2 The Array Editor, showing the values of array A.

Command Window: This is the window in which all MATLAB commands are entered and results are displayed. It is used to enter variables, execute functions, and run M-files.

In addition to the above, the following windows may be opened:

Editor: The MATLAB Editor (see Fig. A.3) is used to create, edit, debug, and execute MATLAB programs (M-files). The command `edit` in the Command Window opens the MATLAB Editor.

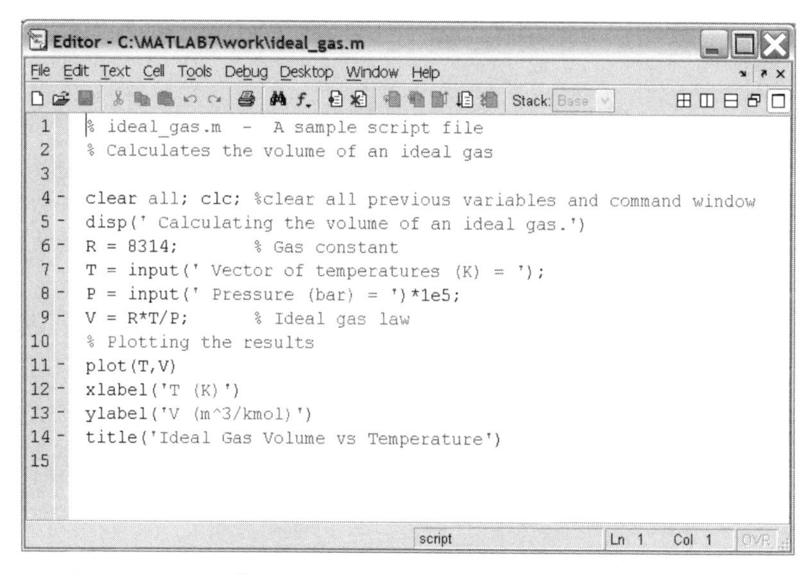

Figure A.3 The MATLAB Editor.

Profiler: MATLAB includes the Profiler to help you improve the performance of your M-files. Run a MATLAB statement or an M-file in the Profiler and it produces a report of where the time is being spent. Access the Profiler from the Desktop menu, or use the profile function.

Start button: The MATLAB Start button, located at the lower left corner of the MATLAB window (Fig. A.1), has a menu interface that provides easy access to all of the above items (and more).

A.1.1 Customizing the MATLAB environment

The MATLAB environment may be customized to suit the user's needs. To do so, go to the menu item File/Preferences. This opens the window shown on Fig. A.4, which gives the user a large number of options from adjusting the size and color of fonts to specifying the format used in displaying the numbers and the compactness of the display.

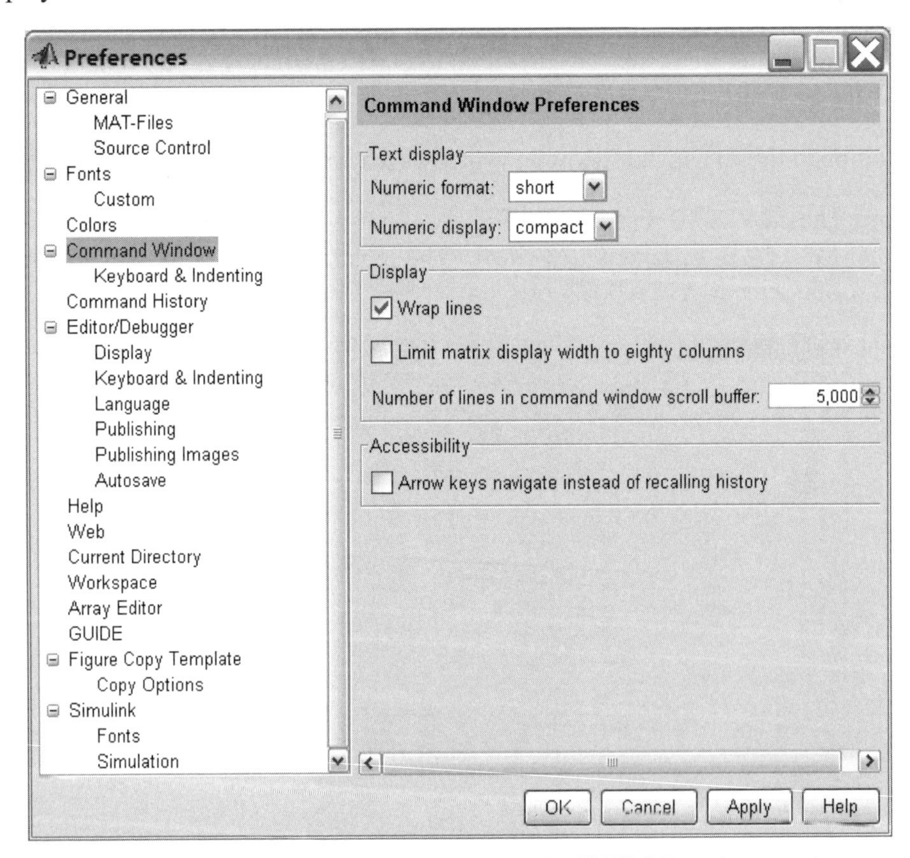

Figure A.4 Preferences for customizing the MATLAB environment.

A.1.2 The MATLAB path

When a command is executed or the name of an M-file is invoked in the Command Window, MATLAB searches through a list of folders and subfolders to locate that command or M-file, in order to execute it. The standard MATLAB search path is set when MATLAB is installed. Users can add additional folders that contain their own files. To add to the path, go to File/Set Path and follow instructions, as shown on Fig. A.5.

Do **not** remove any of the standard MATLAB search path folders. These are needed for proper operation of MATLAB.

Caution should be exercised in naming new M-files. If a file is named with an identical name as one already existing in MATLAB, the file that is found first during the search of the path will be executed. It is good practice to avoid duplicating names, and to add the users' folders to the bottom of the path.

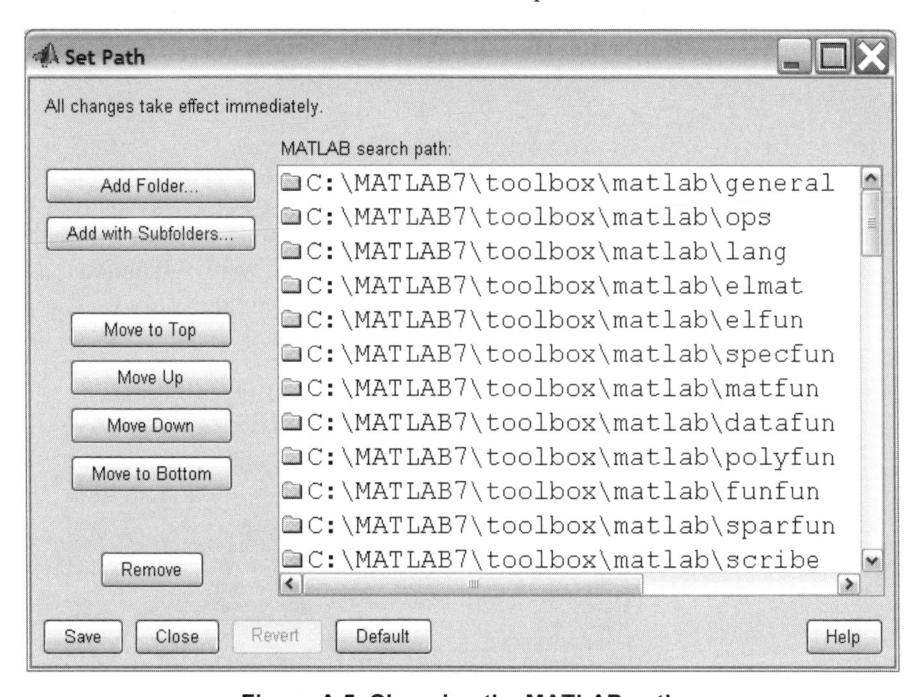

Figure A.5 Changing the MATLAB path.

A.1.3 Where to find help for MATLAB

As a beginner, you may want to see a tutorial about MATLAB. Typing `demo` at the command line will display the available demonstrations. In the MATLAB demo window (Fig. A.6), you may choose the subject you are interested in and then follow the lessons.

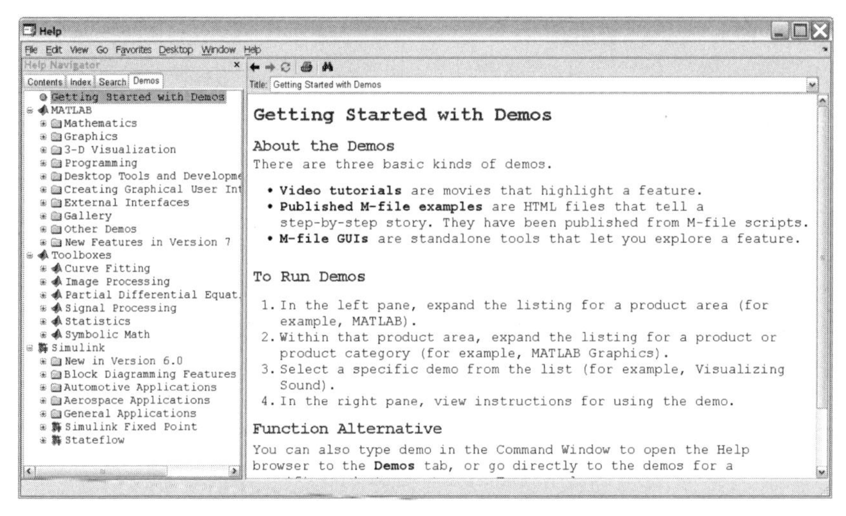

Figure A.6 The MATLAB demo window.

The command

```
>> helpwin
```

opens the window shown on Fig. A.7, and displays the overall organization of all available help. It has links to the general purpose commands; operators and special characters; elementary and specialized math functions; and matrix functions, among others. Single clicking on any of the links will show the available help for the given topic.

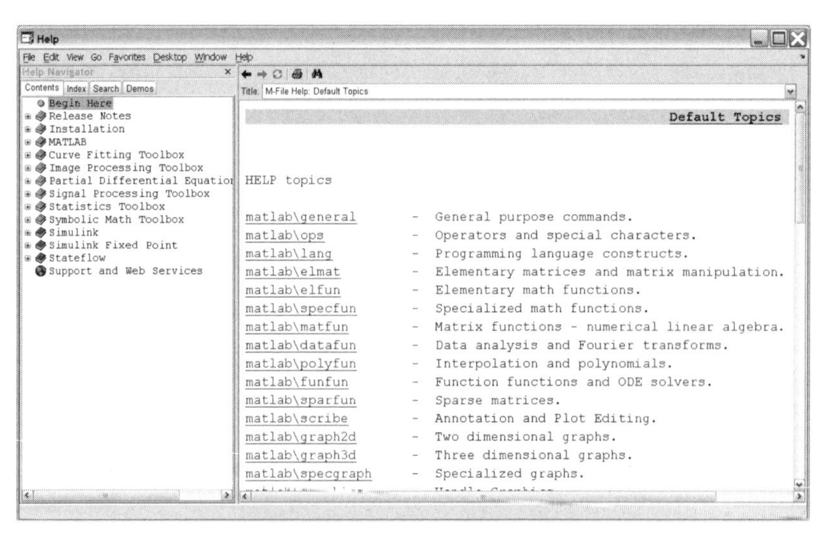

Figure A.7 The default help page.

Typing `help` alone lists the names of all directories in the MATLAB search path with a one-line description for each. Also, if you type a directory name following the command `help`, MATLAB lists the contents of the directory (if the directory is one in the MATLAB path search and contains a `contents.m` file). Try the following commands:

```
>> help
>> help general
>> helpdesk
>> doc
```

If you know the name of the function for which you need help, you can use the `help` command. For example, to get help for the command `sign`:

```
>> help sign

SIGN   Signum function.
    For each element of X, SIGN(X) returns 1 if the element
    is greater than zero, 0 if it equals zero and -1 if it is
    less than zero. For the nonzero elements of complex X,
    SIGN(X) = X ./ ABS(X).
```

The lines printed above are in the first block of comment lines in this function. These are designed to give a brief explanation of the function and its use. The percent symbol (`%`) is used to mark a line that contains comments:

```
% This is a comment statement in a program
```

If you are not sure of the function name, you can try to find the name using the `lookfor` command:

```
>> lookfor absolute
ABS    Absolute value.
IMABSDIFF Compute absolute difference of two images.
CIRCLEPICK Pick bad triangles using an absolute tolerance
MAD Mean/median absolute deviation.
ABS    Absolute value.
 LOCATEFILE Resolve a filename to an absolute location.
applyabsolute.m: % out =
applyabsolute(in,cspace,source_wp,dest_wp)
```

Extensive MATLAB help and manuals may be found on the following Web sites:

http://www.mathworks.com
http://www.mathworks.com/academia/

A.2 Elementary Operations

The four elementary arithmetic operations in MATLAB—addition, subtraction, multiplication, and division—are represented by the operators `+`, `-`, `*`, and `/`, respectively:

```
>> (20+10-4*5)/2
ans =
      5
```

The symbol `^` stands for the power operator:

```
>> 5^2
ans =
      25
```

The operator `\` is for left division, where the denominator is at the left, rather than the right. Compare `/` and `\`:

```
>> 2/4
ans =
      0.5000
>> 2\4
ans =
      2
```

MATLAB can handle complex numbers

```
>> sqrt(-1),
ans =
      0 + 1.0000i
```

and infinite numbers

```
>>1/0
Warning: Divide by zero.
(Type "warning off MATLAB:divideByZero" to suppress this warning.)
ans =
      Inf
```

The letters `i` and `j` represent the complex number $\sqrt{-1}$ unless another value is assigned to them. Also, the variable `pi` represents the ratio of the circumference of a circle to its diameter (i.e., 3.141592653. . .). If an expression cannot be evaluated, MATLAB returns `NaN`, which stands for Not-a-Number:

```
>> 0*log(0)
Warning: Log of zero.
ans =
       NaN
```

The equality sign is used to represent the assignment of values to variables:

```
>> a = 2
a =
       2
>> b=3*a
b =
       6
```

If there is no assignment, then the result of the expression is saved in a variable named `ans`:

```
>> a+b
ans =
       8
```

If you want to suppress the display in the Command Window of the result of the assignment, put a semicolon at the end of the statement:

```
>>c = a+b;
```

You can see the value of the variable by simply typing the variable:

```
>> c
c =
      8
```

MATLAB is case sensitive, meaning that MATLAB distinguishes between variables with upper- and lower-case names:

```
>> A=10;
>> A
A =
      10
>> a
a =
       2
```

All computations in MATLAB are done in double precision (see Chapter 3). The precision of the calculation is independent of the formatted display. The results of calculation are normally displayed, or printed, with only five significant digits. The `format` command may be used to switch between different display formats:

```
>> d = exp(pi)
d =
     23.1407
>> format long, d
d =
   23.14069263277927
```

```
>> format short e, d
d =
  2.3141e+001
>> format long e, d
d =
    2.314069263277927e+001
>> format short, d
d =
   23.1407
```

All of the different variations of the `format` command may be viewed by getting the help for `format`:

```
>> help format

 FORMAT Set output format.
    All computations in MATLAB are done in double precision.
    FORMAT may be used to switch between different output
    display formats as follows:
      FORMAT         Default. Same as SHORT.
      FORMAT SHORT   Scaled fixed point format with 5 digits.
      FORMAT LONG    Scaled fixed point format with 15 digits.
      FORMAT SHORT E Floating point format with 5 digits.
      FORMAT LONG E  Floating point format with 15 digits.
      FORMAT SHORT G Best of fixed or floating point format with 5
                     digits.
      FORMAT LONG G  Best of fixed or floating point format with
                     15 digits.
      FORMAT HEX     Hexadecimal format.
      FORMAT +       The symbols +, - and blank are printed
                     for positive, negative and zero elements.
                     Imaginary parts are ignored.
      FORMAT BANK    Fixed format for dollars and cents.
      FORMAT RAT     Approximation by ratio of small integers.

    Spacing:
      FORMAT COMPACT Suppress extra line-feeds.
      FORMAT LOOSE   Puts the extra line-feeds back in.
```

In order to delete a variable from the memory, use the `clear` command:

```
>>clear a
```

Using `clear` or `clear all` command deletes all the variables from the workspace:

```
>> clear
```

or

```
>> clear all
```

Check to verify that all variables have been deleted from the workspace by looking at the Workspace window.

The `clc` command clears the Command Window and brings the cursor to the top of the window (home):

```
>> clc
```

The clf command clears the current figure:

```
>> clf
```

Remember that by using the up-arrow (or down-arrow) key you can scroll backwards (or forwards) through all the commands you have entered so far in each session. If you need to recall a specific command that has been used already, just type its first letter (or first few letters) and then use the up arrow key to locate that command.

Several operating system navigational commands may be executed from the Command Window, such as:

`cd` (for change directory),
`dir` (for list the files in directory),
`mkdir` (for make new directory),
`pwd` (for print current working directory),
`ls` (for list directory).

For example, to change to the work directory of MATLAB:

```
>> cd c:\matlab704\work
```

To change to Chapter1 directory of the Biosystems program directory:

```
>> cd 'C:\Program Files\Biosystems\Chapter1'
```

The single quotation mark (') is needed in the above command because of the presence of blank spaces in the name of the directory. For this reason, the student is strongly advised not to place blanks in names of directories and in the names of M-files. The underscore character (_) should be used to separate words in names of files.

A.3 Vectors and Matrices

The word MATLAB is an abbreviation for Matrix Laboratory: MATLAB was designed to make operations on matrices as easy as possible. All variables in MATLAB are represented as arrays. A scalar number is a (1×1) array, a row vector is a $(1 \times n)$ array, and a column vector is a $(n \times 1)$ array. The standard matrix notation M(row, column) is used by MATLAB for two-dimensional arrays. Multidimensional matrices (i.e., matrices with more than two dimensions) may also be used in

MATLAB. The notation used is M(row, column, page, …). The third dimension is called *page*, but no generic names are given to the higher dimensions.

Creating a matrix is also done by assigning values to a MATLAB variable, where elements of the matrix must be enclosed in square brackets:

```
>> M = [1, 2, 3; 4, 5, 6]
M =
      1      2      3
      4      5      6
```

or

```
>> M = [1, 2, 3
  4 5 6]
M =
      1      2      3
      4      5      6
```

The elements of a row may be separated either by a comma or a space, and the rows may be separated by a semicolon or carriage return (i.e., a new line). Elements of a matrix can be accessed or replaced individually, as in M(row, column):

```
>> M(1,3)
ans =
      3

>> M(2,1) = 7
M =
      1      2      3
      7      5      6
```

The `row` and `column` variables are used for indexing the position of an element within the matrix. Array indices must be integer numbers greater than or equal to 1. Zero is not acceptable as an index for arrays in MATLAB.

Matrices may be combined together to form new matrices:

```
>> N = [M; M]
N =
      1      2      3
      7      5      6
      1      2      3
      7      5      6

>> O = [M, M]
O =
      1      2      3      1      2      3
      7      5      6      7      5      6
```

The transposition of a matrix results from interchanging its rows and columns. This can be done by putting a single quote after a matrix:

```
>> M_trans=M'
M_trans =
      1      7
      2      5
      3      6
```

A convenient shorthand notation for a sequence is the colon (:) operator. A colon placed between two numbers generates the vector of numbers from the left limit to the right limit:

```
>> v = [-1:4]
v =
    -1     0     1     2     3     4
```

The default increment is 1, but the user can change it, if required:

```
W = [-1:0.5:2; 6:-1:0; 1:7]
W =
   -1.0000   -0.5000         0    0.5000    1.0000    1.5000    2.0000
    6.0000    5.0000    4.0000    3.0000    2.0000    1.0000         0
    1.0000    2.0000    3.0000    4.0000    5.0000    6.0000    7.0000
```

A very common use of the colon notation is to refer to rows, columns, or a part of the matrix:

```
>> W(:,5)   % This command refers to all rows of the 5th column
ans =
     1
     2
     5

>> W(1,:)   % This refers to the 1st row and all the columns
ans =
-1.0000  -0.5000        0  0.5000  1.0000  1.5000  2.0000

>> W(2:3,4:7)   % This refers to rows 2 to 3 and column 4 to 7
ans =
     3     2     1     0
     4     5     6     7

>> W(2,5:end)   % This refers to row 2 and columns 5 to last column
ans =
     2     1     0
```

Multidimensional matrices (i.e., matrices with more than two dimensions) may also be handled in MATLAB. Let us add a third dimension to the matrix `W` to make it into `W(row, column, page)`:

```
>> W(:,:,2) = [2:2:14; 0.1 0.2 0.3 0.4 0.5 0.6 0.7; 2 2 2 2 2 2 2]
W(:,:,1) =
   -1.0000   -0.5000         0    0.5000    1.0000    1.5000    2.0000
    6.0000    5.0000    4.0000    3.0000    2.0000    1.0000         0
    1.0000    2.0000    3.0000    4.0000    5.0000    6.0000    7.0000
W(:,:,2) =
    2.0000    4.0000    6.0000    8.0000   10.0000   12.0000   14.0000
    0.1000    0.2000    0.3000    0.4000    0.5000    0.6000    0.7000
    2.0000    2.0000    2.0000    2.0000    2.0000    2.0000    2.0000
```

As mentioned earlier, the third dimension is called *page*.

A.3.1 MATLAB construction functions for special arrays

MATLAB has many built-in array construction functions. For example:

```
>>ones(2)              % generates a (2 × 2) matrix of ones
ans =
     1     1
     1     1

>>ones(2,3)            % generates a (2 × 3) matrix of ones
ans =
     1     1     1
     1     1     1

>>zeros(2,3,2)         % generates a (2 × 3 × 2) array of zeros
ans(:,:,1) =
     0     0     0
     0     0     0
ans(:,:,2) =
     0     0     0
     0     0     0

>>eye(3)               % generates a (3 × 3) identity matrix
ans =
     1     0     0
     0     1     0
     0     0     1

>>rand(4,2)            % generates a (4 × 2) matrix of random
                       % variables uniformly distributed on the
                       % interval (0.0,1.0)
ans =
    0.9501    0.8913
    0.2311    0.7621
    0.6068    0.4565
    0.4860    0.0185

>>linspace(-1,5,7)     % generates a 7-element row vector of
                       % equally spaced numbers between -1 and 5
ans =
    -1     0     1     2     3     4     5
```

```
>>logspace(-1,2,8)        % generates an 8-element row vector of
                          % logarithmically equally spaced points
                          % between 10^-1 and 10^2
ans =
    0.1000    0.2683    0.7197    1.9307    5.1795   13.8950
37.2759  100.0000
```

Two very useful array functions are `size`, which gives the dimensions of the array, and `length`, which gives the maximum length of the array:

```
>> size(W)
ans =
     3     7     2
>> length(W)
ans =
     7
```

A.3.2 Array arithmetic

Multiplying an array by a scalar multiplies each element of the array by the scalar:

```
>> A= [1, 2, 6; 2:4; 10:0.5:11], 2*A
A =
    1.0000    2.0000    6.0000
    2.0000    3.0000    4.0000
   10.0000   10.5000   11.0000
ans =
     2     4    12
     4     6     8
    20    21    22
```

Only arrays of the same size may be added or subtracted:

```
>> B = [4, 7, 14; 1, 5, 7; 7, 7, 5]
B =
     4     7    14
     1     5     7
     7     7     5
ans =
>> A-B
   -3.0000   -5.0000   -8.0000
    1.0000   -2.0000   -3.0000
    3.0000    3.5000    6.0000

>> c = [5; 2; 4];
>> A+c
??? Error using ==> +
Matrix dimensions must agree.
```

Adding a scalar to an array results in adding the scalar to each element of the array:

```
>> A+5
ans =
    6.0000    7.0000   11.0000
    7.0000    8.0000    9.0000
   15.0000   15.5000   16.0000
```

Vector and matrix multiplication requires that they are conformable (see Appendix C):

```
>> A*B    % conformable because both matrices are (3 × 3)
ans =
   48.0000   59.0000   58.0000
   39.0000   57.0000   69.0000
  127.5000  199.5000  268.5000

>> A*c    % conformable because A is (3 × 3) and c is (3 × 1)
ans =
    33
    32
   115

>> c*A    % nonconformable because c is (3 × 1) and A is (3 × 3)
??? Error using ==> *
Inner matrix dimensions must agree.

>> c'*A   % conformable because c' is (1 × 3) and A is (3 × 3)

ans =
    49    58    82
```

To perform element-by-element operations on matrices and vectors, use a period (.) before the operator symbol:

```
>> A.*B
ans =
    4.0000   14.0000   84.0000
    2.0000   15.0000   28.0000
   70.0000   73.5000   55.0000

>> A.^2
ans =
    1.0000    4.0000   36.0000
    4.0000    9.0000   16.0000
  100.0000  110.2500  121.0000
```

```
>> A^2          % compare this with A.^2 above
ans =
    65    71    80
    48    55    68
   141   167   223

>> 1./A
ans =
    1.0000    0.5000    0.1667
    0.5000    0.3333    0.2500
    0.1000    0.0952    0.0909
```

Some useful matrix functions are:

```
>> det(A)          % determinant of a square matrix
ans =
    -27

>> inv(A)          % inverse of a square matrix
ans =
    0.3333   -1.5185    0.3704
   -0.6667    1.8148   -0.2963
    0.3333   -0.3519    0.0370

>>rank(A)          % rank of a matrix (see Appendix C)
ans =
     3

>> [V,D]=eig(A)  % eigenvectors and eigenvalues of a square matrix
V =
   -0.3457   -0.7603    0.6014
   -0.2817   -0.1527   -0.7737
   -0.8951    0.6314    0.1993
D =
   18.1660         0         0
         0   -3.5811         0
         0         0    0.4150
```

The columns of matrix `V` above are the eigenvectors of `A`, and the diagonal elements of matrix `D` are the corresponding eigenvalues of `A`.

```
>>poly(A)          % coefficients of characteristic polynomial of A
ans =
    1.0000  -15.0000  -59.0000   27.0000
```

The above result indicates that the characteristic polynomial, $\det(\mathbf{A} - \lambda\mathbf{I})$, is

$$\lambda^3 - 15\lambda^2 - 59\lambda + 27$$

A.4 MATLAB Built-in Functions

MATLAB provides standard elementary and advanced mathematical and visualization functions, such as `abs`, `sqrt`, `exp`, `sin`, `cos`, `eig`, `plot`, `plot3`.

To see a listing of all the elementary functions, give the command:

```
>> help elfun
  Elementary math functions.

  Trigonometric.
    sin         - Sine.
    sind        - Sine of argument in degrees.
    sinh        - Hyperbolic sine.
    asin        - Inverse sine.
    asind       - Inverse sine, result in degrees.
    asinh       - Inverse hyperbolic sine.
    cos         - Cosine.
    cosd        - Cosine of argument in degrees.
    cosh        - Hyperbolic cosine.
    acos        - Inverse cosine.
    acosd       - Inverse cosine, result in degrees.
    acosh       - Inverse hyperbolic cosine.
    tan         - Tangent.
    tand        - Tangent of argument in degrees.
    tanh        - Hyperbolic tangent.
    atan        - Inverse tangent.
    atand       - Inverse tangent, result in degrees.
    atan2       - Four quadrant inverse tangent.
    atanh       - Inverse hyperbolic tangent.
    sec         - Secant.
    secd        - Secant of argument in degrees.
    sech        - Hyperbolic secant.
    asec        - Inverse secant.
    asecd       - Inverse secant, result in degrees.
    asech       - Inverse hyperbolic secant.
    csc         - Cosecant.
    cscd        - Cosecant of argument in degrees.
    csch        - Hyperbolic cosecant.
    acsc        - Inverse cosecant.
    acscd       - Inverse cosecant, result in degrees.
    acsch       - Inverse hyperbolic cosecant.
    cot         - Cotangent.
    cotd        - Cotangent of argument in degrees.
    coth        - Hyperbolic cotangent.
    acot        - Inverse cotangent.
    acotd       - Inverse cotangent, result in degrees.
    acoth       - Inverse hyperbolic cotangent.
```

```
Exponential.
  exp         - Exponential.
  expm1       - Compute exp(x)-1 accurately.
  log         - Natural logarithm.
  log1p       - Compute log(1+x) accurately.
  log10       - Common (base 10) logarithm.
  log2        - Base 2 logarithm and dissect floating point
                number.
  pow2        - Base 2 power and scale floating point number.
  realpow     - Power that will error out on complex result.
  reallog     - Natural logarithm of real number.
  realsqrt    - Square root of number greater than or equal to
                zero.
  sqrt        - Square root.
  nthroot     - Real n-th root of real numbers.
  nextpow2    - Next higher power of 2.

Complex.
  abs         - Absolute value.
  angle       - Phase angle.
  complex     - Construct complex data from real and imaginary
                parts.
  conj        - Complex conjugate.
  imag        - Complex imaginary part.
  real        - Complex real part.
  unwrap      - Unwrap phase angle.
  isreal      - True for real array.
  cplxpair    - Sort numbers into complex conjugate pairs.

Rounding and remainder.
  fix         - Round towards zero.
  floor       - Round towards minus infinity.
  ceil        - Round towards plus infinity.
  round       - Round towards nearest integer.
  mod         - Modulus (signed remainder after division).
  rem         - Remainder after division.
  sign        - Signum.
```

To obtain a listing of the specialized math functions, type the command:

```
>> help specfun
```

For a listing of elementary matrices and matrix manipulation functions, type:

```
>> help elmat
```

For a listing of the graphics functions, give the command:

```
>> help graphics
```

For a listing of all the operators and special characters, give the command:

```
>> help ops
```

For a help with any of the Toolboxes give the command `help` followed by the name of the Toolbox:

```
>> help symbolic math
```

or

```
>> help simulink
```

A.5 Graphics

MATLAB has a variety of visualization tools: 2-D graphics, 3-D graphics, pie charts, bar charts, histograms, and contour plots that may be used to represent data and functions.

2-D graphs

Functions of one independent variable can easily be visualized in MATLAB (see Fig. A.8):

```
>> x = linspace(0,2,30);          % create a vector of 30 values
                                  % equally spaced between 0 and 2
>> y = x.*exp(-x);                % calculate the y values
                                  % corresponding to the vector x
>> plot(x,y)                      % plot y versus x
>> grid                           % add grid lines to the current axes
>> xlabel('Distance')             % label the x-axis
>> ylabel('Concentration')        % label the y-axis
>> title('Figure 1: y = xe^-^x')      % add the title of the graph
>> gtext('anywhere')              % place text with mouse
>> text(1,0.2,'(1,0.2)')          % place text at the specific point
```

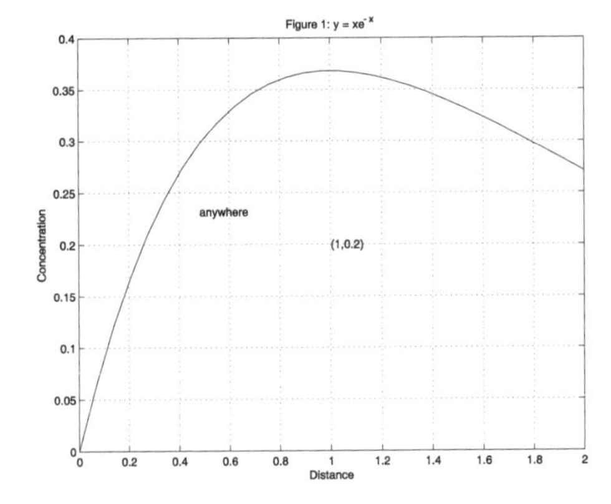

Figure A.8 Sample of 2-D graph.

You can use symbols instead of lines. You can also plot more than one function on the same figure (see Fig. A.9):

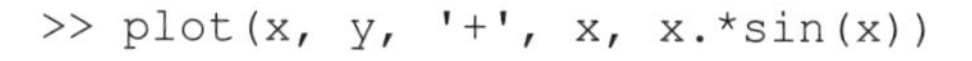

```
>> plot(x, y, '+', x, x.*sin(x))
```

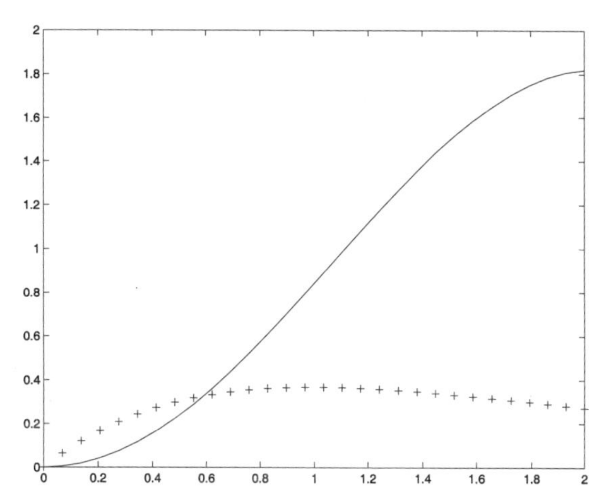

Figure A.9 Multiple functions plotted on the same figure.

If desired, more than one graph may be shown in different frames of the same figure. The following commands create a (2×1) array of frames and place the first plot in frame 1 and the second in frame 2 (see Fig. A.10):

```
>> subplot(2,1,1), plot(x,x.*cos(x))
>> subplot(2,1,2), plot(x,x.*sin(x))
```

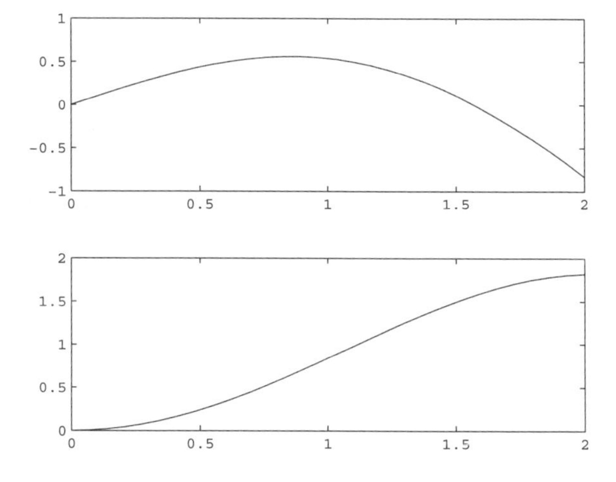

Figure A.10 Multiple graphs on the same figure.

Axis limits can be seen and modified using the `axis` command:

```
>> axis
ans =
     0     2     0     2
```

Try the following command and observe what happens to the plot:

```
>> axis([0, 1.5, 0, 1.5])
```

Before continuing, clear the graphics window:

```
>> clf
```

Now let us see the comet-like trajectory of the function:

```
>> shg, comet(x,y)
```

The `shg` command shows the current graphics window. It is possible to use more than one graphics window by using the `figure(n)` command, where `n` is a positive integer.

Another easy way to plot a function is to use `fplot(FUN, LIMS)` that plots the function `FUN` between the x-axis limits `LIMS = [XMIN XMAX]`:

```
>> figure(2)              % creates a second figure window
>> fplot('x*exp(-x)',[0, 2])
```

The function to be plotted may also be a user-defined function (see Section A.6).

Other useful two-dimensional plotting facilities are (the student is encouraged to try all of these):

```
>> plotyy(x1,y1,x2,y2) % plots x1 versus y1 with y-axis labeling
                       % on the left and plots x2 versus y2 with
                       % y-axis labeling on the right
>> semilogx(x,y) % semilogarithmic plot, x-axis is logarithmic
>> semilogy(x,y) % semilogarithmic plot, y-axis is logarithmic
>> loglog(x,y)   % full logarithmic plot, both axes logarithmic
>> area(x,y)     % filled area plot
>> polar(x,y)    % polar coordinate plot
>> bar(x,y)      % bar graph of x on the horizontal axis
                 % and y on the vertical axis shown as columns
```

3-D graphs

MATLAB has several commands for visualizing three-dimensional functions. A 3-D curve can be shown by the `plot3` command:

```
>> t = 0:0.01:3*pi;
>> plot3(t, sin(t), cos(t))
>> xlabel('t'), ylabel('sin t'), zlabel('cos t')
```

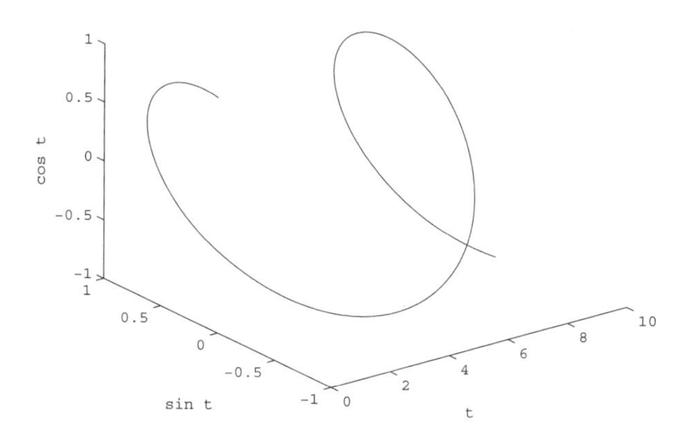

Figure A.11 Example of a 3-D graph.

Surfaces can be shown in many ways. Here are two that may be of interest:

```
>> [x, y] = meshgrid(-pi:pi/10:pi,-pi:pi/10:pi);
>> z = cos(x).*cos(y);
>> surf(x,y,z)
```

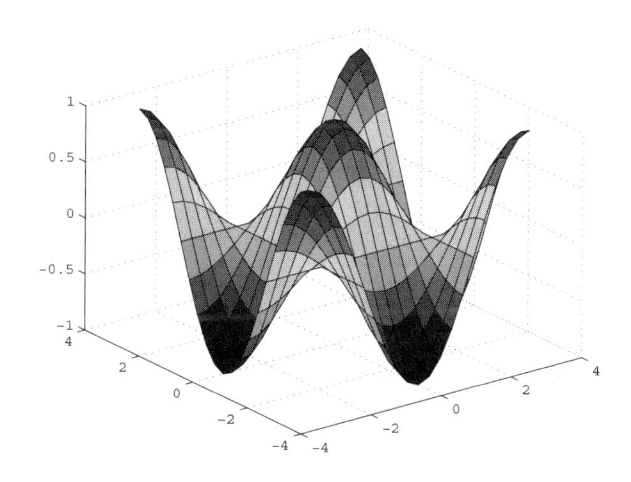

Figure A.12 Example of a 3-D surface graph.

Also try the following commands:

```
>> mesh(x,y,z)
>> view(30,60)
```

You may make your graph look better by using the `shading` command

```
>> shading interp      %controls the color shading of surface
```

To see the color scale, use `colorbar`

```
>> colorbar
```

2½-D Graphs

The so-called 2½-D graph is used for visualizing a 3-D graph on a 2-D system of coordinates. Let us use x, y, and z variables from the previous section. We can show different z-levels on an x-y system of coordinates by its contour lines (Fig. A.13):

```
>> contour(x,y,z)
```

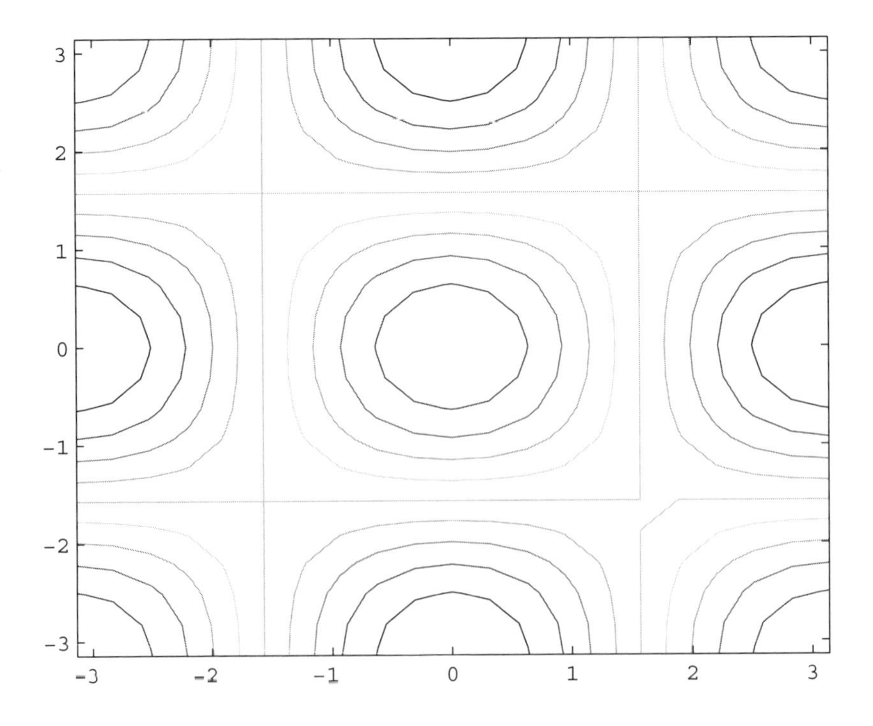

Figure A.13 Contour plots.

It is possible to show only specific contour lines, as required:

```
>> contour(x,y,z,[-0.9:0.3:0.9])  % shows contours between -0.9
                                  % and 0.9 in intervals of 0.3
```

and to print the level values on the contour lines:

```
>> [c,h] = contour(x,y,z,[-0.9:0.2:0.9]);
>> clabel(c,h)
```

Another method is to look at the graph as a pseudocolor plot (Fig. A.14) that assigns different colors to different z-values:

```
>> pcolor(x,y,z)
>> colorbar
>> shading interp
```

The `quiver` command can also be useful in visualizing vector fields such as velocity profiles.

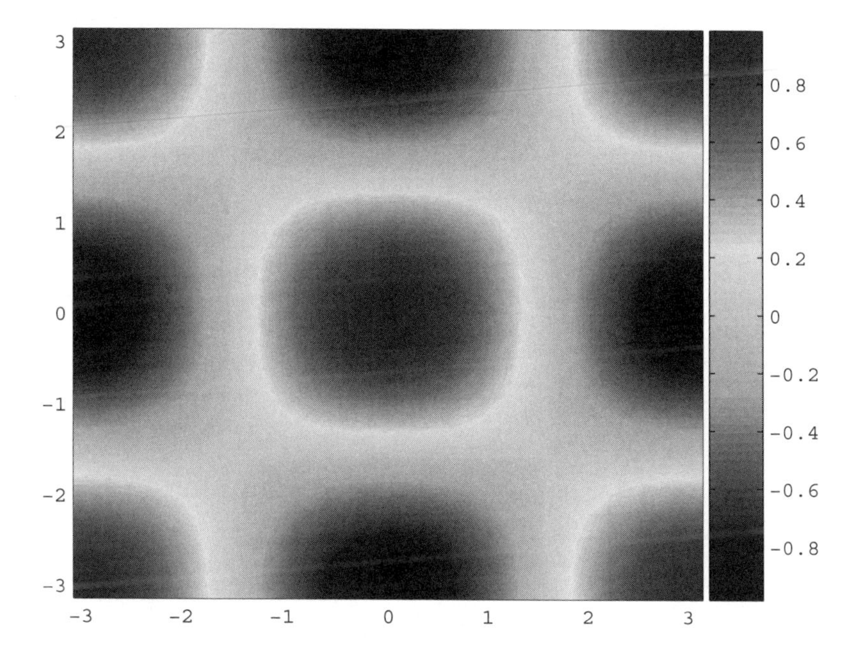

Figure A.14 Pseudocolor plot.

The figures in MATLAB may be edited, rotated, zoomed, and modified using the menus in the graphics window, as described in the next section.

Interactive Plot Creation

MATLAB 7 introduces a new set of tools to let you interactively create and edit plots without typing any MATLAB code. You can create plots from scratch and visualize your data quickly. You can also automatically generate the code to create the plot again with new data. Interactive plot creation and editing is performed by using the plot tools, which are launched from the figure window toolbar, as shown in Fig. A.15. You can launch the plot tools after opening a blank figure window. However, it is more likely that you will turn them on after creating an initial plot by selecting and plotting from the workspace browser or array editor, from a plotting function entered at the command window, or even from an existing .fig file. The plot tools provide point-and-click functionality that enables you to:

- Drag and drop new data sets onto the figure
- Add new subplot axes
- Change object properties
- Add annotations and draw shapes

In MATLAB 7, you can generate an M-code function from the graphic you created with plot tools (or with plotting functions entered at the command line). You do this by choosing Generate M-file... from the figure's File menu.

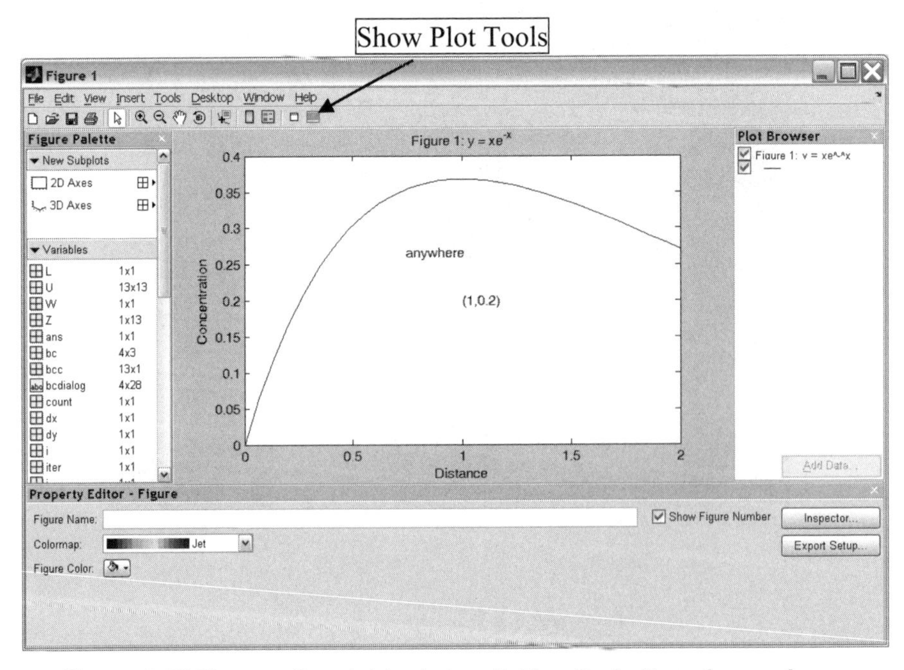

Figure A.15 Turn on the plot tools by clicking the button shown above.

A.6 Scripts and Functions

Use the MATLAB Editor to write programs in the MATLAB language and save them for future use. The editor saves your files with the `m` extension, which is where the name of *M-file* originates. The editor may also be used to execute and debug the programs. M-files can be in the form of *scripts*, such as a sequence of commands that perform a certain task and serve the purpose of a main program, or *functions*, which perform specific tasks, such as repetitive calculations. The M-files, once debugged, may be executed in the MATLAB workspace. Scripts may call functions written by the user, or built-in MATLAB functions, to perform certain tasks and return results. Scripts may also call on other scripts, as shown on Fig. A.16.

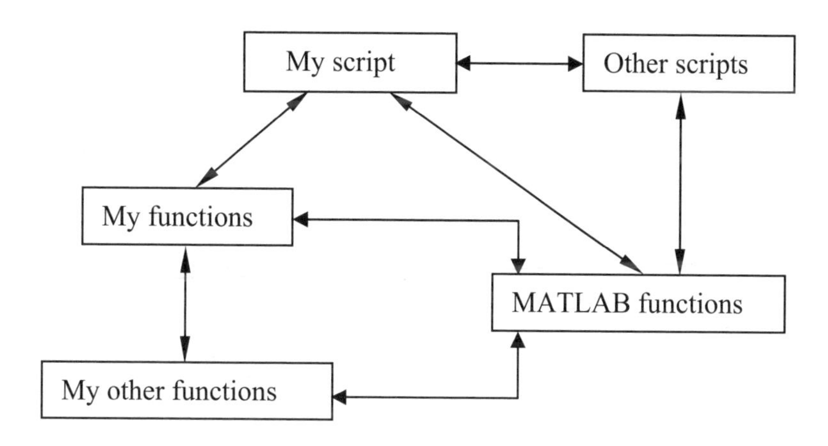

Figure A.16 Scripts, functions, and MATLAB functions.

A script is a series of MATLAB commands that are assembled together into a program using the editor. When typing the name of the script, the commands will be executed sequentially as if they were typed in the same order in the MATLAB Command Window. For example, let us calculate the volume of an ideal gas as a function of pressure and temperature. Type the following commands in the editor:

```
% ideal_gas.m  -  A sample script file
% Calculates the volume of an ideal gas
%
clear all; clc; %clear all previous variables and command window
disp(' Calculating the volume of an ideal gas.')
R = 8314;          % Gas constant
T = input(' Vector of temperatures (K) = ');
P = input(' Pressure (bar) = ')*1e5;
V = R*T/P;         % Ideal gas law
```

```
% Plotting the results
plot(T,V)
xlabel('T (K)')
ylabel('V (m^3/kmol)')
title('Ideal Gas Volume vs Temperature')
```

Save the above script as `ideal_gas.m`. Return to the MATLAB command window and type its name:

```
>> ideal_gas
```

Input the required data and observe the results.

A practical method for beginners to create a script is using `diary`. You can start creating a diary by typing:

```
>> diary mydiary
```

Then you start typing your statements in the Command Window, one by one. For example, you can type the statements of the script developed above, see the results at each step, and make corrections if necessary. When you get your desired results, close the diary:

```
>> diary off
```

Now you can develop a script by editing the file `mydiary` (no extension is added by MATLAB). Delete the unnecessary commands and output, and save it as an M-file.

You can develop your own functions and execute them just like any other function in MATLAB. A function takes in some data as input, performs the required calculations, and returns the results of calculations back to the command that invoked the function. As an example, let us write a function to do the ideal gas volume calculations that we have already done in a script. We can make this function more general so that it will be able to calculate the volume at multiple pressures and multiple temperatures:

```
function V = ideal_gas_func(T, P)
% This function calculates the specific volume of an ideal gas
R = 8314;                  % Gas constant
for k = 1:length(P)
   V(k,:) = R*T/P(k);   % Ideal gas law
end
```

The first line of a function is the function declaration line and begins with the word `function` followed by the output argument(s), equality sign, name of the function, and input argument(s), as illustrated in the example above. The first set of continuous comment lines immediately after the function declaration line is the help for the function and can be reviewed separately:

```
>> help ideal_gas_func
```

This function must be saved as `ideal_gas_func.m`. You can now use this function in the workspace, in a script, or in another function. For example:

```
>> P = [1:10]; T = [300:10:400];
>> vol = ideal_gas_func(T, P);
>> surf(T, P, vol)
>> view(135,45), colorbar
```

The results of these commands are shown on Fig. A.17.

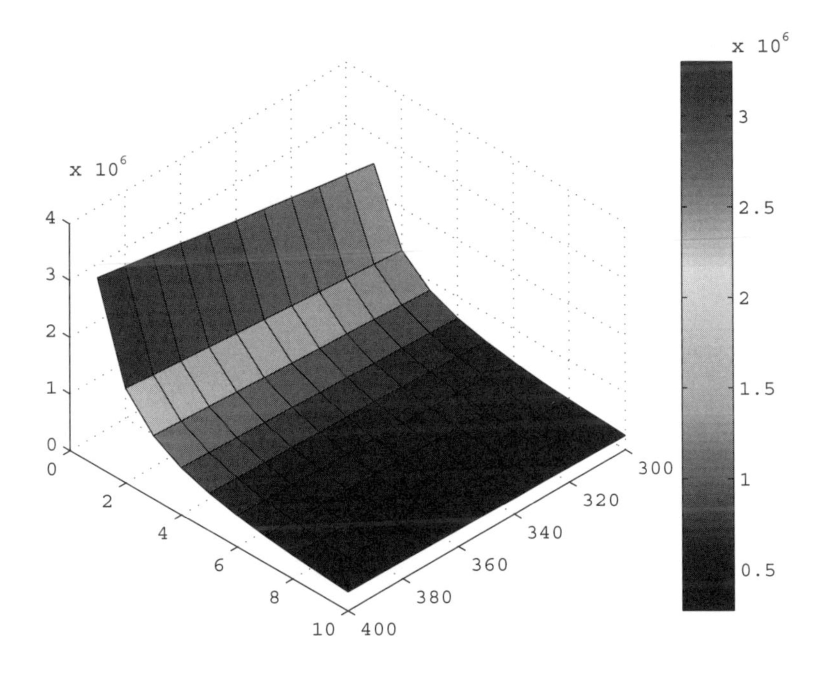

Figure A.17 Plot of ideal gas volume as function of temperature and pressure

Functions do not necessarily require input, and may not return the results of calculations back to the command. For example, the script `ideal_gas.m` listed above may also be saved in the form of a function:

```
function ideal_gas
% Calculates the volume of an ideal gas
%
clear all; clc;  %clear all previous variables and command window
```

```
disp(' Calculating the volume of an ideal gas.')
R = 8314;          % Gas constant
T = input(' Vector of temperatures (K) = ');
P = input(' Pressure (bar) = ')*1e5;
V = R*T/P;         % Ideal gas law
% Plotting the resultsa
plot(T,V)
xlabel('T (K)')
ylabel('V (m^3/kmol)')
title('Ideal Gas Volume vs Temperature')
```

In this case, there is little difference between the script and the function. Either one may be executed from the Command Window (or from another script) by invoking it by name.

A.7 Flow Control

MATLAB has several flow control structures that allow the program to execute different sequences of statements depending on the value of data. These structures are `if`, `switch`, `for`, and `while`, which we describe briefly below. We do not show the results of each operation. It is left as an exercise to the student to use these commands in sample programs and observe the results:

`if . . . (else . . .) end` –The `if` control structure enables the program to make decision about what commands to execute. In the following sequence, the line `y = x^2` executes only if `x` input is greater than or equal to zero:

```
x = input(' x = ');
if x >= 0
   y = x^2
end
```

You can also add an `else` clause with statements that are executed if the condition in the `if` statement is not true:

```
x = input(' x = ');
if x >= 0
   y = x^2
else
   y = -x^2
end
```

`switch . . . case . . . end` –Another form of conditional execution is to execute a different block of statements for different value of a variable. The `switch . . . case . . . end` structure is easier to use than a nested `if` structure. For example:

```
a = input('Give a value of a (1, 2, or 3) = ');
switch a
case 1
   disp('One')
case 2
   disp('Two')
case 3
   disp('Three')
end
```

`for ... end` –This control structure is used to repeat the execution of a block of statements a fixed number of times:

```
k = 0;
for x = 0:0.2:1
   k = k+1
   y(k) = exp(-x)
end
```

`while ... end` –This control structure is used to repeat the execution of a block of statements until a condition of the data exists:

```
x = 0;
while x<1
   y = sin(x)
   x = x+0.1;
end
```

There are three other flow control commands: `break`, `pause`, and `return`. The `break` command is used to stop executing an iteration before it has completed. The `pause` command will cause the program to wait for a key to be pressed (on the keyboard) before continuing:

```
k = 0;
for x = 0:0.2:1
   if k>3
      break
   end
   k = k+1
   y(k) = exp(-x)
   pause
end
```

The `return` command terminates a function invocation, returning to either the calling script or to the command line.

A.8 Display, Export, and Import of Data

A.8.1 Displaying data and results

There are many different ways that data and other information may be displayed on the screen, sent to the printer, saved as files, or exported to other formats. We begin with the most elementary ones: `disp` and `fprintf`.

```
>> X=[1 2 3; 4 5 6];
>> disp(X)
     1     2     3
     4     5     6
```

The `disp` command displays the array without printing the array name. Only one variable may be displayed at a time. In all other ways the `disp` command is the same as leaving the semicolon off an expression, except that empty arrays do not display. The `disp` command may also be used to display text:

```
>> disp('This is the X array:'); disp(X)
This is the X array:
     1     2     3
     4     5     6
```

The `fprintf` command writes formatted data to the screen or to a file. For example, to write text and data to the screen:

```
>> fprintf('\n The value of position (2,2) of X = %g',X(2,2))

 The value of position (2,2) of X = 5
```

The special format control characters `\n`, `\r`, `\t`, `\b`, and `\f` can be used to produce new line, carriage return, tab, backspace, and formfeed characters, respectively. Use `\\` to produce a backslash character and `%%` to produce the percent character.

The `%g` above is an example of a format specification string that controls how the data will be exhibited. This string contains C language conversion specifications that involve the character `%`, optional flags, optional width and precision fields, optional subtype specifier, and conversion characters `c`, `d`, `e`, `E`, `f`, `g`, `G`, `i`, `o`, `s`, `u`, `x`, and `X`. The uses of these characters are described in Table A.1.

Several examples of using the `fprintf` command are given below. The students should format the commands in a way that presents their data and results in an informative and easy-to-read display, appropriate for the problem at hand.

```
>> fprintf('This is a test of the printing commands')
This is a test of the printing commands
>> a=-5.23 ;   b=3;   c=0.000000015;   d = 'Test';
```

```
>> fprintf('Value of a = %7.4f   b = %2i  c = %10g  d = %s \n', a,
b, c, d)
Value of a = -5.2300   b =  3  c =    1.5e-008  d = Test

>> fprintf('%12.2e', pi)
   3.14e+000

>> fprintf('%0.5e', pi)
3.14159e+000

>> fprintf('%0.5f', pi)
3.14159

>> fprintf('%10.5f', pi)
   3.14159

>> fprintf('%15.5g', eps)
    2.2204e-016

>> y=[1 2 3 4];
>> fprintf('%5.2f \n' ,y)
 1.00
 2.00
 3.00
 4.00
```

Table A.1 List of format symbols for printing in MATLAB

Specifier	Description
%c	Single character
%d	Decimal notation (signed)
%e	Exponential notation (using a lowercase e as in 3.1415e+00)
%E	Exponential notation (using an uppercase E as in 3.1415E+00)
%f	Fixed-point notation
%g	The more compact of %e or %f. Insignificant zeros do not print.
%G	Same as %g, but using an uppercase E
%i	Decimal notation (signed)
%o	Octal notation (unsigned)
%s	String of characters
%u	Decimal notation (unsigned)
%x	Hexadecimal notation (using lowercase letters a-f)
%X	Hexadecimal notation (using uppercase letters A-F)

A.8.2 Saving and loading data

There are several ways in which you can save your data in MATLAB. Let us first clear all other variables from the workspace and generate some new variables:

```
>> clear
>> a = magic(3),  b = magic(4)  %magic is a special MATLAB matrix
a =
       8     1     6
       3     5     7
       4     9     2
b =
      16     2     3    13
       5    11    10     8
       9     7     6    12
       4    14    15     1
```

The following command saves all the variables in the MATLAB workspace in the file `f1.mat`:

```
>> save f1
```

If you need to save only specific variables from the workspace, list their names after the file name. The following saves only the variable `a` in the file `f2.mat`:

```
>> save f2 a
```

The files generated above have the extension "`.mat`" and can be retrieved only by MATLAB. To use the data elsewhere you may want to save them in text format:

```
>> save f3 b -ascii
```

Here, the file `f3` is a text file with no extension in its name. To see all the different ways that `save` may be used, give the command `help save`.

You can load data into the MATLAB workspace using the `load` command. If the file to be loaded was generated by MATLAB (carrying the `.mat` extension), the variables will appear in the workspace with the same names they had at the time they were saved:

```
>> clear
>> load f1
>> whos        % verifies the presence of these data

  Name       Size                       Bytes  Class

  a          3x3                           72  double array
```

```
  b         4x4                         128  double array

Grand total is 25 elements using 200 bytes
```

However, if the file is a text file, the variables will appear in the workspace under the name of the file:

```
>> clear
>> load f3
>> whos
  Name      Size                      Bytes  Class

  f3        4x4                         128  double array

Grand total is 16 elements using 128 bytes
```

Data may also be created using the MATLAB editor. For example, open the Editor and enter the values of the vector variable `y` as a column of numbers:

```
0.8345
0.0381
0.0163
0.0287
0.0220
0.0434
```

Save the file as `y.dat`.

To load the data from the file just saved, use the `load` command from the MATLAB Command Window (or from within a script):

```
load y.dat
```

The workspace now contains the vector variable `y`.

Another way of reading the data is by using the `open` command:

```
open y.dat
```

The Import Wizard (Fig. A.18) opens up automatically when the open command is used. It gives you the opportunity to specify how the data are formatted before they are read into the workspace.

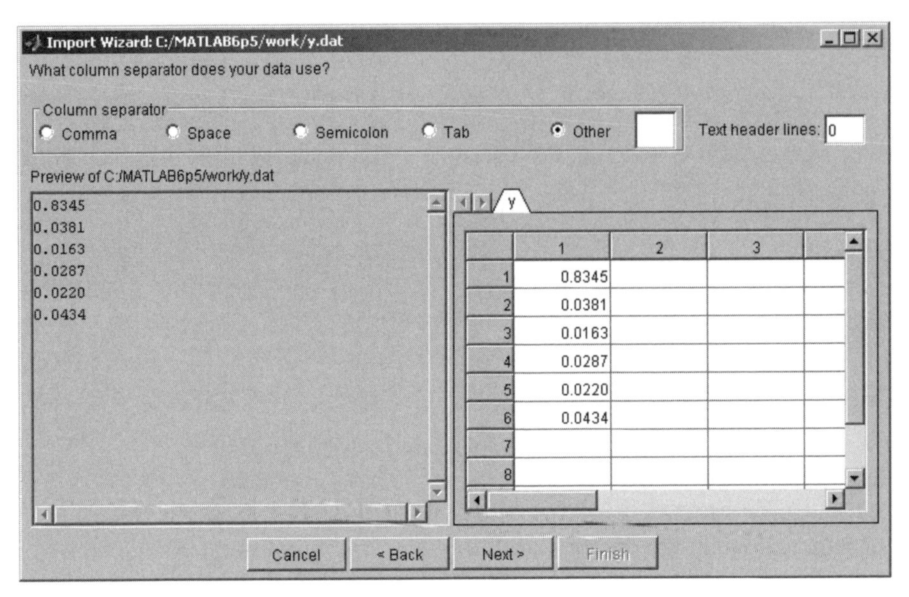

Figure A.18 Import Wizard.

Press the `Next` button,to see the following window (Fig. A.19), which gives options for formatting multidimensional data:

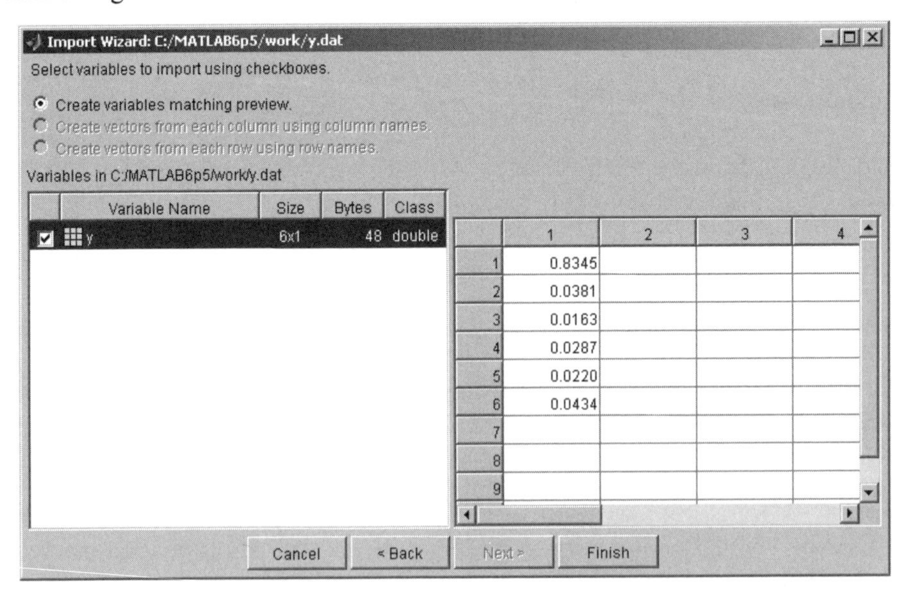

Figure A.19 Import Wizard: options for formatting multidimensional data.

Pressing the `Finish` button will complete the task of loading `y`. Spreadsheet files, such as Excel files, may also be read this way, including column and row headings.

Alternatively, from the MATLAB Command Window (or from within a script), open the file and read it:

```
fidy = fopen('y.dat');
% fidy is the identification number for this file
y=fscanf(fidy, '%f')
y =
     0.8345
     0.0381
     0.0163
     0.0287
     0.0220
     0.0434
```

Always close the file when finished reading it:

```
fclose(fidy)
```

Similarly, you can create a file with the values of `x` and `y` and name it `xy.dat`:

```
0.0    0.8345
0.1    0.0381
0.2    0.0163
0.3    0.0287
0.4    0.0220
0.5    0.0434
```

Since the first column is `x` and the second column is `y`, we need to read each data in the sequence `x(1), y(1), x(2), y(2), x(3), y(3)`,. . . etc. To do so let us write a script using the MATLAB Editor:

```
% Opening and reading xy.dat file with two columns of data
%
clear
fidxy=fopen('xy.dat');
for i=1:6
   x(i) = fscanf(fidxy,'%f',1);  % reads one value of x at a time
   y(i) = fscanf(fidxy,'%f',1);  % reads one value of y at a time
end
fclose(fidxy);
x, y
```

The `fscanf` command reads formatted data from a file. Save this script as `read_xy.m` and use it from the Command window:

```
>> read_xy
x =
```

```
  Columns 1 through 5
         0    0.1000    0.2000    0.3000    0.4000
  Column 6
    0.5000
y =
  Columns 1 through 5
    0.8345    0.0381    0.0163    0.0287    0.0220
  Column 6
    0.0434
```

Please note that x and y are read as row vectors (rather than column vectors).

Also, try using the commands `open` and `load` and notice the difference in how these commands read the file:

```
open xy.dat
```

or

```
load xy.dat
```

A.8.3 Generating data in a program and saving into a file

The program below generates some data, creates and opens a text file named `expon.dat`, uses the `fprintf` command to write the data into a file using the desired format, and closes the file `expon.dat` that contains a short table of the exponential function:

```
x = 0:0.1:1;
y = [x; exp(x)];
fid = fopen('expon.dat','w');
fprintf(fid,'%6.2f  %12.8f\n',y);
fclose(fid);
```

Use the command `load` to read the file and place all the data into one variable (matrix):

```
clear
load expon.dat
expon
expon =
         0    1.0000
    0.1000    1.1052
    0.2000    1.2214
    0.3000    1.3499
    0.4000    1.4918
    0.5000    1.6487
    0.6000    1.8221
    0.7000    2.0138
```

```
0.8000     2.2255
0.9000     2.4596
1.0000     2.7183
```

If necessary, use the method described in Section A.8.2 to open the `expon.dat` file and read it as two variables.

A.9 Symbolic Computation

The Symbolic Math Toolbox, an integral part of the MATLAB Student Version, incorporates analytic computation into MATLAB's numeric environment. This toolbox enables the student to perform computations using symbolic mathematics and variable-precision arithmetic. The student is strongly encouraged to walk through the Symbolic Math demos that are provided (see Help/Demos/Toolboxes/Symbolic Math). Only a few examples of the symbolic math commands will be given in this Appendix.

To create a symbolic variable use the command `sym( )`:

```
>> y= sym('y')
y =
y
>> om= sym('omega')
om =
omega
```

The above commands created a symbolic variable `y` that prints as `y`, and a symbolic variable `om` that prints as `omega`. Variables may also be declared as symbolic by the command `syms`:

```
>>syms e f g
```

A.9.1 Symbolic solution of algebraic equations

Now, let us use the symbolic function `solve` to find the solution of the quadratic equation:

$$ax^2 + bc + c = 0$$

Several ways of using the `solve` function will be demonstrated below:

(a) Find the zeroes of the function, assuming that the unknown is `x`.

```
>>clear
>> x = solve('a*x^2+b*x+c')
x =
[ 1/2/a*(-b+(b^2-4*a*c)^(1/2))]
[ 1/2/a*(-b-(b^2-4*a*c)^(1/2))]
```

(b) The user specifies that the function = 0, and the unknown is `x`.

```
>> x = solve('a*x^2 + b*x + c=0','x')
x =
[ 1/2/a*(-b+(b^2-4*a*c)^(1/2))]
[ 1/2/a*(-b-(b^2-4*a*c)^(1/2))]
```

(c) The function is defined symbolically as `S`.

```
>> syms x a b c S
S = a*x^2 + b*x + c;
x = solve(S)
x =
[ 1/2/a*(-b+(b^2-4*a*c)^(1/2))]
[ 1/2/a*(-b-(b^2-4*a*c)^(1/2))]
```

(d) The function is = 5, and the unknown is `x`.

```
>> x = solve('a*x^2 + b*x + c=5','x')
x =
[ 1/2/a*(-b+(b^2-4*a*c+20*a)^(1/2))]
[ 1/2/a*(-b-(b^2-4*a*c+20*a)^(1/2))]
```

(e) The function is solved for `a` instead of `x`.

```
>> a = solve('a*x^2 + b*x + c=5','a')
a =
-(b*x+c-5)/x^2
```

Solve two simultaneous algebraic equations:

$$\alpha N_1 - \beta N_1 N_2 = 0$$
$$-\gamma N_2 + \delta N_1 N_2 = 0$$

```
>> [N1,N2] = solve('alpha*N1 - beta*N1*N2 = 0',...
                 '-gamma*N2 + delta*N1*N2 = 0','N1,N2')
N1 =
[             0]
[ 1/delta*gamma]
N2 =
[            0]
[ 1/beta*alpha]
```

Note: Three or more points at the end of a line (...) indicate continuation of a MATLAB statement into the next line.

A.9.2 Symbolic solution of differential equations

Solve the differential equation:

$$\frac{dy}{dt} + 4y = e^{-t}$$

```
>> y = dsolve('Dy + 4*y = exp(-t)')
y =
(1/3*exp(3*t)+C1)*exp(-4*t)
```

Solve the same differential equation with the initial condition `y(0)=2`:

```
>> y = dsolve('Dy + 4*y = exp(-t)', 'y(0)=2')
y =
(1/3*exp(3*t)+5/3)*exp(-4*t)
```

Set the range of the independent variable *tt*, evaluate the dependent variable `y`, that was obtained above, for this range, and plot the profile (Fig. A.20):

```
>> tt = [0:.1:5];
>> for i=1:length(tt)
t=tt(i);
yy(i)=eval(y);
end
```

Note that the symbolic solution of `y` contains `t`, as a scalar, therefore `t` had to be created as such.

```
>> plot(tt,yy)
```

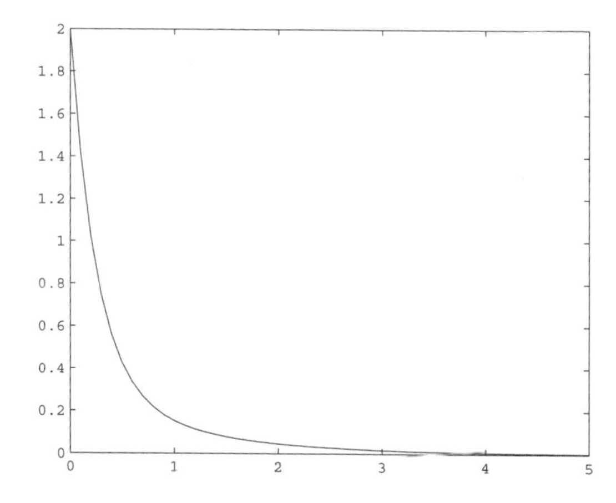

Figure A.20

Solve two simultaneous differential equations:

$$\frac{dy_1}{dt} + 4y_2 = e^{-t} \quad \text{with initial condition } y_1(0) = 1$$

and

$$\frac{dy_2}{dt} = -y_1 \quad \text{with initial condition } y_2(0) = 2$$

```
>> [y1,y2]=dsolve('Dy1 + 4*y2 = exp(-t)','Dy2=-y1','y1(0)=1, y2(0)=2')
y1 =
-4/3*exp(2*t)+2*exp(-2*t)+1/3*exp(-t)
y2 =
2/3*exp(2*t)+exp(-2*t)+1/3*exp(-t)
```

Verify the answers by substituting them into the differential equations. The function `diff` differentiates the symbolic expression (See Section A.9.3)

```
>> diff(y1) + 4*y2
ans =
exp(-t)
>> diff(y2)
ans =
4/3*exp(2*t)-2*exp(-2*t)-1/3*exp(-t)
```

Set the range of the independent variable, evaluate the dependent variable for this range, and plot the profile:

```
>> t=[0:.01:1];  yy1=eval(y1);  yy2=eval(y2);
>>plot(t,yy1,':', t, yy2)
```

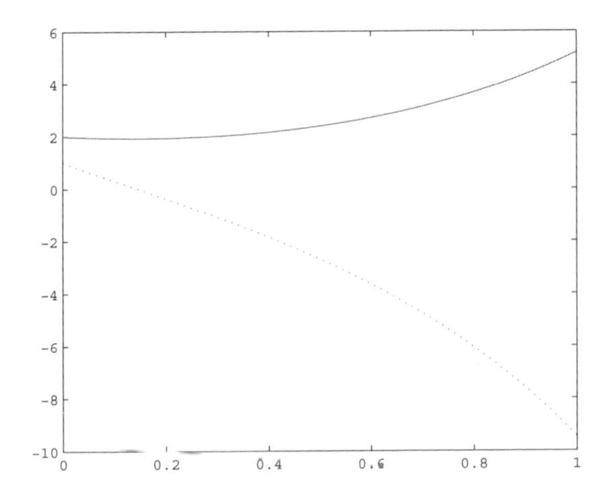

Figure A.21

A.9.3 Symbolic differentiation

Clear all previous variables and define the new symbolic variables:

```
>> clear
>> syms x a f Re e
```

Differentiate:

```
>> diff(x^2)
ans =
2*x
>> diff(sin(x))
ans =
cos(x)

>> diff(exp(x^2))
ans =
2*x*exp(x^2)

>> diff(log(x))
ans =
1/x

>> diff(acos(x))
ans =
-1/(1-x^2)^(1/2)

>> diff(acos(x))
ans =
-1/(1-x^2)^(1/2)

>> diff(1/sqrt(f)+.86*log(e/3.7+2.51/Re/sqrt(f)), f)
ans =
-1/2/f^(3/2)-10793/10000/Re/f^(3/2)/(10/37*e+251/100/Re/f^(1/2))
```

A.9.4 Symbolic integration

In the following vector, the first two elements involve integration with respect to `x`, while the second two are with respect to `a`.

```
>> [int(x^a), int(a^x), int(x^a, a), int(a^x, a)]
ans =
[ x^(a+1)/(a+1),  1/log(a)*a^x,  1/log(x)*x^a, a^(x+1)/(x+1)]
```

Another example of integration:

```
>> int('a*T+b*T^2+c*T^4','T')
ans =
1/2*a*T^2+1/3*b*T^3+1/5*c*T^5
```

Limits of integration may be used:

```
>> int('a*T+b*T^2+c*T^4','T',298,500)
ans =
80598*a+98536408/3*b+28899927176032/5*c
```

A.10 MATLAB Toolboxes

Toolboxes are specialized collections of M-files built specifically for solving particular classes of problems. These toolboxes are optional, and may be licensed separately from MathWorks, Inc. Toolboxes related to the topics of this textbook are listed in Table A.2.

Table A.2 Selected MATLAB Toolboxes

Toolbox	Description
Curve Fitting	Perform model fitting and analysis
Image Processing	Perform image processing, analysis, and algorithm development
Partial Differential Equation	Solve and analyze partial differential equations
Signal Processing	Perform signal processing, analysis, and algorithm development
Simulink	Model, simulate, and analyze dynamic systems
Statistics	Apply statistical algorithms and probability models
Symbolic Math	Perform computations using symbolic mathematics and variable-precision arithmetic

The MATLAB Student Version contains Simulink and Symbolic Math toolboxes in addition to the basic MATLAB.

A.11 References

Hanselman, D., and Littlefield, B. 2005. *Mastering MATLAB 7.* Upper Saddle River, NJ: Prentice Hall.

Hahn, B. D. 2002. *Essential MATLAB for Scientists and Engineers.* Oxford, UK: Butterworth- Heinemann Publications.

Lyshevski, S. E. 2003. *Engineering and Scientific Computations Using MATLAB.* Hoboken, NJ: John Wiley & Sons, Inc..

Appendix B: Introduction to Simulink

Simulink is a graphical extension to MATLAB for modeling and simulating systems. Simulink is a visual programming language and environment: functions either built in or user defined, are represented as blocks and the data passed between functions are represented as lines. Input and output devices are also available, in addition to the broad range of functions. Simulink is integrated with MATLAB, and data can be easily transferred between the programs. In this tutorial, we will introduce the basics of using Simulink to model and simulate a system using some of the numerical methods from this text. Simulink is supported on UNIX, Macintosh, and Windows environments, and it is included in the MATLAB Student Version for personal computers. For more information on Simulink, contact The MathWorks, Inc.

B.1 Dynamic System Simulation

Engineers now have sufficient computational power readily available to simulate (or mimic) dynamic systems without requiring an actual physical setup. Simulation of dynamic systems has proved to be immensely useful when it comes to control design, saving time and money that would otherwise be spent in prototyping a physical system. The main advantage of Simulink is that it is a visual programming environment allowing the user to rapidly prototype a simulation and immediately see the results.

Concept of signal and logic flow

In Simulink, the data flow between blocks (functions) is specified by lines connecting the output of one block to the next. The final output can then be dumped into sinks, such as scopes and displays, or saved to a file. Data can be connected from one block to another, branched, or multiplexed. During a simulation, data is processed and transferred only at discrete simulation time steps (otherwise called integration time steps). The selection of a time step is determined by the fastest dynamics in the simulated system.

In the following sections, the different blocks that are available are explained.

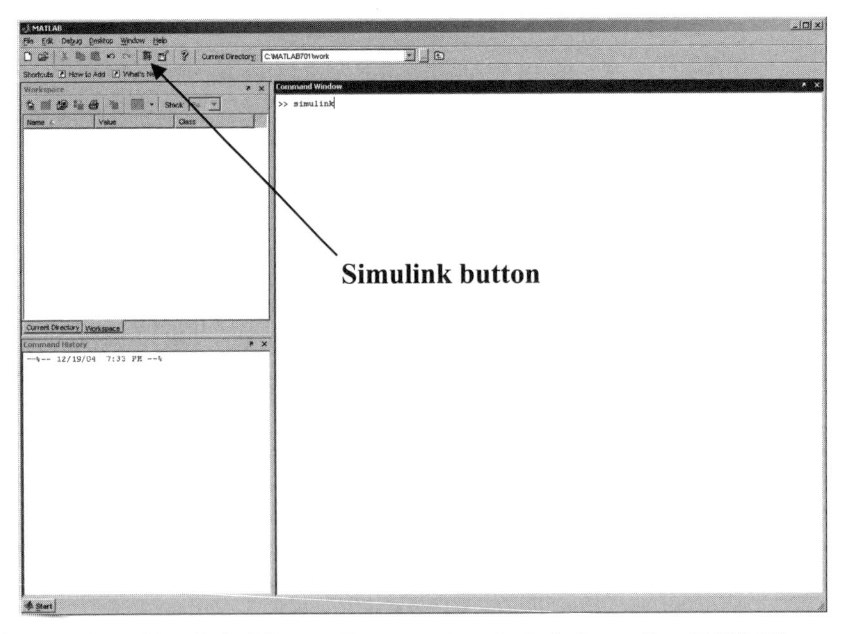

Figure B.1 **The Simulink Library Browser is started from the MATLAB command window by typing the command `simulink` at the command prompt, or by pressing the Simulink Library Browser button (black arrow) on the toolbar.**

B.2 Getting Started

Simulink may be started from the MATLAB command window in one of two ways: Either type the command `simulink` at the command prompt or press the Simulink button on the toolbar as shown in Fig. B.1.

The Simulink Library Browser will appear in a separate window. This window has three panes identified as A, B, and C in Fig. B.2. Pane A is a directory-like listing of the Simulink blocks available in your system; for each category listing, the blocks available will appear in pane B. The definition of a block that has been selected will appear in pane C that is above panes A and B.

Simulink block diagrams are created and edited in a worksheet; to open a new worksheet, press the new worksheet button on the Simulink Library Browser toolbar, as shown in Fig. B.2.

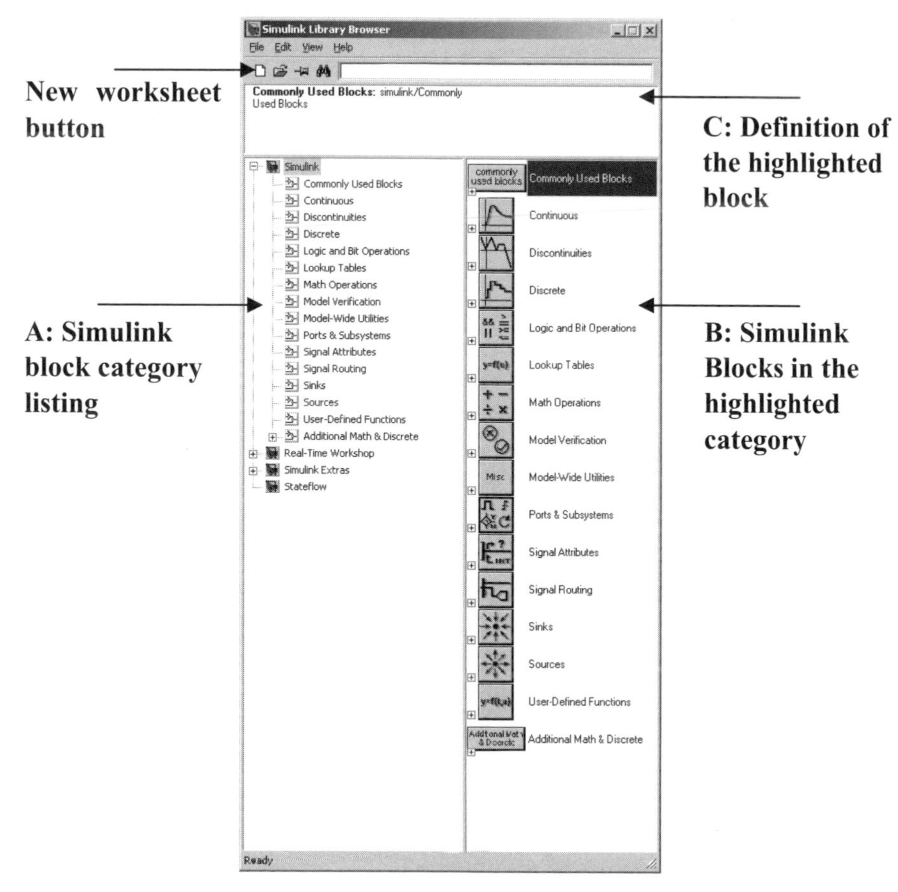

Figure B.2 The Simulink Library Browser.

B.2.1 A Simulink model of a sine wave generator

Copying the blocks to a model

Fig. B.3 shows an empty worksheet. A simple, but illustrative example block diagram can be created using only three blocks. A model is created by dragging blocks from the browser and dropping them into the worksheet.

To create a model that displays a sinusoid with period 2π, drag the following three blocks from the browser to the empty worksheet:

1. The **Sine Wave** block from the **Sources** category.
2. The **Scope** block from the **Sinks** category.
3. The **Gain** block from the **Math Operations** category.

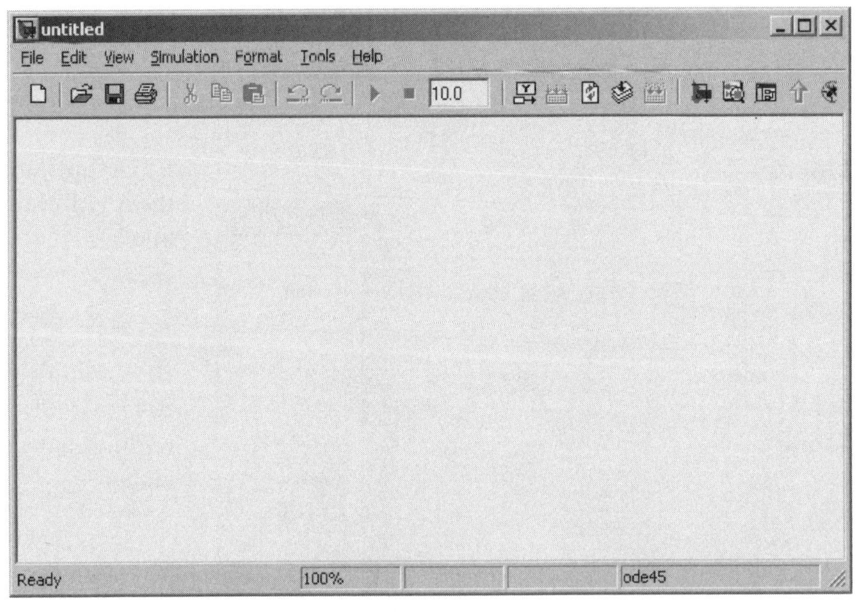

Figure B.3 An empty Simulink worksheet.

Connecting the blocks

For the example system, to display a sinusoid with period 2π, the signal output by the Sine Wave block is connected to the input of the Gain block. The Gain block amplifies this signal and outputs the amplified signal to the Scope block, which displays the signal, as a function of time, in a separate figure. Lines connecting the output of the Sine Wave block to the input of the Gain block, and from the output of the Gain block to the input of the Scope block (Fig. B.5) must now be added to the diagram. Lines are drawn by dragging the mouse from where a signal starts (output terminal of a block) to where it ends (input terminal of another block). When drawing lines, it is important to make sure that the signal reaches each of its intended terminals. Simulink will turn the mouse pointer into a crosshair when it is close enough enough to an output terminal to begin drawing a line, and the pointer will change into a double crosshair when it is close enough to snap to an input terminal. A signal is properly connected if its arrowhead is filled in. If the arrowhead is open, the signal is not connected to both blocks. To fix an open signal, treat the open arrowhead as an output terminal and continue drawing the line to an input terminal in the same manner as explained before. The completed block diagram is shown in Fig. B.5, and the scope output appears in Fig. B.6 (see running the simulation, below).

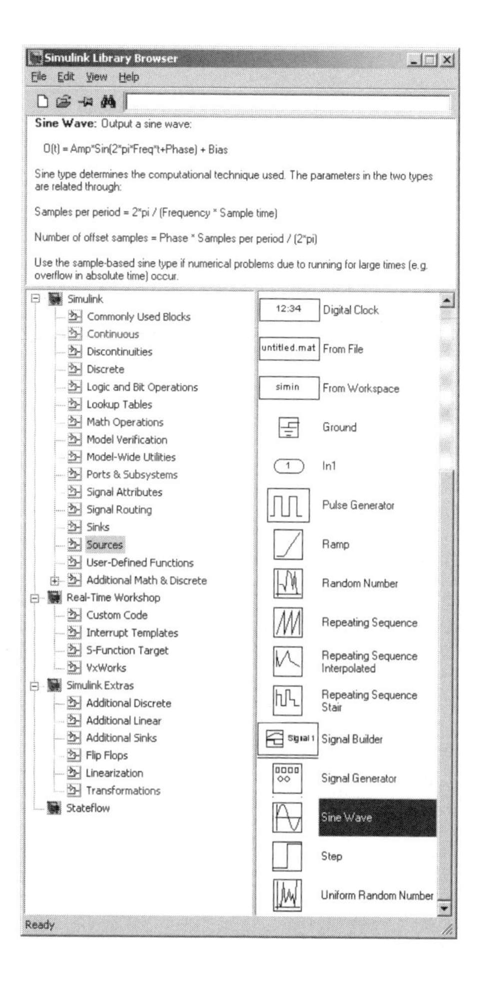

Figure B.4 Creating a Simulink model, beginning with a sine wave generator. The block is dragged to a Simulink worksheet.

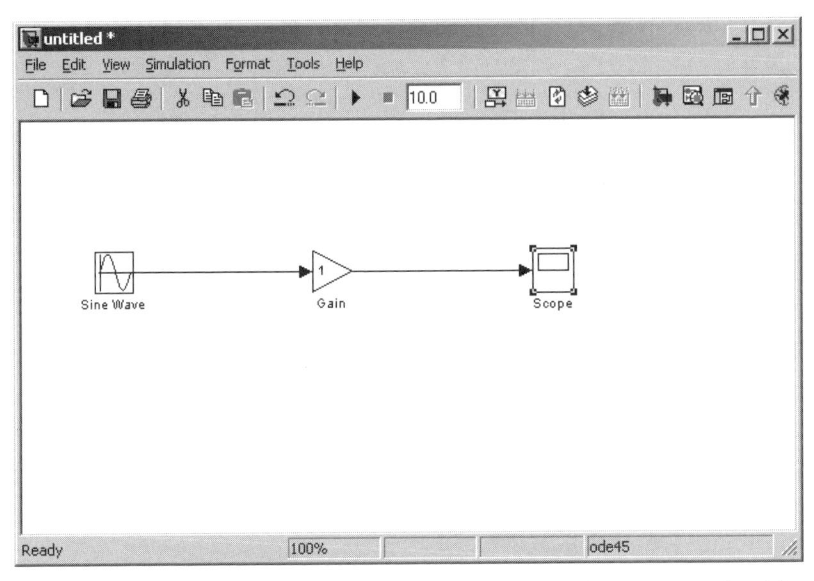

Figure B.5 Simulink blocks are connected by dragging the output of a block to its destination, the input of the next block in the pipeline.

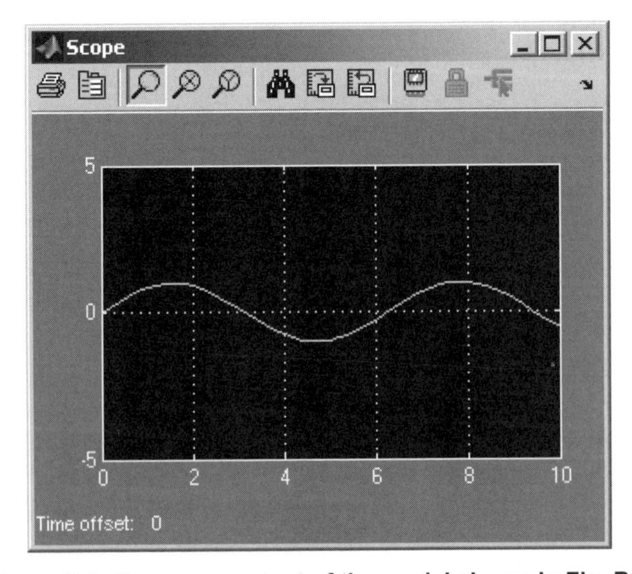

Figure B.6 The scope output of the model shown in Fig. B.5.

Table B.1 is a summary of editing operations that can be performed to edit a Simulink model.

Table B.1 Editing operations in Simulink

Operation	Keystrokes or Mouse Actions
Copy a block from a library	Drag the block to the model window with the left button on the mouse, or use COPY and PASTE from the EDIT menu.
Duplicate block(s) in a model	Hold down the CTRL key and select the block(s) with the left mouse button, then drag the block(s) to a new location.
Display block parameters	Double-click on the block
Flip a block	CTRL-F
Rotate a block (clockwise 90° with each keystroke)	CTRL-R
Change block names	Click on block's label and position the cursor to desired place.
Disconnect a block	Hold down the SHIFT key and drag the block to a new location
Draw a diagonal line	Hold down the SHIFT key while dragging the mouse with the left button
Divide a line	Move the cursor to the line to where you want to create the vertex. Use the left button on the mouse to drag the line while holding down the SHIFT key

Running the simulation

To start the simulation, pull down the simulation menu on the worksheet and press the start menu entry, or type control-T as indicated. Because the example is a simple model, its simulation runs almost instantaneously. With more complicated systems, however, you will be able to see the progress of the simulation by observing its running time in the lower box of the model window. When the simulation has ended, the result can be seen by double-clicking on the scope to view the output of the Gain block. The output of the simulation is amplitude as a function of time. Once the Scope window appears, click the "Autoscale" button in the toolbar (looks like a pair of binoculars) to scale the graph to better fit the window. The output appears in a separate window, as shown in Fig. B.6.

B.2.2 Modifying Simulink models

Properly connected signals

When drawing lines, you do not need to worry about the path you follow, as the lines will be routed automatically. Once the blocks are connected, they can be repositioned for a neater appearance. This is done by clicking on and dragging each block to its desired location (signals will stay properly connected and will re-route themselves).

In some models, it will be necessary to branch a signal so that it is transmitted to two or more different input terminals. This is done by first placing the mouse cursor at the location from where the signal is to branch. Then, using either the CTRL key in conjunction with the left mouse button, or just the right mouse button, drag the new line to its intended destination. This method was used to construct the branch in the Sine Wave output signal shown in Fig. B.7. The Scope blocks were renumbered as Scope 1 and Scope 2 by clicking the mouse on the name of the block and making the appropriate changes.

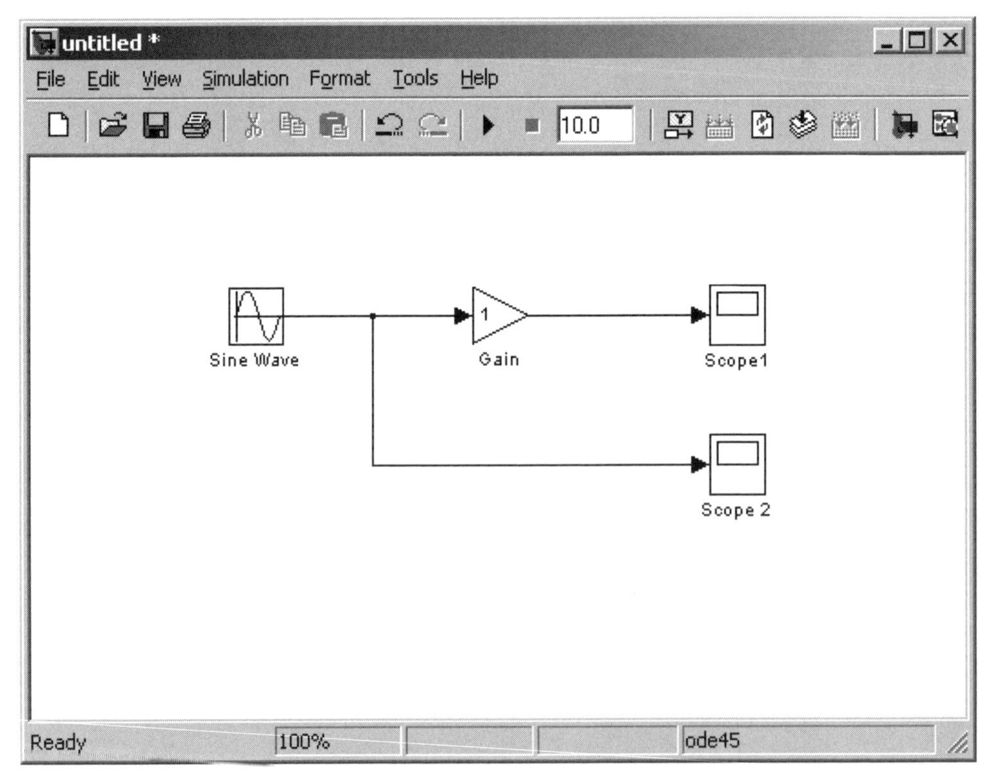

Figure B.7 Branches are added by right-clicking the mouse on a line or using CTRL-key with the left mouse button.

Modifying blocks: Entering block parameters

The parameters of a block are displayed when the block is double-clicked. To change the gain of the Gain block above, first double-click on the Gain block and change the gain to be two, as shown in Fig. B.8.

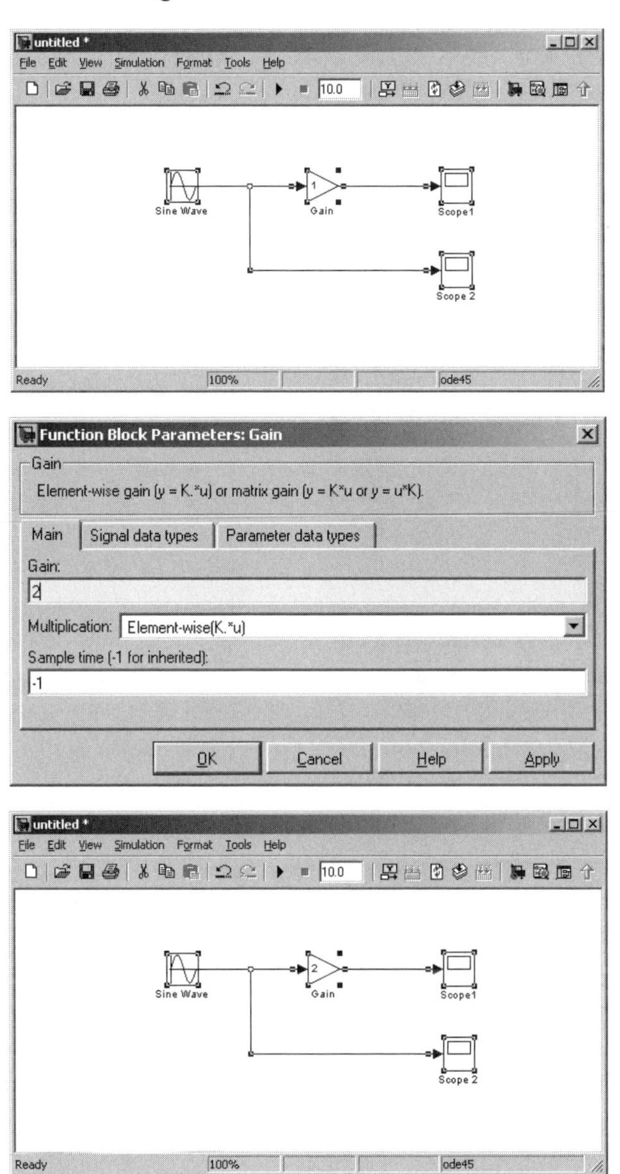

Figure B.8 Modifying the parameters of the Gain block.

Simulation configurations

Simulation time: The simulation of the system shown in Fig. B.8 can be started in one of three ways: go to the Simulation menu and click on Start, click the "Start/Pause" button in the model window toolbar, or type Control-T. The output is shown in Fig. B.9. Notice that the amplitude of the sine wave shown in Scope 1 is 2, the result of having the gain block in the path from the generator. The signal shown in Scope 2 has amplitude 1, since it is directly the output of the signal generator.

Notice that the simulation has been run for 10 seconds, as shown on both the model worksheet (to the right of the "Start/Pause" button) and the x-axis of the scope plots. The length of the simulation can be changed by changing the simulation time entry on the model toolbar or by changing the configuration parameters (Simulation/ Configuration Parameters). The scope output of a 20 second simulation is shown in Fig. B.10.

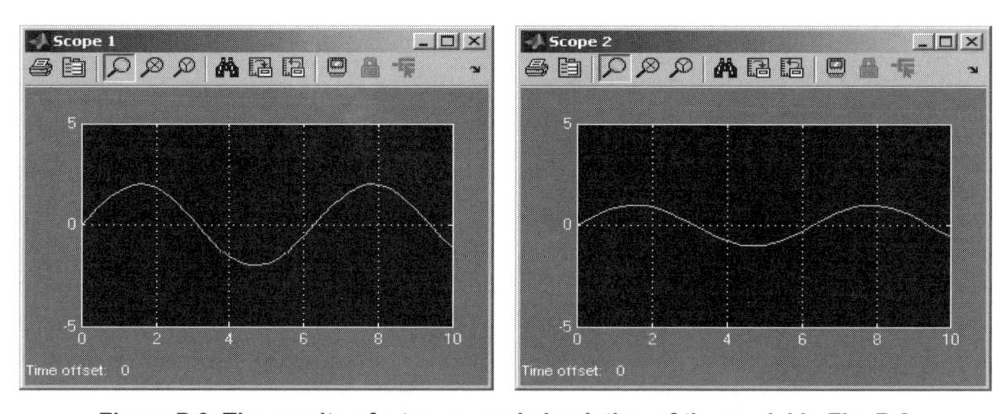

Figure B.9 The results of a ten second simulation of the model in Fig. B.8.

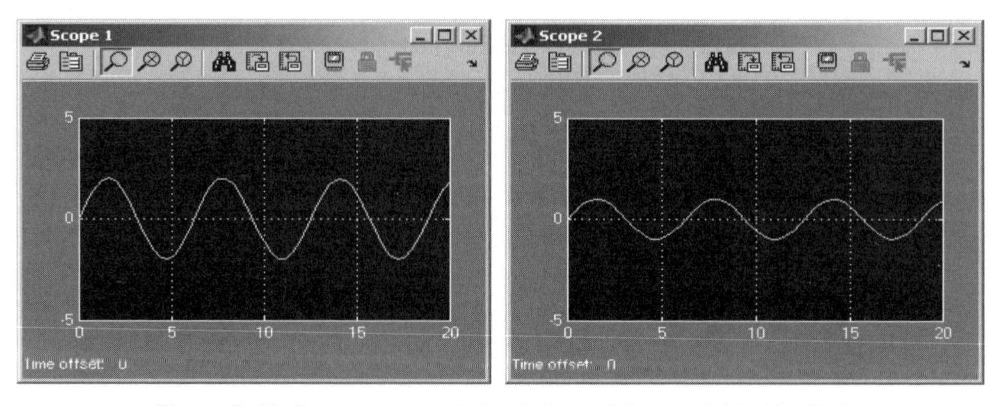

Figure B.10 Twenty second simulation of the model in Fig. B.8.

States and solvers

Many, if not all, of the models of biomedical systems are structured such that the outputs are functions of the previous values of the output variables (which are, in turn, typically functions of time). In Simulink, as well as other simulations packages, such variables are referred to as *states*. Computing the outputs requires saving the state at each time step in order to compute the next output value(s).

In MATLAB, the programmer would have to save the state(s) in variables in the MATLAB workspace and manage their use to compute the next output value(s). In Simulink, a block that needs some or all of its previous outputs to compute its current output implies that there are states that need to be saved. Simulink performs these tasks automatically if the model has state variables.

Understanding the notion of states and state solvers becomes important when selecting a numerical method for simulating a model (or equivalently, solving an ordinary differential equation, or ODE). There are two types of states that can occur in a Simulink model: *continuous* or *discrete* states. If the model has continuous states, then one of the ODE solvers described in Chapter 7 can be used; if the model has only discrete states, then the discrete solver can be used in the simulation.

A Simulink model with a *continuous state* is one where some computation is performed to compute a derivative, and then this derivative is integrated. The output of the integrator is used to compute the next value of the derivative. Simulink models will use integrator blocks (continuous menu), as well as a set of blocks to compute the derivative. This approach to building a model is sometime referred to as the *state-space* approach. The chain of blocks that are the input to the integrator block represent an ODE. The accuracy of the simulation performed by Simulink will depend on the step size and the ODE solver used. See Chapter 7 for more details on ODE solvers and their implementation in MATLAB; these MATLAB functions are used directly in Simulink to solve models with continuous states.

A *discrete state* is one where the output can be computed at finite intervals. Simulink includes two discrete solvers: a fixed-step and a variable-step discrete solver. The former is used to update the state periodically, and the latter is used to update the state aperiodically, or in a manner that ensures that all discrete states are updated. As the discrete step size approaches zero, the discrete state approaches a continuous state.

From the same menu where the simulation duration is changed, the ODE solver used in the simulation can be changed as well. The choices for ODE solvers for continuous state systems are given in Table 7.2; recall that in Simulink there are also two solvers for discrete state system.

Prior to simulation, Simulink examines the model to determine whether it has continuous or discrete states. An error would result if a discrete solver were chosen for a model when continuous states were required, such as a model that included an integrator, a state-space, or a transfer function block. Since the model in Fig. B.8 does not have an integrator, a discrete space solver may be used. Fig. B.11 is the solution of the model in Fig. B.8, but using a variable-step discrete solver.

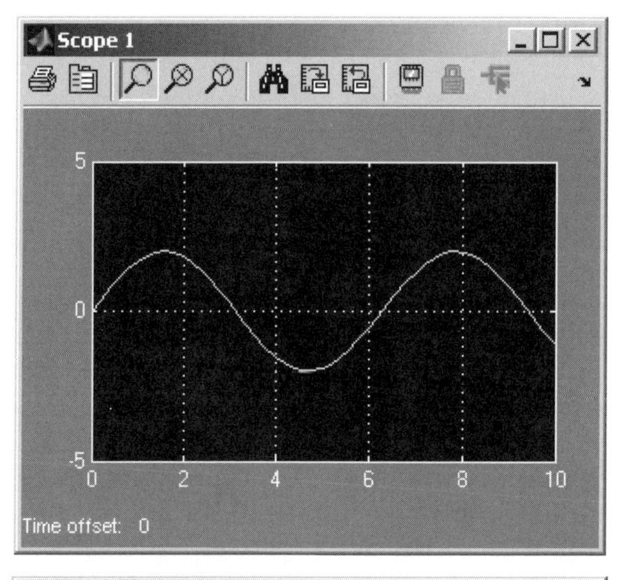

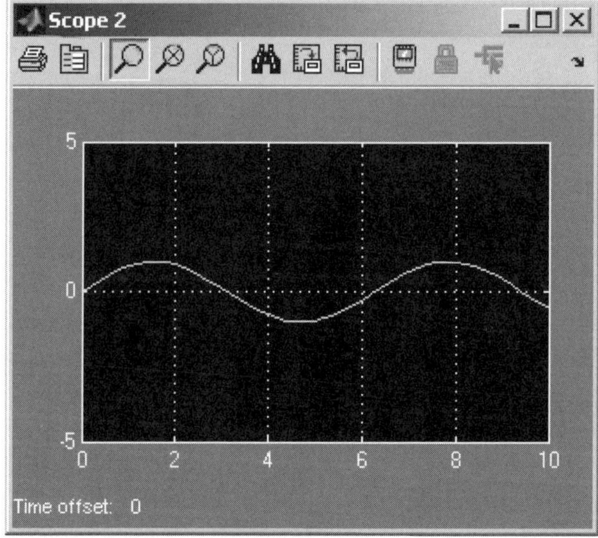

Figure B.11 The simulation results of the model in Fig. B.8, using a discrete solver instead of the continuous solver ode45. As expected, there is no difference, since there is no continuous state in the model.

There are a number of other simulation parameters, such as data input/output and optimization that are beyond the scope of this brief introduction. The reader is referred to the Simulink help for information on these parameters.

B.3 The Simulink Block Libraries

When the Simulink library browser is opened, the lower left-hand pane has a list of the categories of blocks available in Simulink. This pane is organized much like a directory browser; if additional Simulink toolboxes are available in your installation, the toolbox name will appear in the same list as the standard Simulink library. The blocks are categorized in subheading below the toolbox listing. The blocks in each category are displayed in the lower right-hand pane and shown in the figures below for each category. The main categories of blocks available are:

The *Sources* category, shown in Fig. B.12, contains blocks that are sources of data (the model and subsystem inputs) or the sources of signals (the signal generators). The signal generators include constants, periodic functions, noise and time stamps.

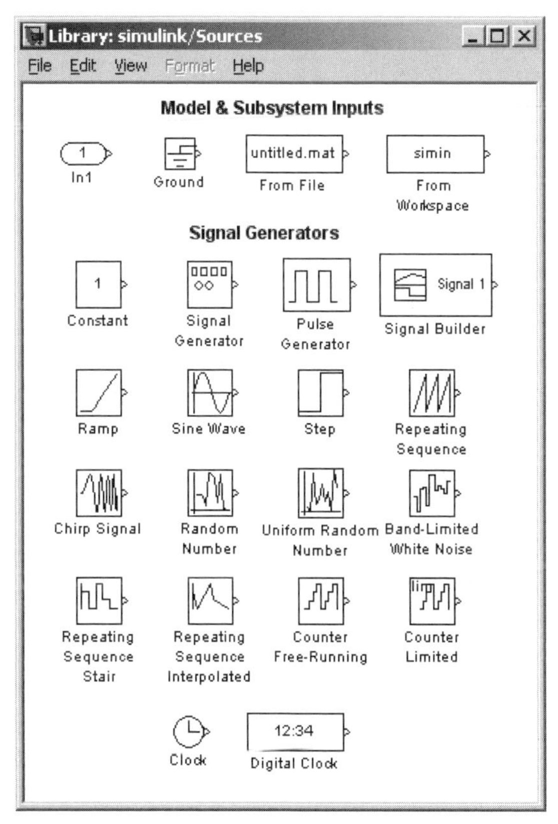

Figure B.12 The Simulink blocks in the sources library.

The *Sinks* are those blocks that use signals or terminate a simulation. There are three subcategories: Model and Subsystem Outputs (transferring signals to other models); Data Viewers (including scopes); and Simulation Control (stopping the simulation).

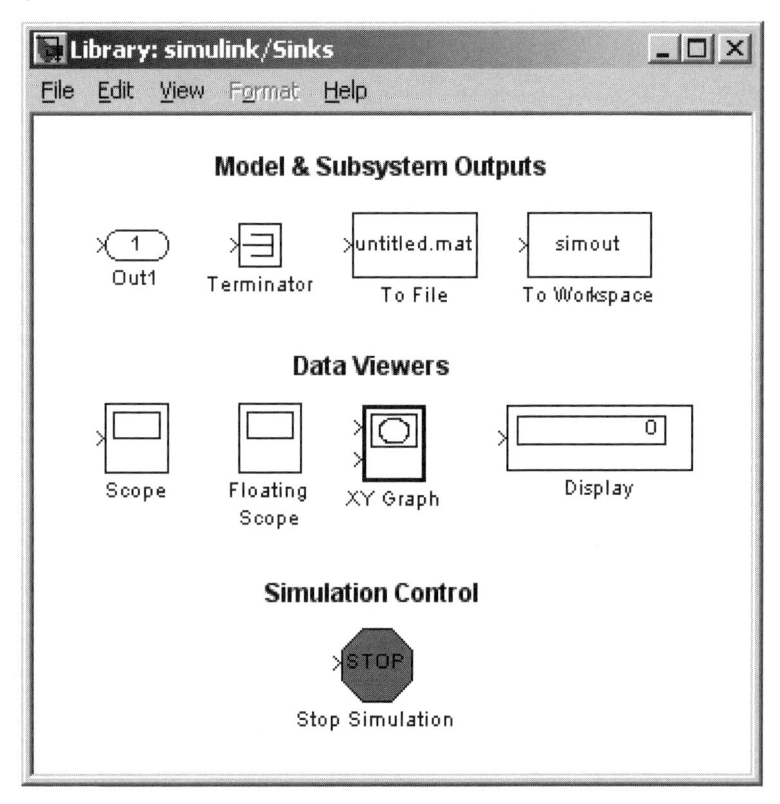

Figure B.13 The Simulink blocks in the sinks library.

The other category that we have already used are the *Math Operations*. Fig. B.14 shows the large list of functions available, including algebraic operations on scalars and matrices and conversions to and from complex numbers.

To understand the impact of changing the ODE solver, it was necessary to introduce the concepts of continuous and discrete states in modeled systems. Fig. B.15 are those blocks in the *Continuous* category, including integration and differentiation. Any model that includes one or more of these blocks will require a continuous state ODE solver (Table 7.2).

The category of *Commonly Used Blocks* is handy to use and it contains many of the blocks that can be used to build even moderately sophisticated models. Fig. B.16 list several commonly used blocks.

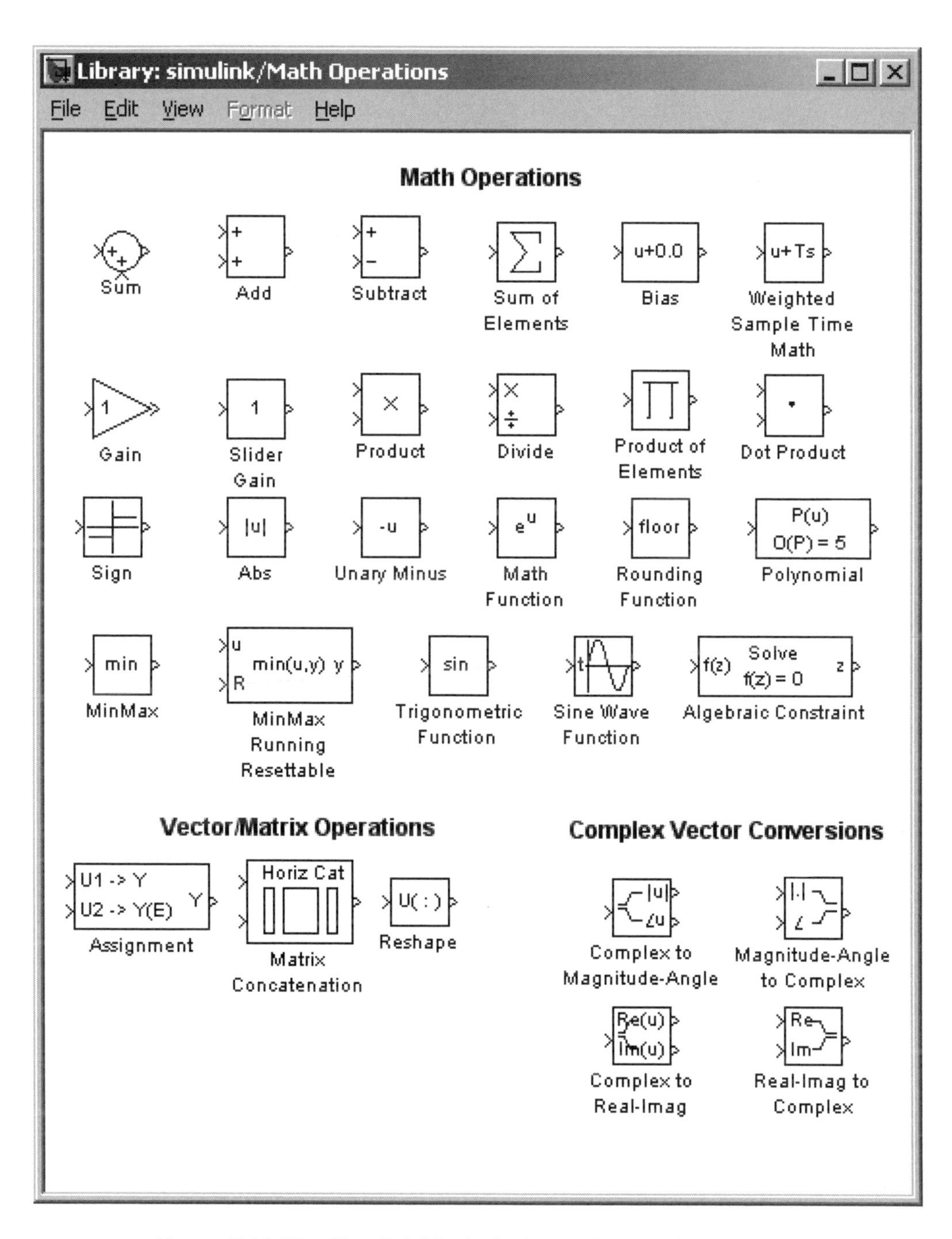

Figure B.14: The Simulink blocks in the math operations library.

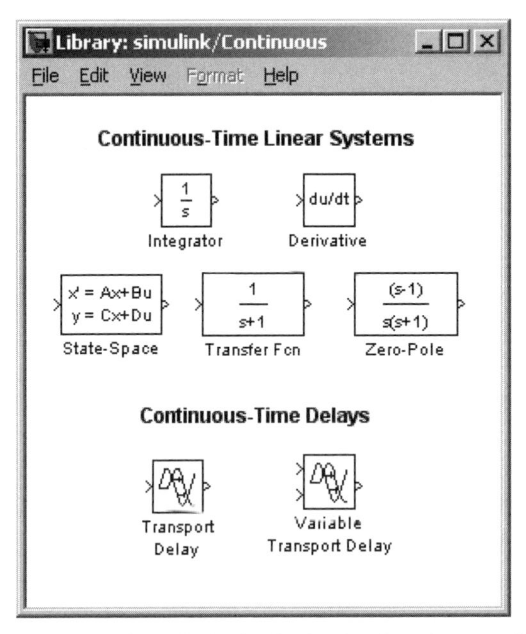

Figure B.15 The Simulink blocks in the Continuous library.

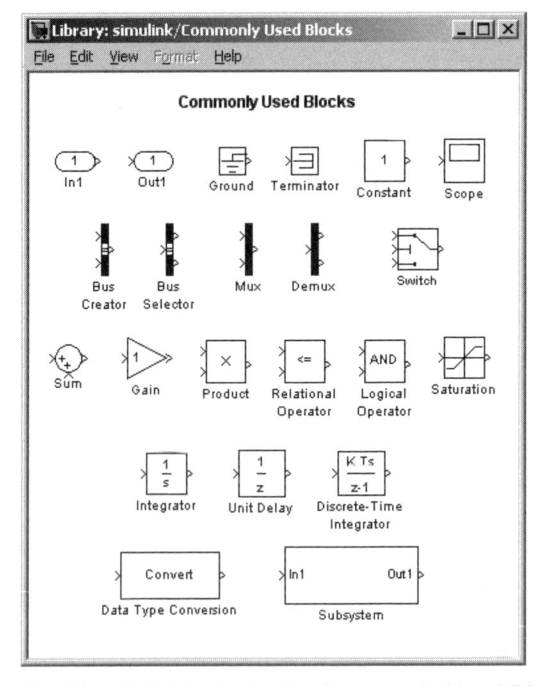

Figure B.16: The Simulink blocks in the Commonly Used Blocks library.

These categories include most of the blocks that will be needed for the majority of the models of biomedical systems. Other categories contain blocks that may be needed for more detailed models, including:

1. *Ports & Subsystems*: This category includes those blocks that are needed to define and connect models. Just like dividing a MATLAB program into scripts and functions, large Simulink models should be divided into subsystems for easier management. The Ports & Subsystems category includes blocks for defining subsystems and passing signals between models.
2. *Signal Routing*: This category of blocks is used to manage signal flow in a model. The most often used blocks will be in pairs:
    a. *Buses* are used to reduce the number of lines within a model.
    b. *Multiplexers* (the Mux/Demux pair) are used to combine (or separate) individual signals (to/from) a single signal that is the composition of a number of signals.
    c. The *Goto* and *From* blocks are used to pass a signal from one block to another without actually connecting them. They are most often used when a signal needs to be communicated between two models.
3. *User-Defined Functions*: These provide the ability to create blocks that perform custom operations. The user-defined operation can be expressed in MATLAB, in C, or as an algebraic expression using C language–style syntax.

Some of the categories in the standard Simulink toolbox will not be shown here; more detail can be found in the standard Simulink help. These categories include:

1. *Discontinuities* are blocks that introduce nonlinearities into the model, such as saturations and dead zones.
2. The *Discrete* category include blocks for filtering and modeling transfer functions in terms of the z-transform (a transform to a discrete signal space);
3. *Logic and Bit Operations* include blocks for logic operations on single bits as well as words.
4. *Lookup Tables* is a category that includes blocks for implementing functions that map a discrete domain to a discrete range.
5. *Model Verification* blocks are those used to create self-validating models, and can be turned off or on as the need arises.
6. *Model-Wide Utilities* include blocks that are used to annotate a model.
7. The *Signal Attributes* category includes blocks used to manage data types within a model.
8. *Additional Math & Discrete* blocks are used to manage real-world representations of signal magnitudes including overflows and fixed-point representations of floating-point numbers.

There is also a second library of *Simulink Extras*. The most important of these categories are the *Additional Sinks* (such as spectrum analyzers) and the blocks in the *Transformations* (changes in coordinate systems and units of measure).

B.4 Constructing Models

The rudimentary examples shown in Section B.2 are meant to illustrate the operation of Simulink but not meant to be an exhaustive tutorial on using Simulink to model biomedical systems. This section has several examples of constructing Simulink models based on the examples and blocks already introduced.

B.4.1 Algebraic operations, signal routing and MATLAB variables

Section B.2 ended with an example of a sine wave generator and two scopes; Scope 2 displayed the output of the sine wave generator and Scope 1 displayed the output of the sine wave generator after the output amplitude had been multiplied by two.

Suppose now that a second sine wave generator is introduced; the amplitude of both sine waves is set to three, the frequency of each sine wave is 0.5 Hz, but the second has a phase difference of .25 rad/sec with respect to the first. The difference between the two signals is computed, integrated, and displayed on a scope along with the two generated signals. Write out the two derived signals to MATLAB for further computation. Set the length of the simulation to be 15 seconds. The analytical equation is:

$$\int [\sin(.5t) - \sin(.5t + .25)]\, dt$$

In this example, two new blocks, the bus and the integrator, are introduced. Signals can be routed together throughout a model using blocks called *buses*, which combine multiple signals and route them to blocks that can take multiple inputs. Scopes are an instance of such blocks, as shown in this example. The integrator outputs the continuous integral of the input signal. Double-click on the integrator to see the properties, the most important of which is the starting value, which defaults to zero. In the following sequence of steps, when a block is to be added to the model, the block category is specified in parenthesis. The steps to create this model are:

1. Create two sine wave generators (Sources). Set the properties by double-clicking on the block. Change the names of the sine wave generators by double-clicking in the name field and setting the names to be Sine Wave 1 and Sine Wave 2, respectively.
2. Add a subtraction box (Math Operations), and set the two inputs to be the output of the two sine wave generators.
3. Add an integrator (Continuous) and connect the input to the output of the subtraction box.

4. Add a second scope and differentiate the two scopes by changing the names as was done in Step 1.
5. Add two bus creators (Signal Routing) to the model. The outputs of theses bus creators, the buses, are the inputs to the scopes. The inputs of the bus to the first scope are Sine Wave 1, Sine Wave 2, and the difference signal. The inputs to the second bus are Sine Wave 1, Sine Wave 2, and the integral of the difference signal.
6. Add two "To Workspace" blocks (Sinks) to output the computed signals to MATLAB to perform further computation. Connect the first to the difference signal, and the input to the second is the integrated difference signal. The MATLAB workspace variable names can be set in the block properties.
7. A third scope, showing the output of Sine Wave 1, is added for reference.

After these blocks are added to the model, connect the inputs and outputs as specified in the problem statement above. The complete model is shown in Fig. B.17, and the outputs from the three scopes are shown in Fig. B.18.

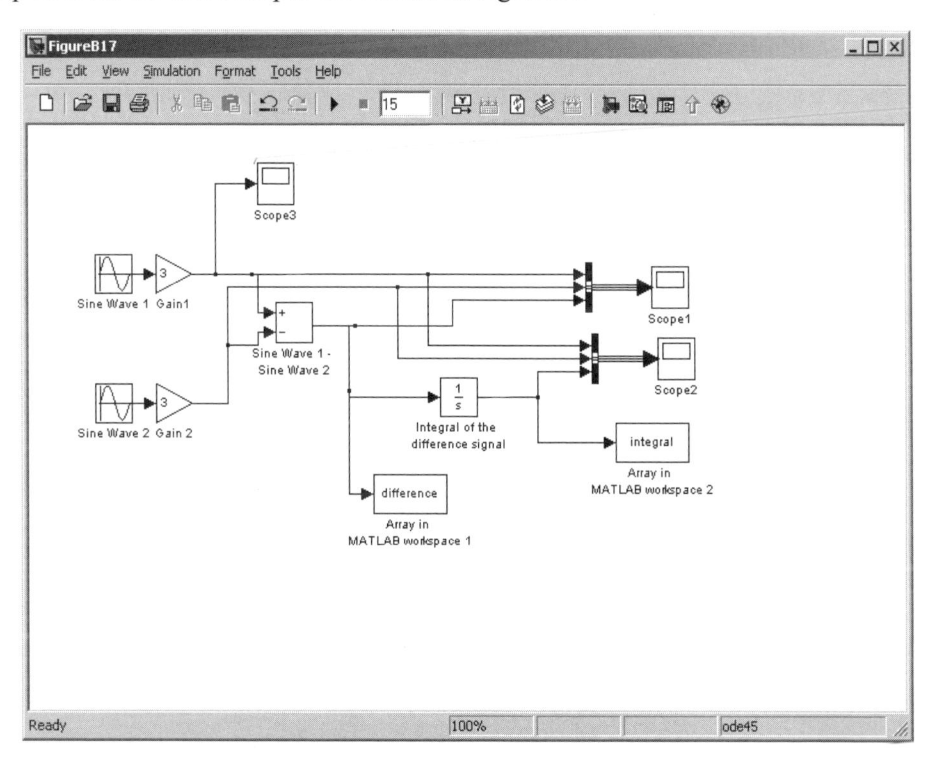

Figure B.17 A model that illustrates the use of integrators, buses and MATLAB output. The difference of two sine waves, with a small phase difference, is computed, displayed in a Simulink scope and output to MATLAB.

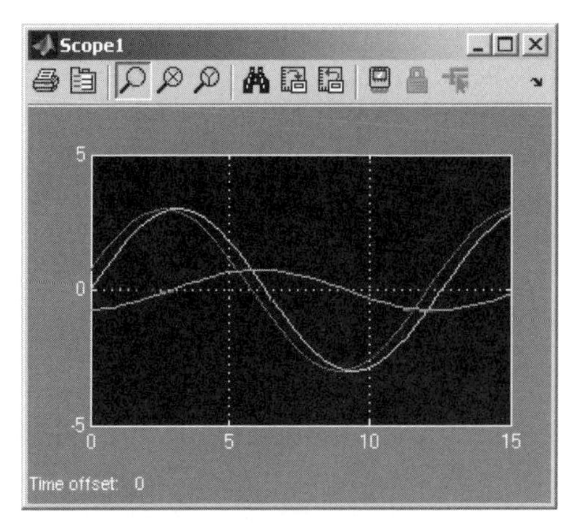

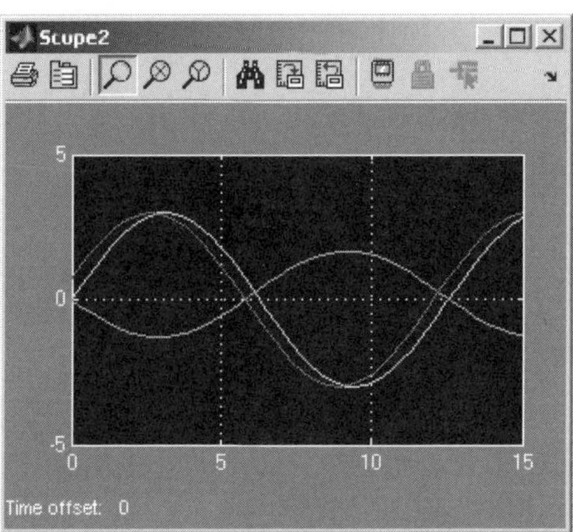

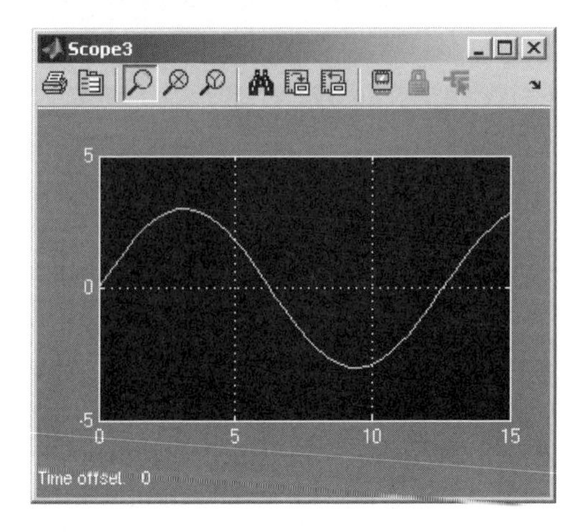

Figure B.18 The scope outputs for the model of Fig. B.17.

Comments

The integral of the analytical expression above is

$$-2\cos(.5t) + 2\cos(.5t + .25)$$

After running the simulation, two variables will appear in the MATLAB workspace, and the traces on all three scopes show approximately 1¼ period of their respective outputs. The outputs of scopes with bus inputs, such as Scopes 1 and 2, are differentiated by color on the actual computer screen: The signal on the first input to the bus would be shown in yellow, the second in pink, and the third in turquoise.

As expected, the instantaneous difference between the two signals (third signal in Scope1 of Fig. B.18) is small by comparison to the two signals. If the phase difference were larger, the difference would be commensurately larger (the reader is encouraged to try this). The actual magnitude and statistics of the signals can be computed from the MATLAB variable output. The MATLAB code

```
max(difference)
max(integral)
```

will yield the following results:

```
>> max(difference)
ans =
    0.7467

>> max(integral)
ans =
    1.6824
```

B.4.2 Simultaneous differential equations

Example 7.2, in Chapter 7 and was solved using the Runge-Kutta fourth-order method of solution (ode45). The example is repeated here to illustrate the power of Simulink by showing how to model simultaneous ordinary differential equations.

Example 7.2 simulates the action of an enzyme, *E*, that catalyzes the conversion of a substrate, *S*, to form a product, *P*, via the formation of an intermediate complex, *ES*, as shown below:

$$S + E \underset{k_{-1}}{\overset{k_1}{\rightleftarrows}} ES \xrightarrow{k_2} P + E$$

In this problem, the reader is asked to determine the time required for the enzyme catalysis reaction to reach 99.9% conversion of a substrate. The mathematical model of this reaction resulted in the following system of four differential equations:

$$\frac{d[S]}{dt} = -k_1[S][E] + k_{-1}[ES] \qquad [S]_0 = 1.0$$

$$\frac{d[E]}{dt} = -k_1[S][E] + k_{-1}[ES] + k_2[ES] \qquad [E]_0 = 0.1$$

$$\frac{d[ES]}{dt} = k_1[S][E] - k_{-1}[ES] - k_2[ES] \qquad [ES]_0 = 0$$

$$\frac{d[P]}{dt} = k_2[ES] \qquad [P]_0 = 0$$

with rate constants: $k_1 = 0.1\ (\mu\text{M})^{-1}\text{s}^{-1}$ $k_{-1} = 0.1\ \text{s}^{-1}$ $k_2 = 0.3\ \text{s}^{-1}$

The Simulink solution shown below is based on a paradigm for solving differential equations, called the *state-space solution*. For a single initial value problem ODE, there are three parts to the Simulink state-space solution: a scope to display the output; an integrator; and a set of blocks that compute the differential equation (Fig. B.19). The initial values are set as the initial conditions by double-clicking on the integrator box.

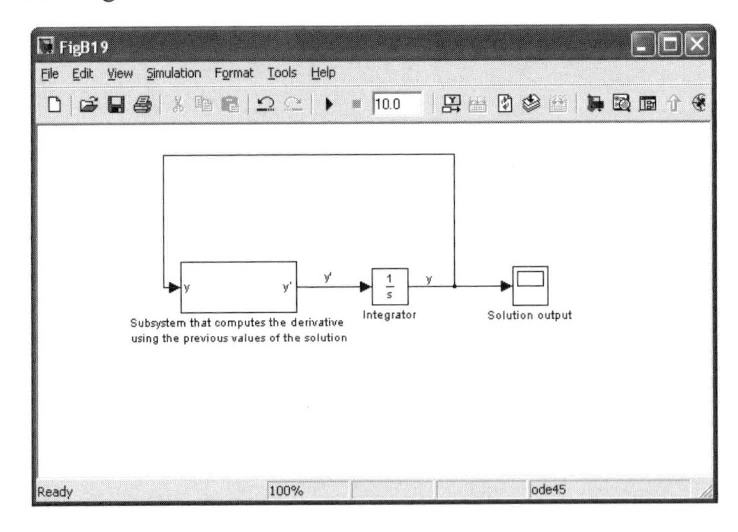

Figure B.19 The general structure of a state-space solution of an ODE. The initial condition is set as the initial value in the integrator block.

In this problem, there are four differential equations so the model will be built from the state-space solutions to the four differential equations. The four "subsystems" that compute the derivatives (and are the inputs to the integrators) are linear combinations of the four outputs. In the MATLAB solution given in Chapter 7, pairs of variables are plotted on one of two graphs; in this Simulink version, two scopes are used to produce graphs similar to those in Example 7.2; and the signals are combined using a bus as input to each scope. The complete Simulink model is shown in Fig. B.20.

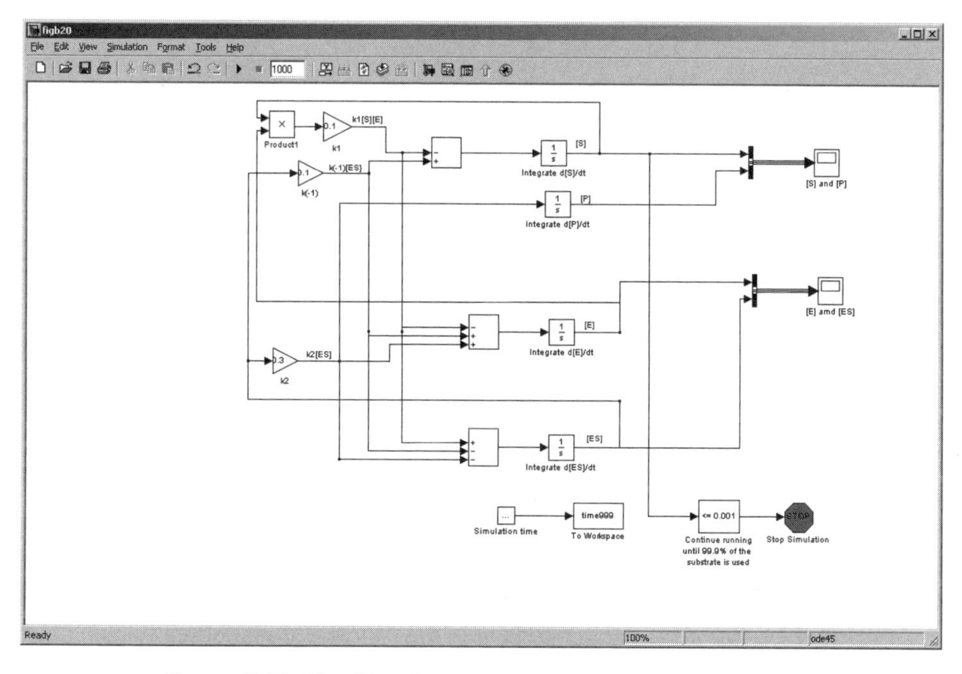

Figure B.20 The Simulink model that solves Example 7.2.

The blocks in the lower right-hand corner of the model are used to determine the time at which 99.9% of the substrate is converted. There are two parts: first, the output [S] is compared to 0.001 (*Compare to Constant* in the *Logic and Bit Operations* category). Once this limit is reached, the simulation is stopped; the Boolean output signals the Stop Simulation block (Sinks). In parallel, the simulation time is stored in a MATLAB workspace variable called time999. When the simulation stops, the last value in the array will be the time at which 99.9% of the substrate has been converted. The results are the same as given in Example 7.2

B.4.3 PHYSBE and subsystems

PHYSBE is a Simulink model of the circulatory system for simulating the flow of oxygen, nutrients, heat, or chemical tracers within the bloodstream (McCleod, 1966, 1968). Although the majority of work on PHYSBE was done in the 1960s, PHYSBE has since been implemented in Simulink, making it accessible for more educational and research programs in biomedical engineering. PHYSBE is introduced here as an example of a biomedical engineering application in Simulink and to demonstrate the use of subsystems, `from` and `goto` blocks, and other signal routing tools in Simulink.

The Simulink implementation of PHYSBE can be downloaded from the MathWorks MATLAB central Web site (www.mathworks.com); all eight files are stored in the single zip file `physbe.zip` (see Fig. B.21). Regardless of which zip file tool is used (WinZip, 7-zip, the explorer in Windows XP, or ZipIt for Mac OS X) it is easiest to organize the files if the pathname given is used when extracting the files and the `physbe` directory is made as a subdirectory of the MATLAB work directory.

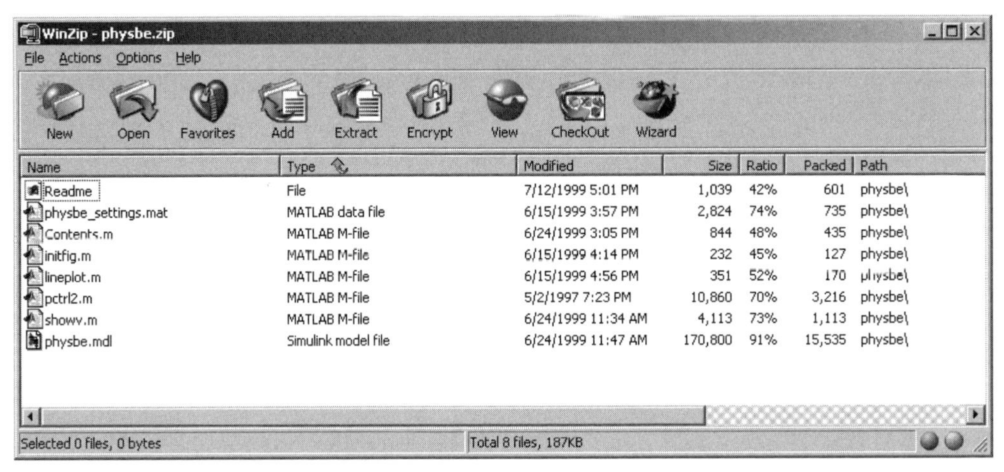

Figure B.21 The .m and the .mdl files of the PHYSBE transport simulator.

After installing PHYSBE and starting MATLAB, start the Simulink model by typing the following commands:

```
>> cd physbe
>> physbe
```

at the MATLAB command prompt.

Five windows will appear: the PHYSBE model, the PHYSBE control center, two scopes for heart pressure and volume, and a floating scope. The highest level of the PHYSBE model of the circulatory system is shown in Fig. B.22. The circulatory system is represented as nine subsystems; and in addition there is an analysis subsystem (lower right). Each subsystem can be expanded by double-clicking on the respective box. For example, the model of the lungs can be seen by double-clicking on the subsystem box labeled LUNGS; the detailed model is shown in Fig. B.23.

The Simulink model of the lungs shown in Fig. B.23 is an example of a *subsystem*. A subsystem is itself a Simulink model, with inputs and outputs, as well as other signals and functions that may be communicated between the current subsystem and other subsystems of the complete model. A subsystem is added to a model by adding a subsystem block (Ports & Subsystems) to the model that is

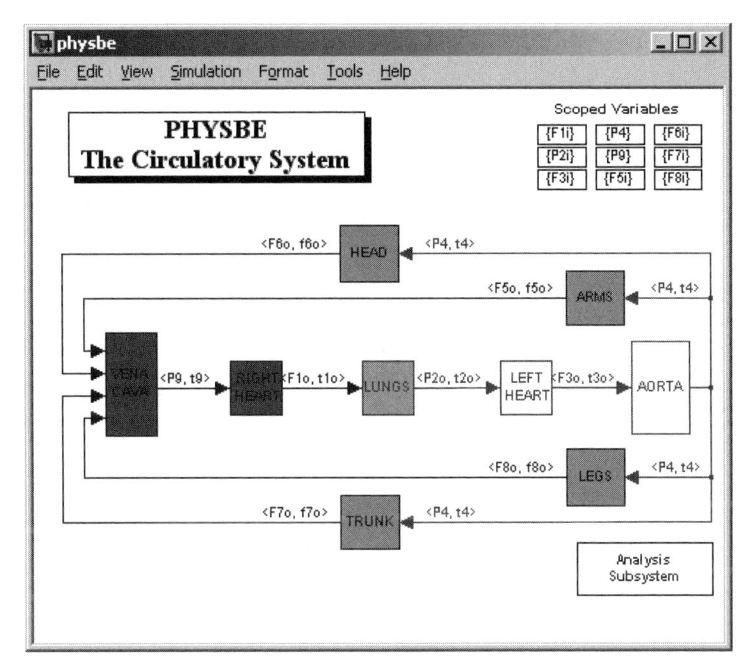

Figure B.22 The Simulink model of PHYSBE. This model is composed of ten subsystems, shown above.

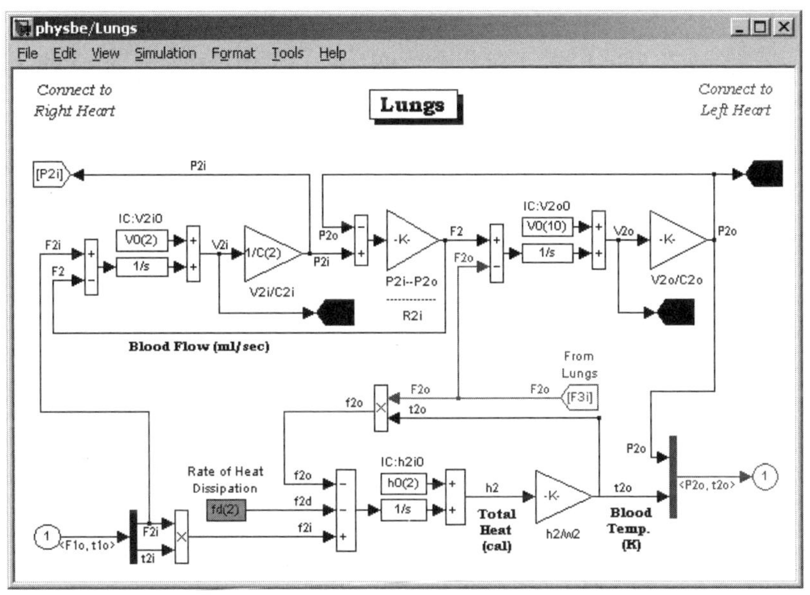

Figure B.23 The PHYSBE model of the Lungs. This is one subsystem of the complete model. The inputs are P2i and <F1o, t1o>; the outputs are V2i, V2o, P2o and <P2o, t2o>.

currently being designed. Double-clicking on the subsystem block will open another Simulink window to display the contents of the subsystem. As shown in Fig. B.23, the initial configuration of a new subsystem is a connection between the input and output signals. A subsystem can be built by first deleting the signal path between the input and output and then inserting the block(s) that perform the desired computation. Examples of detailed subsystems are available by adding the Subsystems Examples block (Ports & Subsystems) to a model.

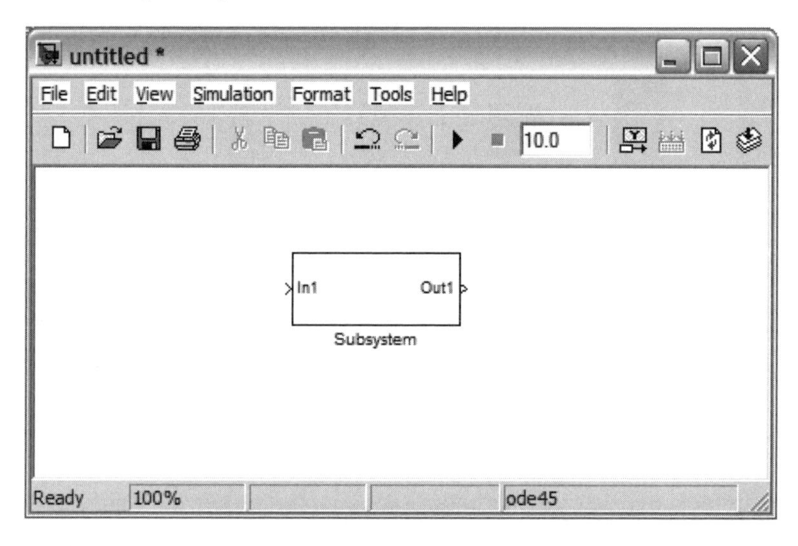

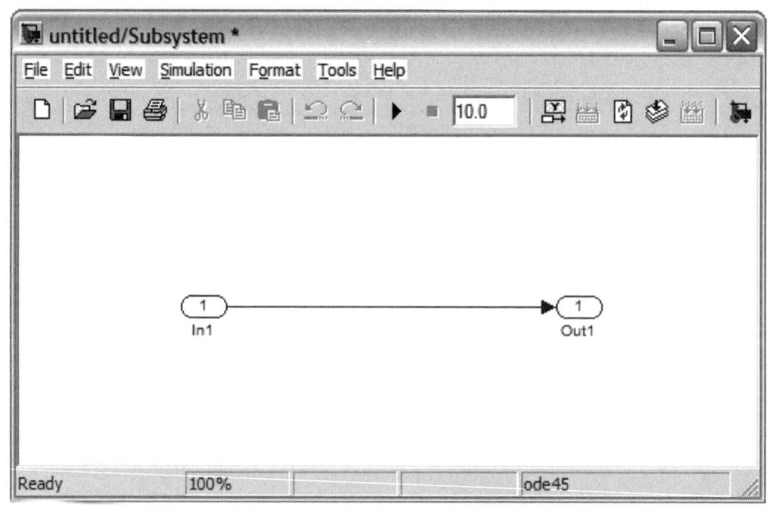

Figure B.24 The Model at the top contains a single subsystem that, when expanded, simply connects the input to the output of the subsystem.

The PHYSBE model parameters are explained in the original papers (McCleod, 1966, 1968) or on the The MathWorks, Inc., PHYSBE Web site. The signal paths between each subsystem are labeled with names of one or more of the parameters. The labels on the signal paths are shown in brackets, e.g., < Bn(i/o), Hn(i/o)>, where the first parameter B is a parameter of blood flow and the second parameter H is a parameter of heat flow. The bracket notation is used to indicate that the parameters are on a *bus*, where two signals are combined as the output of one block and they are separated as the input to the second block using the input and output notation of a subsystem (Fig. B.24).

For example, the output of the RIGHT HEART (Fig. B.22) is a bus that carries the two signals <F1o, t1o>, the output flow rate and the output temperature from the RIGHT HEART. These two signals are separated in the subsystem that models the LUNGS, where the signals are referred to as F2i and t2i; respectively, the Blood Flow Rate into the LUNGS and temperature of the heat flow into the LUNGS. The number in the designation refers to the row in the PHYSBE Control Center; i.e., F2i is the flow rate into the LUNGS (Fig. B.25). The PHYSBE Control Center is a graphical user interface that was designed and developed using GUIDE, the graphical user interface layout editor that is part of MATLAB. Although GUIDE is a very powerful tool for creating user interfaces to complex models, its use is beyond the scope of this book.

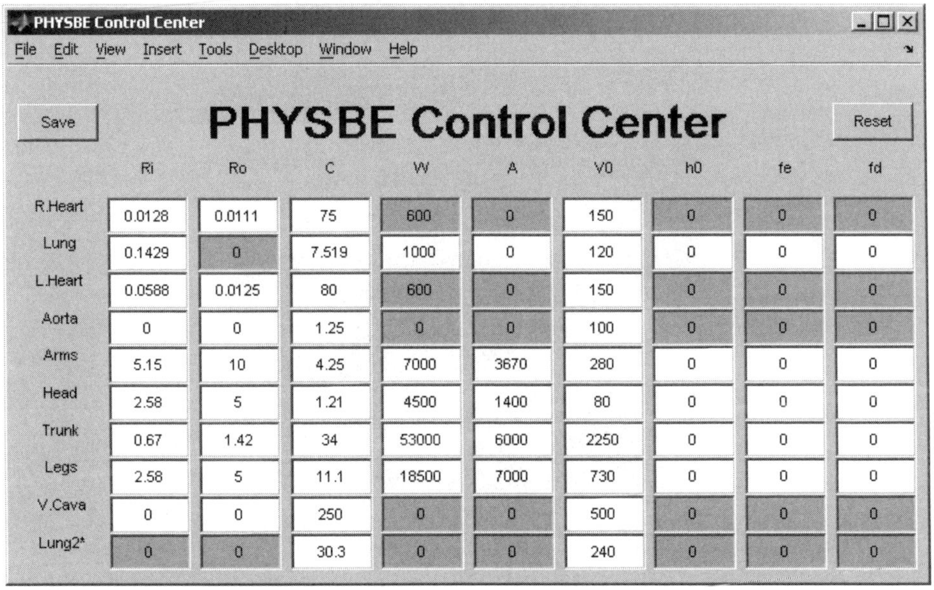

PHYSBE Control Center

File Edit View Insert Tools Desktop Window Help

Save **PHYSBE Control Center** Reset

	Ri	Ro	C	W	A	V0	h0	fe	fd
R.Heart	0.0128	0.0111	75	600	0	150	0	0	0
Lung	0.1429	0	7.519	1000	0	120	0	0	0
L.Heart	0.0588	0.0125	80	600	0	150	0	0	0
Aorta	0	0	1.25	0	0	100	0	0	0
Arms	5.15	10	4.25	7000	3670	280	0	0	0
Head	2.58	5	1.21	4500	1400	80	0	0	0
Trunk	0.67	1.42	34	53000	6000	2250	0	0	0
Legs	2.58	5	11.1	18500	7000	730	0	0	0
V.Cava	0	0	250	0	0	500	0	0	0
Lung2*	0	0	30.3	0	0	240	0	0	0

Figure B.25: The PHYSBE Control Center. Parameters can be changed or reset before a simulation is run.

In each PHYSBE subsystem, the signal flow proceeds (visually) from left to right. For example, in the LUNGS subsystem (Fig. B.23), the input (the circle labeled with a 1) is the bus that carries the two model parameters F1o and t2o. The output is generated on the right, where a bus is created carrying P2o and t2o. In addition, other signals are passed between the subsystems using From and Goto: For example, P2i, the blood pressure in the lungs, is sent to the right heart using a Goto (from the LUNGS subsystem) and a From (in the RIGHT HEART subsystem).

The PHYSBE model is run as any other Simulink model, from the Simulation pull-down menu or by typing control-T at the keyboard. The output, for a simulation using the predefined parameters is shown on the two scopes labeled Heart Pressure and Heart Volume in Fig. B.26.

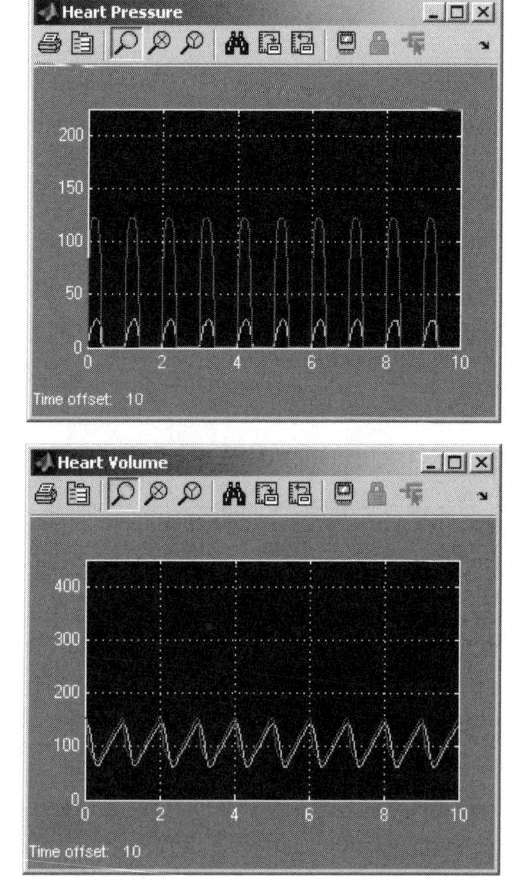

Figure B.26 The PHYSBE output shown on the Heart Pressure scope (above) and the Heart Volume scope (below). No parameters where changed in this simulation.

The inputs to these two scopes are shown in the analysis subsystem: [P1] and [P3] to the Heart Pressure scope and [V1] and [V3] to the Heart Volume scope. The origins of the four signals can be found in the RIGHT HEART and the LEFT HEART subsystems.

PHYSBE also starts with a floating scope, i.e., a scope that can display signals carried on more than one line and without using buses, as used for the Heart Pressure and Heart Volume scopes. The floating scope is used by opening the scope, opening the signal selector, and checking the box to the left of each signal to be displayed. Fig. B.27 shows both windows; the signal selector has two panes that list the subsystems and the signals within each subsystem.

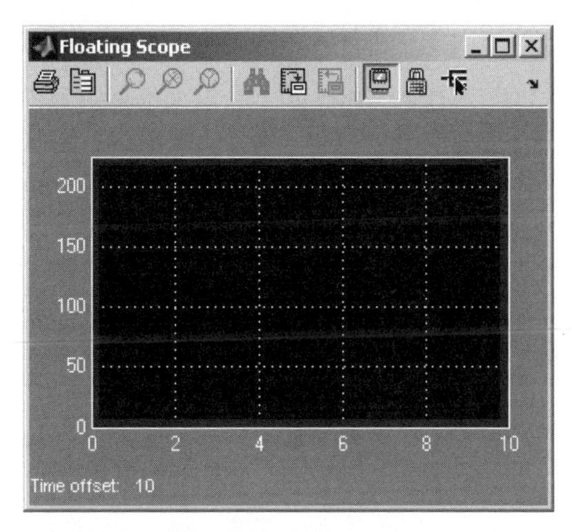

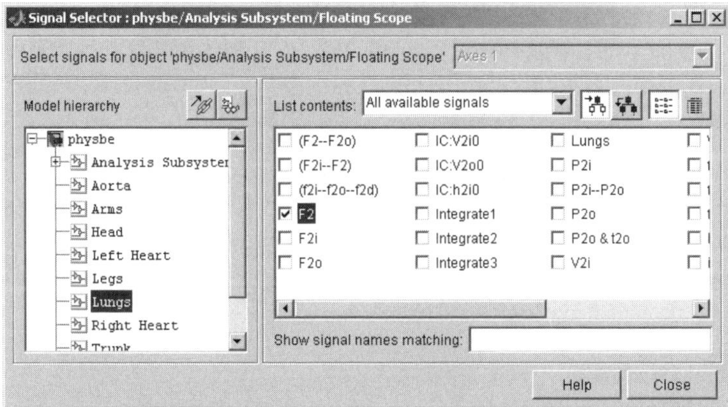

Figure B.27 The PHYSBE Floating Scope (above) and its Signal Selector (below). The left pane of the Signal Selector shows the hierarchy of models, and the right pane shows the signals within the subsystem selected. The box F2 is checked to display signal F2 on the floating scope.

For example, use the floating scope to display the blood pressure in the HEAD (P6), the RIGHT HEART (P1), and the LEGS (P8) all on the same floating scope. Open the floating scope and its signal selector and, in turn, go to each of the three subsystems and select the pressure signals. Close the Signal Selector and run the simulation. The Floating Scope output is shown in Fig. B.28.

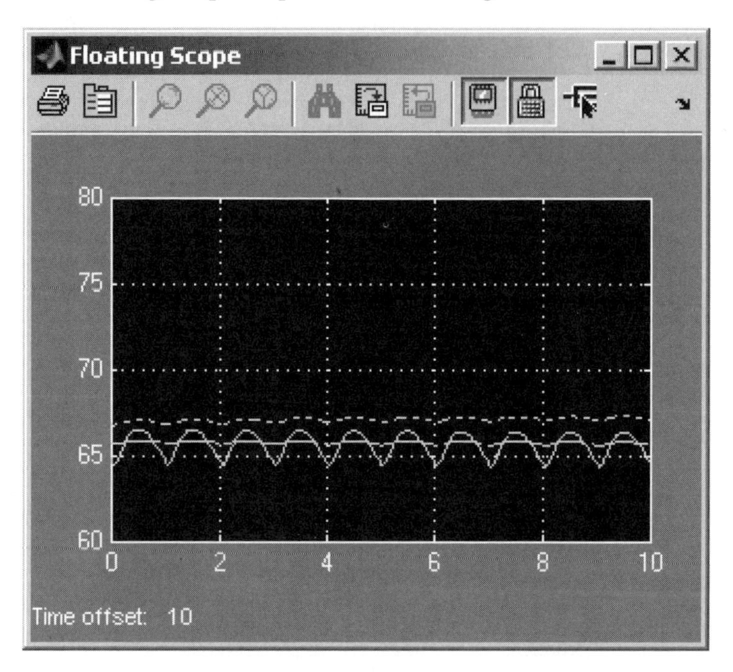

Figure B.28: The PHYSBE Floating Scope showing three blood pressure signals: The head (P6, sinusoid), the trunk (P7, dotted small segments) and the legs (P8, line segments). The blood pressure in the trunk is relatively constant and higher than that in the head or legs.

The purpose of this Appendix is to introduce components of Simulink models and to show how the most elementary can be used to solve even relatively complex problems such as simulating the transport in the cardiovascular system. The reader is referred to Chapter 10 for some examples of how PHYSBE can be used to model pathologies of the cardiovascular system.

B.5 References

McLeod J, (1966). PHYSBE…a physiological simulation benchmark experiment. *SIMULATION*, **7(6)**:324-329.

McLeod J, (1968). PHYSBE...a year later. *SIMULATION*, **10(1)**:37-45.

Appendix C: Review of Linear Algebra and Related MATLAB Commands

C.1 Matrix and Vector Operations

Matrix **A** is an $(m \times n)$ array of elements arranged in rows and columns, as shown below:

$$\mathbf{A} = \begin{bmatrix} a_{11} & a_{12} & \cdots & a_{1n} \\ a_{21} & a_{22} & \cdots & a_{2n} \\ \cdots & \cdots & \cdots & \cdots \\ a_{m1} & a_{m2} & \cdots & a_{mn} \end{bmatrix}$$

In MATLAB, elements in a row are separated by commas, while rows are separated by semicolons or by line breaks. Thus,

```
>> B=[1,3,5;2,0,4;1,5,6]
B =
     1     3     5
     2     0     4
     1     5     6
```

The *trace* of a matrix is the sum of the elements on the main diagonal of a square matrix. The MATLAB command for calculating the trace is

```
>> trace(B)
ans = 7
```

A matrix of one column of n elements is referred to as an n-dimensional *column vector*,

```
>> x=[1; 3; 5; 2; 0]
x =
     1
     3
     5
     2
     0
```

A matrix of only one row of n elements is referred to as an n-dimensional *row vector*,

```
>> y=[5, 6, 7 , 8, 9]
y =
     5     6     7     8     9
```

Matrix addition and subtraction can be performed if matrices have the same number of rows and the same number of columns.

Matrix multiplication between **A** and **B** is allowed if **A** has the same number of columns as the number of rows of **B**. When this condition is satisfied, then **A** and **B** are said to be *conformable*. The multiplication of the matrix **A,** that has ($m \times n$) elements, with matrix **B**, of ($n \times p$) elements, results in a matrix with `m` rows and `p` columns that has an order of ($m \times p$). For example,

$$\mathbf{AB} = \underset{(4\times 3)}{\begin{bmatrix} a_{11} & a_{12} & a_{13} \\ a_{21} & a_{22} & a_{23} \\ a_{31} & a_{32} & a_{33} \\ a_{41} & a_{42} & a_{43} \end{bmatrix}} \underset{(3\times 2)}{\begin{bmatrix} b_{11} & b_{12} \\ b_{21} & b_{22} \\ b_{31} & b_{32} \end{bmatrix}}$$

$$= \underset{(4\times 2)}{\begin{bmatrix} (a_{11}b_{11} + a_{12}b_{21} + a_{13}b_{31}) & (a_{11}b_{12} + a_{12}b_{22} + a_{13}b_{32}) \\ (a_{21}b_{11} + a_{22}b_{21} + a_{23}b_{31}) & (a_{21}b_{12} + a_{22}b_{22} + a_{23}b_{32}) \\ (a_{31}b_{11} + a_{32}b_{21} + a_{33}b_{31}) & (a_{31}b_{12} + a_{32}b_{22} + a_{33}b_{32}) \\ (a_{41}b_{11} + a_{42}b_{21} + a_{43}b_{31}) & (a_{41}b_{12} + a_{42}b_{22} + a_{43}b_{32}) \end{bmatrix}}$$

Matrix multiplication is not commutative, that is, $\mathbf{AB} \neq \mathbf{BA}$. It should be noted that to an *element-by-element* product of two matrices is denoted by the (.*) operator:

$$\mathbf{A}.*\mathbf{B} = \begin{bmatrix} a_{11} & a_{12} \\ a_{21} & a_{22} \end{bmatrix} .* \begin{bmatrix} b_{11} & b_{12} \\ b_{21} & b_{22} \end{bmatrix} = \begin{bmatrix} a_{11}b_{11} & a_{12}b_{12} \\ a_{21}b_{21} & a_{22}b_{22} \end{bmatrix}$$

The transpose of the matrix $\mathbf{A}$, $\mathbf{A}^{\mathrm{T}}$, is defined as an interchange of the rows and columns of $\mathbf{A}$, and is denoted by `A'` in MATLAB. If $\mathbf{A} = \mathbf{A}^{\mathrm{T}}$, then $\mathbf{A}$ is called a *symmetric* matrix.

An *identity*, or *unit* matrix, is one whose diagonal elements are all unity and the rest of the elements are zero. An identity matrix is always symmetric and square. Thus, a (3×3) order identity matrix is given by:

$$\mathbf{I} = \begin{bmatrix} 1 & 0 & 0 \\ 0 & 1 & 0 \\ 0 & 0 & 1 \end{bmatrix}$$

The MATLAB function `eye(n)` returns an $(n \times n)$ identity matrix.

```
>> eye(3)
ans =
     1     0     0
     0     1     0
     0     0     1
```

The inverse of a matrix, $\mathbf{A}^{-1}$, is that matrix which when multiplied by the matrix $\mathbf{A}$ gives the identity matrix,

$$\mathbf{AA}^{-1} = \mathbf{I}$$

A matrix has an inverse if it is square and *nonsingular*, or equivalently, if the determinant of the matrix is non-zero. To check if a matrix $\mathbf{B}$ is nonsingular, the determinant of the matrix can be readily computed with a MATLAB command, as shown below:

```
>> B=[1,3,5;2,0,4;1,5,6]
B =
     1     3     5
     2     0     4
     1     5     6

>> det(B)
ans =
     6
```

Similarly, the inverse of matrix can be easily computed in MATLAB, as shown below for the matrix **B** defined above:

```
>> inv(B)
ans =
   -3.3333    1.1667    2.0000
   -1.3333    0.1667    1.0000
    1.6667   -0.3333   -1.0000
```

In MATLAB, the division of two matrices, **A** divided by **B**, is equivalent to the product of matrix **A** and the inverse of **B**. Thus, $\mathbf{A/B} = \mathbf{A\ B}^{-1}$, while $\mathbf{A \backslash B}$ is equivalent to $\mathbf{A}^{-1}\mathbf{B}$. Note also that the inverse of the product of two matrices is the reverse product of the inverses of the two matrices:

$$(\mathbf{AB})^{-1} = \mathbf{B}^{-1}\mathbf{A}^{-1}$$

The *rank* of a matrix **A** is defined as the order of the largest nonsingular square matrix within **A**. Consider the $(m \times n)$ matrix, where $n \geq m$. Then, the rank of the largest square submatrix within **A** is m, provided the $(m \times m)$ submatrix is nonsingular, that is, the determinant of the $(m \times m)$ submatrix is nonzero. The MATLAB command `rank(A)` gives the rank of the matrix:

```
>> A=[1,2,1,0;1,0,0,1;2,1,0,1]
A =
     1     2     1     0
     1     0     0     1
     2     1     0     1
>> rank(A)
ans =
     3
```

Each square matrix has its characteristic scalar properties, called the eigenvalues, and corresponding vectors, called the eigenvectors. The matrix **A** is said to have eigenvalue λ and eigenvector **x** if and only if:

$$\mathbf{Ax} = \lambda\mathbf{x}$$

which can also be expressed as:

$$(\mathbf{A} - \lambda\mathbf{I})\mathbf{x} = 0$$

To find λ and the corresponding vector **x** requires the solution of the homogeneous set of equations shown above. The homogeneous problem above has nontrivial solutions if the determinant of $(\mathbf{A} - \lambda\mathbf{I})$, also called the characteristic matrix, equals zero. For example, for a (2×2) matrix, **A**, shown below, the values of λ can be found readily:

$$\mathbf{A} = \begin{bmatrix} 1 & 0 \\ 1 & 1 \end{bmatrix}$$

Since the determinant of the characteristic matrix equals zero:

$$|A - \lambda I| = \begin{vmatrix} 1-\lambda & 0 \\ 1 & 1-\lambda \end{vmatrix} = (1-\lambda)^2 = 0 \quad \Rightarrow \quad \lambda_1 = 1, \quad \lambda_2 = 1$$

In MATLAB, `eig(A)` finds the values of λ. The output vector of eigenvalues can be named, as `m=eig(A)`.

The command `[Z,D]=eig(A)` generates a diagonal matrix `D` of eigenvalues and a full matrix `Z` whose columns contain the corresponding eigenvectors:

```
>> B
B =
       1      3      5
       2      0      4
       1      5      6
>> m=eig(B)
m =
      9.6540
     -0.2596
     -2.3944
>> [Z,D]=eig(B)
Z =
      0.5587     0.8176     0.0067
      0.4136     0.3791    -0.8595
      0.7189    -0.4334     0.5111
D =
      9.6540          0          0
           0    -0.2596          0
           0          0    -2.3944
```

C.2 Matrix Factorization

The purpose of LU factorization of a matrix **A** is to find a lower triangular matrix, **L**, and an upper triangular matrix **U**, such that **A** = **LU**. For any matrix, **A** (certain conditions may apply; see text for discussion), the problem is to find **L** and **U** such that **LU** = **A**:

$$\begin{bmatrix} l_{11} & 0 & 0 \\ l_{21} & l_{22} & 0 \\ l_{31} & l_{32} & l_{33} \end{bmatrix} \begin{bmatrix} u_{11} & u_{12} & u_{13} \\ 0 & u_{22} & u_{23} \\ 0 & 0 & u_{33} \end{bmatrix} = \begin{bmatrix} a_{11} & a_{12} & a_{13} \\ a_{21} & a_{22} & a_{23} \\ a_{31} & a_{32} & a_{33} \end{bmatrix}$$

A lower triangular matrix **L** with *1*s on the diagonals can be constructed from the multipliers used in Gaussian elimination (reviewed in Chapter 4). The elimination process transforms the original matrix **A** into an upper triangular matrix **U**. The lower triangular matrix **L** is formed by placing the negatives of the multipliers in the appropriate positions, as illustrated below.

LU factorization from basic Gaussian elimination (no pivoting)

```
Input
    A                              n-by-n matrix to be factored
    n                              dimension of A
Initialize
    L = I                          n-by-n identity matrix
    U = A
Operate
For k = 1 to n-1
    For i = k + 1 to n             each row of matrix U after the kth row
    m(i,k) = -U(i,k)/U(k,k)
        For j = k to n             transform row i
            U(i,j) = U(i,j) + m(i,k)*U(k,j)
        End
        L(i,k) = -m(i,k)           update L matrix
    End
End
    L                              lower triangular matrix
    U                              upper triangular matrix
```

Direct LU factorization

There are two alternative approaches to Gaussian elimination for finding the **LU** factorization of matrix **A**: Doolittle factorization, wherein the diagonal elements of matrix **L** are *1*s; and Cholesky factorization for symmetric matrix **A**, wherein the upper triangular matrix **U** is the transpose of the lower triangular matrix **L**.

Doolittle factorization

For a three-by-three matrix **A**, the problem is to find matrices **L** and **U** so that

$$\begin{array}{ccc} \mathbf{L} & \mathbf{U} & = \quad \mathbf{A} \end{array}$$

$$\begin{bmatrix} 1 & 0 & 0 \\ l_{21} & 1 & 0 \\ l_{31} & l_{32} & 1 \end{bmatrix} \begin{bmatrix} u_{11} & u_{12} & u_{13} \\ 0 & u_{22} & u_{23} \\ 0 & 0 & u_{33} \end{bmatrix} = \begin{bmatrix} a_{11} & a_{12} & a_{13} \\ a_{21} & a_{22} & a_{23} \\ a_{31} & a_{32} & a_{33} \end{bmatrix}$$

One begins by finding $u_{11} = a_{11}$ and then solving for the remaining elements in the first row of **U** and the first column of **L**. At the second stage, we find u_{22} and then the remainder of the second row of **U** and the second column of **L**. This procedure is continued till all the elements of **U** and **L** are determined. A general algorithm is summarized below: in what follows, the MATLAB colon (:) notation is used to represent a row or column.

Input

A	n-by-n matrix to be factorized
n	dimension of **A**

Initialize

U = zeros(n)	initialize **U** to n-by-n zero matrix
L = **I**(n)	initialize **L** to identity matrix

Operate

For $k = 1$ to n

$$U(k,k) = A(k,k) - L(k,1:k-1)*U(1:k-1,j)$$

For $j = k+1$ to n

$$U(k,j) = A(k,j) - L(k,1:k-1)*U(1:k-1,j)$$

$$L(j,k) = (A(j,k) - L(j,1:k-1)*U(1:k-1,k))/U(k,k)$$

End

End

Cholesky LU factorization

If the matrix **A** is symmetric, there is a convenient form of LU factorization called the *Cholesky* factorization. In this method, the upper triangular matrix **U** is the transpose of the lower triangular matrix **L**; that is, $\mathbf{A} = \mathbf{LL}^{\mathrm{T}}$. The basic steps for the Cholesky factorization are summarized below:

Input

A	symmetric n-by-n matrix ($A = A^T$)

Initialize

L=**I**(n)	n-by-n identity matrix

Operate

For $k = 1$ to n

$x = L(k,1:k-1)$ columns 1 to k-1 of the kth row of **L**

$$L(k,k) = \sqrt{A(k,k) - xx^T}$$

For $j = k+1$ to n

$y = L(j,1:k-1)$ columns 1 to k-1 of the jth row of **L**

```
        End
End
Transpose
U=L^T
```

MATLAB contains several convenient commands for LU factorization.

The MATLAB function `lu` can be used to find the LU decomposition of a square matrix **A**. The function call `[L, U] = lu(A)` returns an upper triangular matrix in `U`, and a product of lower triangular matrices in `L` such that `A = LU`. For example:

```
>> a=[4,12,8,4;1,6,25,9;2,9,30,30;3,11,20,18]
a =
        4    12     8     4
        1     6    25     9
        2     9    30    30
        3    11    20    18
>>
>> [L, U]=lu(a)
L =
    1.0000         0         0         0
    0.2500    1.0000         0         0
    0.5000    1.0000    1.0000         0
    0.7500    0.6667   -0.4444    1.0000

U =
    4.0000   12.0000    8.0000    4.0000
         0    3.0000   23.0000    8.0000
         0         0    3.0000   20.0000
         0         0         0   18.5556
```

The MATLAB function for the Cholesky factorization is `chol`. This function uses only the diagonal and upper triangular part of **A**, as demonstrated below. Note that the Cholesky factorization only works on positive definite matrices: A symmetric matrix **A** is positive definite if $\mathbf{x}^T\mathbf{A}\mathbf{x} > 0$; all eigenvalues of **A** are positive and all diagonal elements are positive. In general, **A** is positive definite if it is nonsingular, diagonally dominant ($|a_{ii}| > \sum_{j \neq i} |a_{ij}|$), and each diagonal element is positive ($a_{ii} > 0$ for $i = 1,..n$).

```
>> A=[1,4,5;4,20,40;5,40,128]
A =
        1     4     5
        4    20    40
        5    40   128
>>
```

```
>> chol(A)
ans =
    1.0000    4.0000     5.0000
         0    2.0000    10.0000
         0         0     1.7321
```

This concludes our brief review of selected topics from linear algebra. The interested student is encouraged to read more on the subject in any of the numerous texts available on linear algebra.

Appendix D: Analytical Solutions of Differential Equations

This appendix is a summary of methods used to obtain analytical solutions to ordinary and partial differential equations. It is not meant to be a complete treatise of such methods, but rather a brief review for the benefit of the student. It is assumed that the student has already had a calculus course in ordinary differential equations, where methods for obtaining analytical solutions to such differential equations have been covered in detail. It is unlikely, however, that the undergraduate student will be familiar with analytical methods for partial differential equations. For this reason, the student is encouraged to review the material in this Appendix. For a thorough discussion of these topics, the student is referred to Boyce and DePrima (2001), O'Neil (2003), and Zill and Cullen (2000).

Examples will be used to demonstrate each of the methods discussed. Whenever possible, solutions will be obtained using the Symbolic Math Toolbox of MATLAB. This will provide the student with some experience in using the symbolic logic capabilities of the MATLAB environment.

D.1 Ordinary Differential Equations of First Order

The standard form of a first-order ordinary differential equation is:

$$M(t,y)dt + N(t,y)dy = 0 \tag{D.1}$$

Alternatively, the standard form may be shown as:

$$\frac{dy}{dt} = f(t,y) \tag{D.2}$$

In the discussion that follows, we will work with either one of these standard forms. In this Appendix, and throughout this book, we will use the symbol y to represent the dependent variable, and the symbol t to designate the independent variable in ordinary differential equations. However, when the geometry of the system requires it (i.e., when the independent dimension is space rather than time) we will designate the independent variable with the symbol x.

D.1.1 Equations with separable variables

If the function $f(t,y)$ in Eq. (D.2) can be written as

$$f(t,y) = g(t)h(y) \tag{D.3}$$

where $g(t)$ is a function of t only, and $h(y)$ is a function of y only, then the differential equation

$$\frac{dy}{dt} = g(t)h(y) \tag{D.4}$$

has variables that can be separated. The solution of this equation may be obtained by rearranging

$$\frac{dy}{h(y)} = g(t)dt \tag{D.5}$$

and integrating both sides:

$$\int \frac{dy}{h(y)} = \int g(t)dt \tag{D.6}$$

For example, a simple first-order linear differential equation is

$$\frac{dy}{dt} = \lambda y \tag{D.7}$$

Separation of the variables and integration of both sides gives:

$$\begin{aligned} \int \frac{dy}{y} &= \int \lambda dt \\ \ln y &= \lambda t + \ln c \end{aligned} \tag{D.8}$$

For convenience in obtaining the final solution, the constant of integration was entered in the logarithmic form. Applying exponentials to both sides of the above equation, we obtain the solution in its final form:

$$y = ce^{\lambda t} \tag{D.9}$$

We will demonstrate the utility of this form of solution as a possible solution for other differential equations in this Appendix.

Example D.1 Solve the differential equation

$$\frac{dy}{dt} = \frac{y}{1+t}$$

Separate the variables and integrate both sides

$$\int \frac{dy}{y} = \int \frac{dt}{1+t}$$

to obtain the solution:

$$\ln(y) = \ln(1+t) + \ln(c)$$

The solution may be simplified by exponentiating both sides

$$y = c(1+t)$$

MATLAB solution

```
>> Y = dsolve('Dy=y/(1+t)')
Y = C1*(1+t)
```

D.1.2 Equations with homogeneous coefficients

If $M(t,y)$ and $N(t,y)$ in Eq. (D.1) are homogeneous functions of the same degree, then replacing one of the two variables, say $y = vt$, will result in an equation whose variables may be separated.

A function $f(t,y)$ is homogeneous of degree k in t and y if and only if $f(mt,my) = m^k f(t,y)$ for a constant m.

Example D.2 Solve the differential equation

$$\left(t^2 - ty + y^2\right)dt - tydy = 0$$

Replace y with vt

$$\left(t^2 - t^2v + t^2v^2\right)dt - t^2v\left(vdt + tdv\right) = 0$$

Simplify by dividing through by t^2

$$\left(1 - v + v^2\right)dt - v\left(vdt + tdv\right) = 0$$

Rearrange the terms to separate the variables:

$$\frac{dt}{t} + \frac{vdv}{v-1} = 0$$

Noting that

$$\frac{v}{v-1} = 1 + \frac{1}{v-1}$$

the differential equation becomes:

$$\frac{dt}{t} + \left[1 + \frac{1}{v-1}\right]dv = 0$$

The solution is obtained as[1]

$$\ln t + v + \ln(v-1) = -c_1$$

which simplifies to:

$$t(v-1)e^{v} = e^{-c_1}$$

and, in terms of the original variable y,

$$t\left(\frac{y}{t}-1\right)e^{y/t} = e^{-c_1}$$

This solution is implicit in y. It will be solved below using the MATLAB `solve` command.

MATLAB explicit solution

```
>> YY=solve('t*(y/t-1)*exp(y/t)=exp(-C1)', 'y')
YY =
(lambertw(1/t*exp(-C1-1))+1)*t
```

The integration of the differential equation by MATLAB is:

```
>> Y=dsolve('t*y*Dy=(t^2-t*y+y^2)')
Y =
t+exp(-lambertw(1/exp(C1)*exp(-1)/t)-C1-1)
>> simplify(Y)
ans =
(lambertw(1/t*exp(-C1-1))+1)*t
```

This solution is identical to the one obtained above. The explicit solution of y is given in terms of the Lambert's W function, which is defined as: w = lambertw(x) is the solution to w*exp(w) = x.

[1] The constant of integration was chosen as $-c_1$ to conform with MATLAB's choice of constant.

D.1.3 Exact equations

If there exists a function $F(t,y)=c$, whose total differential is exactly equal to $M(t,y)dt+N(t,y)dy$, then Eq. (D.1) is an *exact equation.* In that case, the following relations are true:

$$\frac{\partial M}{\partial y}=\frac{\partial N}{\partial t} \tag{D.10}$$

and

$$\frac{\partial F}{\partial t}=M(t,y) \tag{D.11}$$

$$\frac{\partial F}{\partial y}=N(t,y) \tag{D.12}$$

The solution of the exact differential equation (Eq. (D.1)) is *F*, and may be obtained by integrating Eq. (D.11) with respect to *t*:

$$F(t,y)=\int Mdt+h(y) \tag{D.13}$$

and integrating Eq. (D.12) with respect to *y*:

$$F(t,y)=\int Ndy+g(t) \tag{D.14}$$

The functions $h(y)$ and $g(t)$ may be identified by comparing (D.13) and (D.14).

Example D.3 Solve the differential equation

$$3t(ty-2)dt+(t^3+2y)dy=0$$

First verify that this is an exact equation:

$$\frac{\partial M}{\partial y}=3t^2=\frac{\partial N}{\partial t}$$

Using Eqs. (D.11) and (D.12):

$$\frac{\partial F}{\partial t} = M = 3t(ty - 2)$$
$$\frac{\partial F}{\partial y} = N = t^3 + 2y$$

Since it is an exact equation, the solution can be determined using Eqs. (D.13) and (D.14). Integrating the first of the above two equations with respect to t, and the second with respect to y, we obtain the functions:

$$F = t^3 y - 3t^2 + h(y)$$
$$F = t^3 y + y^2 + g(t)$$

For the right-hand sides to be equal to each other, the following must be true:

$$h(y) = y^2$$
$$g(t) = -3t^2$$

Therefore, the function F is

$$F = t^3 y - 3t^2 + y^2$$

and the solution of the differential equation is

$$t^3 y - 3t^2 + y^2 = c$$

MATLAB solution

```
>>y=dsolve('Dy=3*t*(t*y-2)/(t^3+2*y)')
Warning: Explicit solution could not be found.
```

MATLAB could not obtain an explicit solution for this equation.

D.1.4 Linear equations and the integrating factor

The following equation

$$\frac{dy}{dt} + P(t)y = f(t) \tag{D.15}$$

is a linear nonhomogeneous ordinary differential equation (See Section 7.2 for a complete definition of ordinary differential equations). To determine the solution to the differential equation, multiply both sides of Eq. (D.15) by the integrating factor

$$\left[e^{\int P(t)dt}\right]$$

to obtain

$$e^{\int P(t)dt}\frac{dy}{dt} + e^{\int P(t)dt}P(t)y = e^{\int P(t)dt}f(t) \tag{D.16}$$

The left side of (D.16) is the derivative of $\left[e^{\int P(t)dt}y\right]$ with respect to t, therefore:

$$\frac{d}{dt}\left[e^{\int P(t)dt}y\right] = e^{\int P(t)dt}f(t) \tag{D.17}$$

Integrate both sides of (D.17)

$$e^{\int P(t)dt}y = c + \int e^{\int P(t)dt}f(t)\,dt \tag{D.18}$$

and rearrange to obtain the solution:

$$y = ce^{-\int P(t)dt} + e^{-\int P(t)dt}\int e^{\int P(t)dt}f(t)\,dt \tag{D.19}$$

Example D.4 Solve the differential equation

$$\frac{dy}{dt} - t^2 y = t^3 e^t$$

Evaluate the integrating factor

$$e^{\int P(t)dt} = e^{\int -t^2 dt} = e^{-t^3/3}$$

Multiply both sides of the differential equation by the integrating factor

$$e^{-t^3/3}\frac{dy}{dt} - e^{-t^3/3}t^2y = e^{-t^3/3}t^3e^t$$

Note that the left side of the above equation is the derivative of $[e^{-t^3/3}y]$

$$\frac{d}{dt}[e^{-t^3/3}y] = e^{-t^3/3}t^3e^t$$

Integrate both sides

$$e^{-t^3/3}y = c + \int e^{-t^3/3}t^3e^t\,dt$$

Divide through by $e^{-t^3/3}$

$$y = ce^{t^3/3} + \int t^3e^t\,dt$$

Evaluate the integral, using the method of integration by parts $\left(\int udv = uv - \int vdu\right)$, to obtain the solution:

$$y = ce^{t^3/3} + t^3e^t - 3t^2e^t + 6te^t - 6e^t$$

MATLAB solution

```
>> dsolve('Dy-t^2*y=t^3*exp(t)')
ans =
(Int(t^3*exp((1-1/3*t^2)*t),t)+C1)*exp(1/3*t^3)
```

This result is equivalent to

$$y = ce^{t^3/3} + \int t^3e^t\,dt$$

To evaluate it further, separate out the term that contains the constant and evaluate the integral:

```
>> syms t C1
>> Int(t^3*exp((1-1/3*t^2)*t)*exp(1/3*t^3))+C1*exp(1/3*t^3)
ans =
t^3*exp(t)-3*t^2*exp(t)+6*t*exp(t)-6*exp(t)+C1*exp(1/3*t^3)
```

The MATLAB solution is identical to the analytical solution obtained above.

D.1.5 Nonlinear equations and the integrating factor

The Bernoulli equation,

$$\frac{dy}{dt} + P(t)y = f(t)y^n \tag{D.20}$$

where $n \neq 1$ is a nonlinear nonhomogeneous ordinary differential equation. The appropriate substitution of variables will convert this Bernoulli equation to a linear equation of the form of Eq. (D.15). This may then be solved by an integrating factor.

To determine the form of the substitution, we first multiply both sides of Eq. (D.20) by y^{-n}

$$y^{-n}\frac{dy}{dt} + P(t)y^{1-n} = f(t) \tag{D.21}$$

We note that the differential of y^{1-n} is $(1-n)y^{-n}$, i.e.,

$$\frac{d(y^{1-n})}{dt} = (1-n)y^{-n}\frac{dy}{dt} \tag{D.22}$$

The equation may be simplified by making the substitution

$$z = y^{1-n} \tag{D.23}$$

which yields:

$$\frac{dz}{dt} = (1-n)y^{-n}\frac{dy}{dt} \tag{D.24}$$

With this substitution, the Bernoulli equation becomes

$$\left(\frac{1}{1-n}\right)\frac{dz}{dt} + P(t)z = f(t) \tag{D.25}$$

which is certainly of the form of (D.15) and may be solved by the integrating factor of Section D.1.4.

D.2 Ordinary Differential Equations of Higher Order

The general form of a linear ordinary differential equation of order n may be written as

$$b_n(t)\frac{d^n y}{dt^n}+b_{n-1}(t)\frac{d^{n-1}y}{dt^{n-1}}+\ldots+b_1(t)\frac{dy}{dt}+b_0(t)y=R(t) \qquad \text{(D.26)}$$

If $R(t)=0$, the equation is called *homogeneous*. If $R(t)\neq 0$, the equation is *nonhomogeneous*. The coefficients $\{b_i \mid i=n,\ldots,1\}$ are called *variable coefficients* when they are functions of t, and *constant coefficients* when they are scalars. A differential equation is *autonomous* if the independent variable does not appear explicitly in that equation.

D.2.1 Linear homogeneous equations with constant coefficients

The following equation is a third-order, linear, homogeneous, autonomous, differential equation

$$b_3\frac{d^3 y}{dt^3}+b_2\frac{d^2 y}{dt^2}+b_1\frac{dy}{dt}+b_0 y=0 \qquad \text{(D.27)}$$

This equation may be written in terms of the differential operator D (where $D\equiv\frac{d}{dt}$):

$$b_3D^3y+b_2D^2y+b_1Dy+b_0y=0 \qquad \text{(D.28)}$$

Since y postmultiplies each of the terms in Eq. (D.28), it may be factored out

$$\left(b_3D^3+b_2D^2+b_1D+b_0\right)y=0 \qquad \text{(D.29)}$$

The bracketed term in Eq. (D.29) is a linear combination of differential operators, and is itself a differential operator that operates on y. For Eq. (D.29) to be true for any value of y, the bracketed term must be equal to zero; i.e.,

$$\left(b_3D^3+b_2D^2+b_1D+b_0\right)=0 \qquad \text{(D.30)}$$

This is called the *characteristic* or *auxiliary equation* of the differential equation. It is a polynomial of degree n (n = 3, in this example), and possesscs n roots $(\lambda_1,\lambda_2,\lambda_3)$.

The roots, which are also called *eigenvalues*, may be real and distinct, real and repeated, complex, or any combination of these. Complex roots always appear in conjugate pairs.

For the case of real and distinct roots $(\lambda_1 \neq \lambda_2 \neq \lambda_3)$, the solution of the differential equation may be constructed from these roots, using the form of solution obtained in Eq. (D.9) as follows:

$$y = c_1 e^{\lambda_1 t} + c_2 e^{\lambda_2 t} + c_3 e^{\lambda_3 t} \tag{D.31}$$

Each part of the solution that depends on an eigenvalue is called an *eigenfunction*.

For the case of real and repeated roots $(\lambda_1 = \lambda_2 \neq \lambda_3)$, the solution may be constructed as follows:

$$y = c_1 e^{\lambda_1 t} + c_2 t e^{\lambda_1 t} + c_3 e^{\lambda_3 t} \tag{D.32}$$

For the case of one pair of complex roots and one real root:

$$\lambda_1 = \alpha + \beta i, \quad \lambda_2 = \alpha - \beta i, \quad \lambda_3 = \text{real}, \ \text{where } i = \sqrt{-1}$$

the solution is

$$y = c_1 e^{(\alpha+\beta i)t} + c_2 e^{(\alpha-\beta i)t} + c_3 e^{\lambda_3 t} \tag{D.33}$$

Utilizing the Euler identities,

$$\begin{aligned} e^{\alpha+\beta i} &= e^{\alpha}(\cos\beta + i\sin\beta) \\ e^{\alpha-\beta i} &= e^{\alpha}(\cos\beta - i\sin\beta) \end{aligned} \tag{D.34}$$

we convert Eq. (D.33) to:

$$y = c_4 e^{\alpha t}\cos(\beta t) + c_5 e^{\alpha t}\sin(\beta t) + c_3 e^{\lambda_3 t} \tag{D.35}$$

where the new constants are

$$\begin{aligned} c_4 &= c_1 + c_2 \\ c_5 &= i(c_1 - c_2) \end{aligned}$$

Example D.5 Determine the solution of the differential equation

$$\frac{d^3y}{dt^3} - 9\frac{d^2y}{dt^2} + 26\frac{dy}{dt} - 24y = 0$$

Introduce the differential operators:

$$D^3y - 9D^2y + 26Dy - 24y = 0$$

Obtain the characteristic equation:

$$\left(D^3 - 9D^2 + 26D - 24\right) = 0$$

Factor it in order to calculate the roots:

$$(D-2)(D-4)(D-3) = 0$$

Therefore, the roots of the characteristic equation are:

$$\lambda_1 = 2, \quad \lambda_2 = 4, \quad \lambda_3 = 3$$

and the solution is constructed as follows:

$$y = c_1e^{2t} + c_2e^{4t} + c_3e^{3t}$$

This solution may be verified by substituting it into the original differential equation.

MATLAB solution

```
>> Y=dsolve('D3y-9*D2y+26*Dy-24*y=0')
Y =
C1*exp(2*t)+C2*exp(4*t)+C3*exp(3*t)
```

The constants of integration may be evaluated when initial conditions are given for the differential equation. For example, for $y(0) = 1$, $\left.\frac{dy}{dt}\right|_0 = 2$, and $\left.\frac{d^2y}{dt^2}\right|_0 = 3$, the constants are obtained algebraically or with a MATLAB command:

```
>>Y=dsolve('D3y-9*D2y+26*Dy-24*y=0','y(0)=1','Dy(0)=2','D2y(0)=3')
Y =
1/2*exp(2*t)-1/2*exp(4*t)+exp(3*t)
```

D.2.2 Linear nonhomogeneous equations (constant coefficients)

The equation

$$\frac{d^3y}{dt^3} - 9\frac{d^2y}{dt^2} + 26\frac{dy}{dt} - 24y = e^t \tag{D.36}$$

is a nonhomogeneous ordinary differential equation. The solution of this category of equations is composed of a complementary and a particular solution:

$$y = y_c + y_p \tag{D.37}$$

The complementary solution, y_c, is the solution of the homogeneous part of Eq. (D.36),

$$\frac{d^3y}{dt^3} - 9\frac{d^2y}{dt^2} + 26\frac{dy}{dt} - 24y = 0$$

that was found in Example D.5 to be $y_c = c_1e^{2t} + c_2e^{4t} + c_3e^{3t}$. The particular solution, y_p, corresponds to the nonhomogeneous part of the equation (the right-hand side of (D.36)). By inspecting the right-hand side of (D.36), we predict that e^t resulted from a root $\lambda_4 = 1$, i.e.,

$$(D-1)R(t) = 0 \tag{D.38}$$

Multiplying both sides of (D.36) by $(D-1)$ forces the right-hand side to become zero:

$$(D-1)(D^3 - 9D^2 + 26D - 24)y = (D-1)e^t = 0 \tag{D.39}$$

Eq. (D.39) is a homogeneous equation and has roots:

$$\lambda_1 = 2, \quad \lambda_2 = 4, \quad \lambda_3 = 3, \quad \text{and} \quad \lambda_4 = 1$$

The solution is

$$y = c_1e^{2t} + c_2e^{4t} + c_3e^{3t} + Ce^t \tag{D.40}$$

The last term in (D.40) is the result of the particular solution, and is shown with constant C to indicate that C must have a value such that the particular solution satisfies (D.36). To calculate the value of C, we take the first, second, and third derivatives of y_p, substitute these into Eq. (D.36), and solve for C. This operation yields C = -1/6, therefore the complete solution is:

$$y = c_1 e^{2t} + c_2 e^{4t} + c_3 e^{3t} - \frac{1}{6} e^{t} \tag{D.41}$$

MATLAB solution:

```
>> Y=dsolve('D3y-9*D2y+26*Dy-24*y=exp(t)')
Y =
-1/6*exp(t)+C1*exp(2*t)+C2*exp(4*t)+C3*exp(3*t)
```

Example D.6 Determine the solution of the nonhomogeneous differential equation

$$\frac{d^3y}{dt^3} - 6\frac{d^2y}{dt^2} + 11\frac{dy}{dt} - 24y = \sin t + \cos t$$

Introduce the differential operators

$$D^3y - 6D^2y + 11Dy - 6y = 0$$

Obtain the characteristic equation

$$\left(D^3 - 6D^2 + 11D - 6\right) = 0$$

Factor it in order to obtain the roots

$$(D-1)(D-2)(D-3) = 0$$

Therefore, the roots of the characteristic equation are:

$$\lambda_1 = 1, \quad \lambda_2 = 2, \quad \lambda_3 = 3$$

and the complementary solution is:

$$y_c = c_1e^t + c_2e^{2t} + c_3e^{3t}$$

The particular solution is of the form:

$$y_p = C_1 \cos t + C_2 \sin t$$

To calculate the values of C_1 and C_2, we take the first, second, and third derivatives of y_p, and substitute these into the differential equation. The results are

$$C_1 = -\frac{1}{10} \quad \text{and} \quad C_2 = \frac{1}{10}$$

Therefore, the complete solution is:

$$y_c = c_1e^t + c_2e^{2t} + c_3e^{3t} - \frac{1}{10}\cos t + \frac{1}{10}\sin t$$

MATLAB solution

```
>> Y=dsolve('D3y-6*D2y+11*Dy-6*y=cos(t)+sin(t)')
Y =
-1/10*cos(t)+1/10*sin(t)+C1*exp(t)+C2*exp(2*t)+C3*exp(3*t)
```

The constants of integration (in the complementary solution) may be evaluated when initial conditions are given for the differential equation. For example, for $y(0) = 1$, $\left.\frac{dy}{dt}\right|_0 = 2$, and $\left.\frac{d^2y}{dt^2}\right|_0 = 3$, the constants are obtained algebraically, or with the MATLAB command:

```
>> Y=dsolve('D3y-6*D2y+11*Dy-6*y=cos(t)+sin(t)', 'y(0)=1',
'Dy(0)=2', 'D2y(0)=3')
Y =
-1/10*cos(t)+1/10*sin(t)+7/5*exp(2*t)-3/10*exp(3*t)
```

D.3 Partial Differential Equations with Separable Variables

The general classification of partial differential equations is given in Section 8.2. These equations may be linear or nonlinear, homogeneous or nonhomogeneous, of first, second, third, or fourth order. The most commonly encountered partial differential equations in engineering and physics are of the second order. For this reason, we will concentrate our treatment of analytical solutions to second order linear partial differential equations.

Partial differential equations can have solutions involving arbitrary functions and solutions involving an unlimited number of arbitrary constants. The general solution of a partial differential equation may be defined as a solution involving n arbitrary functions.

D.3.1 The diffusion equation

The classic example of a partial differential equation that has separable variables is the one-dimensional unsteady-state diffusion equation:

$$\frac{\partial u}{\partial t} = \alpha \frac{\partial^2 u}{\partial x^2} \tag{D.42}$$

When u is concentration and α is the diffusivity, then Eq. (D.42) represents the mechanism of diffusion of a solute through a permeable membrane. On the other hand, if u is temperature and α is thermal diffusivity, then this equation describes the diffusion of energy across a slab.

For the sake of discussion, let us assume that we have a membrane of finite thickness, L, that initially has an internal concentration of solute that is a function of position, i.e., $f(x)$. At the start of the experiment, the membrane is immersed in a large quantity of liquid that is maintained at a concentration of zero. Both sides of the membrane are exposed to the liquid. The initial and boundary conditions for Eq. (D.42) are

$$\begin{aligned} &\text{at } t = 0, \quad u(x,0) = f(x) \quad \text{for} \quad 0 \le x \le L \\ &\text{at } t > 0, \quad u(0,t) = 0 \quad \text{and} \quad u(L,t) = 0 \end{aligned} \tag{D.43}$$

Let us assume that u is a solution to the partial differential equation that can be expressed as the product of two functions: $X(x)$, which is a function of x alone, and

$T(t)$, which is a function of t alone:

$$u(x,t) = X(x)T(t) \tag{D.44}$$

When we substitute Eq. (D.44) into (D.42), we have:

$$X(x)T'(t) = \alpha X''(x)T(t) \tag{D.45}$$

The variables in this equation may be separated:

$$\frac{X''(x)}{X(x)} = \frac{T'(t)}{\alpha T(t)} \tag{D.46}$$

Since the left side of (D.46) is a function of x only, and the right side is a function of t only, it follows that in order for these to be equal to each other, they must be equal to a constant, so:

$$\frac{X''(x)}{X(x)} = \frac{T'(t)}{\alpha T(t)} = -\lambda^2 \tag{D.47}$$

For reasons that will become evident later, the constant is chosen to be $-\lambda^2$. We can now split Eq. (D.47) into two ordinary differential equations:

$$\begin{aligned} X'' + \lambda^2 X &= 0 \\ T' + \alpha\lambda^2 T &= 0 \end{aligned} \tag{D.48}$$

The solutions of these equations may be determined by the methods of Section D.2:

$$\begin{aligned} X &= c_1 \cos(\lambda x) + c_2 \sin(\lambda x) \\ T &= c_3 e^{-\alpha\lambda^2 t} \end{aligned} \tag{D.49}$$

Using first the boundary conditions:

$$\left.\begin{aligned} u(0,t) &= X(0)T(t) = 0 \\ u(L,t) &= X(L)T(t) = 0 \end{aligned}\right\} \quad \text{for} \quad t > 0 \tag{D.50}$$

we conclude that $X(0)=0$ and $X(L)=0$. Applying these to the first part of (D.49) shows that $c_1=0$ and

$$X(L)=c_2\sin(\lambda L)=0$$

For a nontrivial solution, c_2 and λ may not be zero, therefore

$$\sin(\lambda L)=0$$

This implies that $\lambda L=n\pi$ for $n=1,2,3,...,\infty$, which enables us to express λ as

$$\lambda=\frac{n\pi}{L} \quad \text{for} \quad n=1,2,3,...,\infty \tag{D.51}$$

The $X(x)$ and $T(t)$ solutions are now shown as $X_n(x)$ and $T_n(t)$:

$$\left.\begin{aligned} X_n(x) &= c_{2_n}\sin\left(\frac{n\pi x}{L}\right) \\ T_n(t) &= c_{3_n}e^{-\alpha\left(\frac{n^2\pi^2}{L^2}\right)t} \end{aligned}\right\} \quad \text{for} \quad n=1,2,3,...,\infty \tag{D.52}$$

Combining these solutions into (D.44), we obtain

$$u_n = A_n e^{-\alpha\left(\frac{n^2\pi^2}{L^2}\right)t}\sin\left(\frac{n\pi x}{L}\right) \quad \text{for} \quad n=1,2,3,...,\infty \tag{D.53}$$

The product of the two constants, $(c_{2_n}c_{3_n})$, is replaced by the new constant A_n, which is a function of n. The *superposition principle* states that if u_1, u_2, …, u_n are solutions of a linear homogeneous partial differential equation in some region, then the linear combination of any of these solutions is also a solution of that equation in that region. In order to obtain the solution that satisfies any boundary conditions, we consider the infinite series combination of all of the solutions

$$u(x,t)=\sum_{n=1}^{\infty}u_n=\sum_{n=1}^{\infty}A_n e^{-\alpha\left(\frac{n^2\pi^2}{L^2}\right)t}\sin\left(\frac{n\pi x}{L}\right) \tag{D.54}$$

We now apply the initial condition (at $t = 0, u(x,0) = f(x)$):

$$u(x,0) = f(x) = \sum_{n=1}^{\infty} A_n \sin\left(\frac{n\pi x}{L}\right) \tag{D.55}$$

To evaluate A_n, we multiply both sides of Eq. (D.55) by $\sin\left(\frac{m\pi x}{L}\right)$, where m is any integer $m = 1,2,3,...,\infty$, and integrate over the interval (0, L):

$$\int_0^L f(x)\sin\left(\frac{m\pi x}{L}\right)dx = \sum_{n=1}^{\infty} A_n \int_0^L \left(\sin\left(\frac{m\pi x}{L}\right)\sin\left(\frac{n\pi x}{L}\right)\right)dx \tag{D.56}$$

The sine functions are orthogonal functions in the interval (0, L) with respect to a weighting function of unity. Therefore,

$$\int_0^L \left(\sin\left(\frac{m\pi x}{L}\right)\sin\left(\frac{n\pi x}{L}\right)\right)dx = 0 \quad \text{for} \quad n \neq m \tag{D.57}$$

But for $n = m$, the integral takes a value of $L/2$, as shown below (with the aid of integral tables):

$$\int_0^L \sin^2\left(\frac{n\pi x}{L}\right)dx = \left[\frac{1}{2}x - \frac{1}{4\frac{n\pi}{L}}\sin\left(\frac{2n\pi x}{L}\right)\right]_0^L = \frac{L}{2} \tag{D.58}$$

Utilizing (D.58), we can now solve Eq. (D.56) for A_n, since only one term in the summation series remains, i.e., when $n = m$,

$$A_n = \frac{2}{L}\int_0^L f(x)\sin\left(\frac{n\pi x}{L}\right)dx \tag{D.59}$$

Thus, the solution of Eq. (D.42), with the initial and boundary conditions given by Eq. (D.43), is

$$u(x,t) = \frac{2}{L}\sum_{n=1}^{\infty}\left(\int_0^L f(x)\sin\left(\frac{n\pi x}{L}\right)dx\right)e^{-\alpha\left(\frac{n^2\pi^2}{L^2}\right)t}\sin\left(\frac{n\pi x}{L}\right) \tag{D.60}$$

Example D.7 Evaluate the concentration profiles of the membrane diffusion problem, discussed above, using the following values:

$$u(x,0) = 1.0,\ L = \pi,\ \text{and}\ \alpha = 1$$

Substituting the above values into Eq. (D.59) gives the value of A_n as

$$A_n = \frac{2}{\pi}\left[\frac{1-(-1)^n}{n}\right]$$

so that the solution (D.54) becomes

$$u(x,t) = \frac{2}{\pi}\sum_{n=1}^{\infty}\left[\frac{1-(-1)^n}{n}\right]e^{-n^2 t}\sin(nx)$$

MATLAB solution

The following MATLAB program evaluates and plots the solution in the range of the independent variable, $0 < t < 1.5$:

```
% Evaluating the analytical solution of the one-dimensional
% unsteady-state diffusion equation
clear, clc

L=pi; alpha=1;
space_steps=50;
dx=L/space_steps;
time_steps=15;
dt=1.5/time_steps;
u=zeros(space_steps,time_steps);
u(1:space_steps+1,1)=1;
syms tt x
for n=1:2:13
    f(n)=((1-(-1)^n)/n)*exp(-n^2*tt)*sin(n*x);
end
sumf=sum(f);
tt=0;
for t=1:time_steps+1
    tt=tt+dt;
    x=0;
      for i=2:space_steps
        x=x+dx;
        u(i,t+1)=(2/pi)*eval(sum(f));
    end
end
```

```
figure(1)
xx=[0:dx:L];
plot(xx,u(:,1), xx,u(:,ceil(0.1/dt)+1), xx,u(:,ceil(0.2/dt)+1),...
     xx,u(:,ceil(0.3/dt)+1), xx,u(:,ceil(0.5/dt)+1),...
     xx,u(:,ceil(1/dt)+1),xx,u(:,ceil(1.5/dt)+1))
axis tight
title('Figure D.1: One-dimensional unsteady-state diffusion')
xlabel('Membrane thickness, x'); ylabel('Concentration, y')
text(L/30, 0.97,'t = 0.0'); text(L/7,0.93,'t = 0.1');
text(L/2.2,0.95,'t = 0.2'); text(L/2.2,0.88,'t = 0.3');
text(L/2.2,0.72,'t = 0.50'); text(L/2.2,0.42,'t = 1.00');
text(L/2.2,0.23,'t = 1.50')
```

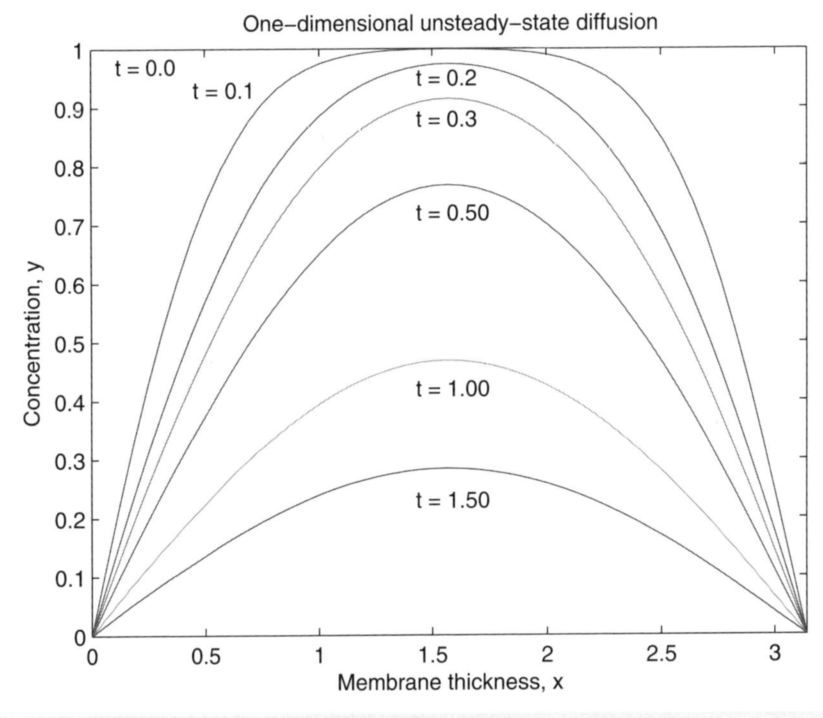

Figure D.1 MATLAB solution: concentration profiles of the membrane diffusion problem

Discussion of results

The concentration profile at $t = 0$ is $u(x,0) = 1$, which is the initial condition. At $t > 0$, the outside surfaces of the membrane (at $x = 0$ and $x = L$) that are in contact with the liquid have $u = 0$, which are the boundary conditions. As t increases, the solute from inside the membrane diffuses into the surrounding liquid, thus the concentration profile decreases towards zero.

It should be noted that when n is even, the coefficient $A_n = 0$, therefore, only the odd terms in the series need to be evaluated. Also, we have determined that the series converges to within $10^{-6}\%$ of itself when $n = 13$. This is a consequence of the term $e^{-n^2 t}$ in the solution.

D.3.2 The potential equation

Eq. (D.61) is the *potential* equation, better known as Laplace's equation. It simulates the potentials of force fields in mechanics, or electromagnetic and gravitational fields. It also models the steady-state heat conduction, or diffusion, in a two-dimensional body, such as a rectangular plate.

$$\frac{\partial^2 u}{\partial x^2} + \frac{\partial^2 u}{\partial y^2} = 0 \tag{D.61}$$

When u is concentration, Eq. (D.61) represents the steady-state mechanism of diffusion of a solute across a rectangular plate, and when u is temperature, this equation describes the steady-state diffusion of energy over a rectangular solid plate.

The Laplace equation is also amenable to the method of separation of variables. A variety of boundary conditions may be assigned to Eq. (D.61) to complete the statement of the problem. Chapter 8 of this book describes in detail the possible conditions that may apply to this problem.

For this discussion, we will solve the heat transfer problem of a rectangular plate of dimensions, $(a \times b)$, as shown on Fig. D.2.

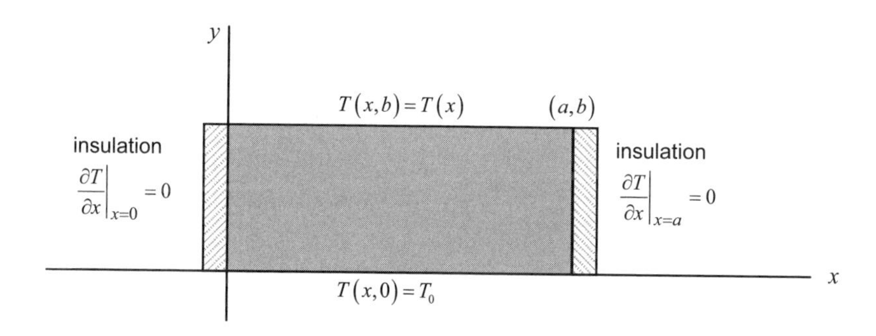

Figure D.2 Heat transfer in a rectangular plate.

The Laplace equation in terms of temperature is

$$\frac{\partial^2 T}{\partial x^2}+\frac{\partial^2 T}{\partial y^2}=0 \tag{D.62}$$

The plate has perfect insulation on the left and right sides (Neumann conditions), and has temperature equal to T_0 on the bottom side, and equal to a function of x on the top side (Dirichlet conditions). These boundary conditions are expressed in mathematical terms as follows:

$$\left.\frac{\partial T}{\partial x}\right|_{x=0}=0 \qquad \left.\frac{\partial T}{\partial x}\right|_{x=a}=0 \qquad \text{for} \quad 0 \le y \le b$$

$$T(x,0)=T_0 \qquad T(x,b)=T(x) \qquad \text{for} \quad 0<x<a \tag{D.63}$$

For convenience in obtaining the solution of the problem, and to maintain the generality of the solution, we transform the variable T to the variable u as follows:

$$u=\frac{T-T_0}{T_0} \tag{D.64}$$

and

$$f(x)=\frac{T(x)-T_0}{T_0} \tag{D.65}$$

This transformation causes the boundary conditions to change to:

$$\left.\frac{\partial u}{\partial x}\right|_{x=0}=0 \qquad \left.\frac{\partial u}{\partial x}\right|_{x=a}=0 \qquad \text{for} \quad 0 \le y \le b$$

$$u(x,0)=0 \qquad u(x,b)=f(x) \qquad \text{for} \quad 0<x<a \tag{D.66}$$

and Eq. (D.62) to become identical to Eq. (D.61). The transformed problem is shown on Fig. D.3.

The solution of Eq. (D.61) with the boundary conditions of Eq. (D.66) is given by Zill and Cullen (2000) and is described below, with some modifications.

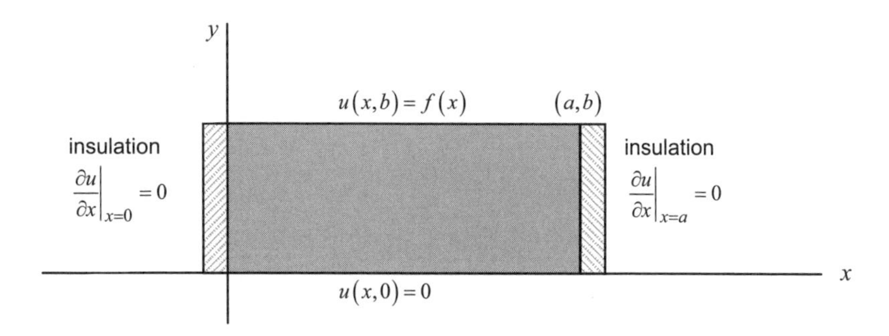

Figure D.3 Heat transfer in a rectangular plate with transformed variables.

Let us assume that u is a solution to the partial differential equation that can be expressed as the product of two functions: $X(x)$ which is function of x alone, and $Y(y)$ which is a function of y alone. Thus,

$$u(x,y) = X(x)Y(y) \tag{D.67}$$

Differentiating Eq. (D.67) and substituting into (D.61) enables us to separate the variables:

$$\frac{X''(x)}{X(x)} = -\frac{Y''(y)}{Y(y)} = -\lambda^2 \tag{D.68}$$

This yields two ordinary differential equations:

$$X'' + \lambda^2 X = 0 \tag{D.69}$$

$$Y'' - \lambda^2 Y = 0 \tag{D.70}$$

The solution of the first equation is:

$$X = c_1 \cos(\lambda x) + c_2 \sin(\lambda x) \tag{D.71}$$

The solution of the second equation is:

$$Y = c_3 e^{\lambda y} + c_4 e^{-\lambda y} \tag{D.72}$$

The boundary conditions will be used to determine the constants, c_1 to c_4. The first boundary condition is equivalent to $X'(0) = 0$. To apply this, we differentiate X, given by Eq. (D.71), and evaluate it at $x = 0$:

$$X'(0) = -c_1\lambda \sin(0) + c_2\lambda \cos(0) = 0 \tag{D.73}$$

Since $\sin(0) = 0$ and $\cos(0) = 1$, then c_2 must be zero, and Eq. (D.71) simplifies to

$$X = c_1 \cos(\lambda x) \tag{D.74}$$

The second boundary condition is equivalent to $X'(a) = 0$. We differentiate X given by Eq. (D.74) and evaluate it at $x = a$:

$$X'(a) = -c_1\lambda \sin(\lambda a) = 0 \tag{D.75}$$

This is satisfied when $\lambda = 0$ and when $\lambda a = n\pi$, or $\lambda = \dfrac{n\pi}{a}$, $n = 1,2,3,\ldots,\infty$. For the case when $\lambda = 0$, $n = 0$, Eq. (D.69) becomes

$$X'' = 0 \tag{D.76}$$

Integrating (D.76) twice, we obtain the equation of a straight line:

$$X = c_1 + c_2 x \tag{D.77}$$

The boundary conditions $X'(0) = 0$ and $X'(a) = 0$ applied to (D.77) require that $X = c_1$, therefore the solutions of Eq. (D.69) are:

$$X = c_{1_0} \quad \text{for} \quad n = 0, \qquad \text{and} \qquad X = c_{1_n} \cos\left(\frac{n\pi x}{a}\right) \quad \text{for} \quad n = 1,2,3,\ldots,\infty \tag{D.78}$$

The third boundary condition, $u(x,0) = 0$ is equivalent to $Y(0) = 0$. Applying this condition to Eq. (D.72):

$$\begin{aligned} Y(0) &= c_3 e^{\lambda 0} + c_4 e^{-\lambda 0} = 0 \\ &= c_3 + c_4 = 0 \end{aligned} \tag{D.79}$$

Eq. (D.79) requires that $c_4 = -c_3$ when $\lambda > 0$. This converts Eq. (D.72) to

$$Y = c_3\left(e^{\lambda y} - e^{-\lambda y}\right) \tag{D.80}$$

The following relation between exponentials and the hyperbolic sine function:

$$\frac{1}{2}\left(e^{\alpha} - e^{-\alpha}\right) = \sinh\alpha \tag{D.81}$$

may be used to transform Eq. (D.80) to

$$Y = \bar{c}_3 \sinh(\lambda y) \tag{D.82}$$

However, when $\lambda = 0$, Eq. (D.70) becomes $Y'' = 0$, and its solution is the equation of the straight line

$$Y = c_3 + c_4 y \tag{D.83}$$

The boundary condition $Y(0) = 0$ implies that $c_3 = 0$. Therefore, the solutions corresponding to Eq. (D.70) are:

$$Y = c_{4_0} y \quad \text{for} \quad n = 0, \qquad \text{and} \qquad Y = \bar{c}_{3_n} \sinh\left(\frac{n\pi y}{a}\right) \quad \text{for} \quad n = 1,2,3,\ldots,\infty \tag{D.84}$$

The combined solutions for $X(x)Y(y)$ are

$$u_0 = A_0 y \;\text{ for }\; n = 0, \quad \text{and} \quad u_n = A_n \sinh\left(\frac{n\pi y}{a}\right)\cos\left(\frac{n\pi x}{a}\right) \;\text{ for }\; n = 1,2,3,\ldots,\infty \tag{D.85}$$

Using the superposition principle, we create the linear combination of all these solutions:

$$u(x,y) = A_0 y + \sum_{n=1}^{\infty} A_n \sinh\left(\frac{n\pi y}{a}\right)\cos\left(\frac{n\pi x}{a}\right) \tag{D.86}$$

We finally utilize the last boundary condition of (D.66):

$$u(x,b) = f(x) = A_0 b + \sum_{n=1}^{\infty} A_n \sinh\left(\frac{n\pi b}{a}\right)\cos\left(\frac{n\pi x}{a}\right) \tag{D.87}$$

To evaluate A_0 and A_n, we multiply both sides of Eq. (D.87) by $\cos(m\pi x/a)$, where m is any integer, $m = 1,2,3,\ldots,\infty$, and integrate over the interval (0, a)

$$\int_0^a f(x)\cos\left(\frac{m\pi x}{a}\right)dx = A_0 b\int_0^a \cos\left(\frac{m\pi x}{a}\right)dx + \sum_{n=1}^{\infty} A_n \sinh\left(\frac{n\pi b}{a}\right)\int_0^a \cos\left(\frac{m\pi x}{a}\right)\cos\left(\frac{n\pi x}{a}\right)dx \tag{D.88}$$

The cosine functions are orthogonal functions in the interval (0, a) with respect to a weighting function of unity. Therefore

$$\int_0^a \cos\left(\frac{m\pi x}{a}\right)\cos\left(\frac{n\pi x}{a}\right)dx = 0 \quad \text{for} \quad n \neq m \tag{D.89}$$

But for $n = m$, the integral takes the value of $(a/2)$, as shown below (with the aid of integral tables):

$$\int_0^a \cos^2\left(\frac{n\pi x}{a}\right)dx = \left[\frac{1}{2}x + \frac{1}{4\frac{n\pi}{a}}\sin\left(\frac{2n\pi x}{a}\right)\right]_0^a = \frac{a}{2} \tag{D.90}$$

When (D.89) and (D.90) are used in Eq. (D.88), the latter simplifies to

$$\int_0^a f(x)\cos\left(\frac{n\pi x}{a}\right)dx = A_0 b\int_0^a \cos\left(\frac{n\pi x}{a}\right)dx + A_n\frac{a}{2}\sinh\left(\frac{n\pi b}{a}\right) \tag{D.91}$$

Note that the summation sign has disappeared, because only one term remains in the series when $n = m$. For $n = 0$, $\cos(0) = 1$ and $\sinh(0) = 0$, therefore, Eq. (D.91) simplifies further to

$$\int_0^a f(x)\,dx = A_0 ab \tag{D.92}$$

This enables us to evaluate A_0 as

$$A_0 = \frac{1}{ab}\int_0^a f(x)\,dx \tag{D.93}$$

For $n > 0$, the integral $\int_0^a \cos(n\pi x/a)\, dx$ vanishes; therefore, A_n may be evaluated from (D.91) as

$$A_n = \frac{2}{a \sinh\left(\frac{n\pi b}{a}\right)} \int_0^a f(x) \cos\left(\frac{n\pi x}{a}\right) dx \tag{D.94}$$

So, the solution of this problem consists of Eq. (D.86) with A_0 and A_n defined by (D.93) and (D.94), respectively. To convert u back to T, we rearrange Eq. (D.64) and apply

$$T = T_0 u + T_0 \tag{D.95}$$

Example D.8 Show that when $f(x) = C$ (constant), the temperature profiles given by Eq. (D.87) are straight lines in the *y* direction.

When $f(x) = C$, the integral in Eq. (D.94) is zero, therefore A_n is zero:

$$A_n = \frac{2C}{a \sinh\left(\frac{n\pi b}{a}\right)} \int_0^a \cos\left(\frac{n\pi x}{a}\right) dx = \frac{2C}{a \sinh\left(\frac{n\pi b}{a}\right)} \left[\frac{a}{n\pi} \sin\left(\frac{n\pi x}{a}\right)\right]_0^a = 0$$

The integral in (D.93) is equal to Ca, therefore A_0 is C/b:

$$A_0 = \frac{1}{ab} \int_0^a C\, dx = \frac{Ca}{ab} = \frac{C}{b}$$

Eq. (D.86) simplifies to

$$u(x,y) = \frac{C}{b} y$$

Applying Eq. (D.95):

$$T = T_0 \frac{C}{b} y + T_0$$

This is the equation of a straight line in the *y* direction. The physical explanation of this result is: the left and right sides of the plate are perfectly insulated, and the top and bottom sides are kept at constant temperatures.

D.3.3 Periodic functions and the Fourier series

The solutions developed in Sections D.3.1 and D.3.2 may also have been obtained by making use of *Fourier series* expansions.

A function $f(x)$ is periodic if there exists a constant $2p$ with the property that

$$f(x+2p) = f(x) \qquad \text{for all } x \tag{D.96}$$

If $2p$ is the smallest number for which this identity holds, it is called the period of the function. It is well known that the sine and cosine functions are periodic:

$$\left.\begin{aligned}\cos\left(\frac{n\pi(x+2p)}{p}\right) &= \cos\left(\frac{n\pi x}{p}\right)\\ \sin\left(\frac{n\pi(x+2p)}{p}\right) &= \sin\left(\frac{n\pi x}{p}\right)\end{aligned}\right\} \quad \text{for all values of } x \tag{D.97}$$

An arbitrary function $f(x)$ of period $2p$ can be expressed by a series of the form

$$\begin{aligned}f(x) = \frac{a_0}{2} &+ a_1\cos\left(\frac{\pi x}{p}\right) + a_2\cos\left(\frac{2\pi x}{p}\right) + \cdots + a_n\cos\left(\frac{n\pi x}{p}\right) + \cdots\\ &+ b_1\sin\left(\frac{\pi x}{p}\right) + b_2\sin\left(\frac{2\pi x}{p}\right) + \cdots + b_n\sin\left(\frac{n\pi x}{p}\right) + \cdots\end{aligned} \tag{D.98}$$

where

$$a_0 = \frac{1}{p}\int_d^{d+2p} f(x)\,dx \qquad \text{for all values of } d \tag{D.99}$$

$$a_n = \frac{1}{p}\int_d^{d+2p} f(x)\cos\left(\frac{n\pi x}{p}\right)dx \qquad \text{for all values of } d \tag{D.100}$$

$$b_n = \frac{1}{p}\int_d^{d+2p} f(x)\sin\left(\frac{n\pi x}{p}\right)dx \qquad \text{for all values of } d \tag{D.101}$$

Eqs. (D.98) to (D.100) constitute the Fourier series expansion of the function $f(x)$.

D.3.3.1 Even and odd symmetric functions

The function $f(x)$ is an *even* function if

$$f(-x) = f(x) \qquad \text{for all } x$$

The graph of $f(x)$ is symmetric around the vertical axis, as shown on Fig. D.4a. The function $f(x)$ is an *odd* function if

$$f(-x) = -f(x) \qquad \text{for all } x$$

Geometrically, the graph of $f(x)$ is symmetric about the origin, as shown on Fig. D.4b. The Fourier series for even functions in the interval $(-p, p)$ is given by

$$f(x) = \frac{a_0}{2} + \sum_{n=1}^{\infty} a_n \cos\left(\frac{n\pi x}{p}\right) \tag{D.102}$$

where

$$a_0 = \frac{2}{p} \int_0^p f(x)\, dx \tag{D.103}$$

$$a_n = \frac{2}{p} \int_0^p f(x) \cos\left(\frac{n\pi x}{p}\right) dx \tag{D.104}$$

The Fourier series for odd functions in the interval $(-p, p)$ is given by

$$f(x) = \sum_{n=1}^{\infty} b_n \sin\left(\frac{n\pi x}{p}\right) \tag{D.105}$$

where

$$b_n = \frac{2}{p} \int_0^p f(x) \sin\left(\frac{n\pi x}{p}\right) dx \tag{D.106}$$

Eq. (D.105) is identical to Eq. (D.55) in Section D.3.1 when $p = L$; therefore, the constant A_n may be obtained from Eq. (D.106). Similarly, Eqs. (D.102), (D.103), and (D.104) may be used with Eq. (D.86) in Section D.3.2 to determine the constants A_0 and A_n.

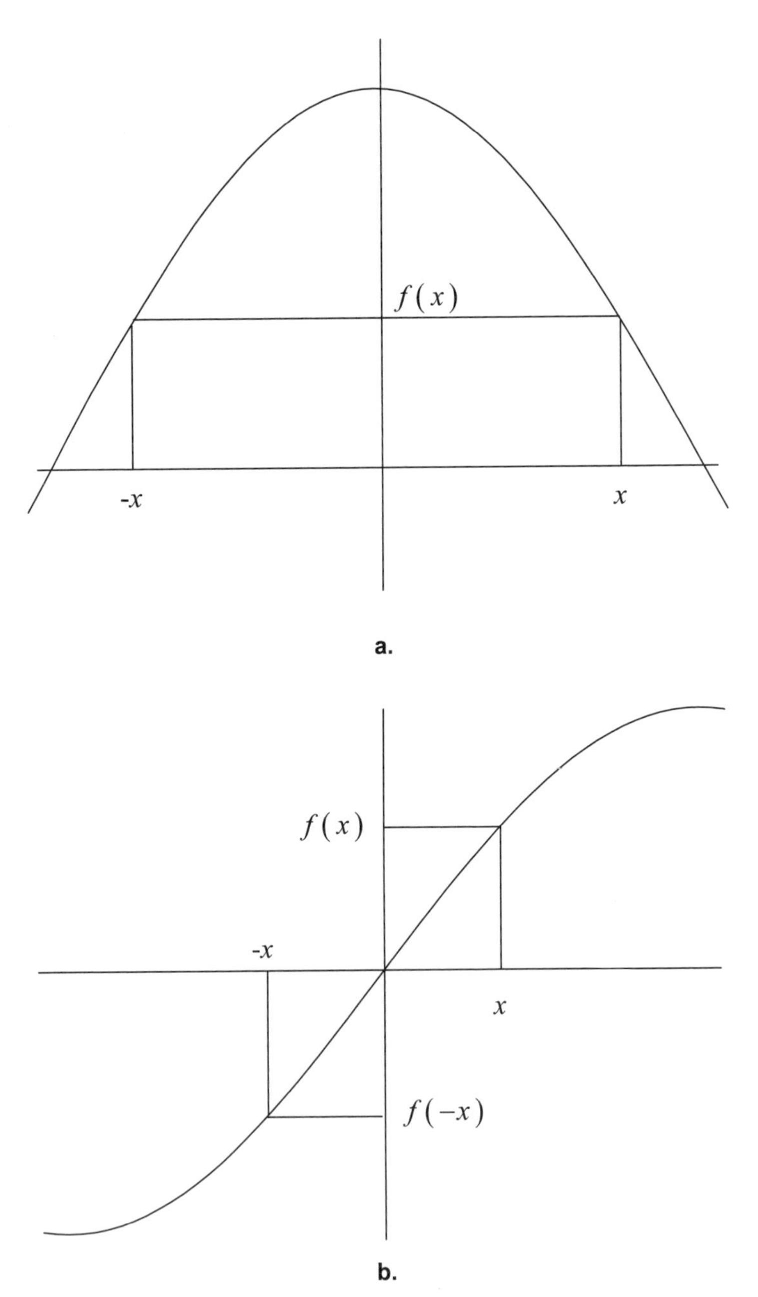

Figure D.4 a. Even symmetric function; b. odd symmetric function.

D.4 Laplace Transform Methods

The method of Laplace transforms is a powerful technique for solving linear ordinary and partial differential equations of dynamic systems. The steps involved in solving linear *ordinary* differential equations using Laplace transforms are shown schematically in Fig. D.5, where the symbol $\mathscr{L}$ represents the operation of the Laplace transformation, and the symbol $\mathscr{L}^{-1}$ depicts the inverse Laplace operation. The mathematical definition of these operations is given in Section D.4.1. Applying the Laplace transformation to an ordinary differential equation whose independent variable is the time domain (or space domain) transforms it to an algebraic equation in the Laplace domain (s domain). Solving the latter algebraically and then applying the inverse Laplace transformation yields the solution of the original problem in the time domain.

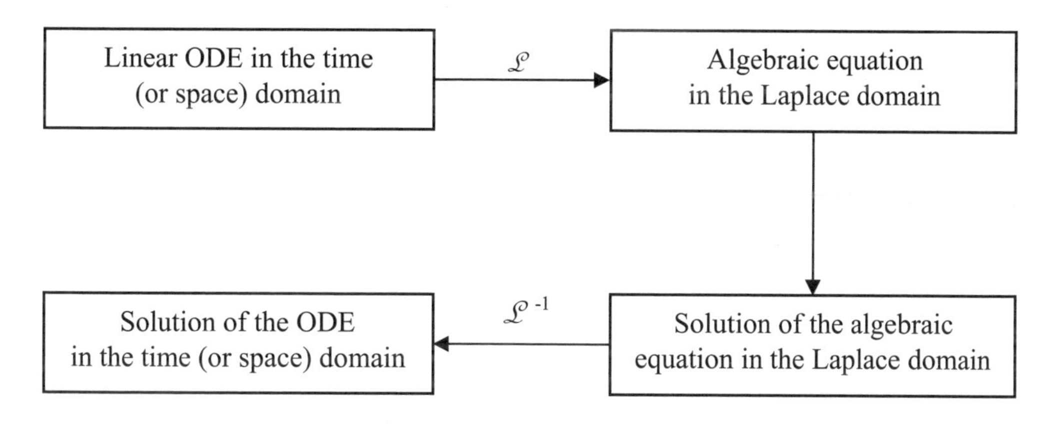

Figure D.5 Method of solution of a linear ordinary differential equation using Laplace transforms.

The steps involved in solving linear *partial* differential equations using Laplace transforms are shown in Fig. D.6. When Laplace transforms are applied to a partial differential equation whose independent variables are the time and space domains, the equation is transformed to an ordinary differential equation in the Laplace domain. Solving the latter in the Laplace domain using solution methods applicable to ordinary differential equations, and then applying the inverse Laplace transformation, results in the solution of the original partial differential equation in the time/space domains.

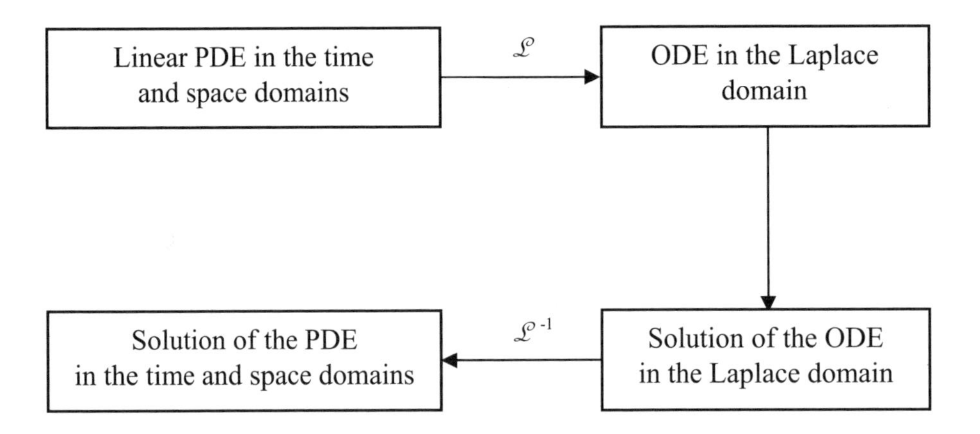

Figure D.6 Method of solution of a linear partial differential equations using Laplace transforms.

D.4.1 The Laplace transform

The Laplace transform maps a function $f(t)$ from the time (or space) domain to the s domain, where s is a variable defined in the complex plane $(s = \alpha + \beta i)$. The operation of the Laplace transformation is defined as:

$$\mathscr{L}[f(t)] \equiv \bar{f}(s) = \int_0^\infty f(t)e^{-st}dt \qquad \text{(D.107)}$$

The transform is an improper integral and may not exist for all continuous functions and all values of s. The *sufficient conditions* for the existence of the Laplace transform for a function $f(t)$ are stated as follows:

a. The function $f(t)$ must be piecewise continuous on every interval $0 < t < T$. A function is said to be piecewise continuous on an interval if it has only a finite number of finite discontinuities.
b. The function $f(t)$ must be of exponential growth at infinity. A function f on $0 < t < \infty$ is said to be of exponential growth at infinity if there exist constants M and α such that $|f(t)| \le Me^{\alpha t}$ for sufficiently large t.
c. The integral $\int_0^\delta |f(t)dt|$ must exist (is finite) for every finite $\delta > 0$.

The inverse of the Laplace transform maps a function $\bar{f}(s)$ from the Laplace domain (s domain) back to the time domain:

$$\mathcal{L}^{-1}\left[\bar{f}(s)\right] = f(t) = \frac{1}{2\pi i}\int_{\gamma-i\infty}^{\gamma+i\infty} \bar{f}(s)e^{st}ds \tag{D.108}$$

Evaluation of this integral requires knowledge of complex variables, the theory of residues, and contour integration.

In the paragraphs that follow, we will derive the Laplace transform of selected functions, tabulate these and other such transforms in Table D.1, and then use them to facilitate the solution of ordinary and partial differential equations. The Laplace transform and its inverse are rarely computed directly from the integrals of Eqs. (D.107) and (D.108), respectively. Instead, extensive tables of Laplace transforms (Roberts and Kaufman, 1966) are used to obtain the transforms and their inverses. Computer software, such as MATLAB, has the ability to generate Laplace transforms and inverse transforms. For a more complete derivation and discussion of Laplace transforms, the student is referred to Bequette (1998) and Stephanopoulos (1984).

The Laplace transform of the exponential function $f(t) = e^{-at}$ is given by

$$\begin{aligned}\mathcal{L}\left[e^{-at}\right] &= \int_0^\infty e^{-at}e^{-st}dt = \int_0^\infty e^{-(s+a)t}dt \\ &= -\frac{1}{s+a}\left[e^{-(s+a)t}\right]_0^\infty = \frac{1}{s+a}\end{aligned} \tag{D.109}$$

The Laplace transform of the first-order derivative $df(t)/dt$ is determined by

$$\mathcal{L}\left[\frac{df(t)}{dt}\right] = \int_0^\infty \frac{df(t)}{dt}e^{-st}dt \tag{D.110}$$

Using integration by parts $\left(\int udv = uv - \int vdu\right)$

$$\begin{aligned}\mathcal{L}\left[\frac{df(t)}{dt}\right] &= \int_0^\infty \frac{df(t)}{dt}e^{-st}dt \\ &= \left[e^{-st}f(t)\right]_0^\infty - \int_0^\infty f(t)(-s)e^{-st}dt \\ &= \left[0 - f(0)\right] + s\int_0^\infty f(t)e^{-st}dt \\ &= s\bar{f}(s) - f(0)\end{aligned} \tag{D.111}$$

The Laplace transform of the nth order derivative $d^n f(t)/dt^n$ is given as

$$\mathcal{L}\left[\frac{d^n f(t)}{dt^n}\right] = s^n \bar{f}(s) - s^{n-1} f(0) - s^{n-2} f'(0) - \ldots - s f^{(n-2)}(0) - f^{(n-1)}(0) \quad \text{(D.112)}$$

where n initial conditions, $f(0), f'(0), \ldots, f^{(n-1)}(0)$, are needed for the evaluation of Eq. (D.112). The Laplace transform of the ramp function $f(t) = bt$ is determined as follows:

$$\begin{aligned} \mathcal{L}[bt] &= \int_0^\infty bte^{-st}\,dt \\ &= \left[-\frac{bt}{s}e^{-st}\right]_0^\infty + \int_0^\infty \frac{b}{s}e^{-st}\,dt \\ &= (-0+0) + \frac{b}{s}\left[-\frac{1}{s}e^{-st}\right]_0^\infty = \frac{b}{s^2} \end{aligned} \quad \text{(D.113)}$$

The Laplace transform of the sine function is obtained as:

$$\begin{aligned} \mathcal{L}[\sin(\omega t)] &= \int_0^\infty \sin(\omega t)e^{-st}\,dt \\ &= \int_0^\infty \frac{e^{i\omega t} - e^{-i\omega t}}{2i}e^{-st}\,dt \\ &= \frac{1}{2i}\left[-\frac{e^{-(s-i\omega)t}}{s-i\omega} + \frac{e^{-(s+i\omega)t}}{s+i\omega}\right]_0^\infty \\ &= \frac{1}{2i}\left[\frac{1}{s-i\omega} - \frac{1}{s+i\omega}\right] = \frac{\omega}{s^2+\omega^2} \end{aligned} \quad \text{(D.114)}$$

In a similar fashion, the Laplace transform of the cosine function is established as:

$$\begin{aligned} \mathcal{L}[\cos(\omega t)] &= \int_0^\infty \cos(\omega t)e^{-st}\,dt \\ &= \int_0^\infty \frac{e^{i\omega t} + e^{-i\omega t}}{2}e^{-st}\,dt \\ &= \frac{1}{2}\left[-\frac{e^{-(s-i\omega)t}}{s-i\omega} - \frac{e^{-(s+i\omega)t}}{s+i\omega}\right]_0^\infty \\ &= \frac{1}{2}\left[\frac{1}{s-i\omega} + \frac{1}{s+i\omega}\right] = \frac{s}{s^2+\omega^2} \end{aligned} \quad \text{(D.115)}$$

Table D.1 Laplace transforms for selected time-domain functions

$f(t)$	$\mathscr{L}[f(t)]$
Unit step: $S(t)=\begin{cases}0 & \text{for} \quad t<0 \\ 1 & \text{for} \quad t>0\end{cases}$	$\frac{1}{s}$
Constant: b	$\frac{b}{s}$
Ramp: bt	$\frac{b}{s^2}$
t^n	$\frac{n!}{s^{n+1}}$
e^{-at}	$\frac{1}{s+a}$
$t^n e^{-at}$	$\frac{n!}{(s+a)^{n+1}}$
$\frac{1}{a_1-a_2}\left(e^{-a_2 t}-e^{-a_1 t}\right)$	$\frac{1}{(s+a_1)(s+a_2)}$
1st order derivative: $\frac{df(t)}{dt}$	$s\overline{f}(s)-f(0)$
nth order derivative: $\frac{d^n f(t)}{dt^n}$	$s^n\overline{f}(s)-s^{n-1}f(0)-s^{n-2}f'(0)-\ldots$ $-sf^{(n-2)}(0)-f^{(n-1)}(0)$
$\sin(\omega t)$	$\frac{\omega}{s^2+\omega^2}$
$\cos(\omega t)$	$\frac{s}{s^2+\omega^2}$
$e^{-at}\sin(\omega t)$	$\frac{\omega}{(s+a)^2+\omega^2}$
$e^{-at}\cos(\omega t)$	$\frac{s+a}{(s+a)^2+\omega^2}$
$\int_0^t f(t)\,dt$	$\frac{1}{s}\overline{f}(s)$
$erfc\left(\frac{a^{\frac{1}{2}}}{2t^{\frac{1}{2}}}\right)$	$\frac{1}{s}e^{-a^{\frac{1}{2}}s^{\frac{1}{2}}}$

Two important theorems related to Laplace transformation are the *initial-value theorem* and the *final-value theorem*. These are stated below, without proof:

Initial-value theorem:

$$\lim_{t\to 0} f(t) = \lim_{s\to\infty}\left[s\bar{f}(s) \right] \tag{D.116}$$

Final-value theorem:

$$\lim_{t\to \infty} f(t) = \lim_{s\to 0}\left[s\bar{f}(s) \right] \tag{D.117}$$

where

$$\bar{f}(s) = \int_0^{\infty} f(t) e^{-st} dt = \text{Laplace transform of } f(t)$$

D.4.2 Solution of ordinary differential equations

We consider the following first-order nonhomogeneous ordinary differential equation:

$$\frac{dy}{dt} + 3y = e^t \qquad \text{with} \qquad y(0) = 0 \tag{D.118}$$

Take the Laplace transform of each term in the equation:

$$\begin{aligned} \mathcal{L}\left[\frac{dy}{dt}\right] + \mathcal{L}[3y] &= \mathcal{L}[e^t] \\ s\bar{y}(s) - y(0) + 3\bar{y}(s) &= \frac{1}{s-1} \end{aligned} \tag{D.119}$$

Use the initial condition $y(0) = 0$ to simplify, and rearrange to solve for $\bar{y}(s)$:

$$\bar{y}(s) = \frac{1}{(s+3)(s-1)} \tag{D.120}$$

The right side of Eq. (D.120) may be expanded into partial fractions:

$$\frac{1}{(s+3)(s-1)} = \frac{C_1}{(s+3)} + \frac{C_2}{(s-1)} \tag{D.121}$$

To calculate the values of the constants c_1 and c_2, we first multiply Eq. (D.121) by $(s+3)$

$$\frac{1}{(s-1)} = c_1 + \frac{c_2(s+3)}{(s-1)} \tag{D.122}$$

and, since this equation must be true for any value of s, we set $s=-3$. This simplifies Eq. (D.122) and enables us to evaluate c_1:

$$c_1 = \frac{1}{(s-1)} = \frac{1}{(-3-1)} = -\frac{1}{4} \tag{D.123}$$

In a similar procedure, we calculate c_2 by multiplying Eq. (D.121) by $(s-1)$, setting $s=1$, and solving for c_2:

$$c_2 = \frac{1}{(s+3)} = \frac{1}{(1+3)} = \frac{1}{4} \tag{D.124}$$

We construct the solution in the s domain by combining Eqs. (D.120), (D.121), (D.123), and (D.124):

$$\bar{y}(s) = \frac{-1/4}{(s+3)} + \frac{1/4}{(s-1)} \tag{D.125}$$

We finally obtain the solution of the differential equation in the time domain by taking the inverse Laplace transform of Eq. (D.125):

$$\begin{aligned} \mathcal{L}^{-1}[\bar{y}(s)] &= \mathcal{L}^{-1}\left[\frac{-1/4}{(s+3)}\right] + \mathcal{L}^{-1}\left[\frac{1/4}{(s-1)}\right] \\ y(t) &= -\frac{1}{4}e^{-3t} + \frac{1}{4}e^{t} \end{aligned} \tag{D.126}$$

The inversion was accomplished by reading Table D.1 from the right column to the left column.

The student is encouraged to verify this solution in two ways: (a) substitute the solution into the original differential equation (D.118) and show that it satisfies it; (b) obtain the solution using the methods developed in Sections D.1 and D.2.

We now consider the second-order nonhomogeneous ordinary differential equation:

$$\frac{d^2y}{dt^2} - 5\frac{dy}{dt} + 6y = \sin t \tag{D.127}$$

with the following initial conditions:

$$y(0) = 0 \qquad \text{and} \qquad \left.\frac{dy}{dt}\right|_{t=0} = 0 \tag{D.128}$$

We take the Laplace transform of each side of the differential equation.

$$\begin{gathered}\mathscr{L}\left[\frac{d^2y}{dt^2}\right] - \mathscr{L}\left[5\frac{dy}{dt}\right] + \mathscr{L}[6y] = \mathscr{L}[\sin t] \\ s^2\overline{y}(s) - sy(0) - y'(0) - 5(s\overline{y}(s) - y(0)) + 6\overline{y}(s) = \frac{1}{s^2+1}\end{gathered} \tag{D.129}$$

We use the initial conditions (D.128) to simplify the above and solve for $\overline{y}(s)$:

$$\overline{y}(s) = \frac{1}{(s^2+1)(s^2-5s+6)} \tag{D.130}$$

We note that (s^2+1) can be factored into $(s+i)(s-i)$, and (s^2-5s+6) into $(s-3)(s-2)$; thus, we can expand the right side of Eq. (D.130) into partial fractions:

$$\frac{1}{(s+i)(s-i)(s-3)(s-2)} = \frac{c_1}{(s+i)} + \frac{c_2}{(s-i)} + \frac{c_3}{(s-3)} + \frac{c_4}{(s-2)} \tag{D.131}$$

By using the method employed earlier in this section, we obtain the values of the constants:

$$c_1 = \frac{1}{10}\frac{1}{(1-i)}, \qquad c_2 = \frac{1}{10}\frac{1}{(1+i)}, \qquad c_3 = \frac{1}{10}, \qquad c_4 = -\frac{1}{5} \tag{D.132}$$

Combining Eqs. (D.130), (D.131), and (D.132), we obtain the solution in the s domain:

$$\overline{y}(s) = \frac{\frac{1}{10}\frac{1}{(1-i)}}{(s+i)} + \frac{\frac{1}{10}\frac{1}{(1+i)}}{(s-i)} + \frac{\frac{1}{10}}{(s-3)} + \frac{-\frac{1}{5}}{(s-2)} \tag{D.133}$$

To simplify the solution, we note that

$$\frac{1}{1-i}\frac{1+i}{1+i}=\frac{1+i}{2} \quad \text{and} \quad \frac{1}{1+i}\frac{1-i}{1-i}=\frac{1-i}{2}$$

With these identities, Eq. (D.133) converts to

$$\bar{y}(s)=\frac{1}{10}\left(\frac{1+i}{2}\right)\frac{1}{(s+i)}+\frac{1}{10}\left(\frac{1-i}{2}\right)\frac{1}{(s-i)}+\frac{1}{10}\frac{1}{(s-3)}+\left(-\frac{1}{5}\right)\frac{1}{(s-2)} \tag{D.134}$$

We take the inverse Laplace transform of (D.134):

$$y(t)=\frac{1}{10}\left(\frac{1+i}{2}\right)e^{-it}+\frac{1}{10}\left(\frac{1-i}{2}\right)e^{it}+\frac{1}{10}e^{3t}-\frac{1}{5}e^{2t} \tag{D.135}$$

Making use of the Euler identities (Eq. (D.34)), we show the solution in trigonometric form:

$$y(t)=\frac{1}{10}\cos t+\frac{1}{10}\sin t+\frac{1}{10}e^{3t}-\frac{1}{5}e^{2t} \tag{D.136}$$

We verify this solution using MATLAB:

```
>> y=dsolve('D2y-5*Dy+6*y=sin(t)','y(0)=0', 'Dy(0)=0')
y =
1/10*exp(3*t)-1/5*exp(2*t)+1/10*cos(t)+1/10*sin(t)
```

We leave it as an exercise for the student to obtain the solution of Eq. (D.127) with $\cos t$ on the right side.

D.4.3 Solution of partial differential equations

The one-dimensional unsteady-state heat conduction problem is modelled by Eq. (D.42) with the variable u replaced by T for temperature:

$$\frac{\partial T}{\partial t}=\alpha\frac{\partial^2 T}{\partial x^2} \tag{D.137}$$

The initial and boundary condition for this problem are

$$\begin{aligned} &\text{at } t = 0, \quad T(x,0) = T_0 \quad \text{for} \quad 0 \leq x \leq L \\ &\text{at } t > 0, \quad T(0,t) = T_0 \quad \text{and} \quad T(L,t) = T_1 \end{aligned} \tag{D.138}$$

For convenience in obtaining the solution of the problem, we transform the variable T to the variable u as follows:

$$u = \frac{T - T_0}{T_1 - T_0} \tag{D.139}$$

The differential equation reverts back to Eq. (D.42):

$$\frac{\partial u}{\partial t} = \alpha \frac{\partial^2 u}{\partial x^2} \tag{D.42}$$

and the initial and boundary conditions become homogeneous

$$\begin{aligned} &\text{at } t = 0, \quad u(x,0) = 0 \quad \text{for} \quad 0 \leq x \leq L \\ &\text{at } t > 0, \quad u(0,t) = 0 \quad \text{and} \quad u(L,t) = 1 \end{aligned} \tag{D.140}$$

Take the Laplace transform of the left side of (D.42)

$$\mathcal{L}\left[\frac{\partial u}{\partial t}\right] = s\bar{u} - u(x,0) \tag{D.141}$$

Take the Laplace transform of the right side of (D.42)

$$\mathcal{L}\left[\frac{\partial^2 u}{\partial x^2}\right] = \frac{\partial^2}{\partial x^2}(\mathcal{L}[u]) = \frac{\partial^2 \bar{u}}{\partial x^2} = \frac{d^2 \bar{u}}{dx^2} \tag{D.142}$$

We were able to obtain (D.142) because x and t are independent variables, and for a continuous function, the order of transformation and differentiation may be reversed. Using the initial condition $u(x,0) = 0$, and combining Eqs. (D.141), (D.142), and (D.42):

$$\alpha \frac{d^2 \bar{u}}{dx^2} = s\bar{u} \tag{D.143}$$

The boundary conditions are also transformed to the s domain:

$$\begin{aligned} &\text{at } x=0, \quad u(0,t)=0, \quad \mathscr{L}[u(0,t)]=\bar{u}(0,s)=0 \\ &\text{at } x=L, \quad u(L,t)=1, \quad \mathscr{L}[u(L,t)]=\bar{u}(L,s)=\frac{1}{s} \end{aligned} \tag{D.144}$$

Eq. (D.143) is a second-order ordinary differential equation in $\bar{u}$, whose solution is obtained by the methods of Section D.2:

$$\bar{u}=c_1 e^{\sqrt{\frac{s}{\alpha}}x}+c_2 e^{-\sqrt{\frac{s}{\alpha}}x} \tag{D.145}$$

Utilizing the boundary conditions, we determine the constants c_1 and c_2:

$$c_1=\frac{1}{s}\frac{1}{\left(e^{\sqrt{\frac{s}{\alpha}}L}-e^{-\sqrt{\frac{s}{\alpha}}L}\right)} \quad \text{and} \quad c_2=-\frac{1}{s}\frac{1}{\left(e^{\sqrt{\frac{s}{\alpha}}L}-e^{-\sqrt{\frac{s}{\alpha}}L}\right)} \tag{D.146}$$

With these values of the constants, Eq. (D.145) becomes:

$$\bar{u}=\frac{1}{s}\frac{\left(e^{\sqrt{\frac{s}{\alpha}}x}-e^{-\sqrt{\frac{s}{\alpha}}x}\right)}{\left(e^{\sqrt{\frac{s}{\alpha}}L}-e^{-\sqrt{\frac{s}{\alpha}}L}\right)} \tag{D.147}$$

The right side of Eq. (D.147) may be expanded by using partial fractions:

$$\frac{1}{s}\frac{\left(e^{\sqrt{\frac{s}{\alpha}}x}-e^{-\sqrt{\frac{s}{\alpha}}x}\right)}{\left(e^{\sqrt{\frac{s}{\alpha}}L}-e^{-\sqrt{\frac{s}{\alpha}}L}\right)}=\frac{A}{s}+\frac{B}{\left(e^{\sqrt{\frac{s}{\alpha}}L}-e^{-\sqrt{\frac{s}{\alpha}}L}\right)} \tag{D.148}$$

To determine A, multiply both sides of (D.148) by s, and then set $s=0$:

$$A=\frac{\left(e^{\sqrt{\frac{s}{\alpha}}x}-e^{-\sqrt{\frac{s}{\alpha}}x}\right)}{\left(e^{\sqrt{\frac{s}{\alpha}}L}-e^{-\sqrt{\frac{s}{\alpha}}L}\right)} \tag{D.149}$$

But A is indeterminate because $s=0$, therefore, L'Hôpital's rule must be used to obtain the value of A:

$$A=\lim_{s\to 0}\frac{\left(e^{\sqrt{\frac{s}{\alpha}}x}-e^{-\sqrt{\frac{s}{\alpha}}x}\right)}{\left(e^{\sqrt{\frac{s}{\alpha}}L}-e^{-\sqrt{\frac{s}{\alpha}}L}\right)}=\frac{x}{L} \tag{D.150}$$

To determine B, multiply both sides of (D.148) by $\left(e^{\sqrt{\frac{s}{\alpha}}L} - e^{-\sqrt{\frac{s}{\alpha}}L}\right)$, and then set this quantity to zero:

$$B = \frac{1}{s}\left(e^{\sqrt{\frac{s}{\alpha}}x} - e^{-\sqrt{\frac{s}{\alpha}}x}\right) \tag{D.151}$$

From trigonometric identities:

$$\left(e^{\sqrt{\frac{s}{\alpha}}L} - e^{-\sqrt{\frac{s}{\alpha}}L}\right) = 2\sinh\left(\sqrt{\frac{s}{\alpha}}L\right) \tag{D.152}$$

For (D.152) to be equal to zero, the following must be true:

$$\sqrt{\frac{s}{\alpha}}L = in\pi \qquad \text{or} \qquad s_n = -\frac{\alpha n^2\pi^2}{L^2}, \quad n = 1,2,\ldots,\infty \tag{D.153}$$

Making use of Eqs. (D.148), (D.150), (D.151), and (D.152), we reconstruct $\bar{u}$ in Eq. (D.147):

$$\bar{u} = \frac{1}{s}\frac{x}{L} + \frac{1}{s}\frac{\left(e^{\sqrt{\frac{s}{\alpha}}x} - e^{-\sqrt{\frac{s}{\alpha}}x}\right)}{2\sinh\left(\sqrt{\frac{s}{\alpha}}L\right)} \tag{D.154}$$

Taking the inverse Laplace transform of Eq. (D.154):

$$u = \mathcal{L}^{-1}[\bar{u}] = \mathcal{L}^{-1}\left[\frac{1}{s}\frac{x}{L}\right] + \mathcal{L}^{-1}\left[\frac{1}{s}\frac{\left(e^{\sqrt{\frac{s}{\alpha}}x} - e^{-\sqrt{\frac{s}{\alpha}}x}\right)}{2\sinh\left(\sqrt{\frac{s}{\alpha}}L\right)}\right] \tag{D.155}$$

Using the tables (Roberts and Kaufman, 1966, p. 284) we obtain the solution:

$$u = \frac{x}{L} + \sum_{n=1}^{\infty}\frac{2}{n\pi}(-1)^n \sin\left(\frac{n\pi x}{L}\right)e^{-\frac{\alpha n^2\pi^2 t}{L}} \tag{D.156}$$

Finally, we transform the variable u to T by rearranging Eq. (D.139)

$$T = u(T_1 - T_0) + T_0 \tag{D.157}$$

Example D.9 The mechanism of immune blood cell (leukocyte) migration on prosthetic materials is an important aspect of material biocompatibility. This mechanism, in its simplest form (no convection), has been modeled by the diffusion equation:

$$\frac{\partial c}{\partial t} = D\frac{\partial^2 c}{\partial x^2}$$

Initially, the concentration of immune blood cells in the prosthetic material is zero:

$$\text{at } t = 0, \quad c(x,0) = 0 \qquad \text{for all } x$$

At t > 0, the concentration of immune blood cells at the surface is c_A:

$$\text{at } \mathrm{x} = 0, \quad c(0,t) = c_A \qquad \text{for } t > 0$$

For the second boundary condition, we assume that the diffusing immune blood cells penetrate only a small distance into the material in comparison with the depth of the material:

$$\text{at } x = \infty, \quad c(\infty,t) = 0 \qquad \text{for all } t$$

Solve the differential equation with the above initial and boundary conditions and plot the migration profiles of the leukocytes.

The analytical solution is simplified by redefining the concentration variable to

$$u = \frac{c}{c_A}$$

This converts the differential equation to

$$\frac{\partial u}{\partial t} = D\frac{\partial^2 u}{\partial x^2}$$

and the initial and boundary conditions become homogeneous

$$\text{at } t = 0, \qquad u(x,0) = 0 \qquad \text{for all } x$$

$$\text{at } x = 0, \qquad u(0,t) = 1 \qquad \text{for } t > 0$$

$$\text{at } x = \infty, \qquad u(\infty,t) = 0 \qquad \text{for all } t$$

Take the Laplace transform of the diffusion equation and use the initial condition to obtain:

$$D\frac{d^2\bar{u}}{dx^2} = s\bar{u}$$

This result is completely analogous to Eq. (D.143). Transform the boundary conditions:

$$\text{at } x = 0, \quad u(0,t) = 1, \quad \mathscr{L}[u(0,t)] = \bar{u}(0,s) = \frac{1}{s}$$

$$\text{at } x = \infty, \quad u(\infty,t) = 0, \quad \mathscr{L}[u(\infty,t)] = \bar{u}(\infty,s) = 0$$

The solution of the transformed differential equation is

$$\bar{u} = c_1 e^{\sqrt{\frac{s}{D}}x} + c_2 e^{-\sqrt{\frac{s}{D}}x}$$

The boundary condition at $x = \infty$ requires the integration constant c_1 to be zero. The boundary condition at $x = 0$ gives the value of the integration constant c_2:

$$c_2 = \frac{1}{s}$$

Therefore, the solution of the differential equation becomes:

$$\bar{u} = \frac{1}{s} e^{-\sqrt{\frac{s}{D}}x}$$

The inverse Laplace transform of the above is listed in Table D.1:

$$u = \text{erfc}\left(\frac{x}{2\sqrt{Dt}}\right) = 1 - \text{erf}\left(\frac{x}{2\sqrt{Dt}}\right)$$

where erf( ) is the *error function* defined as:

$$\text{erf}(\theta) = \frac{2}{\sqrt{\pi}} \int_0^{\theta} e^{-t^2} dt$$

Finally, the solution in terms of the normalized concentration is:

$$\frac{c}{c_A} = 1 - \text{erf}\left(\frac{x}{2\sqrt{Dt}}\right)$$

MATLAB solution

We use the following MATLAB program to plot the migration profiles of leukocytes.

```
% Plotting the migration profiles of the immune blood cells
clear; clc

D=0.01;
x=[0:0.01:1];
nc=0;
for t=0:0.5:5
nc=nc+1;
    c(nc,:)=(1-erf(x/(2*sqrt(D*t))));
end
plot(x,c)
title('Figure D.7: Migration profiles of immune blood cells')
xlabel('Position, x'); ylabel('Normalized Concentration')
text(0.4, 0.4,'Time');
```

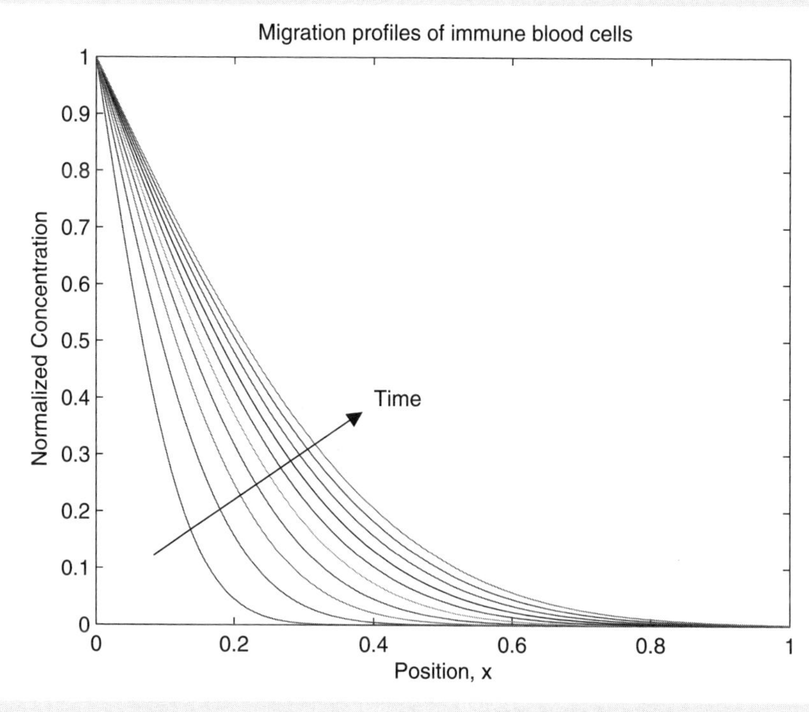

Figure D.7 Migration profiles of immune blood cells.

This concludes our discussion of the solution of partial differential equations using Laplace transforms.

D.5 References

Bequette, B. W. 1998. *Process Dynamics: Modeling, Analysis, and Simulation.* Upper Saddle River, NJ; Prentice Hall, Inc.

Boyce, W. E. and DiPrima, R. C. 2001. *Elementary Differential Equations.* New York: John Wiley & Sons, Inc.

O'Neil, P. V. 2003. *Advanced Engineering Mathematics,* 5th ed. Pacific Grove, CA: Brooks/Cole-Thompson Learning, Inc.

Roberts, G.E and Kaufman, H. 1966. *Table of Laplace Transforms.* Philadelphia, PA: W. B. Saunders Company.

Stephanopoulos, G. 1984. *Chemical Process Control: An Introduction to Theory and Practice.* Upper Saddle River, NJ; Prentice Hall, Inc.

Zill, D. G. and Cullen, M. R. 2000. *Advanced Engineering Mathematics.* Boston, MA: James & Bartlett Publishers.

Appendix E: Numerical Stability and Other Topics

Three topics of great importance in the numerical integration of differential equations equations are error propagation, stability, and convergence of the solutions. Two types of stability considerations enter into the solution of ordinary differential equations: inherent stability (or instability) of the analytical solution and numerical stability (or instability). Inherent stability is determined by the mathematical formulation of the problem and is dependent on the eigenvalues of the Jacobian matrix of the differential equations, as was shown in Section 7.6. On the other hand, numerical stability is a function of the error propagation in the numerical integration method. The behavior of the error propagation depends on the values of the characteristic roots of the difference equations that yield the numerical solution. In this Appendix, we systematically examine the error propagation and stability of several numerical integration methods and suggest ways of reducing these errors by the appropriate choice of step size and integration algorithm.

E.1 Stability of the Euler Methods

Consider the initial-value differential equation in the linear form:

$$\frac{dy}{dt} = \lambda y \tag{E.1}$$

where the initial condition is given as

$$y(t_0) = y_0 \tag{E.2}$$

Furthermore, assume that λ is real and y_0 is finite. The analytical solution of this differential equation

$$y(t) = y_0 e^{\lambda t} \tag{E.3}$$

is *inherently stable* for $\lambda < 0$, that is:

$$\lim_{t \to \infty} y(t) = 0 \tag{E.4}$$

Consider now the stability of the numerical solution of this problem, obtained using the explicit Euler method, and momentarily ignore the truncation and roundoff errors. Applying Eq. (7.25), we obtain the recurrence equation:

$$y_{n+1} = y_n + h\lambda y_n \tag{E.5}$$

which is, after rearrangement, the following *first-order homogeneous difference equation*:

$$y_{n+1} - (1 + h\lambda) y_n = 0 \tag{E.6}$$

Using the definition of the shift operator, $Ey_n = y_{n+1}$, we obtain

$$Ey_n - (1 + h\lambda) y_n = 0 \tag{E.7}$$

which yields the characteristic equation

$$E - (1 + h\lambda) = 0 \tag{E.8}$$

whose root is

$$\mu_1 = (1 + h\lambda) \tag{E.9}$$

From this, we obtain the solution of the difference equation (E.6) as

$$y_n = C(1 + h\lambda)^n \tag{E.10}$$

The constant C is calculated from the initial condition, at $t = t_0$:

$$n = 0 \qquad y_n = y_0 = C \tag{E.11}$$

Therefore, the final form of the solution is

$$y_n = y_0 (1 + h\lambda)^n \tag{E.12}$$

The differential equation is an initial-value problem; therefore, n can increase without bound. Because the solution y_n is a function of $(1 + h\lambda)^n$, its behavior is determined by the value of $(1 + h\lambda)$.

A numerical solution is said to be *absolutely stable* if

$$\lim_{n \to \infty} y_n = 0 \tag{E.13}$$

The numerical solution of the differential equation (E.1) using the explicit Euler method is absolutely stable if

$$|1 + h\lambda| \le 1 \tag{E.14}$$

Because $(1 + h\lambda)$ is the root of the characteristic equation (E.8), an alternative definition of absolute stability is

$$|\mu_i| \le 1 \qquad i = 1, 2, \ldots, k \tag{E.15}$$

where more than one root exists in the multi-step numerical methods.

The inequality (E.14) is equivalent to:

$$-2 \le h\lambda \le 0 \tag{E.16}$$

which sets the limits of the integration step size for a stable solution as follows: Because h is positive, $\lambda < 0$ and

$$h \le \frac{2}{|\lambda|} \tag{E.17}$$

Inequality (E.17) is a finite *general stability boundary*, and the explicit Euler method is called *conditionally stable*. Any method with an infinite general stability boundary is called *unconditionally stable*.

At the outset, we assumed that λ was real in order to simplify the derivation. This assumption is not necessary; λ can be a complex number. A solution is stable, converging with damped oscillations, when complex roots are present $(\alpha \pm \beta i)$ and the moduli of the roots $\left(r = \sqrt{\alpha^2 \pm \beta^2}\right)$ are less than or equal to unity:

$$|r| \leq 1 \tag{E.18}$$

The inequalities, (E.16) and (E.18), describe the circle with a radius of unity on the complex plane. Fig. E.1 shows the regions of numerical stability for the Euler and Runge-Kutta methods. The set of values of $h\lambda$ inside the circle yields stable numerical solutions of Eq. (E.1) using the Euler, and equivalently, first-order Runge-Kutta integration method.

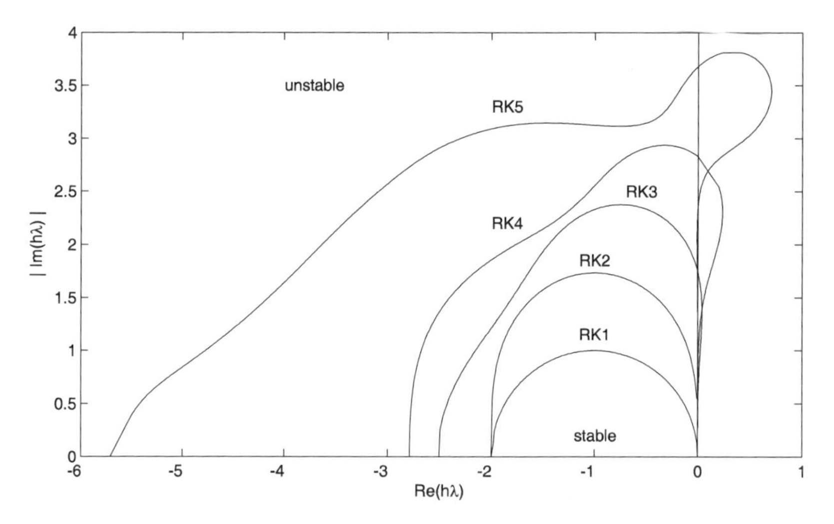

Figure E.1 Stability regions in the complex plane for Runge-Kutta methods of order 1 (explicit Euler), 2, 3, 4, and 5.

We now return to the consideration of the truncation and roundoff errors of the Euler method. The propagation of these errors in the numerical solution will be represented by a difference equation. Beginning with the nonlinear form of the initial-value problem:

$$\frac{dy}{dt} = f(t, y) \tag{E.19}$$

where the initial condition is given by

$$y(t_0) = y_0 \tag{E.20}$$

we define the accumulated error of the numerical solution at step $(n+1)$ as

$$\varepsilon_{n+1} = y_{n+1} - y(t_{n+1}) \tag{E.21}$$

where $y(t_{n+1})$ is the *exact* value of *y*, and y_{n+1} is the *calculated* value of *y* at t_{n+1}. We then write the exact solution, $y(t_{n+1})$, as a Taylor series expansion, showing as many terms as needed for the Euler method:

$$y(t_{n+1}) = y(t_n) + hf(t_n, y(t_n)) + T_{E,n+1} \tag{E.22}$$

where $T_{E,\,n+1}$ is the local truncation error for step $(n+1)$. We also write the calculated value y_{n+1} obtained from the explicit Euler formula:

$$y_{n+1} = y_n + hf(t_n, y_n) + R_{E,n+1} \tag{E.23}$$

where $R_{E,n+1}$ is the roundoff error introduced by the computer in step $(n+1)$. Combining Eqs. (E.21), (E.22), and (E.23):

$$\varepsilon_{n+1} = y_n - y(t_n) + h\left[f(t_n, y_n) - f(t_n, y(t_n))\right] - T_{E,n+1} + R_{E,n+1} \tag{E.24}$$

which simplifies to:

$$\varepsilon_{n+1} = \varepsilon_n + h\left[f(t_n, y_n) - f(t_n, y(t_n))\right] - T_{E,n+1} + R_{E,n+1} \tag{E.25}$$

The mean-value theorem:

$$f(t_n, y_n) - f(t_n, y(t_n)) = \left.\frac{\partial f}{\partial y}\right|_{\alpha, x_n} \left[y_n - y(t_n)\right] \qquad y_n < \alpha < y(t_n) \tag{E.26}$$

can be used to simplify Eq. (E.25) to

$$\varepsilon_{n+1} - \left[1 + h\left.\frac{\partial f}{\partial y}\right|_{\alpha, x_n}\right]\varepsilon_n = -T_{E,n+1} + R_{E,n+1} \tag{E.27}$$

This is a *first-order nonhomogeneous difference equation with varying coefficients,* which can be solved only by iteration. However, by making the following simplifying assumptions:

$$T_{E,n+1} = T_E = \text{constant} \tag{E.28}$$

$$R_{E,n+1} = R_E = \text{constant} \tag{E.29}$$

$$\left.\frac{\partial f}{\partial y}\right|_{\alpha,x_n} = \lambda = \text{constant} \tag{E.30}$$

Eq. (E.27) simplifies to

$$\varepsilon_{n+1} - (1+h\lambda)\varepsilon_n = -T_E + R_E \tag{E.31}$$

the solution of which is given by the sum of the homogeneous and particular solutions:

$$\varepsilon_n = C_1(1+h\lambda)^n + \frac{-T_E + R_E}{1-(1+h\lambda)} \tag{E.32}$$

Comparison of Eqs. (E.6) and (E.31) reveals that the characteristic equations for the solution y_n and the error ε_n are identical. The truncation and roundoff error terms in Eq.(E.31) determine the particular solution. The constant C_1 is calculated by assuming that the initial condition of the differential equation has no error; that is, $\varepsilon_0 = 0$. The final form of the equation that describes the behavior of the propagation error is

$$\varepsilon_n = \frac{-T_E + R_E}{h\lambda}\left[(1+h\lambda)^n - 1\right] \tag{E.33}$$

A great deal of insight can be gained by thoroughly examining Eq. (E.33). As expected, the value of $(1+h\lambda)$ is the determining factor in the behavior of the propagation error. Consider first the case of a fixed finite step size h, with the number of integration steps increasing to a very large n. The limit on the error as $n \to \infty$ is:

$$\lim_{n\to\infty}|\varepsilon_n| = \frac{-T_E + R_E}{h\lambda} \qquad \text{for} \quad |1+h\lambda| < 1 \tag{E.34}$$

$$\lim_{n\to\infty}|\varepsilon_n| = \infty \qquad \text{for} \quad |1+h\lambda| > 1 \tag{E.35}$$

In the first situation (Eq. (E.34)), $\lambda < 0$, $0 < h < 2/|\lambda|$, the error is bounded and the numerical solution is stable. The numerical solution differs from the exact solution by only the finite quantity $(-T_E + R_E)/h\lambda$, which is a function of the truncation error, the roundoff error, the step size, and the eigenvalue of the differential equation.

In the second situation (Eq. (E.35)), $\lambda > 0$, $h > 0$, the error is unbounded and the numerical solution is unstable. For $\lambda > 0$, however, the exact solution itself is *inherently unstable*. For this reason we introduce the concept of *relative error*, defined as:

$$\text{relative error} = \frac{\varepsilon_n}{y_n} \tag{E.36}$$

Utilizing Eqs. (E.12) and (E.33), we obtain the relative error as:

$$\frac{\varepsilon_n}{y_n} = \frac{-T_E + R_E}{y_0 h\lambda}\left[1 - \frac{1}{(1+h\lambda)^n}\right] \tag{E.37}$$

The relative error is bounded for $\lambda > 0$ and unbounded for $\lambda < 0$. So we conclude that for inherently stable differential equations, the absolute error is the pertinent criterion for numerical stability, whereas for inherently unstable differential equations, the relative error must is the right criterion.

Let us now consider a fixed interval of integration, $0 < t < \alpha$, so that

$$h = \frac{\alpha}{n} \tag{E.38}$$

and we increase the number of integration steps to a very large *n*. This, of course, causes $h \to 0$.

A numerical method is said to be *convergent* if

$$\lim_{h\to 0}|\varepsilon_n| = 0 \tag{E.39}$$

In the absence of roundoff error, the Euler method, and most other integration methods, would be convergent because

$$\lim_{h\to 0} T_E = 0 \tag{E.40}$$

therefore, Eq. (E.39) would be true. However, roundoff error is *never* absent in numerical calculations.

As $h \to 0$, the truncation error goes to zero and the roundoff error is the crucial factor in the propagation of error:

$$\lim_{h\to 0}|\varepsilon_n| = R_E \lim_{h\to 0}\frac{(1+h\lambda)^n - 1}{h\lambda} \tag{E.41}$$

By L'Hôpital's rule, the roundoff error propagates unbounded as the number of integration steps becomes very large:

$$\lim_{h \to 0} \varepsilon_n = R_E [\infty] \tag{E.42}$$

This is the "catch-22" of numerical methods: a smaller integration step size reduces the truncation error but requires more computation, thereby increasing the roundoff error.

A similar analysis of the *implicit Euler method* (backward Euler) results in the following equations, for the solution:

$$y_{n+1} = \frac{y_0}{(1 - h\lambda)^n} \tag{E.43}$$

and the propagation error:

$$\varepsilon_{n+1} = \frac{-T_E + R_E}{h\lambda}(1 - h\lambda)\left[\frac{1}{(1 - h\lambda)^n} - 1\right] \tag{E.44}$$

For $\lambda < 0$ and $0 < h < \infty$, the solution is stable:

$$\lim_{n \to \infty} y_n = 0 \tag{E.45}$$

and the error is bounded:

$$\lim_{n \to \infty} \varepsilon_n = -\frac{-T_E + R_E}{h\lambda}(1 - h\lambda) \tag{E.46}$$

No limitation is placed on the step size; therefore, the implicit Euler method is *unconditionally stable* for $\lambda < 0$. On the other hand, when $\lambda > 0$, the following inequality must be true for a stable solution:

$$|1 - h\lambda| \leq 1 \tag{E.47}$$

This imposes the limit on the step size:

$$-2 \leq h\lambda < 0 \tag{E.48}$$

It can be concluded that the implicit Euler method has a wider range of stability than the explicit Euler method (see Table E.1).

E.2 Stability of the Runge-Kutta Methods

Using methods parallel to those of the previous section, the recurrence equations and the corresponding roots for the Runge-Kutta methods can be derived (Lapidus and Sienfeld, 1971). For the differential equation (E.1), these are:

Second-order Runge-Kutta:

$$y_{n+1} = \left(1 + h\lambda + \frac{1}{2}h^2\lambda^2\right)y_n \tag{E.49}$$

$$\mu_1 = 1 + h\lambda + \frac{1}{2}h^2\lambda^2 \tag{E.50}$$

Third-order Runge-Kutta:

$$y_{n+1} = \left(1 + h\lambda + \frac{1}{2}h^2\lambda^2 + \frac{1}{6}h^3\lambda^3\right)y_n \tag{E.51}$$

$$\mu_1 = 1 + h\lambda + \frac{1}{2}h^2\lambda^2 + \frac{1}{6}h^3\lambda^3 \tag{E.52}$$

Fourth-order Runge-Kutta:

$$y_{n+1} = \left(1 + h\lambda + \frac{1}{2}h^2\lambda^2 + \frac{1}{6}h^3\lambda^3 + \frac{1}{24}h^4\lambda^4\right)y_n \tag{E.53}$$

$$\mu_1 = 1 + h\lambda + \frac{1}{2}h^2\lambda^2 + \frac{1}{6}h^3\lambda^3 + \frac{1}{24}h^4\lambda^4 \tag{E.54}$$

Fifth-order Runge-Kutta:

$$y_{n+1} = \left(1 + h\lambda + \frac{1}{2}h^2\lambda^2 + \frac{1}{6}h^3\lambda^3 + \frac{1}{24}h^4\lambda^4 + \frac{1}{120}h^5\lambda^5 + \frac{0.5625}{720}h^6\lambda^6\right)y_n \tag{E.55}$$

$$\mu_1 = 1 + h\lambda + \frac{1}{2}h^2\lambda^2 + \frac{1}{6}h^3\lambda^3 + \frac{1}{24}h^4\lambda^4 + \frac{1}{120}h^5\lambda^5 + \frac{0.5625}{720}h^6\lambda^6 \tag{E.56}$$

The last term in the right-hand side of Eqs. (E.55) and (E.56) is specific to the fifth-order Runge-Kutta, which appears in Table 5.2 of Constantinides and Mostoufi (1999) and varies for different fifth-order formulas.

The condition for absolute stability

$$|\mu_i| \le 1 \qquad i = 1, 2, \ldots, k \tag{E.57}$$

applies to all the above methods. The absolute real stability boundaries for these methods are listed in Table E.1, and the regions of stability in the complex plane are shown in Fig. E.2. In general, as the order increases, so do the stability limits.

Table E.1 Real stability boundaries

Method	**Boundary**
Explicit Euler	$-2 \le h\lambda < 0$
Implicit Euler	$\begin{cases} 0 < h < \infty & \text{for} \quad \lambda < 0 \\ -2 \le h\lambda < 0 & \text{for} \quad \lambda > 0 \end{cases}$
Modified Euler (predictor-corrector)	$-1.077 \le h\lambda < 0$
Second-order Runge-Kutta	$-2 \le h\lambda < 0$
Third-order Runge-Kutta	$-2.5 \le h\lambda < 0$
Fourth-order Runge-Kutta	$-2.785 \le h\lambda < 0$
Fifth-order Runge-Kutta	$-5.7 \le h\lambda < 0$
Adams	$-0.546 \le h\lambda < 0$
Adams Moulton	$-1.285 \le h\lambda < 0$

E.3 Stability of Multistep Methods

Using methods parallel to those of the previous section, the recurrence equations and the corresponding roots for the modified Euler, Adams, and Adams-Moulton methods can be derived (Lapidus and Sienfeld, 1971). For the differential equation (E.1), these are:

Modified Euler (combination of predictor and corrector):

$$y_{n+1} = \left(1 + h\lambda + h^2\lambda^2\right) y_n \tag{E.58}$$

$$\mu_1 = 1 + h\lambda + h^2\lambda^2 \tag{E.59}$$

Adams:

$$y_{n+1} = \left(1 + \frac{23}{12}h\lambda\right) y_n - \left(\frac{4}{3}h\lambda\right) y_{n-1} + \left(\frac{5}{12}h\lambda\right) y_{n-2} \tag{E.60}$$

$$\mu^3 - \left(1 + \frac{23}{12}h\lambda\right)\mu^2 + \left(\frac{4}{3}h\lambda\right)\mu - \left(\frac{5}{12}h\lambda\right) = 0 \tag{E.61}$$

Adams-Moulton (combination of predictor and corrector):

$$\begin{aligned} y_{n+1} = {} & \left(1 + \frac{7}{6}h\lambda + \frac{55}{64}h^2\lambda^2\right) y_n - \left(\frac{5}{24}h\lambda + \frac{59}{64}h^2\lambda^2\right) y_{n-1} \\ & + \left(\frac{1}{24}h\lambda + \frac{37}{64}h^2\lambda^2\right) y_{n-2} - \left(\frac{9}{64}h^2\lambda^2\right) y_{n-3} \end{aligned} \tag{E.62}$$

$$\begin{aligned} \mu^4 - {} & \left(1 + \frac{7}{6}h\lambda + \frac{55}{64}h^2\lambda^2\right)\mu^3 + \left(\frac{5}{24}h\lambda + \frac{59}{64}h^2\lambda^2\right)\mu^2 \\ & - \left(\frac{1}{24}h\lambda + \frac{37}{64}h^2\lambda^2\right)\mu + \left(\frac{9}{64}h^2\lambda^2\right) = 0 \end{aligned} \tag{E.63}$$

The condition for absolute stability,

$$\left|\mu_i\right| \le 1 \qquad i = 1, 2, \ldots, k \tag{E.57}$$

applies to all the above methods. The absolute real stability boundaries for these methods are also listed in Table E.1, and the regions of stability in the complex plane are shown in Fig. E.2.

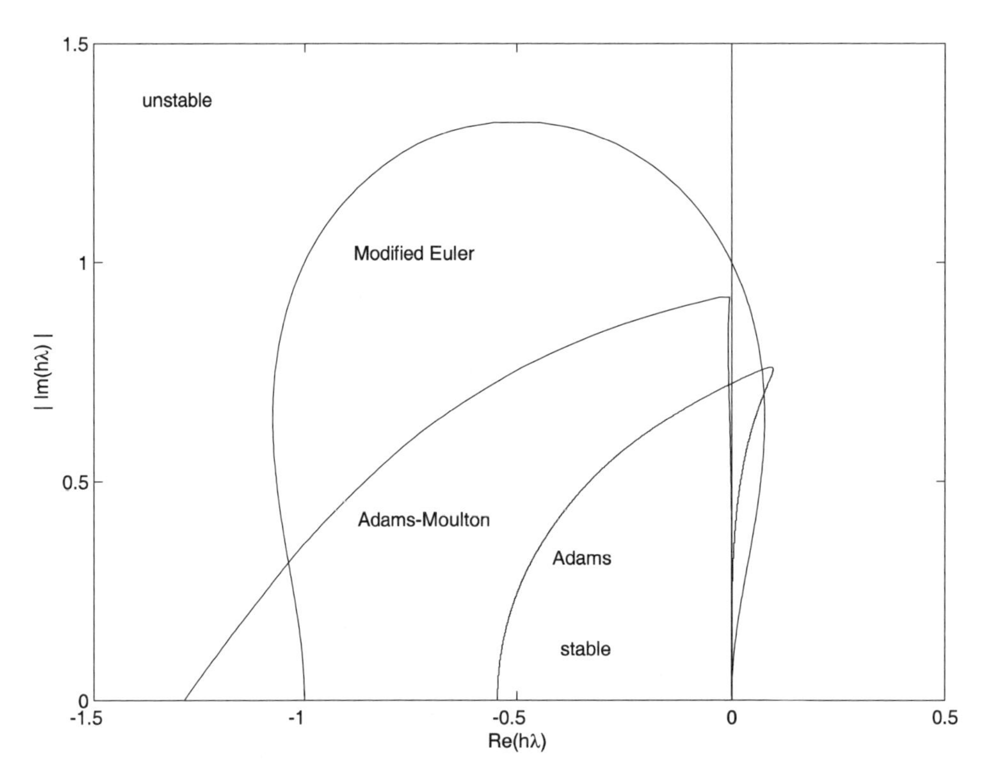

Figure E.2 Stability regions in the complex plane for the modified Euler (Euler predictor-corrector), Adams, and Adams-Moulton methods.

E.4 Stability of Methods for Partial Differential Equations

In this section, we discuss the stability of finite difference approximations of partial differential equations using the well-known von Neumann procedure. This method introduces an initial error represented by a finite Fourier series and examines how this error propagates during the solution. The von Neumann method applies to initial-value problems; for this reason it is used to analyze the stability of the explicit method for parabolic equations developed in Section 8.5.2 and the explicit method for hyperbolic equations developed in Section 8.5.3.

We define the error $\varepsilon_{m,n}$ as the difference between the solution of the finite difference approximation, $u_{m,n}$, and the exact solution of the differential equation, $\overline{u}_{m,n}$, at step (m, n):

$$\varepsilon_{m,n} \equiv u_{m,n} - \overline{u}_{m,n} \tag{E.64}$$

The explicit finite difference solution (8.61) of the parabolic partial differential equation (8.22) can be written for $u_{m,n+1}$ and $\overline{u}_{m,n+1}$ as follows:

$$u_{m,n+1} = \left(\frac{\alpha \Delta t}{\Delta x^2}\right) u_{m+1,n} + \left(1 - 2\frac{\alpha \Delta t}{\Delta x^2}\right) u_{m,n} + \left(\frac{\alpha \Delta t}{\Delta x^2}\right) u_{m-1,n} + R_{E_{m,n+1}} \tag{E.65}$$

$$\bar{u}_{m,n+1} = \left(\frac{\alpha \Delta t}{\Delta x^2}\right) \bar{u}_{m+1,n} + \left(1 - 2\frac{\alpha \Delta t}{\Delta x^2}\right) \bar{u}_{m,n} + \left(\frac{\alpha \Delta t}{\Delta x^2}\right) \bar{u}_{m-1,n} + T_{E_{m,n+1}} \tag{E.66}$$

where $R_{E_{m,n+1}}$ and $T_{E_{m,n+1}}$ are the roundoff and truncation errors, respectively, at step $(m, n+1)$. Combining Eqs. (E.64)–(E.66), we obtain

$$\varepsilon_{m,n+1} - \left(\frac{\alpha \Delta t}{\Delta x^2}\right) \varepsilon_{m+1,n} - \left(1 - 2\frac{\alpha \Delta t}{\Delta x^2}\right) \varepsilon_{m,n} - \left(\frac{\alpha \Delta t}{\Delta x^2}\right) \varepsilon_{m-1,n} = R_{E_{m,n+1}} + T_{E_{m,n+1}} \tag{E.67}$$

This is a *nonhomogeneous finite difference equation in two dimensions*, representing the propagation of error during the numerical solution of the parabolic partial differential equation (8.22). The solution of this finite difference equation is rather difficult to obtain. For this reason, the von Neumann analysis considers the *homogeneous* part of Eq. (E.67):

$$\varepsilon_{m,n+1} - \left(\frac{\alpha \Delta t}{\Delta x^2}\right) \varepsilon_{m+1,n} - \left(1 - 2\frac{\alpha \Delta t}{\Delta x^2}\right) \varepsilon_{m,n} - \left(\frac{\alpha \Delta t}{\Delta x^2}\right) \varepsilon_{m-1,n} = 0 \tag{E.68}$$

The above equation represents the propagation of the error introduced at the initial point only, i.e., at $n = 0$, and ignores truncation and roundoff errors that enter the solution at $n > 0$.

The solution of the homogeneous finite difference equation may be written in the following separable form:

$$\varepsilon_{m,n} = c e^{\gamma n \Delta t} e^{i \beta m \Delta x} \tag{E.69}$$

where $i = \sqrt{-1}$ and c, γ, and β are constants. At $n = 0$,

$$\varepsilon_{m,0} = c e^{i \beta m \Delta x} \tag{E.70}$$

which is the error at the initial point. Therefore, the term $e^{\gamma \Delta t}$ is the *amplification factor* of the initial error. In order for the original error not to grow as n increases, the amplification factor must satisfy the *von Neumann condition for stability*:

$$\left| e^{\gamma \Delta t} \right| \le 1 \tag{E.71}$$

The amplification factor can have complex values. In that case, the modulus of the complex numbers must satisfy the above inequality; that is,

$$|r| \le 1 \tag{E.72}$$

Therefore the stability region in the complex plane is a circle of radius = 1, as shown in Fig. E.3.

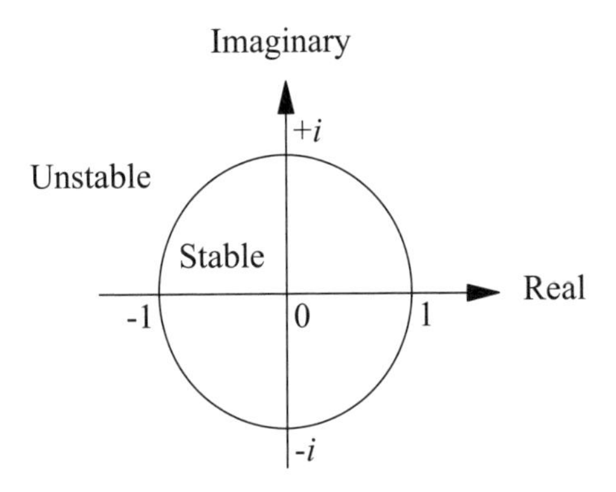

Figure E.3 Stability region in the complex plane.

The amplification factor is determined by substituting Eq. (E.69) into Eq. (E.68) and rearranging to obtain:

$$e^{\gamma \Delta t} = \left(1 - 2\frac{\alpha \Delta t}{\Delta x^2}\right) + \frac{\alpha \Delta t}{\Delta x^2}\left(e^{i\beta \Delta x} + e^{-i\beta \Delta x}\right) \tag{E.73}$$

Using the trigonometric identities,

$$\frac{e^{i\beta \Delta x} + e^{-i\beta \Delta x}}{2} = \cos(\beta \Delta x) \tag{E.74}$$

and

$$1 - \cos(\beta \Delta x) = 2\sin^2\left(\frac{\beta \Delta x}{2}\right) \tag{E.75}$$

Eq. (E.73) becomes

$$e^{\gamma \Delta t} = 1 - \left(4\frac{\alpha \Delta t}{\Delta x^2}\right)\left[\sin^2\left(\frac{\beta \Delta x}{2}\right)\right] \tag{E.76}$$

Combining this with the von Neumann condition for stability, we obtain the stability bound:

$$0 \le \left(\frac{\alpha \Delta t}{\Delta x^2}\right)\left[\sin^2\left(\frac{\beta \Delta x}{2}\right)\right] \le \frac{1}{2} \tag{E.77}$$

The $\sin^2(\beta\Delta x/2)$ term has its highest value equal to unity; therefore,

$$0 \le \left(\frac{\alpha \Delta t}{\Delta x^2}\right) \le \frac{1}{2} \tag{E.78}$$

is the limit for conditional stability for this method. It should be noted that this limit is identical to that obtained by using the positivity rule (Section 8.5.2).

The stability of the explicit solution (8.84) of the hyperbolic equation (8.82) can be similarly analyzed using the von Neumann method. The homogeneous equation for the error propagation of that solution is:

$$\varepsilon_{m,n+1} - 2\left(1 - \frac{\alpha^2 \Delta t^2}{\Delta x^2}\right)\varepsilon_{m,n} - \frac{\alpha^2 \Delta t^2}{\Delta x^2}\left(\varepsilon_{m+1,n} + \varepsilon_{m-1,n}\right) + \varepsilon_{m,n-1} = 0 \tag{E.79}$$

Substitution of the solution (E.69) into (E.79) and use of the trigonometric identities (E.74) and (E.75) give the amplification factor as:

$$e^{\gamma \Delta t} = \left[1 - 2\frac{\alpha^2 \Delta t^2}{\Delta x^2}\sin^2\left(\frac{\beta \Delta x}{2}\right)\right] \pm \sqrt{\left[1 - 2\frac{\alpha^2 \Delta t^2}{\Delta x^2}\sin^2\left(\frac{\beta \Delta x}{2}\right)\right]^2 - 1} \tag{E.80}$$

The above amplification factor satisfies inequality (E.72) in the complex plane, that is, when

$$\left[1 - 2\frac{\alpha^2 \Delta t^2}{\Delta x^2}\sin^2\left(\frac{\beta \Delta x}{2}\right)\right]^2 - 1 \le 0 \tag{E.81}$$

which converts to the following inequality:

$$\frac{\alpha^2 \Delta t^2}{\Delta x^2} \le \frac{1}{\sin^2\left(\dfrac{\beta \Delta x}{2}\right)} \tag{E.82}$$

The $\sin^2(\beta\Delta x/2)$ term has its highest value equal to unity; therefore,

$$\frac{\alpha^2\Delta t^2}{\Delta x^2} \le 1 \tag{E.83}$$

is the conditional stability limit for this method.

In a similar manner, the stability of other explicit and implicit finite difference methods may be examined. This has been done by Lapidus and Pinder (1982), who conclude that "most explicit finite difference approximations are conditionally stable, whereas most implicit approximations are unconditionally stable."

E.5 Step Size Control

The discussion of stability analysis in the previous sections relied on the simplifying assumption that the value of λ remains constant throughout the integration. This is true for linear equations such as Eq. (E.1); however, for nonlinear equations, the value of λ may vary considerably over the interval of integration. The minimum integration step size must be chosen using the maximum possible value of λ, which will guarantee stability at the expense of computation time. For problems in which computation time becomes excessive, it is possible to develop strategies for automatically adjusting the step size at each step of the integration.

A simple test for checking the step size is to do the calculations at each interval twice: do them once with the full step size, and then repeat the calculations over the same interval with a smaller step size, usually half of the first one. If, at the end of the interval, the difference between the predicted values of y by both approaches is less than a specified convergence criterion, the step size may be increased. Otherwise, a larger than acceptable difference between the two calculated y values suggests that the step size is large, and it should be shortened in order to achieve an acceptable truncation error.

Another method of controlling the step size is to obtain an estimation of the truncation error at each interval. A good example of such an approach is the Runge-Kutta-Fehlberg method (see Table 5.2 of Constantinides and Mostoufi, 1999), which provides an estimate of the local truncation error. This error estimate can be easily introduced into the computer program, and you can let the program automatically change the step size at each point until the desired accuracy is achieved.

As mentioned before, the optimum number of correctors is two. Therefore, in the case of using a predictor-corrector method, if the convergence is achieved before the second corrected value, the step size may be increased. On the other hand, if the convergence is not achieved after the second application of the corrector, the step size should be reduced.

E.6 Stiff Differential Equations

In Section E.1, we showed that the stability of the numerical solution of differential equations depends on the value of $h\lambda$, and that λ, together with the stability boundary of the method, determines the step size of integration. In the case of the linear differential equation

$$\frac{dy}{dt} = \lambda y \tag{E.1}$$

λ is the eigenvalue of that equation, and it remains constant throughout the integration. The nonlinear differential equation:

$$\frac{dy}{dt} = f(t, y) \tag{7.11}$$

can be linearized at each step using the mean-value theorem (E.26), so that λ can be determined from the partial derivative of the function with respect to y:

$$\lambda = \left.\frac{\partial f}{\partial y}\right|_{\alpha, t_n} \tag{E.84}$$

The value of λ is no longer a constant but varies in magnitude at each step of the integration.

This analysis can be extended to a set of simultaneous nonlinear differential equations:

$$\begin{aligned} \frac{dy_1}{dt} &= f_1(t, y_1, y_2, \ldots, y_n) \\ \frac{dy_2}{dt} &= f_2(t, y_1, y_2, \ldots, y_n) \\ &\vdots \\ \frac{dy_n}{dt} &= f_n(t, y_1, y_2, \ldots, y_n) \end{aligned} \tag{7.60}$$

Linearizing the set produces the Jacobian matrix

$$\mathbf{J} = \begin{bmatrix} \frac{\partial f_1}{\partial y_1} & \cdots & \frac{\partial f_1}{\partial y_n} \\ \vdots & \ddots & \vdots \\ \frac{\partial f_n}{\partial y_1} & \cdots & \frac{\partial f_n}{\partial y_n} \end{bmatrix} \tag{E.85}$$

The eigenvalues $\{\lambda_i \mid i = 1, 2, \ldots, n\}$ of the Jacobian matrix are the determining factors in the stability analysis of the numerical solution. The step size of integration is determined by the stability boundary of the method and the maximum eigenvalue.

When the eigenvalues of the Jacobian matrix of the differential equations are all of the same order of magnitude, no unusual problems arise in integrating the set. However, when the maximum eigenvalue is several orders of magnitude larger than the minimum eigenvalue, the equations are said to be *stiff*. The *stiffness ratio* (SR) of such a set is defined as

$$\mathrm{SR} = \frac{\max\limits_{1 \le i \le n} \left|\mathrm{Real}\left(\lambda_i\right)\right|}{\min\limits_{1 \le i \le n} \left|\mathrm{Real}\left(\lambda_i\right)\right|} \tag{E.86}$$

The integration step size is determined by the largest eigenvaluc, and the final time of integration is usually dictated by the smallest eigenvalue; since the variable most affected by the smallest eigenvalue changes very slowly; therefore, integrating stiff differential equations using explicit methods may be computer time intensive.

The MATLAB functions `ode23s` and `ode15s` are solvers suitable for the solution of stiff ordinary differential equations (see Table 7.2).

E.7 References

Constantinides, A. and Mostoufi, N. 1999. *Numerical Methods for Chemical Engineers with MATLAB Applications*. Upper Saddle River, NJ: Prentice Hall PTR.

Lapidus, L. and Sienfeld J. H. 1971. *Numerical Solution of Ordinary Differential Equations*. New York: Academic Press.

Lapidus, L. and Pinder G. F. 1982. *Numerical Solution of Partial Differential Equations in Science and Engineering*, New York: J. Wiley & Sons, Inc.

Index

A

C

D

F

G

H

I

M

N

O

Q

R

S

T

U

V

W

Y

Z